Ecological Studies, Vol. 77

Analysis and Synthesis

Ecological Studies

Volume 60
Amazonian Rain Forests Ecosystem Disturbance and Recovery (1987)
Edited by C. F. Jordan

Volume 61
Potentials and Limitations of Ecosystem Analysis (1987)
Edited by E.-D. Schulze and H. Zwölfer

Volume 62
Frost Survival of Plants (1987)
By A. Sakai and W. Larcher

Volume 63
Long-Term Forest Dynamics of the Temperate Zone (1987)
By P. A. Delcourt and H. R. Delcourt

Volume 64
Landscape Heterogeneity and Disturbance (1987)
Edited by M. Goigel-Turner

Volume 65
Community Ecology of Sea Otters (1987)
Edited by G. R. van Blaricom and J. A. Estes

Volume 66
Forest Hydrology and Ecology at Coweeta (1987)
Edited by W. T. Swank and D. A. Crossley, Jr.

Volume 67
Concepts of Ecosystem Ecology A Comparative View (1988)
Edited by L. R. Pomeroy and J. J. Alberts

Volume 68
Stable Isotopes in Ecological Research (1989)
Edited by P. W. Rundel, J. R. Ehleringer and K. A. Nagy

Volume 69
Vertebrates in Complex Tropical Systems (1989)
Edited by M. L. Harmelin-Vivien and F. Bourliere

Volume 70
The Northern Forest Border in Canada and Alaska (1989)
By J. A. Larsen

Volume 71
Tidal Flat Estuaries: Simulation and Analysis of the Ems Estuary (1988)
Edited by J. Baretta and P. Ruardij

Volume 72
Acidic Deposition and Forest Soils (1989)
By D. Binkley, C. T. Driscoll, H. L. Allen, P. Schoeneberger, and D. McAvoy

Volume 73
Toxic Organic Chemicals in Porous Media (1989)
Edited by Z. Gerstl, Y. Chen, U. Mingelgrin, and B. Yaron

Volume 74
Inorganic Contaminants in the Vadose Zone (1989)
Edited by B. Bar-Yosef, N. J. Barrow, and J. Goldshmid

Volume 75
The Grazing Land Ecosystems of the African Sahel (1989)
By H. N. Le Houérou

Volume 76
Vascular Plants as Epiphytes (1989)
Edited by U. Lüttge

Volume 77
Forest Decline and Air Pollution (1989)
Edited by E.-D. Schulze, O. L. Lange, and R. Oren

Forest Decline and Air Pollution

A Study of Spruce (*Picea abies*) on Acid Soils

Edited by

E.-D. Schulze O.L. Lange R. Oren

With 186 Figures

Springer-Verlag Berlin Heidelberg New York
London Paris Tokyo Hong Kong

Prof. Dr. ERNST-DETLEF SCHULZE
Lehrstuhl für Pflanzenökologie
der Universität Bayreuth
Universitätsstraße 30
8580 Bayreuth, FRG

Prof. Dr. OTTO L. LANGE
Lehrstuhl für Botanik II
der Universität Würzburg
Mittlerer Dallenbergweg 64
8700 Würzburg, FRG

Prof. Dr. RAM OREN
School of Forestry and
Environmental Studies
Duke University
Durham, North Carlina 27706
USA

Design of the cover illustration by Roswitha Asche.

ISBN 3-540-50793-0 Springer-Verlag Berlin Heidelberg New York
ISBN 0-387-50793-0 Springer-Verlag New York Berlin Heidelberg

Library of Congress Cataloging-in-Publication Data. Air pollution and forest decline : a study of spruce (Picea abies) on acid soils / E.-D. Schulze, O.L. Lange, R. Oren, editors. p. cm. – (Ecological studies ; vol. 77) Includes bibliographical references. 1. Norway spruce – Fichtelgebirge (Germany and Czechoslovakia) – Effect of air pollution on. 2. Norway spruce – Fichtelgebirge (Germany and Czechoslovakia) – Ecology. 3. Forest declines – Fichtelgebirge (Germany and Czechoslovakia) 4. Forest ecology – Fichtelgebirge (Germany and Czechoslovakia) 5. Acid soils – Fichtelgebirge (Germany and Czechoslovakia) 6. Soil ecology – Fichtelgebirge (Germany and Czechoslovakia) I. Schulze, E.-D. (Ernst-Detlef), 1941– II. Lange, O.L. (Otto Ludwig) III. Oren, Ram. IV. Series: Ecological studies ; v. 77.
QK494.5.P66A47 1989 574.5'2642'0943315 – dc20 89-21705

Printed in Germany

Typesetting: International Typesetters Inc., Makati, Philippines
Printing: Druckhaus Beltz, Hemsbach
Binding: J. Schäffer, Grünstadt
2131/3145-543210 – Printed on acid-free paper

Contents

Part 1: Air Pollution: Transport and Deposition

Institutions Cooperating in the Fichtelgebirge-Project

University of Bayreuth, Postfach 10 12 51, D-8580 Bayreuth, FRG

1) Lehrstuhl Bodenkunde und Bodengeographie
2) Lehrstuhl Bodenkunde, Abt. Bodenphysik
3) Abteilung Elektronenmikroskopie
4) Lehrstuhl Hydrologie
5) Lehrstuhl Hydrologie, Abt. Meteorologie
6) Lehrstuhl Ökologische Chemie und Geochemie
7) Lehrstuhl Pflanzenökologie
8) Lehrstuhl Pflanzenphysiologie
9) Lehrstuhl Tierökologie

Bundesanstalt für Land- und Forstwirtschaft Berlin, Königin-Luise-Str. 19, D-1000 Berlin 33, FRG

10) Abteilung Mikrobiologie

University of Bonn, Venusbergweg 22, D-5300 Bonn 1, FRG

11) Institut für Pflanzenkrankheiten

Biolog. Bundesanstalt für Land- und Forstwirtschaft Braunschweig, Messeweg 11/12, D-3300 Braunschweig, FRG

12) Institut für Pflanzenschutz im Forst

University of Erlangen-Nürnberg, Schloßplatz 4, D-8520 Erlangen, FRG

13) Institut für Hydromechanik
14) Lehrstuhl für Strömungsmechanik

University of Munich

15) Institut für Meteorologie, Theresienstr. 37, D-8000 München 2, FRG
16) Lehrstuhl Bioklimatologie und Angewandte Meteorologie, Amalienstr. 52, D-8000 München 40, FRG

Technical University of Munich

17) Lehrstuhl für Botanik, Postfach 20 24 20, D-8000 München 2, FRG
18) Lehrstuhl Bodenkunde, 8050 Freising-Weihenstephan, FRG

Bayerisches Landesamt für Umweltschutz, Rosenkavalierplatz 22, D-8000 München, FRG

19) Abteilung Luftreinhaltung

Zentrale Analytik der Gesellschaft für Strahlen- und Umweltforschung, München-Neuherberg, Ingolstädter Landstr. 31, D-8042 Neuherberg, FRG

20) Zentrale Analytik der Projektgruppe Bayern zur Erforschung der Wirkung von Umweltschadstoffen (PBWU)
21) Institut für Hydrologie

University of Regensburg

22) Institut für Botanik, Universitätsstr. 31, D-8400 Regensburg, FRG

University of Würzburg, Institut für Botanik und Pharmazeutische Biologie, Mittlerer Dallenbergweg 64, D-8700 Würzburg, FRG

23) Lehrstuhl für Botanik I
24) Lehrstuhl für Botanik II
25) Lehrstuhl für Pharmazeutische Biologie

Scientists Involved in the Fichtelgebirge-Project

Contributors to this volume are indicated by bold face
(Numbers refer to the respective institutions, see pp. XV and XVI)

Acker, G	(3)
Barthmann, U.	(7)
Baumgartner, A.	(16)
Beck, E.	(8)
Berger, A.	(7)
Beyschlag, W.	(24)
Bierl, R.	(4)
Bresinsky, A.	(22)
Brietzke, Ch.	(24)
Buchmann, N.	(7)
Butin, H.	(12)
Czygan, F.-Ch.	(25)
Durst, F.	(14)
Egger, J.	(15)
Eiden, R.	(4)
Ernst, F.	(14)
Fischer, R.	(9)
Förster, J.	(4)
Frevert, T.	(4)
Führer, G.	(24)
Gebauer, G.	(7)
Gebel, J.	(24)
Geschke, A.	(6)
Gietl, G.	(19)
Grade, J.	(1)
Graf, J.	(15)
Haderlein, K.	(15)
Hantschel, R.	(1)
Hartmann, P.	(9)
Heber, U.	(23)
Herrmann, R.	(4)
Herterich, R.	(5)
Hock, B.	(17)
Hoess, P.	(24)
Horn, R.	(2)
Hutzinger, O.	(6)
Kaiser, W.	(23)
Kammerbauer, H.	(17)
Katz, C.	(7)
Kaupenjohann, M.	(2)
Klemm, O.	(4)
Köstner, B.	(24,25)
Kuhn, U.	(8)
Kurbasik, K.	(17)
Laisk, A.	(23)
Lange, O. L.	(24)
Lippert, M.	(24)
Ludwig, J.	(4)
Magel, E.	(17)
Martinoia, E.	(23)
Merk, L.	(17)
Meyer, A.	(24)
Meyer, J.	(7)
Mitterhuber, E.	(23)
Nätscher, L.	(18)
Nemmaier, Ch.	(15)
Nienhaus, F.	(11)
Oren, R.	(7)
Paulus, W.	(22)
Peters, K.	(5)
Petzold, H.	(10)
Pfanz, H.	(23,24)
Pfeiffer, S.	(24)
Poss, G.	(13)
Rall, A.	(16)
Reichmann, T.	(14)
Reif, A.	(7)
Reischl, A.	(6)
Reissinger, M.	(6)
Riederer, M.	(17)
Römmelt, R.	(17)
Schaaf, W.	(2)

Scheitler, M.	(9)	Taubner, H.	(1)
Schneider, B. U.	(2)	Tenhunen, J.	(24)
Schönwitz, R.	(17)	Thomas, W.	(4)
Schramel, P.	(20)	**Trautner, F.**	(4)
Schulze, E.-D.	(7)	**Ulrich, W.**	(15)
Schumacher, U.	(2)	Verhoeven, W.	(5)
Schwab, M.	(24)	Völklein, J.	(14)
Schwarzmaier, G.	(16)	Wedler, M.	(24)
Schwertmann, U.	(18)	Weikert, R.	(24)
Stenzel, A.	(4)	**Werk, K. S.**	(7)
Stichler, W.	(21)	**Zech, W.**	(1)
Steudle, E.	(7)	Zellner, H.	(24)
Suske, J.	(3)	**Ziegler, H.**	(17)
Süsser, P.	(18)	**Zimmermann, R.**	(7)

1 Introduction: The Problem of Forest Decline and the Bavarian Forest Toxicology Research Group

W. UHLMANN[1], H. ALTNER[2], E.-D. SCHULZE[3], and O.L. LANGE[3]

Forest decline first came to public attention when the dominant silvicultural tree of Germany showed damage on a large geographic scale. Damaged Norway spruce, *Picea abies* (L.) Karts., displayed symptoms which could not be explained by known pathogens or other factors. In the Fichtelgebirge (northeast Bavaria, Federal Republic of Germany), the earliest observations of the typical decline symptoms of needle chlorosis and needle loss were made in 1978 (Zech, personal communication). When these symptoms appeared over a larger area, research on permanent observation plots began in 1981–82 (Bayerische Staatsforstverwaltung 1982–87). Research first focused on classifying the damage according to the various degrees of chlorosis and needle loss (Table 1).

Slightly damaged trees (Class 1) may be found for many reasons, including temporal or phenological variation of needle chlorosis and loss. Moderately and severely damaged as well as dying trees (Classes 2–4), however, represent that portion of the forest which is in decline for reasons other than the natural variation in tree appearance (Forschungsbeirat Waldschäden 1986). If the damage estimate from the initial 1981–82 observations on the permanent plots in the Fichtelgebirge is taken as a baseline (Fig. 1), tree damage increased during 1983. From 1983–1986, damage stabilized for 80% of the trees on these plots. By 1986, the last year for which data are available, 12% of the permanent observation plots in the Fichtelgebirge contained only dead trees (Kennel 1987). For the rest of Bavaria, a smaller number of the trees on permanent observation plots was damaged during this period (Bayerische Staatsforstverwaltung 1982–87).

By 1986, total inventories of the forest area (Fig. 2) had classified 30% of the Bavarian forests and 20% of the other German forests as moderately or severely damaged from unknown causes. On an European scale, forest damage from causes other than known stress factors was 23% in 1986 (ECE 1987). In Bavaria, the proportion of timber harvested before schedule because of damage by wind, insects, and decline had jumped from 20% in 1981 to 50% in later years. As a result, over the last 5 years a large proportion of damaged forests has been severely thinned (Bayerische Staatsforstverwaltung 1982–87).

[1]Prof. Dr. W. UHLMANN (Würzburg), Head (1980–1986) of the Scientific Advisory Board of the Bavarian Ministry of Education and Cultural Affairs
[2]Prof. Dr. H. ALTNER (Regensburg), Chairman of the Review Committee
[3]Prof. Dr. E.-D. SCHULZE (Bayreuth) and Prof. Dr. O.L. LANGE (Würzburg), Organizers of the Research Group

E.-D. Schulze, O.L. Lange, R. Oren (Eds.)
Ecological Studies Vol. 77

Table 1. Classification of forest damage on the basis of needle chlorosis and needle loss

Damage Class	Vitality	Needle loss (%)	Yellowing (%)
0	Healthy	0–10	0–25
1	Slight damage	11–25	$<$ 60 at 0–10% needle loss $<$ 25 at 11–25% needle loss
2	Moderate damage	26–60	$>$ 25 at 11–25% needle loss $<$ 25 at 26–60% needle loss
3	Severe damage	61–99	$>$ 25 at 26–60% needle loss $>$ 25 at 61–99% needle loss
4	Dying	100	

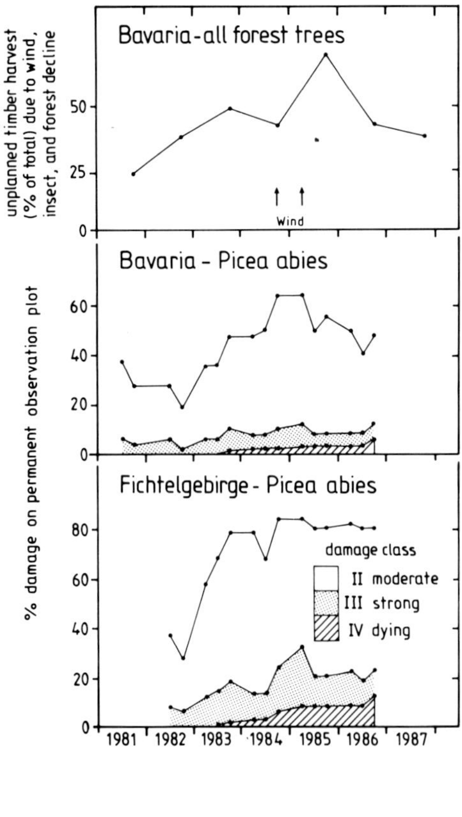

Fig. 1. Damage of spruce (*Picea abies*) and unplanned timber harvest for Bavaria and for the Fichtelgebirge

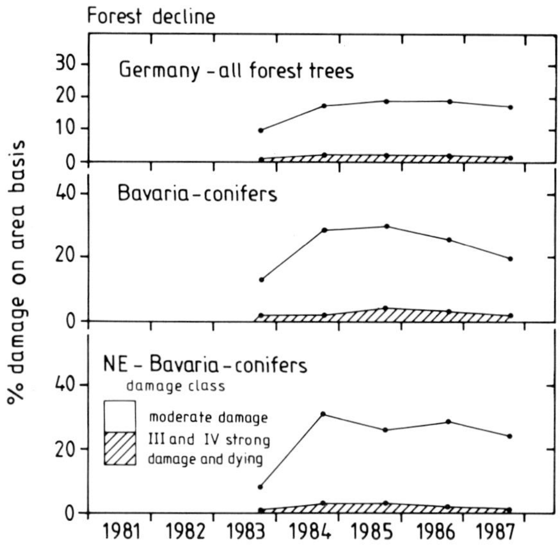

Fig. 2. Damage of all forest trees in the Federal Republic of Germany, and damage of conifers in Bavaria and in NE-Bavaria. In 1988 15.1% of all German forests and 17.4% of Bavarian conifers were damaged

Since the initial increase in damaged forests, numerous scientific hypotheses have been proposed to explain forest decline (Forschungsbeirat Waldschäden 1984). The following four hypotheses account for forest decline: (1) natural causes (climate) and epidemics; (2) direct effects of air pollutants on aboveground plant organs; (3) mineral deficiencies and imbalances as a consequence of acid deposition and soil acidification; or (4) various combinations of these factors (Blank 1985).

Support for these various forest decline hypotheses has come from different sources of evidence. Because the number of fungal infections markedly increased in the early period of forest decline, pathogens were thought to play an important role in explaining the pattern of symptoms and their spread (Kandler 1983; Rehfuess and Rodenkirchen 1984). Similarly, it was known that direct effects from such gaseous pollutants as sulfur dioxide, nitrous oxide, and ozone affected forest health. The severe, large-scale forest damage in Poland and Czechoslovakia indicated that forest decline in Germany might be directly linked to such gaseous pollutants (Mohr 1983), particularly in northeast Bavaria, which borders Czechoslovakia and experiences high atmospheric sulfur dioxide concentrations in winter. In contrast to natural epidemics and possible direct effects of gaseous air pollutants, soil acidification from nutrient cation leaching by acid rain, and the subsequent aluminum enrichment of the soil solution, have been proposed as major mechanisms for explaining forest decline (Ulrich 1980; Likens et al. 1981). Nutrient deficiencies were in fact observed very early in some of the declining stands in the Fichtelgebirge (Zech and Popp 1983). Finally, it has been suggested that forest decline is unrelated to anthropogenic pollution and is a result of climatic variation, e.g., a drought during a sequence of dry years, or unseasonably early or late but severe winter frost events (Cramer and Cramer-Mittendorf 1984).

Such climatic variations have been implicated in predisposing trees to other biological and chemical stress factors.

By 1982, the disturbing increase in forest decline and the lack of scientific consensus regarding its causes had alarmed the Scientific Advisory Board (Beirat für Wissenschafts- und Hochschulfragen) of the Bavarian Ministry of Education and Cultural Affairs (Bayerisches Staatsministerium für Unterricht und Kultus). Established in 1980 by the minister Prof. Dr. H. Maier, the Scientific Advisory Board consisted of members of most of the Bavarian universities. In an advisory capacity, it had the task of suggesting scientific activities to promote scientific development in Bavaria. The head of the board (W. Uhlmann, mathematician) recognized the urgent need for interdisciplinary research on tree damage phenomena. Following discussions with botanists from the universities of Bayreuth and Würzburg, it was arranged for scientists of various disciplines to meet during the winter term of 1982–83. In a first conference, botanists, chemists, forest botanists, hydrologists, silviculturists, and soil scientists discussed the possibilities of concentrated field work in the Bavarian forests. It was impressive to see how willingly the scientists followed the initiative of the Scientific Advisory Board. Thus, the Bavarian Forest Toxicology Research Group (Bayerische Forschungsgruppe Forsttoxikologie) was founded. It encouraged researchers in biology, earth science, forestry, and meteorology to test competing hypotheses on unexplained forest damage by working cooperatively at the same integrated research site, the Fichtelgebirge. Because of its economic importance and the severe damage it was experiencing, Norway spruce was chosen as the study tree. Selection and establishment of the research sites would not have been possible without the advice and the help of the various Bavarian forest departments (Staatsministerium für Landwirtschaft und Forsten: Ltd. Ministerialrat Wild; Oberforstdirektion Bayreuth: Forstdirektor Meier; Forstamt Fichtelberg and Forstamt Goldkronach). All this assistance and the installation of a container for recording atmospheric pollution (Staatsministerium für Landesplanung und Landesentwicklung; Landesamt für Umweltschutz) at one of the measuring sites is gratefully acknowledged.

From within the Fichtelgebirge, two stands of 30-year-old *P. abies* at Wülfersreuth and Oberwarmensteinach were selected as experimental sites because of similar climatic conditions and the same base rock (phyllite). The stand at Wülfersreuth was obviously healthy (damage Class 0), but the stand at Oberwarmensteinach showed great variation in declining trees ranging from damage Classes 0 to 3. In patches at Oberwarmensteinach were normal, full-grown spruce trees with full crowns, growing next to other individuals with heavy signs of damage. Figure 3 illustrates typical symptoms of tree damage in the Fichtelgebirge. In the typical severely damaged case, spruce needles of the older age-classes are chlorotic, particularly on the insulated upper side of the branches, while the current-year terminal flush and the current beginning growth of the second-order branches appear fully green (Fig. 3 above right). Yellowing begins at the tips of the older needles and progresses to their base (Fig. 3 above left). Extensively discolored parts of a tree can be seen in association with needle and

Fig. 3. Typical symptoms of spruce damage in the Fichtelgebirge; *above left:* bleaching of the tips of the needles; *above right:* branches of a damaged tree from below (*left*) and from above (*right*); *below left:* damaged tree with needle and branch loss and with chlorotic needles at the upper side of the branches; *below right:* collapsed stand with dead and heavily damaged trees. (Fot. Köstner, Lange, Reif, Schulze)

branch loss (Fig. 3 below left). In the final stage of damage, the whole stand can collapse, as has been the case at the Schneeberg, the highest mountain in the Fichtelgebirge (Fig. 3 below right).

The Fichtelgebirge project tried to explain patterns of decline by studying pathogens, the relationships between gaseous pollutants, precipitation acidity, and the physiological responses in aboveground tree parts, as well as the correlations between soil factors, soil processes, whole plant growth, and nutrient relationships. Both single-twig and whole-tree studies were made to compare green trees at the two sites and trees ranging from green to yellow at the declining site. The whole-stand decline phenomenon was studied with randomly distributed observation plots, and parts of the study sites were experimentally treated with fertilization.

In 1984, the Bavarian Forest Toxicology Research Group had begun its work with the financial support of the state of Bavaria (Ministerien für Unterricht und Kultus, für Landwirtschaft und Forsten und für Landesplanung und Landesentwicklung), the Federal Government (Bundesministerium für Forschung und Technologie), and the German Science Foundation (Deutsche Forschungsgemeinschaft). A review committee of the Scientific Advisory Board (chaired by H. Altner, zoologist) was in charge of evaluating the proposals and results of the individual projects within the research group. We are grateful to Ms Dr. Dr. h.c. Anita Hoffmann (Deutsche Forschungsgemeinschaft) for her active involvement in helping with project evaluations. Administration was highly facilitated by the competent cooperation of Ministerialrat D. Magerkurth and Ministerialrat H. Grote from the Bavarian Ministry, who also insured that the research group met annually. Their support and personal interest are gratefully acknowledged.

Many other people have facilitated our work with their assistance. We want to thank Ms Rebecca Richards (Logan, Utah) for carefully editing the English rendition of this Volume and Ms Evelyn Amberg for assisting the Bavarian Forest Toxicology Research Group and organizing the manuscripts. Finally, our studies would not have been possible without all the technical help at the measuring sites, in the laboratories, and at the computers and the word processors.

The Bavarian Forest Toxicology Research Group's project has been conducted by researchers from the Bavarian Environmental Protection Agency (Landesamt für Umweltschutz) and 23 different departments at 7 Bavarian universities. This book summarizes the results obtained to date and is presented as a general case study on forest decline. Naturally, the number of authors in such a synthesis has been restricted and many more scientists have been actively involved with the research group; see p. xvii and xviii for a complete listing. Many students have worked at the experimental sites in the Fichtelgebirge for their diploma thesis, their teaching qualification, or their doctoral thesis. We dedicate this book to this young generation who are enthusiastically involved in scientific ecological work. We hope that together we can contribute to solving an urgent and threatening problem for mankind's future.

References

Bayerische Staatsforstverwaltung (1982, 1983, 1984, 1985, 1986, 1987) Information 4/82, 4/83, 4/84, 4/85, 4/86, 4/87

Blank LW (1985) A new type of forest decline in Germany. Nature (Lond) 314:311–314

Cramer HH, Cramer-Mittendorf M (1984) Untersuchungen über Zusammenhänge zwischen Schadensperioden und Klimafaktoren in mitteleuropäischen Forsten seit 1851.Pflanzenschutz-Nachr Bayer 37:208–334

ECE (1987) Forest damage and air pollution. ECE/UNED Project FP/9101–86–05, Geneva 1987

Forschungsbeirat Waldschäden (1984) Zwischenber 1984 des Forschungsbeirats, Literaturabteilung Kernforschungszentrum Karlsruhe, 368 pp

Forschungsbeirat Waldschäden (1986) Ber Literaturabteilung Kernforschungszentrum Karlsruhe 2:229

Kandler O (1983) Emissions- oder Epidemie-Hypothese. Naturwiss Rundsch 11:488–490

Kennel E (1987) Entwicklung der Waldschäden auf den Dauerbeobachtungsflächen in Bayern von 1981 bis 1986. Bayer forstliche Versuch- und Forschungsanst, Sachgebiet III, 57 pp

Likens G, Wright RF, Galloway JN, Butler TJ (1981) Saurer Regen. Spektrum Wiss 1:18–25

Mohr H (1983) Zur Faktorenanalyse des "Baumsterbens". Biol unserer Zeit 13:74–78

Rehfuess KE, Rodenkirchen H (1984) Über die Nadelröte-Erkrankung der Fichte (*Picea abies* Karst.) in Süddeutschland. Forstwiss Centralbl 103:245–262

Ulrich B (1980) Die Wälder in Mitteleuropa: Meßergebnisse ihrer Umweltbelastung, Theorie ihrer Gefährdung, Prognose ihrer Entwicklung. AFZ 35:1198–1202

Zech W, Popp E (1983) Magnesiummangel, einer der Gründe für das Fichten- und Tannensterben in NO-Bayern. Forstwiss Centralbl 102:50–55

2 The Vegetation of the Fichtelgebirge: Origin, Site Conditions, and Present Status

A. REIF

1 Introduction

A small, horseshoe-shaped mountain range in northeastern Bavaria, the Fichtelgebirge, covers an area of about 1000 km^2 (Fig. 1). The highest mountains are Ochsenkopf (1023 m a.s.l.) and Schneeberg (1053 m a.s.l.; Fig. 2). "Fichtelgebirge" is related to the German word Fichte (spruce, *Picea abies*) and means "spruce range".

Consisting mainly of metamorphic gneiss, granite, and schist of the Paleozoic (Vollrath 1955–1957; Rutte 1981), the Fichtelgebirge represents a pluton, situated in the center of three mountain ranges: the Erzgebirge in the northeast on the border between the German Democratic Republic and Czechoslovakia, the Oberpfälzer Wald to the southeast on the border between the Federal Republic of Germany and Czechoslovakia, and the Frankenwald in the northeast (Fig. 1). In the upper Carboniferous period, granite intrusions formed an epizone of mainly metamorphic phyllite. Other substrates occur locally, e.g., basalt, limestone, serpentine, and diabase. During the early Mesozoic, marine sediments are thought to have been deposited and then to have eroded.

The soils of the Fichtelgebirge can be broadly described as acid and poor in nutrients. On rocky sites, entisols are found locally. Most frequent are inceptisols, spodosols, and transitions between these. In valleys and on terraces, hydromorphic soils are typical. On saddles, peaty soils have developed locally, with peat deposits of up to 6 m (Beug 1957).

The higher altitudes of the Fichtelgebirge are characterized by their eastern latitude and relative height, which result in warmer summers and colder winters than other mountain ranges in Germany (Fig. 3). With increasing altitude, annual mean temperatures decrease from ca. 8°C at the foothills to ca. 4–5°C at the highest altitudes (Klimaatlas von Bayern 1952). At the summits, precipitation reaches a maximum of 1200–1400 mm per year, the soil is covered with snow for up to 150 days per year, and frost occurs 150 days per year.

2 Vegetation History

The vegetation history of the Fichtelgebirge after the last glaciation can be reconstructed by pollen profile diagrams from peat deposits of raised bogs (Firbas and v. Rochow 1956; Beug 1957). During the Pleistocene, the Fichtelgebirge area

E.-D. Schulze, O.L. Lange, R. Oren (Eds.)
Ecological Studies Vol. 77

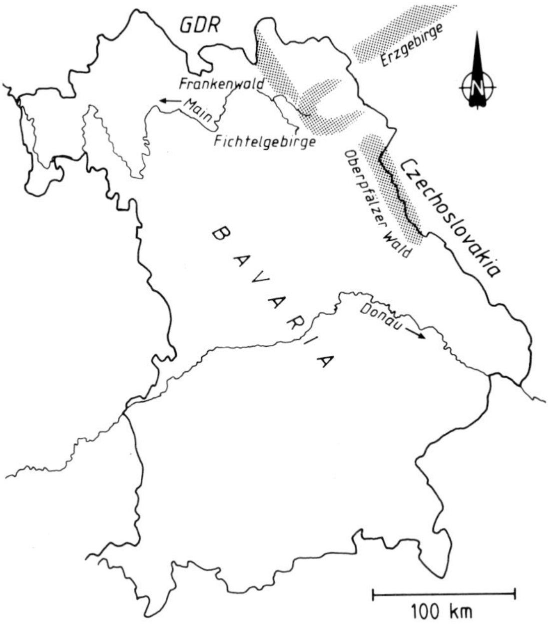

Fig. 1. Location of the Fichtelgebirge in northeast Bavaria, southeastern Germany

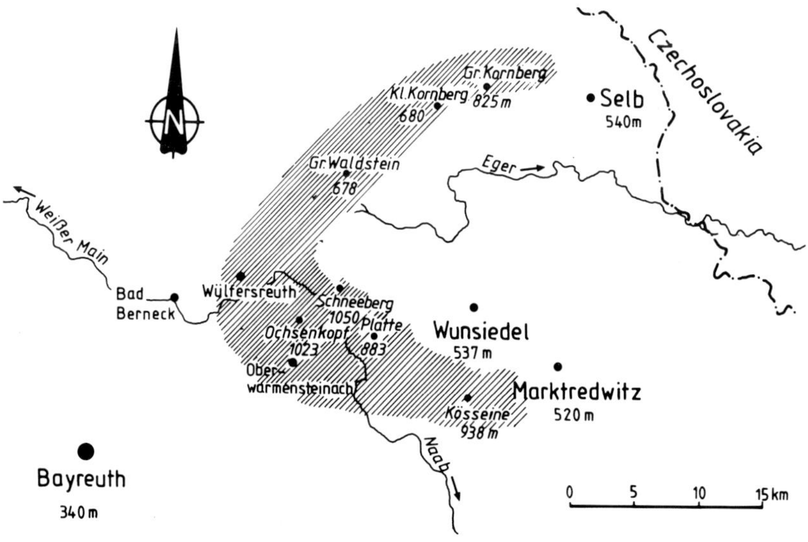

Fig. 2. Localities of the Fichtelgebirge

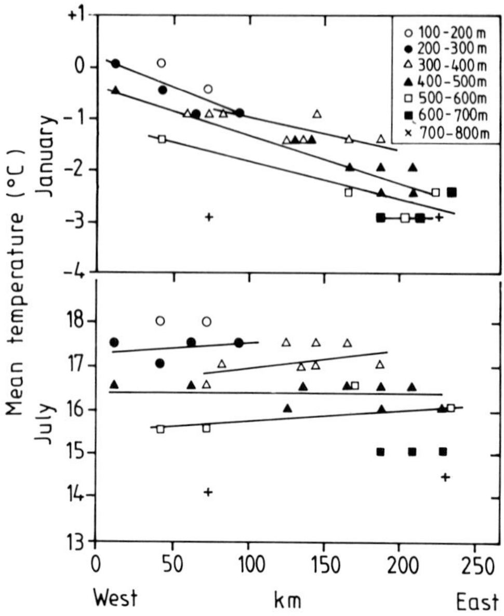

Fig. 3. Mean temperatures of January and July in relation to latitude (from west to east) and altitudinal belts in northern Bavaria. Mean January temperatures decrease from west to east, while mean July temperatures show little difference between the latitudes (Reif 1983)

was ice-free. After the Pleistocene, the climate was still cool, and tundra vegetation developed. Following this tundra period, six periods of forest tree colonization can be distinguished:

1. In the early Betula-Pinus period (IV), before 6800 B.C., the Fichtelgebirge area was colonized by the wind-dispersed woody tree species birch (*Betula pendula*) and Scots pine (*Pinus sylvestris*).
2. Subsequent invaders were hazel (*Corylus avellana*), oak (*Quercus robur*), and elm (*Ulmus glabra*) (Corylus period, V; ca. 6800–5500 B.C.).
3. Increasing temperatures were then accompanied by increased frequency of oak, elm, and such minor tree species as lime (*Tilia* spec.) and ash (*Fraxinus excelsior*; Quercus-robur period, VI; ca. 5500–4000 B.C.). Alder (*Alnus glutinosa*) and spruce (*Picea abies*) were already present.
4. In the following oak-forest-Picea period (VII; ca. 4000–2500 B.C.), spruce became the dominant tree species (Fig. 4). Towards the end of this period, beech (*Fagus sylvatica*), sycamore (*Acer pseudoplatanus*) and hornbeam (*Carpinus betulus*) invaded.
5. In the following Picea-Fagus period (VIII), beech replaced spruce as the dominant canopy species. Fir (*Abies alba*) became a frequent forest species. In the Fagus-Abies period (IX; ca. 700 B.C. to 1200 A.D.), spruce pollen became infrequent and the forests were dominated by beech and fir.
6. The most recent period, the Picea-Pinus-Fagus-Abies period (X), is characterized by a decrease of beech- and fir-pollen, and by an increase of spruce and Scots pine. This shift of species composition can be related to colonization and forest destruction by man over the last 800 years.

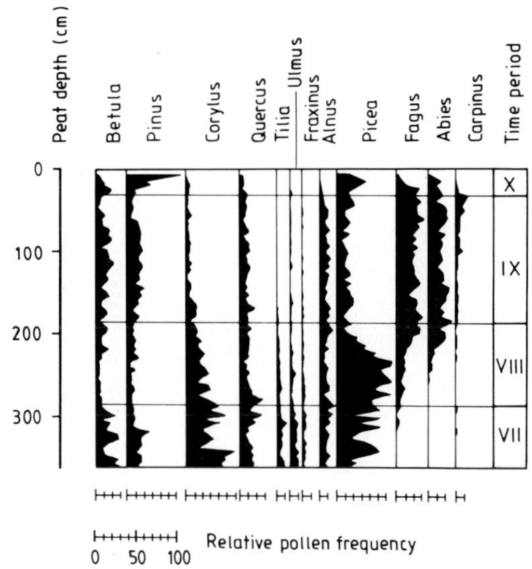

Fig. 4. Pollen profile diagram from "Torflohe", a raised bog area in the Fichtelgebirge. The relative frequency of the pollen of different tree species related to peat depth (and age) indicates the time course of colonization of the Fichtelgebirge (after Firbas and v. Rochow 1956)

3 Natural Forest Vegetation

The Fichtelgebirge has been mostly covered by natural forests (Firbas 1952). In the 16th century, beech (*Fagus sylvatica*) was still the main canopy species of the forest; with other deciduous tree species such as sycamore (*Acer pseudoplatanus*), it comprised about 60% of the canopy composition. Fir (*Abies alba*) was also important, comprising about 40% of the canopy (Seuss 1977). At the highest altitudes, around the edges of bogs, in frost hollows, and on the shallow ranker soils on boulders, spruce is thought to have occurred naturally (Firbas 1952). On exposed sites and on infertile or shallow soils, pioneer species which demanded light occurred, like downy birch (*Betula pubescens* ssp. *carpatica*), rowan (*Sorbus aucuparia*), and Scots pine. On local peat deposits, *Pinus mugo* ssp. *rotundata* formed nearly pure stands (Lutz 1956).

Below ca. 700 m a.s.l., beech is still the most competitive tree species, but sycamore, elm, and oak also occur. On wetter sites near water courses, willows (*Salix triandra, S. fragilis, S. purpurea*), ash (*Fraxinus excelsior*) and alder (*Alnus glutinosa*) grow. On poor sandy soils, mostly weathered granite and limestone and serpentine (Gauckler 1954), Scots pine locally dominates the forests below ca. 700 m a.s.l. The conical and narrow crowns of Scots pine in the Fichtelgebirge are an adaptation to snow damage and represent an indigenous variety (*Selber Höhenkiefer;* Krauss 1952).

4 Present Forest Vegetation

Over the last 400 years, the forests of the Fichtelgebirge have been severely depleted by mining, iron smelting, charcoal-burning and the glass and porcelain

Fig. 5. View from the summit of "Platte". Around the edges of a large boulder field in the foreground, less competitive small tree species like rowan (*Sorbus aucuparia*) and downy birch (*Betula pubescens* ssp. *carpatica*) occur. In the background, planted spruce (*Picea abies*) trees dominate large forested areas. Few of the spruce trees show crown mortality. The farmland in the background is mainly used for grassland and cultivation of cereals

industries. In this last century, reforestation of large areas has resulted in the present forest composition, with 96% spruce, 2% beech and 1% fir (Fig. 5). Today, only a few deciduous forests remain in remote areas (Vollrath 1957). Most forests are planted and are strongly modified by anthropogenic influences (Fig. 6); as a result, many forest species have disappeared. Of the approximately 500 mosses and liverworts which were present in the last century (Funck 1800–1832), 30% have disappeared, mainly because of changing forest management (Hertel 1974). A survey of the forests of the Fichtelgebirge based on 177 releves has been analyzed by numerical methods (Leonhard 1988; Table 1). The releves from extreme sites form well-defined groups and include the "azonal" forest communities. However, the floristic composition of the majority of the forests gradually changes with the environment. Within the forests, the main gradients are altitude and nutrient supply, which is correlated with soil water conditions. The forests of the Fichtelgebirge can thus be divided into eight communities (Table 1):

1. A species-poor beech forest, the Luzulo-Fagetum, is characterized by beech in the canopy and *Luzula luzuloides* in the understory. Other differentiating, but less frequent, species are *Prenanthes purpurea* and the ferns *Dryopteris filix-mas, Thelypteris phegopteris* and *Gymnocarpium dryopteris*. This community mainly occurs on mid-slopes at lower altitudes.

Fig. 6. The ground cover of the spruce forests at altitudes below ca. 800 m a.s.l. is dominated by the grass *Avenella flexuosa*. Because of intensive forest management the spruce trees are more or less even-sized

2. A canyon forest (*Acer pseudoplatanus*-Tilio-Acerion-community) replaces the Luzulo-Fagetum on the bottom of slopes and in gullies where the soil is unstable. On such sites, sycamore, ash, and elm replace the beech trees. Nutrient-demanding forbs like hedge woundwort (*Stachys sylvatica*), herb robert (*Geranium robertianum*), and yellow archangel (*Lamiastrum galeobdolon*) dominate the ground.

3. Adjacent to rivers and streams, alder becomes dominant in a *Stellaria nemorum-Alnus glutinosa*-community on relatively fertile, wet sites. Differentiating species are *Stellaria nemorum, Polygonum bistorta, Galeopsis tetrahit*, and *Chaerophyllum hirsutum*.

4. In the smaller catchments further upstream, ash is locally dominant in a *Carex remota-Fraxinus excelsior*-community on wet valley bottoms near springs. Species characteristic of permanent wet sites comprise the understorey, e.g., *Equisetum sylvaticum, Deschampsia cespitosa, Crepis paludosa, Lysimachia nemorum*, and *Chrysosplenium oppositifolium*.

5. Spruce forests with the dominant understory grass species *Calamagrostis villosa* (*Calamagrostis villosa-Picea abies*-community) are the most frequent community at higher altitudes. Another frequent, differentiating species of this community is chickweed wintergreen (*Trientalis europaea*; Fig. 7); the ferns *Blechnum spicant* and *Thelypteris limbosperma* occur less frequently.

Table 1. Summarized frequency table of the forest communities of the Fichtelgebirge[a]

Community type	1	2	3	4	5c	5b	5a	5d	6a	6b	6c	6d	7	8
Number of releves	14	4	6	7	23	11	21	8	19	14	9	10	16	15
Canopy tree species (T = tree; S = subcanopy; G = ground tier):														
Picea abies (T)	4		2	2	2	4	5	5	5	5	5	5	4	5
Picea abies (S)	1		2		2	3	3	3	2	3	3	4	3	5
Picea abies (G)	4	2	1	2	1	3	3	5	4	5	5	5	3	4
Fagus sylvatica (T + S)	2	3				1	1		2	1	1		1	
Acer pseudoplatanus (T + S)		3	1	1		1			1					
Fraxinus excelsior (T + S)		3	2											
Fraxinus excelsior (G)	1	3	1											
Carpinus betulus (T + S)		4												
Ulmus glabra (T + S)		3												
Alnus glutinosa (T)	1		4	2	1									
Sorbus aucuparia (T + S)	1					1		1	1	1		1	3	2
Betula carpatica (T + S)													1	1
Pinus sylvestris (T + S)	1											2	1	2
Pinus mugo ssp. *rotundata*														1
Differentiating species of the ground tier:														
Luzula luzuloides	5													
Poa nemoralis	1	5												
Geranium robertianum		5							1	1				
Stachys sylvatica		4		1										
Lamiastrum galeobdolon	1	4	1	1										
Chrysosplenium oppositifolium		3												
Melica uniflora	1	3												
Quercus robur (G)	1	3		1					1					
Dryopteris filix-mas	1	3				1								
Phyteuma spicatum		3												
Sambucus racemosa	1	3		1										
Viola reichenbachiana		3												
Hieracium sylvaticum	1	2								1				
Prenanthes purpurea	1	2	1	1						1				
Festuca gigantea		2							1	1				
Calamagrostis arundinacea	1	2												
Cirsium oleraceum		2												
Chelidonium majus		2												
Galium odoratum		2												
Anemone nemorosa		2	5											
Galeopsis tetrahit	1	2	4						1	2				
Polygonum bistorta			4											
Chaerophyllum hirsutum	1	2	3											
Filipendula ulmaria			2											
Stellaria nemorum			4	2										
Phalaris arundinacea			2	1										
Deschampsia cespitosa			3	3	1									
Lysimachia nemorum			3	3	1									
Holcus mollis		2	3	3	1					1				
Crepis paludosa			3	4	1									
Cirsium helenioides			3	2										
Equisetum sylvaticum			3	5	2									

Table 1. *(continued)*

Community type	1	2	3	4	5c	5b	5a	5d	6a	6b	6c	6d	7	8
Number of releves	14	4	6	7	23	11	21	8	19	14	9	10	16	15
Differentiating species of the ground tier *(continued):*														
Cardamine amara				3										
Carex remota	1			3										
Ajuga reptans			1	3										
Galium palustre				3	1									
Juncus effusus	1			3	1		1							1
Glyceria fluitans				2	1									
Scirpus sylvaticus				2										
Caltha palustris				1										
Carex brizoides				1										
Fagus sylvatica (G)	3	3	1	1	1		1	1	3	1	1		1	
Oxalis acetosella	4	3	5	4	2	5			5		1			1
Athyrium filix-femina	3	4	5	5	2	2	1		1	2	1			
Acer pseudoplatanus (G)	2	3	2	2					2	1				
Mnium hornum	3		3	4	2	1			1			1	1	1
Rubus idaeus	3		5	3		1	1		2	2		1	2	1
Maianthemum bifolium	2		1	3	1	2	1		1	1	1			1
Rubus fruticosus agg.	1	3	2							1				
Impatiens noli-tangere	1	2	2	2					1					
Gymnocarpium dryopteris	2			1					1					
Thelypteris phegopteris	1			1	1									
Sphagnum fallax				1	5		1						1	1
Polytrichum commune	1			3	4		1							1
Viola palustris	1				2									
Agrostis canina				3	2									
Carex rostrata					1									
Calamagrostis villosa	1		4	3	5	5	5	5	2	2	3	1	2	2
Trientalis europaea	1		1		3	3	3	2	1	2	1		1	1
Ptilidium ciliare												4		
Leucobryum glaucum	1									1	1	3	1	1
Calluna vulgaris												3	1	1
Pleurozium schreberi										1		3	1	
Melampyrum pratense										1		1		1
Huperzia selago													1	
Avenella flexuosa	5		1	2	2	5	5	5	5	5	5	5	5	3
Vaccinium myrtillus	4		1	1	2	4	5	5	5	5	5	5	5	5
Sorbus aucuparia (G)	3		1	1	1	3	2	5	5	5	5	4	3	2
Dicranum scoparium	3	2	1		1	1	2	4	2	3	5	5	3	4
Vaccinium vitis-idaea							1	1		1	1	2	3	4
Bazzania trilobata	1				1	1		2		1	1	1	1	4
Sphagnum girgensohnii	2			1	1	1	1	1			1	1	1	3
Sphagnum spec.	1			3	1		1						1	2
Molinia caerulea					1		1					1		2
Vaccinium uliginosum														2
Eriophorum vaginatum														1
Juncus squarrosus														1
Juncus filiformis														1

Table 1. *(continued)*

Community type	1	2	3	4	5c	5b	5a	5d	6a	6b	6c	6d	7	8
Number of releves	14	4	6	7	23	11	21	8	19	14	9	10	16	15
Species of disturbed sites:														
Galium harcynicum	1				1	1	2	2	4	4	2	2	1	1
Epilobium angustifolium	1				1	1	1	3	2	3		2	1	1
Urtica dioica		3	1	2					2	2		1		
Agrostis capillaris	1						1		1	2	1	1	1	
Rumex acetosella						1	1		1	2	1			
Carex pilulifera	2						1		2	1	1			
Epilobium montanum			1		1		1		1	1		1		
Digitalis purpurea							1		1	1				
Taraxacum officinale							1		2	1		1		
Mycelis muralis		2							1	2				
Anthoxanthum odoratum			1							1		1		
Other species:														
Dryopteris dilatata	5	2	3	3	3	3	4	4	5	5	5	2	3	1
Polytrichum formosum	4	2	3	3	2	3	3	5	2	3	5	3	5	3
Plagiothecium curvifolium	1	2	1	1			1	1	1		1	2	2	
Plagiothecium undulatum	1				1		2	3	1		1	1	1	
Campylopus flexuosus	1				1	1	1	3			2	1	2	2
Pohlia nutans	1						1	1				2	1	
Carex curta (= canescens)	1				1	1	1	1		1				1
Frangula alnus (G)	1						1			1		1		1
Salix caprea (G)	1	2					1			1				
Dryopteris carthusiana	1				1									
Luzula pilosa	1		1						1	1				
Plagiothecium nemorale				1	1		1							
Carex echinata	1				1						1			
Carex leporina					1					1	1			
Hieracium lachenalii										1		1	1	
Betula spec. (G)					1					1			1	1
Abies alba (G)	1									1	1			
Betula pendula (G)	1						1					1		
Blechnum spicant	1						1				1			
Thelypteris limbosperma				1	1		1							
Lophocolea bidentata			1	1										
Ranunculus repens		2		2										

[a](Frequency classes of the species: 1 = occurring in 1–20% of the releves; 2 = in 21–40%; 3 = in 41–60%; 4 = in 61–80%; 5 = in 81–100%). 1, Luzulo-Fagetum (beech forest); 2, *Acer pseudoplatanus*-Tilio-Acerion-community (forest in gullies and on steep slopes); 3, Stellario-Alnetum (forest on river margins); 4, Carici remotae-Fraxinetum (wet forest near springs); 5, *Calamagrostis villosa-Picea abies* community (spruce forests at high altitudes); 5a, Pure type; 5b, *Oxalis acetosella*-type; 5c, *Sphagnum fallax*-type; 5d, *Vaccinium myrtillus*-type; 6, *Avenella flexuosa-Picea abies*-community (spruce forests at low altitudes); 6a, Pure type; 6b, *Oxalis acetosella*-type; 6c, *Vaccinium myrtillus*-type; 6d, *Ptilidium ciliare*-type; 7, *Picea abies-Sorbus aucuparia*-community (forest on boulders); 8, *Bazzania trilobata-Picea abies*-community (on peat).

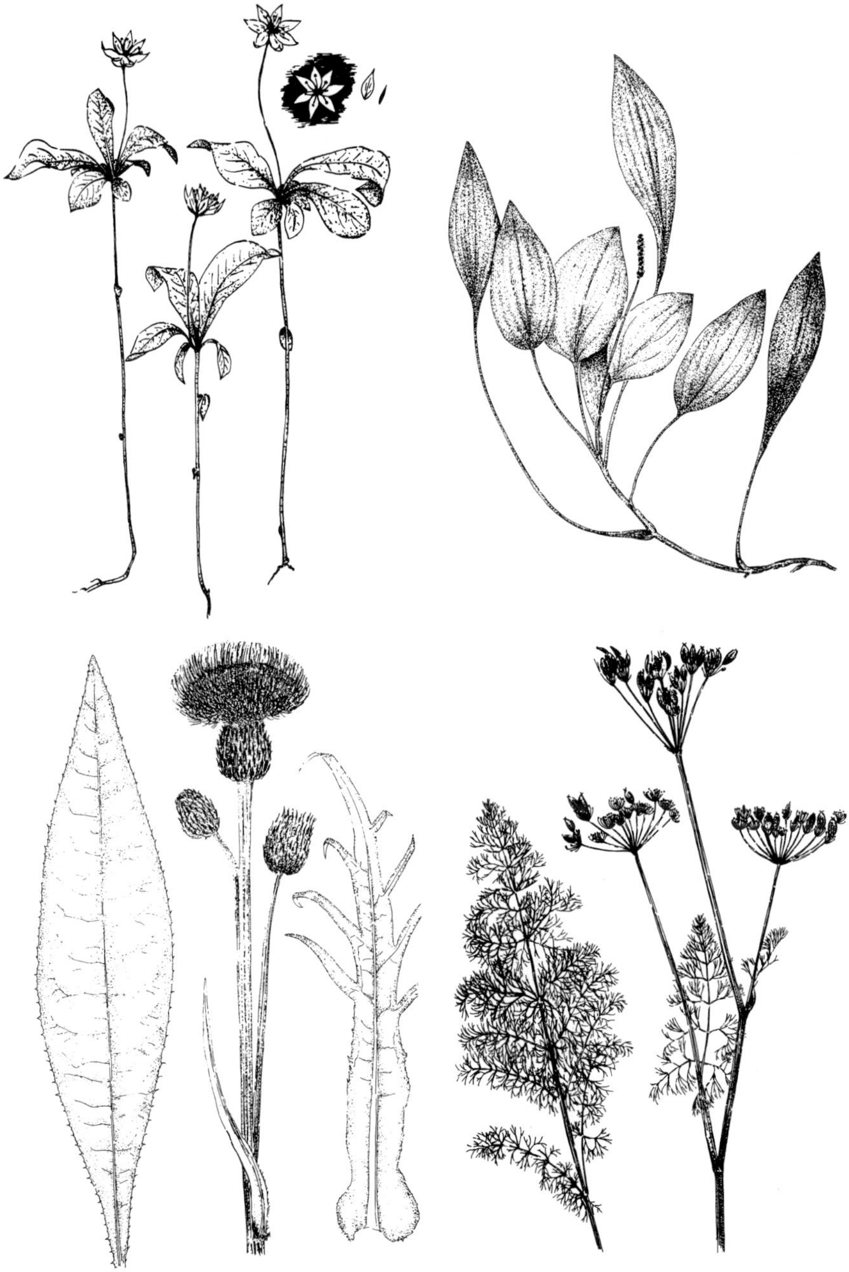

Fig. 7. Four characteristic plant species of the Fichtelgebirge. – Chickweed wintergreen (*Trientalis europaea; top left*) is a common species of the spruce forests at higher altitudes. – Bog pondweed (*Potamogeton polygonifolius; top right*) is a rare and endangered species of clean, acidic streams. The center of its German distribution is in the Fichtelgebirge area in northeast Bavaria. – Melancholy thistle (*Cirsium helenioides; bottom left*) is a typical species of relatively fertile meadows. Its natural sites are in wet forests near rivers. – Spignel (*Meum athamanticum; bottom right*) is a characteristic species of many meadows of less fertile sites at higher altitudes. The Fichtelgebirge populations form the southeastern border of the distribution of this "subatlantic" species (Walter 1982)

Within all the communities studied, forest damage is highest in these high altitude forests, as 70% of all releves with heavily-damaged canopy trees (damage class 3–4) fall into this community. Relatively young stands are particularly affected.

Within this high-altitude forest community, 3 types can be distinguished. A pure type (5a) mainly occurs at higher altitudes above ca. 800 m a.s.l. and has a nearly complete ground cover of *Calamagrostis villosa*. On steep, southerly slopes, *Oxalis acetosella* differentiates an *Oxalis acetosella*-type (5b) between 750 and 990 m a.s.l. A *Sphagnum fallax*-type (5c) is typical of wet, cold sites with long snow cover. Differentiating species are *Sphagnum fallax, Polytrichum commune, Viola palustris*, and *Agrostis canina*. Within this type, *Calamagrostis villosa* descends to lower (ca. 600 m) altitudes. A *Vaccinium myrtillus*-type (5d) is found on sites with a mosaic of earth and boulders. *Calamagrostis villosa* is still the dominant species on bare ground, but on the raw humus of the boulders differentiating vascular species like *Avenella flexuosa* and blueberry (*Vaccinium myrtillus*) occur. On bare rocks, epilithic mosses are frequent, e.g., *Paraleucobryum longifolium, Grimmia hartmanii*, and *Hypnum cupressiforme* (Hertel 1974).

6. Spruce forests with the dominant understory grass species *Avenella flexuosa* (*Avenella flexuosa-Picea abies*-community) are the most frequent community at altitudes between 600 and 800 m a.s.l. Beside *Avenella flexuosa*, only a few species comprise the low ground cover. Because of disturbance, mainly from forest management, species like *Galium harcynicum, Epilobium angustifolium, Urtica dioica, Carex pilulifera*, and *Rumex acetosella* occur. Within this community, four types can be distinguished. A pure type (6a) contains forests with older spruce trees and a more open canopy. *Avenella flexuosa* covers the ground nearly completely. An *Oxalis acetosella*-type (6b) is differentiated by *Oxalis acetosella* and seedlings of beech, which indicate more fertile site conditions. A *Vaccinium myrtillus*-type (6c) with such species as blueberry and rock- and raw-humus mosses, e.g., *Paraleucobryum longifolium, Dicranum scoparium* and *Polytrichum formosum*, are found in forests with some rocks and boulders. On relatively dry, shallow soils, a *Ptilidium ciliare*-type (6d) occurs. Differentiating species are heather (*Calluna vulgaris*), common cow-wheat (*Melampyrum pratense*), the mosses *Leucobryum glaucum* and *Pleurozium schreberi*, and the liverwort *Ptilidium ciliare*. On some releves, Scots pine is found in the canopy.

7. On upper slopes and ridges, e.g., the Haberstein and Platte, large boulder fields occur locally. Their margins are colonized by an open forest with spruce, the less-competitive tree species rowan, and the less-frequent downy birch (*Betula pubescens* ssp. *carpatica*) in the canopy (*Picea abies-Sorbus aucuparia*-community). In the ground cover, this community is differentiated by vascular plant species typical of raw humus deposits on boulders and rocks. Blueberry (*Vaccinium myrtillus*) and cowberry (*Vaccinium vitis-idaea*) are found on the boulders, while the fern *Dryopteris dilatata* grows between the gaps. Mineralization of raw humus in crevices and high light conditions have favored

species like *Epilobium angustifolium* and *Rubus idaeus*. Nonvascular plants are frequent (Hertel 1974). In this community, spruce would naturally be the dominant canopy species. These forests represent a transition to open vegetation on boulder streams.

8. In spruce forests near the margins of bog areas and on drained peat deposits, blueberry and cowberry are still dominant understory species. Differentiating species are cotton-grass (*Eriophorum vaginatum*) and the mosses *Bazzania trilobata* and *Sphagnum girgensohnii* (*Bazzania trilobata-Picea abies*–community). In the canopy, Scots pine and *Pinus mugo* ssp. *rotundata* are infrequently found. This community resembles the Vaccinio uliginosi-Pinetum sylvestris sensu Neuhäusl (1969).

5 Vegetation of Raised Bogs

A few sites in the Fichtelgebirge are naturally free of forests; these include raised bogs and boulder streams. Raised bogs are typically established on saddles. Dominant species are *Sphagnum* mosses, including *Sphagnum fuscum, S. cuspidatum,* and *S. rubellum* (Schwab 1908). Vascular plants include the Ericaceae Northern bilberry (*Vaccinium uliginosum*), cranberry (*Vaccinium oxycoccus*), bog rosemary (*Andromeda polifolia*), crowberry (*Empetrum nigrum*), the carnivorous sundew (*Drosera rotundifolia*), and sedges like *Eriophorum vaginatum*. Today, most raised bogs are drained and the peat is dug. The most intact remaining bog area is the *Fichtelseemoor* which is situated in a depression of 45 ha in the saddle between the two highest mountains of the Fichtelgebirge, Ochsenkopf and Schneeberg (Fig. 2). Some 9000 years ago, a ca. 6.5 m-thick peat deposit began to form from *Sphagnum* mosses, but was almost completely dug during a period of only 9 years (1840–1849; Kronberger 1940). Nearby, an artificial lake, the "Fichtelsee", was also built ca. 200 years ago to maintain a continuous water supply to the adjacent mining area. Today, peat accumulation is hindered, and many rare species, e.g., labrador tea (*Ledum palustre*), Rannoch rush (*Scheuchzeria palustris*), *Carex dioica, C. limosa,* and *Rhynchospora alba* have disappeared (Kronberger 1940; Walter 1982).

6 Vegetation of Boulder Streams and Summit Rocks

On a few steep slopes, e.g., the Haberstein and the Platte, large areas are covered by boulder streams (Fig. 5). These were formed under periglacial conditions. They have never been covered by soil and have therefore never been forested. On such sites, less competitive, and thus rare, species, e.g., the fern allies and ferns *Huperzia selago* and *Asplenium septentrionale,* mosses (*Andraea rupestris, Rhacomitrium heterostichum, R. lanuginosum*), liverworts (*Barbilophozia lycopodioides, Cephalozia ambigua*), and lichens (*Umbilicaria polyphylla, Parmelia saxatilis, P. incurva, Stereocaulon vesuvianum*) have not been shaded out by trees and could survive. In humid, shady rock crevices, the moss *Schistostega pennata* occurs (Hertel 1974). With increasing soil cover, an open forest of rowan, downy birch (*Betula pubescens* ssp. *carpatica*), Scots pine and low-growing spruce can develop, e.g., near the margins of boulder fields and on the rocks of the summits. Similar forests can be found in other mountain ranges, e.g., Rhön (Bohn 1981) and Harz (Stöcker 1967).

7 Water Plants and Acidic Swamp Vegetation

The water of creeks, rivers, and (artificial) lakes has naturally low pH- and nutrient-content. Typical species for acid, oligotrophic lakes are *Nymphaea candida*, greater bladderwort (*Utricularia vulgaris*), and broad-leaved pondweed (*Potamogeton natans*) (Paul 1912; Walter 1982). On lake margins, *Juncus bulbosus*, marsh cinquefoil (*Comarum palustre*), and sedges (*Carex lasiocarpa, C. rostrata*) initiate succession. In contact, acidic swamp vegetation can be found and is also dominated by sedges (*Carex nigra, C. echinata, C. curta, Eriophorum angustifolium*) and *Sphagnum* mosses (*S. fallax*). Other typical species are bogbean (*Menyanthes trifoliata*); *Trifolium spadiceum;* common butterwort (*Pinguicula vulgaris*); marsh violet (*Viola palustris*); heath spotted orchid (*Dactylorhiza maculata* ssp. *fuchsii*); bog arum (*Calla palustris*); and *Agrostis canina*. Adjacent to swamps, successional shrub communities with eared willow (*Salix aurita*) and alder buckthorn (*Frangula alnus*) can be found.

In clean creeks, and in occasional association with the freshwater pearl mussel (*Margaritifera margaritifera* L.), reddish pondweed (*Potamogeton alpinus*) occurs. The most intact populations of southern Germany of Bog pondweed (*Potamogeton polygonifolius*; Fig. 7) occur in the Fichtelgebirge (Bauer et al. 1980; Roweck et al. 1986). Typical nonvascular submerged plants are *Fontinalis squamosa, F. antipyretica, Scapania undulata,* and *Chiloscyphus polyanthus* (Hertel 1974; Paul 1912). Adjacent to creeks, stream banks are frequently colonized by dense communities of butterbur (*Petasites hybridus*) and at higher altitudes, locally by white butterbur (*P. albus*).

8 Vegetation of Farmland

Within the last 800 years, man has colonized the Fichtelgebirge and established permanent settlements (Scherzer 1962). Typically, village names end with the syllable *-grün* (= green), *-greuth*, or *-gereuth* (= clearing) in reminder of the forest clearing. Agricultural land use has resulted in a variety of anthropogenic vegetation types. Pasture land for summer animal grazing, regularly mowed hay meadows for winter fodder, and fields for potatoes, oats and rye have been cleared. As a result, a variety of different vegetation types occurs today.

Meadows represent the most common land use at the higher altitudes. Because of regular mowing, grasses are favored and are the dominant life form (Oberdorfer 1983). In the last century, many meadows of the Fichtelgebirge have been systematically irrigated (Paul 1912). This is believed to have compensated for the permanent loss of nutrients from harvest (Klapp 1971). Today, differences in use intensity, fertilization, and soil and water conditions have resulted in a variety of different community types (Reif and Weiskopf 1988). Characteristic species of the meadows of the Fichtelgebirge are bistort (*Polygonum bistorta*), melancholy thistle (*Cirsium helenioides*; Fig. 7), and, on less fertile sites, spignel (*Meum athamanticum*; Fig. 7). On wetter meadows, marsh marigold (*Caltha palustris*), marsh thistle (*Cirsium palustre*) and forget-me-not (*Myosotis palustris* spp. *nemorosa*) occur. In river valleys, abandoned farmland is covered with the dominant meadowsweet (*Filipendula ulmaria*) and hairy chervil (*Chaerophyllum hirsutum*), and less frequently with *Juncus acutiflorus* and *Carex brizoides.*

Pasture land was more common in past centuries. During the day, the animals, mainly cows, were kept in unfertilized areas around the settlements, and in adjacent forests (Walter 1982). The impoverishment of these sites due to grazing has resulted in low-growing, open grasslands with less competitive small forbs and grasses, e.g., *Nardus stricta*; heath bedstraw (*Galium harcynicum*); common milkwort (*Polygala vulgaris*); heath milkwort (*P. serpyllifolia*); arnica (*Arnica montana*); sulphur cinquefoil (*Potentilla erecta*); harebell (*Campanula rotundifolia*); *Hypericum maculatum; Carex pallescens; C. pilulifera; Hieracium lactucella; Gentianella campestris* ssp. *baltica;* and mountain everlasting (*Antennaria dioica*). On wetter sites, *Juncus squarrosus,* lousewort (*Pedicularis sylvatica*), and *Sphagnum* mosses enter these communities and represent a transition towards acidic swamp vegetation. Since livestock raising as a lifestyle is disappearing, most of these pasture lands today have disappeared, and some species restricted to them are in danger of extinction.

9 Conclusion

The Fichtelgebirge is a small mountain range in southwest Germany. In natural forests, beech (*Fagus sylvatica*) was the most frequent canopy tree. In the last few centuries, nearly all natural forests of this area have been cleared and burned by man. The present forest vegetation was planted since the last century, and today, spruce (*Picea abies*) is the dominant tree species. Because of differences in altitude, soil fertility, disturbance, and soil water conditions, 8 plant communities can be distinguished. The forests dominated by beech and the forests of extreme sites, e.g., on wet sites, on boulders, and on peat, are clearly differentiated by a number of species. The spruce forest communities and community types mainly are differentiated by varying dominance of three species of the ground cover: at altitudes below ca. 800 m a.s.l. by the presence of the grass *Avenella flexuosa*, and at higher altitudes by the grass *Calamagrostis villosa*. In spruce forests on boulders, raw humus accumulates, and blueberry (*Vaccinium myrtillus*) becomes dominant. There are differences in canopy tree mortality between these communities. Mostly affected by tree damage are relatively young stands of *Calamagrostis villosa*- spruce forests at higher altitudes.

References

Bauer G, Schrimpff E, Thomas W, Herrmann R (1980) Zusammenhänge zwischen dem Bestandsrückgang der Flu¯perlmuschel (Margaritifera margaritifera) im Fichtelgebirge und Gewässerbelastung. Archiv Hydrobiol 88:505–513

Beug HJ (1957) Untersuchungen zur frühpostglazialen Floren- und Vegetationsgeschichte einiger Mittelgebirge (Fichtelgebirge, Harz und Rhön). Flora 145:167–210

Bohn U (1981) Vegetationskarte der Bundesrepublik Deutschland 1:200 000 – Potentielle natürliche Vegetation – Blatt CC 5518 Fulda. Schriftenr Vegetationskd 15:330 s

Firbas F (1952) Spät- und nacheiszeitliche Waldgeschichte Mitteleuropas nördlich der Alpen. 2. Bd: Waldgeschichte der einzelnen Landschaften. Gustav Fischer, Jena, 256 s

Firbas F, Von Rochow M (1956) Zur Geschichte der Moore und Wälder im Fichtelgebirge. Forstwiss Centralbl 75:367–380

Funck HC (1800–1832) Cryptogamische Gewächse (besonders) des Fichtelgebirg's. Hof Leipzig

Gauckler K (1954) Serpentinvegetation in Nordbayern. Ber Bayer Bot Ges 30:19–26

Hertel E (1974) Epilithische Moose und Moosgesellschaften im nordöstlichen Bayern. Ber Naturwiss Ges Bayreuth, Beih 1:489 s

Klapp E (1971) Wiesen und Weiden. Parey, Berlin Hamburg, 620 s

Klimaatlas von Bayern (1952) Deutscher Wetterdienst in der US Zone, Zentralamt Bad Kissingen 230 s

Krauss GS (1952) Zur Frage der forstlichen Standortsrassen. Forstwiss Centralbl 71:349–359

Kronberger K (1940) Der "weltberuffene Fichtelberger See", jetzt Naturschutzgebiet Torf-, See- und Hüttenlohe. Ber Naturwiss Ges Bayreuth 4:1–27

Leonhard A (1988) Wald- und Forstgesellschaften im Fichtelgebirge. Diplom Thesis, Univ Bayreuth

Lutz JL (1956) Spirkenmoore in Bayern. Ber Bayer Bot Ges 31:58–69

Neuhäusl R (1969) Systematisch-soziologische Stellung der baumreichen Hochmoorgesellschaften Europas. Vegetatio 18:104–121

Oberdorfer E (1983) Süddeutsche Pflanzengesellschaften. Gustav Fischer, Stuttgart, Teil III, 455s

Paul H (1912) Zur Pflanzenwelt des Fichtelgebirges und des angrenzenden oberpfälzischen Keupers. Ber Bayer Bot Ges 2:402–410

Reif A (1983) Nordbayerische Heckengesellschaften. Hoppea, Denkschr Regensb Bot Ges 41:3–204

Reif A, Weiskopf A (1988) Ökologische Untersuchungen an der Verschiedenblättrigen Kratzdistel (Cirsium helenioides L.) in Oberfranken. Teil I: Vergesellschaftung und Standort. Tüxenia 8:101–148

Roweck H, Weiss K, Kohler A (1986) Zur Verbreitung und Biologie von Potamogeton coloratus und P. polygonifolius in Bayern und Baden-Württemberg. Ber Bayer Bot Ges 57:17–52

Rutte E (1981) Bayerns Erdgeschichte. Der geologische Führer durch Bayern. Ehrenwirth, München, 266 s

Scherzer C (1962) Franken. Land, Volk, Geschichte, Kunst und Wirtschaft. Drexel, Nürnberg, Bd 2

Schwab A (1908) Torfmoos-Flora des Fichtelgebirges. Denkschr Kgl Bayer Bot Ges Regensb 10 NF 4:75–92

Seu¯E (1977) Die Pflanzenwelt zwischen Waldstein und Döbraberg. Heimtbeilage zum amtlichen Schulanzeiger des Regierungsbezirks Oberfranken 56:1–43

Stöcker G (1967) Der Karpatenbirken-Fichtenwald des Hochharzes. Eine vegetationskundlich-ökologische Studie. Pflanzensoziologie (Jena) 15:123 s

Vollrath H (1955/1957) Die Pflanzenwelt des Fichtelgebirges und benachbarter Landschaften in geobotanischer Schau. Ber Naturwiss Ges Bayreuth 9:1–250

Walter E (1982) Wildpflanzen in Fichtelgebirge und Steinwald. Hoermann-Verlag Hof, Bayreuth, 162 s

3 Potentials and Limitations of Field Studies on Forest Decline Associated with Anthropogenic Pollution

R. OREN, K.S. WERK, J. MEYER, and E.-D. SCHULZE

1 Introduction

Responses to anthropogenic pollution may occur at various levels of biological organization from organelles to the ecosystem, over periods of time varying from seconds to years. Thus, it is no wonder that single explanations for the decline of forests have ranged from the biochemical to the biogeochemical, and from the microbiological to the ecological. Explaining an observed phenomenon depends on the level of organization immediately below the level of observation, but the significance of the observation lies in the level immediately above (Passioura 1979). Information about different levels of biological organization must be integrated if clearer understanding, rather than greater confusion, is to be achieved (Ashmore 1988). Thus, Ashmore (1988) has emphasized that quantitatively measuring selected tree components may fail to explain the ultimate course of a tree's decline. Only when the whole organism's vitality both above- and belowground is assessed, and the significance of the results on the appropriate spatial and temporal scales evaluated, can a firm conclusion about the causes of decline be reached. These criteria also apply to assessing decline in whole forest ecosystems (Ulrich 1987).

In the following review of approaches to the study of stress effects on ecosystems, we present the appropriate criteria for selecting an approach and choosing study sites in order to ensure relevant results. Because the study of forest ecosystems in relation to forest decline is complicated by high natural variability, natural stress factors, and logistic, practical limitations, the study of such complex systems must be carefully planned so that the objectives are meaningful, attainable, and appropriate for the expected scope of inference. A combination of both reductionist and holistic approaches must be used for a complete diagnosis of forest decline. However, the objectives of forest decline studies have often been too limited in scope, and thus have failed to narrow convincingly the number of stress factors to those most likely responsible for decline. It may be valuable, therefore, to trace the pattern by which scientific knowledge accumulates, relating it to the study of forest decline.

In his treatise on the structure of scientific revolutions, Kuhn (1970) cited the methodological dictum of Francis Bacon that "truth emerges more readily from error than from confusion." Scientific revolutions, argued Kuhn, progress in a fairly defined fashion, which at first is comprised of observation, logical synthesis and experimentation. Together these stages give rise to a "paradigm," or a set of

E.-D. Schulze, O.L. Lange, R. Oren (Eds.)
Ecological Studies Vol. 77

theories and views. The paradigm is usually accepted if it promises to be more successful in solving problems than other paradigms, and not because it solves all the problems that any preceding paradigm failed to solved. Subsequent research is primarily aimed at articulating and elaborating the paradigm, not at falsifying it. Observations that appear to contradict the paradigm are initially ignored, but as evidence of an anomaly mounts, a period of conceptual adjustment begins. When unexpected research results based on the established paradigm become anticipated results, the paradigm is changed. Precise paradigms are sensitive indicators of anomalies and provide a greater chance of a paradigm than ambiguous paradigms.

Recently, Hersher (1988) has argued that Kuhnian revolutions do not occur in biology, but instead, like the germ theory of disease, paradigms result from the accumulation of knowledge. For example, many "theories" of illness have coexisted, but because anomalies can develop only in relation to a single theory, research has been unable to progress from one theory to another as Kuhn suggested. Similarly, for more than a decade, many "theories" of forest decline from anthropogenic pollution have also coexisted (Cowling 1988). Like research on disease etiology, studies of forest decline have not progressed from one paradigm to another but appear to involve "a process of selective retention among a suite of wild conjectures" (Mentis 1988), a process which seems to dominate the development of knowledge in many new research fields (Mentis 1988). A researcher of forest decline may find comfort in this thought, considering the ultimate success of disease research. However, Bacon's dictum may also mean that conducting research based on some acceptable order of rules may reduce the time required to identify the causes of forest decline at different locations.

2 Defining Forest Decline

While the occurrence of forest decline in numerous European countries is hardly disputed (Rehfuess 1987; Cape et al. 1988), it is difficult to identify stands in which to study the phenomena. The lack of specific symptoms may prevent distinguishing those stands suffering from air pollution from those which are experiencing an episodic occurrence of natural stress, or simply following the normal path of stand development.

Declining stands are usually associated with visual symptoms of unusual foliage color, thinning of the crowns, and excessive mortality. Needle chlorosis and premature needle shedding are the most commonly used qualitative indications of decline and its severity (Rehfuess 1987; Cape et al. 1988). These nonspecific indicators of a stress-afflicted forest can result from many biotic (e.g., disease or insect) or abiotic (e.g., water or nutrient limitation, low temperature, or air pollution) factors (Cowling 1988; Ashmore 1988). Moreover, some of these symptoms may indicate a normal state in the phenology of plants and stand development. The foliage color of evergreen conifers can temporarily turn yellow and then re-green within one growing season. This represents a normal seasonal phenomenon (Blinn and Buckner 1987). In addition to the lack of specificity, damage estimates based on these indicators may be greatly affected by observer subjectivity, light conditions, and site and stand characteristics (Innes 1988; Vales and Bunnell 1988).

Reduction in stand growth is a less often used indicator of forest decline (Franz 1983; Oren et al. 1988) because of the time required for quantification and the difficulty of proper measurement. Tree mortality, a more easily measured stand variable, does not necessarily indicate the action of any one agent, but may simply be a self-thinning process, resulting from intraspecific competition in dense plant populations as the average size of the plants increases (Kira et al. 1956; Yoda et al. 1963).

Many other variables have also been identified as indicators of decline. For example, the needle ratio of high-energy compounds in their fully charged state to their less charged form (e.g., ATP/ADP, NADPH/NADP; Keil et al. 1988) indicates changes in the metabolic pathways and accelerated senescence. Still other variables range from changes in leaf photosynthetic characteristics (Zimmermann et al. 1988) to ecosystem variables, such as the change in the nitrate concentration in streamwater (Hauhs and Wright 1986).

In general, morphological, physiological and biochemical parameters specifically relating to stress factors, and showing a consistent and unique association with stress factors, should be selected over nonspecific indicators (Ashmore 1988). Such selection may only be possible for some stresses. Alternatively, a unique sequence or combination of plant symptoms and ecosystem responses may indicate decline from a specific stress factor. Identification of such a sequence or combination would necessarily require a comprehensive study. For optimal use of available resources, the most likely stress factor should be identified. If, for example, ozone is thought to be the cause of decline, specific parameters should be studied. However, if acidic precipitation is the probable cause of forest decline, a comprehensive study is required, because most of the responses are nonspecific and only their sequence or combination can be specifically associated with such pollution.

3 Inference of Causality

Rules to orderly conduct research have been used by plant pathologists for many decades to ascertain a specific relationship between a stress factor and a particular plant response. Such rules are provided by Koch's postulates for the study of human disease (Koch 1884). Koch's postulates require that an agent must be isolated from a diseased plant and reinoculated in healthy plants. Only if the same symptoms and decline in health are found, and the agent is re-isolated from these plants, can causality be proven. Koch's postulates have also been adopted to determine which abiotic stress factor is the cause of decline in complex ecosystem (Cowling 1988). Briefly, an injury or dysfunction in individuals in the ecosystem must be consistently associated with the suspected causal factor and duplicated when a healthy organism is exposed to that factor under controlled conditions. In addition, healthy clones of plants representing the natural range of variation in resistance to the stress factor should also be exposed to the factor, and the same range in symptoms found in their parents in the field, must be duplicated before causality can be inferred.

In practice, the necessary causal linkages between a suspected stress factor and forest decline may be established by a sequence of research approaches. First, observations are made and new knowledge, formulation of specific hyphotheses, and development of general theories based on the specific data result through inductive reasoning (the hypothetico-inductive approach; Mentis 1988). Then data are collected and analyzed in order to test the specific a-priori hyphotheses suggested through deductive reasoning by the new theories (the hypothetico-deductive approach). Each approach has merits and limitations when applied singly to ecological research.

3.1 The Inductive Approach

In ecology, multiple causes and effects and the interdependence of the measured variables result in data which contain many correlates. Because no replication or only pseudo-replication is possible, inferential statistics cannot be used, so that cause and effect relationships often cannot be identified (Mentis 1988). However, some statistical analyses, e.g., regression and correlation analysis, may allow the screening of a large number of variables, so that a few apparently key variables can be identified and subsequently tested for cause-and-effect relationships (Ashmore 1988). Such screening allows patterns in the observational data to be identified and information about ecosystem functioning extracted. The insights and inferences resulting from this approach are largely independent of premises and less constrained by preconceived views than those drawn from the hypothetico-deductive approach. Such inferences, called ampliative inferences (Mentis 1988), may be used as a descriptive-predictive tool or for generating testable hypothesis. Ampliative inferences, however, may not be necessarily true. The larger the data-set on which they are based, the more we tend to believe them. However, their predictive value must be tested with additional data-sets. In addition, ampliative inference and hypothesis-testing may suffer common drawbacks (Mentis 1988). First, there are limitations in obtaining sufficient replication in data-sets to draw any inferences. Secondly, both ampliative inferences and hypothesis-testing are affected by theories, so that the choice of variables to be measured and the analyses to be used are not free from bias.

Another approach to ecosystem research attempts to overcome the difficulty of replication by drastically manipulating the experimental system, e.g., eradicating all vegetation from a watershed, but not altering the reference system (Likens 1985). Descriptive data are then collected to evaluate how the changes in the manipulated system can be compared to those in the reference system, and whether this can be most plausibly explained by the manipulations (Carpenter 1988). This approach does not involve statistics, but rather relies on extreme manipulation to exclude the probability that natural variation could have caused the changes. While the changes in the manipulated system are accepted as evidence of the effect of the manipulation, the value of such extreme manipulations has been questioned (Crowder et al. 1988).

Such an approach may be useful in evaluating the ecosystem effects of high levels of anthropogenic pollution as found near pollution sources. However, there is little scientific uncertainty that pollution is the cause of forest decline near such sources, because the steep pollution gradient has provided strong correlative evidence and decline symptoms have been reproduced in controlled experiments (Cowling 1988; Ashmore 1988). Conversely, great uncertainties still exist at the fringes of areas receiving pollution from a particular source, and in areas where there is chronic exposure to low level pollution, uncertainties about the cause-and-effect relationship between anthropogenic pollution and forest decline, and about the role of natural factors which contribute or predispose ecosystems to stress. These uncertainties stem from the subtlety of visible symptoms, the large spatial and temporal variation in pollutant concentrations, the possibility of pollutant interactions, and the difficulty of distinguishing possible responses to pollution against a background of large natural variation in ecosystems.

The most commonly used method of studying the effects of air pollution in the field is the comparison of two or more sites at seemingly different stages of decline. One should think of the site comparison approach as the first step in following the routine developed from Koch's postulates, if a causal relationship between a stress factor and decline is to be confirmed. However, the approach is often misused in two ways: firstly, sites are normally chosen on the basis of apparent decline symptoms such as yellow foliage, needle loss and tree death. Other sites are apparently healthy because they lack these symptoms. Differences between such sites in terms of measured variables are then used to explain the apparent decline. Such selection ignores the fact that any two sites are likely to differ on several measurable parameters, even when both are either healthy or in decline, so that differences between sites on any parameter may be unrelated to decline. Thus, when direct damage is suspected, damage or injury surveys require accompanying data on whether pollution concentrations were sufficient to cause injury and were consistently present at the time and place where the symptoms were observed. This still presents only correlational, not confirmatory, evidence (Cowling 1988). A second misuse of this approach concerns comparisons between trees apparently in decline (e.g., with yellow needles) and those apparently healthy (e.g., with green needles only); such comparisons may reveal differences which are related, for example, to genetic variations as reflected in phenology, or to plant responses to decline. Site comparison is thus not a study that alone can demonstrate a causal relationship.

Among several approaches which Jarvis (1987) has proposed for studying ecosystems is the use of unplanned experiments, such as pollution damage. Unplanned experiments do not meet the rigorous standards of Koch's postulates. However, careful design of such observational studies can prevent construing false causes as true. Qualitative hypothesis-rejection is a key part of the comparative investigative design.

Designing a study on the effects of air pollution on forest stands first requires a careful definition of decline. Sites must then be selected in which all site and stand characteristics, except for the pollution load, affect plant function equally or are more conducive to proper plant function and growth at the site displaying decline. By either choosing sites and stands which are qualitatively equivalent, or selecting a declining stand of superior growth characteristics, or one in a more conducive environment, the researcher immediately eliminates several non-pollution factors such as stand density, age, and water or temperature as possible explanations for the poor performance of the declining stand. In this way, several hypothetical reasons for the decline can be qualitatively rejected.

If sites are chosen such that a declining stand is at a site receiving high levels of pollution, but the environment is otherwise much more conducive to plant function than at a site receiving low pollution, two problems may arise. First, the pollution effects may be compensated for, so that erroneous conclusions on the role of pollution in decline may result. Secondly, environmental factors are thought to interact with some pollutants. Such interactions may render a pollutant more or less detrimental in a better environment and lead one to an erroneous assessment of the damage caused by the pollutant. Thus, only slight environmental differences in favor of the declining stand may be accepted when declining and healthy stands in equal environments are not available. For best results, the site comparison approach requires some knowledge of the requirements necessary for good performance of the species studied. It also requires measurements of many sites and stands at different stages of decline, so that a few sites in fairly similar environments can be chosen.

Finally, any study should begin with thorough planning and evaluation of the objectives and approach, experimental design, methods, data analysis, interpretation, and the expected scope of inference (Mentis 1988). Pilot studies and data simulation can be used to save time.

3.2 The Deductive Approach

Through the hypothetico-deductive approach, hypotheses can never be proven but may only be rejected or not, depending on the degree to which observations, experience, or other theories conflict (Mentis 1988; Ambrose and Ambrose 1977). The aim of a study is to falsify, not to validate, hypotheses. One problem with the deductive approach is the possibility of generating a true prediction from both true and false hypotheses (Ambrose and Ambrose 1977). This is the reason why a theory which has never been falsified, even through rigorous attempts, may still be replaced by a better theory, or that a refuted theory may be reinstated as conflicts disappear when other related theories have changed (Mentis 1988).

Mentis (1988) has discussed several conceptual problems to applying the hypothetico-deductive approach to ecology. One is the proper way by which falsifiable hypotheses can be formulated and refuted. Mentis has argued and demonstrated that the results of statistical tests are often affected by the design of the experiment via the choice of sampling intensity. In addition, statistical results can be affected by the selection of the power of the test, a selection often done without consideration of the biological meaning of statistical differences, the expected variation in the data, and the expected risk from erroneous judgment.

Some of these problems may be solved by careful experimental design, including an a priori definition of the hypotheses and the degree to which differences can be accepted as biologically important. The variations may be reduced by performing controlled experiments in the laboratory. However, such experiments have been criticized on the grounds that effects may not be unique to the factor under investigation, or that the conditions of the experiment are not representative of the field (Ashmore 1988). In contrast, field experiments with their considerable potential for hypothesis testing in ecology are subject to logistic and ethical limitations (Jarvis 1987; Mentis 1988). Therefore, experimental manipulation of forest stands to study the effects of air pollution on forests (Bååth et al. 1980; Stuanes et al. 1988) are rarely practiced.

The major logistic limitations to ecosystem experiments are the lack of funding committed to long-term experiments and the shortage of sufficient experimental units. For example, because it is difficult to find a sufficient number of lakes in a given region and only a few manipulations can be made, only a limited number of questions can be answered (Carpenter 1988). If the appropriate unit in the study of forest decline is a watershed, even fewer questions can be addressed. Large scale manipulations are expensive, even without replications. However, because of the high natural variability in ecosystems, many replications of extreme, and thus ecologically unrealistic, long-term continuous manipulations (i.e., press experiments) may be required to detect ecosystem responses. Even more replications may be needed if a single perturbation is used (i.e., pulse experiments). Resolving these problems by reducing replications may lead to erroneous results and faulty conclusions.

New analytical techniques such as the randomized intervention analysis (see Carpenter 1988) have been developed to analyze unreplicated experiments on different perturbations of aquatic ecosystem structure and function. These techniques may be adaptable in studying the effects of anthropogenic pollution on terrestrial ecosystems. Such techniques use data which are collected before and after a treatment has been applied to two ecosystems, one of which is experimental, the other serving as a reference. The techniques are designed to detect if the same changes occur in both ecosystems following the treatment; if not, the treatment cannot be considered to have caused the change. To establish that the manipulation could have caused the response, all alternative causes which could have resulted in a similar response must be rejected beyond reasonable doubt. Determination of causality in an unreplicated experiment ultimately relies on ecological, not statistical arguments (Carpenter 1988).

Finally, testing one or a few favored hypotheses of ecosystem functioning may result in an inappropriate corroboration of the experimental hypothesis, simply because of the choice of measured variables, the experimental design, and the choice of data analysis (Mentis 1988). Such a limited perspective may fail to recognize that an ecosystem response emerges from numerous events which involve many interrelated variables and processes, all operating simultaneously, and many of which may vary with a given stress (Trewavas 1986). The limited perspective or "babying the hypothesis" does not meet the assumptions of inferential statistics (Ambrose and Ambrose 1977).

3.3 The Integrated Approach

Wallace (1978) has suggested a synoptic approach to diagnosing complex problems: (1) key factors are identified in an initial survey; (2) hypotheses are generated from a multiple regression analysis of the data; (3) the diagnosis is verified through field tests. Thus, the initial statistical analyses is an inductively-derived multiple regression, not a test of a priori hypothesis. When hypotheses are formed a posteriori after the data have been collected, there is a risk that the hypotheses are only ad hoc descriptors of a particular data set and have no general predictive or explanatory value (Mentis 1988). Legitimization of such post facto hypotheses must be achieved by some independent means.

4 The Study of a Decline in the Fichtelgebirge – a First Step in an Integrated Approach

A descriptive study, like the one summarized in this Volume, can only identify plausible causes of forest decline and generate hypotheses, which in turn require rigorous testing (Ashmore 1988). Nevertheless, because the study in the Fichtelgebirge was comprehensive and the sites and stands were selected so that most natural stress factors were qualitatively excluded, many possible causes of decline could be eliminated. The most probable hypothesis remaining was that forest decline was caused by a nutritional disharmony (Chap. 4-F,H).

Upon first screening the two sites in the Fichtelgebirge, yellow crowns were seen at only one site, while mortality occurred at both sites. In this volume, the stand with yellow crowns at Oberwarmensteinach will be referred to as "declining" to distinguish it from the other "healthy" stand at Wülfersreuth. After establishing five plots at each site, mean tree biomass was compared with tree density in an analysis of the self-thinning in the two stands (Fig. 1; Oren et al. 1988). Although mortality occurred in all plots, plot means of tree biomass at the declining site were lower at any level of stand density than those at the healthy site (Fig. 1). The self-thinning phenomenon may be complicated by stand history (Mohler et al. 1978), but detailed stand history of the two sites was not known. Nevertheless, trees at the declining stand, where some crowns were at different degrees of yellowing and needle loss, were dying for reasons other than density-related mortality.

In the Fichtelgebirge study, a decline was recognized only if needle-yellowing was accompanied by reduced stand growth rate per ground area. Growth reduction may be not only the first consequence of stress, but also the earliest indicator of incipient forest damage. When resources become limited, plants

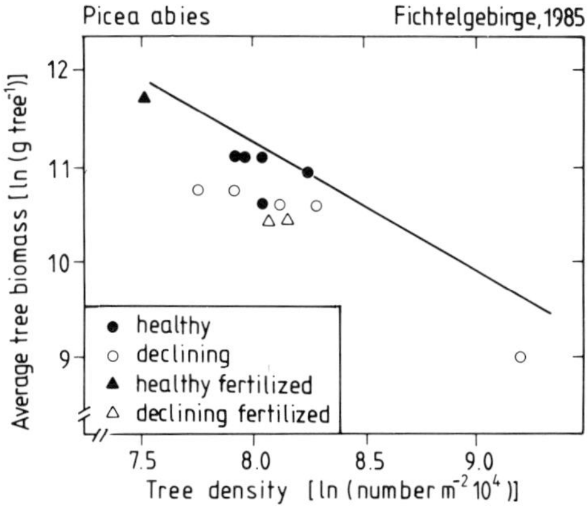

Fig. 1. Relationships between log-log transformed (base e) average tree biomass and plot density among plots in healthy and declining stands. The *line* represents a slope of -3/2. *Triangles* representing three additional plots in which Mg2+ was added at different rates and forms (Hantschel et al. 1988) in a fertilizer experiment are shown for comparison (Oren et al. 1988)

seem to favor maintenance of mature tissue functions (e.g., photosynthesis) over growth processes (McDonald et al. 1986).

Stand growth is highly related to stand density and tree size. The available data (Table 1; Fig. 1) have allowed us to reject current stand density as a reason for possible growth differences. Furthermore, higher maintenance respiration should not have caused a reduction in the growth rate at the declining site. Maintenance respiration should have been higher at the healthy site because respiration is a function of the living biomass, which was greater at the healthy site (Table 1; Oren et al. 1988; see Chap. 4-D) and temperature, which was similar in the air and soil at both sites during the year (Figs. 2 and 6 in Chap. 4-D; Zimmermann 1985).

Similarly, precipitation and light conditions were excluded a priori as reasons for decline by choosing the sites so that the declining stand received 25% more precipitation (Chap. 4-C) and had a more favorable aspect for irradiance input (Table 1; Oren et al. 1988). Although the two stands grew in soils which developed from the same metamorphic parent material (phyllite), the soils were podzolized to different degrees (Hantschel 1987; Chap. 4-A, B) and this difference probably played a role in reducing growth at the declining site even before the symptoms attributed to air pollution began (Chap. 4-F, H, I). Soil podzolization at the declining site may have been enhanced by higher precipitation, accompanied by a 60% greater proton input than at the healthy site, primarily because of a higher sulfate concentration in the precipitation (Hantschel et al. 1988; Chap. 1-B, 4-I).

Biomass estimation and growth analysis at the two sites revealed an important anomaly (Table 1). Although the biomass and annual production of crown components and roots were similar at both sites and the stands were of similar age and canopy leaf areas, the stem biomass and stem growth were lower at the declining site. The two stands are progenies of previous stands that were regenerated ca. 150 years ago from a mixed seed source. The source was drawn from various stands in Germany and introduced into the Fichtelgebirge (Forstamt Fichtelberg, pers. comm.). Even intensive genetic tree-improvement

Table 1. Site and stand description, biomass and growth of healthy and declining stands of *P. abies* located in the Fichtelgebirge, northeast Bavaria, FRG[a]

	Healthy		Declining
Site parameters			
Elevation (m)	675		755
Aspect	NW		SE
Slope (degree)	1		5
Growing season	May-Oct.		May-Oct.
Stand parameters			
Age (yrs)	30		30
Density (trees m^{-2})	0.31		0.44
Basal area (1.3 m; $cm^2 m^{-2}$)	32.2		26.8
Height (m)	10.04		7.93
Leaf-area index (projected; $m^2 m^{-2}$)	10.6		11.9
Biomass (g m^{-2})			
Needle	1767		2009
Branch	1371		1539
Stem	14386	*	7381
Root (< 10 mm)	352		369
Growth (g $m^{-2} yr^{-1}$)			
Needle	216		182
Branch	168		175
Stem	1114	*	737
Growth efficiency ($g_{wood} m^{-2}{}_{leaf\ area} yr^{-1}$)	100.8	*	58.3

[a] Significant difference ($P \leq 0.05$, t-test) between the sites is depicted by an *asterisk* (not tested for site parameters). Additional information on the soils and on the atmospheric conditions and pollution input at the two sites is presented in Chaps. 1-A, 1-B and 4-A to 4-C.

programs with pines can result in an increase in stemwood growth of 12% over one rotation (Zobel and Talbort 1983). This is far less than the 80% difference in stem biomass found between the two study stands. Therefore, it is unlikely that such a growth difference between the two stands was simply a result of genotypic differences between trees at the two sites.

Although the stem biomass was 80% higher, the current annual stem growth was only 55% higher at the healthy site. Thus, the differences in growth rate must have begun many years before even the earliest reports of forest decline in the Fichtelgebirge (Zech and Popp 1983). The difference in current growth rate is probably from the pre-decline differences in site quality (Oren et al. 1988) compounded by the recent addition of a factor or factors responsible for the appearance of the visual symptoms of decline.

When the density of trees in a forest stand is reduced by natural agents or harvesting, the growth of the remaining trees increases and may compensate for the loss of trees, so that no reduction of growth on a stand level occurs (Waring and Pitman 1985; Oren et al. 1985). This study, however, detected such a reduction. Indeed, tree-ring analyses of trees of similar age in central Europe have shown a great reduction in growth, regardless of stand density or tree position in the stand canopy (Schweingruber et al. 1983). The reduction in tree growth in central Europe began about 30 years ago and has been steadily increasing. Since the declining stand in this study grew on a site classified as lower in quality than the site of the healthy stand, it is difficult to separate the effects of pollution from natural site differences (Oren et al. 1988). However, current growth compared with that expected from site quality indicates that growth at the declining site has been reduced, but not growth at the healthy site.

It is tempting to interpret the smaller difference in annual growth rate between the stands compared to the difference in stem biomass as meaning that the differences between the stands have been decreasing. Such a decrease in growth might be an early indication of a decline in the so-called healthy stand. Some tree crowns in stands adjacent to the healthy stand of this study have recently turned yellow, indicating that under such conditions as nutrient-poor soils, stands growing in this site may also be susceptible. However, when growth is expressed on a leaf-area basis so that differences in leaf area and tree size are excluded (Waring et al. 1981), growth at the healthy site has been 73% higher than that at the declining site (Table 1). This indicates that a large difference in the production efficiency of foliage between the sites still exists (Fig. 2). Nonetheless, it is clear that this study only compared relatively healthy and declining stands, and that there is no assurance that either of the stands is not stressed.

Eliminating most factors which could cause one stand to grow more slowly than the other by carefully selecting sites and stands allowed us to isolate inadequate and imbalanced nutrition as the most probable reason for the decline. The unfavorable nutrient supply at the declining site is probably a consequence of a poorer soil impacted by a higher proton load. These conclusions must be verified with independent data-sets, and specific hypotheses suggested by the proposed explanation for the decline must be formed and tested experimentally. This second step in the integrated approach is currently being conducted in field and laboratory studies.

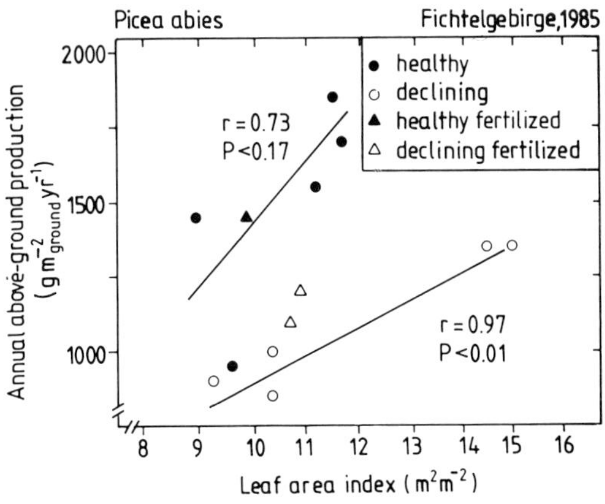

Fig. 2. Relationships between annual aboveground production and leaf-area index among plots in healthy and declining stands. *Triangles* represent three additional plots which were measured in a Mg2+ fertilizer experiment (Hantschel et al. 1988) but are not included in the analyses (Oren et al. 1988)

5 Conclusion

Before a study on forest decline can begin, decline must be defined. Many visual symptoms and ecological, physiological, and biochemical variables have been used as indicators of stress. Reduction in growth rate is an early response to stress, and is a good indicator of decline from air pollution, if all other factors are excluded both qualitatively and quantitatively as hypothetical causes of the growth reduction.

The study of forest ecosystems in relation to forest decline is complicated by the high natural variability, the effects of natural stress factors, and logistic, practical limitations. The most commonly used method of studying the effects of air pollution in the field is the comparison of two or more sites at apparently different stages of decline. For best results, this approach requires some knowledge of the plant species and its requirements for good performance. Also, measurements of many sites and stands at different stages of decline must be taken, so that a few sites in fairly similar environments can be chosen. The initial statistical analysis in such studies is an inductively-derived multiple regression, not a test of a priori hypothesis. When hypothesis are formed a posteriori, they may serve only to empirically describe a particular data set without any general predictive or explanatory value, and must be legitimized by some independent means (Mentis 1988).

When applied singly to ecological research, both inductive and deductive approaches have their limitations. However, an integrated approach, such as the adaptation of Koch's postulates to ecosystem studies, provides a valuable tool for studies such as forest decline from air pollution. First, observations are made through inductive reasoning which result in new knowledge, formulation of specific hypotheses and development of general theories based on specific data. Initial surveys may be used to identify key factors, hypotheses can be generated from multiple regression analyses of the data, and information can be integrated into conceptual models. Then, deductive reasoning is used in which data are collected and analyzed in order to test specific a priori hypotheses about stress mechanisms suggested by the new theories. During this phase of the research, the diagnosis may be verified through field tests. For a complete diagnosis, both reductionist and holistic approaches should be used, and processes at all levels of biological organization whose impairment could result in a decline of the observed magnitude should be studied. Regardless of how scientific knowledge is accumulated in biology (Kuhn 1970; Hersher 1988), theories should be stated unambiguously so they can be judged by the body of accumulated knowledge.

Conclusions based on short-term descriptive studies such as the Fichtelgebirge study described in this Volume must be considered only as a set of plausible hypotheses which can best explain the patterns in the data without major contradictions. To determine whether the pattern identified in our study is consistent, further studies must be carried out over several growing seasons (Mentis 1988). Even with experimental manipulations, symptoms may appear only after several years (Stuanes et al. 1988). The hypotheses must be rigorously tested before causes can be conclusively identified. Nevertheless, because of the comprehensiveness of this study, many possible causes of decline could be eliminated. The most probable cause of the forest decline was a nutritional disharmony (Chaps. 4-F, H). Several hypotheses relating nutritional disharmony to forest decline are currently being tested in controlled experiments in the field and in the laboratory.

References

Ambrose HW III, Ambrose KP (1977) A handbook of biological investigation. Hunter Textbook, Winston-Salem 3rd edn 170 pp

Ashmore MR (1988) Report on the session on the effects of air pollution on forest trees and forest ecosystems. In: Mathy P (ed) Air pollution and ecosystems. Proc Comm Eur Communities Symp Grenoble, France, May 1987, pp 284–287, Reidel Publishing, Dordrecht

Bååth E, Berg B, Lohm U, Lundgren B, Lundkvist H, Rosswall T, Söderström B, Wiren A (1980) Effects of experimental acidification and liming on soil organisms and decomposition in a scots pine forest. Pedobiologia 20:85–100

Blinn CR, Buckner ER (1987) Foliage color and nutrient levels. J For 85:48–49

Cape JN, Paterson IS, Wellburn AR, Wolfenden J, Mehlhorn H, Freer-Smith PH, Fink S (1988) Early diagnosis of forest decline. Inst Terrestrial Ecology report, 68 pp

Carpenter SR (1989) Replication and treatment strength in whole-lake experiments. Ecology (in press)

Cowling EB (1988) Ecosystems and their response to airborne chemicals: the current situation in North America and Europe. In: Mathy P (ed) Air pollution and ecosystems. Proc Comm Eur Communities Symp Grenoble, France, May 1987, pp 18–38, Reidel Publishing, Dordrecht

Crowder LB, Drenner RW, Kerfoot WC, McQueen DJ, Mills EL, Sommer U, Spencer CN, Vanni MJ (1988) Food web interactions in lakes. In: Carpenter SR (ed) Complex interactions in lake communities. Springer, Berlin Heidelberg New York Tokyo, pp 141–160

Franz F (1983) Auswirkungen der Waldkrankheiten auf Struktur und Wuchsleistung von Fichtenbeständen. Forstwiss/centralbl 102:186–200

Hantschel R (1987) Water and element balance of damaged, fertilized Norway spruce ecosystems in the Fichtelgebirge with respect to physical and chemical soil heterogenity. Bayreuther Bodenkundl Ber Bd 3:1–216

Hantschel R, Kaupenjohan M, Horn R, Zech W (1988) Acid rain studies in the Fichtelgebirge (NE-Bavaria). In: Mathy P (ed) Air pollution and ecosystems. Proc Comm Eur Communities Symp Grenoble, France, May 1987, pp 881–886, Reidel Publishing, Dordrecht

Hauhs M, Wright RF (1986) Regional pattern of acid deposition and forest decline along a cross section through Europe. Water Air Soil Pollut 31:463–474

Hersher L (1988) On the absence of revolution in biology. Perspect Biol Med 31:318–323

Innes JL (1988) Forest health surveys: problems in assessing observer objectivity. Can J For Res 18:560–565

Jarvis PG (1987) Water and carbon fluxes in ecosystems. In: Schulze ED, Zwölfer H (eds) Potentials and limitations of ecosystem analysis. Ecol Stud 61, pp 50–67, Springer, Berlin Heidelberg New York Tokyo

Keil R, Benz T, Einig W, Hampp R (1988) Adenine and pyridine nucleotides, and carbohydrates in healthy and deceased needles: Alterations in relation to vegetation period (May to October 1985) and needle age (1979 to 1985). In: Mathy P (ed) Air pollution and ecosystems. Proc Comm Eur Communities Symp Grenoble, France, May 1987, pp 584–589, Reidel Publishing, Dordrecht

Kira T, Ogawa H, Hozumi K, Koyama H, Yoda K (1956) Intraspecific competition among higher plants. V. Supplementary notes on the C-D effect. J Inst Polytechnics, Osaka City Univ, Ser D, Biology 7:1–14

Koch R (1884) Die Aetiologie der Tuberkulose. Mitt Kaiserl Gesundheitsamt 2:1–88

Kuhn TS (1970) The structure of scientific revolutions. Int Encycl Unified Sci 2 (no 2). Univ Chicago press, Chicago, 2nd edn 210 pp

Likens GE (1985) An experimental approach to the study of ecosystems. J Ecol 73:381–396
McDonald AJS, Ericsson A, Lohammar T (1986) Growth response to step-decrease in nutrient availability in small birch (*Betula pendula* Roth.). Plant Cell Environ 9:427–432
Mentis MT (1988) Hypothetico-deductive and inductive approaches in ecology. Funct Ecol 2:5–12
Mohler CL, Marks PL, Sprugel DG (1978) Stand structure and allometry of trees during self-thinning of pure stands. J Ecol 66:599–614
Oren R, Thies WG, Waring RH (1985) Tree vigor and stand growth of Douglas-fir as influenced by laminated root rot. Can J For Res 15:985–988
Oren R, Schulze ED, Werk KS, Meyer J, Schneider BU, Heilmeier H (1988) Performance of two *Picea abies* (L.) Karst. stands at different stages of decline. I. Carbon relations and stand growth. Oecologia (Berl) 75:25–37
Passioura JB (1979) Accountability, philosophy and plant physiology. Search 10:347–350
Rehfuess KE (1987) Perception on forest diseases on central Europe. Forestry 60:1–11
Schweingruber FH, Kontic R, Winkler-Seifert A (1983) Eine jahrringanalytische Studie zum Nadelbaumsterben in der Schweiz. Eidg Anst Forstl Versuchswes, Birmensdorf CH, 253:3–29
Stuanes AO, Abrahamsen G, Tveite B (1988) Effect of artificial rain on soil chemical properties and forest growth. In: Mathy P (ed) Air pollution and ecosystems. Proc Comm Eur Communities Symp Grenoble, France, May 1987, pp 248–253, Reidel Publishing, Dordrecht
Trewavas A (1986) Understanding the control of plant development and the role of growth substances. Aust J Plant Physiol 13:447–457
Ulrich B (1987) Stability, elasticity, and resilience of terrestrial ecosystems with respect to matter malance. In: Schulze ED, Zwölfer H (eds) Potentials and limitations of ecosystem analysis. Ecol Stud 61, pp 11–49, Springer, Berlin Heidelberg New York Tokyo
Vales DJ, Bunnell FL (1988) Comparison of methods for estimating forest overstory cover. I. Observer effects. Can J For Res 18:606–609
Wallace HR (1978) The diagnosis of plant diseases of complex etiology. Annu Rev Phytopathol 16:379–402
Waring RH, Newman K, Bell J (1981) Efficiency of tree crowns and stemwood production at different canopy leaf densities. Forestry 54:15–23
Waring RH, Pitman GB (1985) Modifying lodgepole pine stands to change susceptibility to mountain pine beetle attack. Ecology 66:889–897
Yoda K, Kira T, Ogawa H, Hozumi H (1963) Self-thinning in overcrowded pure stands under cultivated and natural conditions. J Inst Polytech, Osaka City Univ, Biology, Ser D 14:107–129
Zech W, Popp E (1983) Magnesiummangel, einer der Gründe für das Fichten-und Tannensterben in NO-Bayern. Forstwiss Centralbl 102:50–55
Zimmermann R (1985) Untersuchungen zur Photosyntheseleistung, Transpiration und Nadelbiomasseverteilung von zwei geschädigten Fischtenforsten in den höheren Lagen des Fichtelgebirges. Thesis, Univ Bayreuth, FRG
Zimmermann R, Oren R, Schulze ED, Werk KS (1988) Performance of two *Picea abies* (L.) Karst. stands at different stages of decline II. Photosynthesis and leaf conductance. Oecologia (Berl) 76:513–518
Zobel B, Talbort JT (1983) Applied tree improvement. J Wiley and Son NY, 505 pp

Part 1. Air Pollution: Transport and Deposition

Preface

For the study of the effects of air pollution on ecosystems, identifying the source, type, and magnitude of inputs is the first step. Identifying pollutant sources is of particular interest when considering steps to reduce the impact in the future. However, it is quite difficult to distinguish distant and local sources from deposition measurements. Long-range transport models, meteorological observations at the ground, and actual deposition measurements have to complement each other in order to accurately describe atmospheric pollution.

A numerical, long-range transport model (Chap. 1-A) suggests that the study area analyzed in this book is strongly influenced by advection of pollutants originating in industrial regions located approximately 100–200 km west, north and east of the study area. Height in relation to cross-sectional length of the Fichtelgebirge (0.6 km over 50 km) allows air masses to flow smoothly over a mountain range with no major turbulence in the lee. Deposition is great above 750 m elevation in all wind directions. Maximum deposition values occur on crests, on slopes of geomorphological gaps, and in zones of blocked flow. Both experimental sites, at Oberwarmensteinach (declining stand) and Wülfersreuth (healthy stand), are exposed to high pollutant deposition levels transported from afar. However, the declining site is located on a crest, which would suggest greater deposition compared to the healthy site on a smooth westerly slope.

Ground measurements at both study sites confirm the results of the transport model (Chap. 1-B). From comparing general meteorological observations with the deposition of pollutants, it is concluded that gaseous and particulate sulfur pollution loads originate from industrial sources located mainly to the north and east. However, the nitrogen oxide pollution originates in the West. Westerly winds are also associated with the most extreme conditions of acidity in fog and rain. Ammonium appears to originate from neighboring regions and long-distance transport. In addition to these bulk pollutants of sulfur and nitrogen, a large number of organic pollutants are deposited in considerable quantities. The concentrations of hydrogen peroxide are generally low.

Pollutant input occurs as rain, fog, dew, aerosol and gas adsorption, and gas uptake. Chapter 1-B mainly discusses the wet deposition by rain and fog and presents the concentrations of aerosols. The deposition of aerosols and gases is discussed in Chapter 4-I as part of the ecosystem balance; it is shown that dry deposition may contribute as much to pollution deposition as does wet deposition.

E.-D. Schulze, O.L. Lange, R. Oren (Eds.)
Ecological Studies Vol. 77

It is impossible to judge from measurements made at the present time what proportion of the atmospheric inputs in Europe is "natural" and what can be attributed to anthropogenic effects. In an attempt to quantify "natural" concentrations of rain and fog, measurements were made on the western coast of the South Island of New Zealand (Chap. 1-B, Section 4). In such a remote area, the minimum pH of fog water was 5.6 (with the minimum pH in the Fichtelgebirge at 2.2). Concentrations of sulfate and nitrate in New Zealand were 0 and 0.008 mmol l^{-1} respectively. They are thus orders of magnitude below those concentrations found in the Fichtelgebirge (2 to 3 mmol l^{-1} respectively).

Chapter 1-B shows that the mean annual input of sulfate and nitrogen is similar to that measured in other mountainous regions of central Europe. Thus, the conclusions derived from Part 3 on direct damage and from Part 4 on indirect effects of pollutants may be extrapolated to a larger geographical forest area where the chemical soil conditions are similar.

1-A Long-Range Transport and Deposition of Pollutants in the Fichtelgebirge

W. ULRICH

1 Introduction

Air quality measurements in the Fichtelgebirge and elsewhere in Europe have demonstrated that it is inappropriate to expect high air quality away from industrial areas. With the progress in mesoscale modeling, the transport of air pollutants over long distances can be studied using computer simulations. Two- and three-dimensional numerical models are used in this paper to estimate meteorological parameters and transport for situations like that presented in Fig. 1. Air pollution in the Fichtelgebirge region is represented as advection of pollutants from a neighboring industrial area into a mountainous one. Indeed, there are sources of air pollution in the Fichtelgebirge as well, such as a power plant near Arzberg and the local glass industry. However, the output of these seems negligible compared to that of numerous sources in the neighboring industrial areas, such as those around Schweinfurt and Frankfurt in the west, Nürnberg in the southwest, Ingolstadt in the south, Sokolov in Czechoslovakia to the east, and those in the German Democratic Republic to the north. All these industrial regions are approximately 100–200 km away from the Fichtelgebirge.

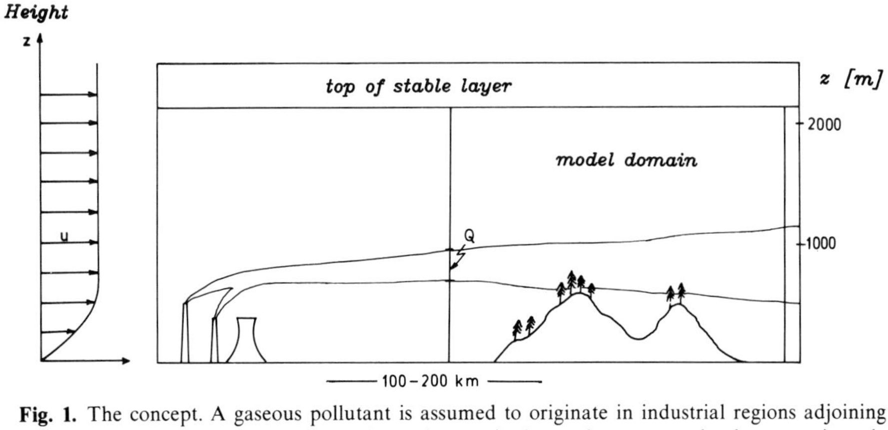

Fig. 1. The concept. A gaseous pollutant is assumed to originate in industrial regions adjoining mountains. Sources and regions where forest damage is observed are assumed to be approximately 100–200 km apart. The model domain does not contain any point sources. The pollutant enters the computational domain as a source Q. On the *top*, we assume an inversion layer. On the *left side*, the wind profile marked an u is shown schematically. The scale on the *right* gives an approximate height above ground; z indicates the elevation in m

E.-D. Schulze, O.L. Lange, R. Oren (Eds.)
Ecological Studies Vol. 77

The meteorological precondition for the scenario in Fig. 1 is a stably stratified atmosphere (e.g., an inversion) in which vertical mixing is so weak that pollutants are transported long distances. Otherwise, substances are dispersed and therefore only nearby sources are important. The models presented do not include moisture or chemical processes so that pollutants can be treated as an ideal gas. In order to obtain a fine spatial resolution in the region concerned, the computational domain is not extended to include the sources. Instead, a plume that enters the computational domain is specified on the inflow boundary; such a plume is denoted by Q in Fig. 1. Two -dimensional simulations with idealized topography and three-dimensional simulations with topographic data of the Fichtelgebirge are presented.

2 Basic Equations for the Time-Dependent Prediction of Pollutant Transport in the Atmosphere

2.1 The Meteorological Model

2.1.1 Equations of Motion

The set of equations in two- and three-dimensions is the same, except that all y-derivatives are set to zero in the two-dimensional cases. All simulations could be performed three-dimensionally as well if the entering plume were specified in three dimensions. These models are described in full detail in Ulrich and Nemmaier (1985) and Ulrich (1987). Only a summary of the mathematical relations is given here.

The conventional Boussinesq-type equations of motion are used, (e.g., Dutton and Fichtel 1969):

$$\frac{d}{dt}\underline{v} = -\alpha_0 \nabla p + g\sigma\underline{k} - f\underline{k} \times (\underline{v} - \underline{v}_g) + \underline{F}^m \quad (1)$$

$\underline{v}$: velocity vector
p : pressure deviation
$\underline{v}_g$: geostrophic (= synoptic) wind
α_0 : specific volume of the basic state
g : gravitational acceleration
f : Coriolis-parameter
$\underline{k}$: unit vector in the vertical direction
$g\sigma\underline{k}$: buoyancy term
$\underline{F}^m$: turbulent terms in the momentum equations

The anelastic approximation is:

$$\nabla \cdot v = 0 \quad (2)$$

The thermodynamic equation reads:

$$\frac{d}{dt}\sigma = -\frac{N^2 w}{g} + F^h \quad (3)$$

where N is the Brunt-Väisälä frequency (measure of stability) and F^h denotes the turbulence terms in the thermodynamic equation.

In contrast to many other mesoscale models, no coordinate transformation is applied to incorporate topography. In the two-dimensional as well as in the three-dimensional model, the topography intersects the rectangular grid. Thus mountain steepness is not restricted by the numerical formulations as is the case with a terrain-following grid. The general method of integrating the meteorological equations is summarized in Appendix I.

2.1.2 The Parameterization of Turbulence

The common splitting of dynamic variables into a space average and a subscale deviation leads to the well-known closure problem. Hence the correlation products have to be parameterized. It is assumed that for the present numerical experiments, a K-parameterization is sufficient; this has the merit of being numerically robust. It is noteworthy that small differences in the calculated turbulent exchange have very little effect on the resulting flow pattern. However, these differences can be significant for the associated transport of a pollutant. Despite some shortcomings the models seem to predict variations of exchange properly. The calculation of the exchange coefficients is then rather standard in both models using the simple profile functions of Businger et al. (1971).

2.2 The Transport Equation

The transport equation in our model is written:

$$\frac{\partial c}{\partial t} = -\nabla \cdot (\underline{v}c) + Q + F^c \tag{4}$$

where c is the concentration of a tracer, Q is the source term, F^c is the turbulent diffusion.

As it is not known at this stage which substances are responsible for the death of trees, we restrict our considerations to an ideal gas. At the present stage the model does not include moisture or chemical processes. As a consequence, processes like wash-out cannot be predicted with this model. It is intended to calculate the near-ground reservoir of pollutants. However, it is likely that wet sinking processes (accretion with water droplets in fog) may be dominant at the locations where the concentration of the pollutant is high.

To calculate the turbulent diffusion F^c, we want to use information from our mesoscale model. The most direct approach is to use calculated K-values which are valid for heat transfer (included in F^h) as well as for transport. Thus we can solve the finite difference approximation of the transport equation, rather than use a particle method. If we had chosen a particle method like "random walk" or "Brownian motion", we would have had to calculate variances (σ-parameters) which are related to the K's by empirical formulae only. This would pose another parameterization problem, e.g. Hanna (1982). Other advantages of finite differences are ease of programming and high formal accuracy. In addition to selecting the Smolarciewicz scheme (1983) as the standard technique, we have also checked all the experiments with the Crowley scheme (1968). The lower boundary condition at the transport equation is specified as a constant value of concentration gradient.

3 Two-Dimensional Prediction of Pollutant Transport

3.1 General Outline of the Experiments

For the present case studies, we specify a geostrophic wind (which is approximately equivalent to the windspeed at the model top) of 5m s^{-1}. The equations of motion are integrated forward in time until a steady state is reached (Appendix I). A concentration is then specified at the inflow boundary using the stationary windfield. The transport equation is integrated forward in time until a steady state is reached. The shape of the topography and the height of the axis of the plume (referred to as plume height in Table 1) entering the integration domain are varied. Special interest is focused on the near-ground gas concentration. Three experiments T1-T3 were performed and are summarized in the following table:

Exp	Plume height (m)	N	Topography	Peak (m)	Fig.
T1	1000	1/100s	Bell-shape	750	2
T2	750	1/100s	Bell-shape	750	3
T3	750	1/100s	Asymmetric	500	4

3.2 Discussion of Results

3.2.1 Idealized Topography

We present only the cases with a stable stratification of N = 10^{-2} s^{-1}. Other stabilities are considered in Ulrich and Nemmaier (1985). The parameters of the experiments presented are summarized in Table 1. The effect of varying plume height above the ground is clearly seen in comparing experiments T1 and T2. The flow over the topography is characterized by a smooth flow over the hill with lee waves as shown in Fig. 2 and Fig. 3 (*top*). The calculated near-ground concentration is depicted in the *frame* under the windvector plot. It should be noted here that the near-ground concentrations are in relative units. The flow-patterns are superimposed with the isolines of concentration. The *thicker* (inner) *lines* correspond to higher concentrations. The maximum concentration is normalized to 1.0. Contours are drawn with intervals of 0.1.

Increased stability implies smaller values of exchange. Because of the presence of thermal effects, waves are induced behind the hill. A pronounced downward motion directly behind the hill is apparent. As noted by Pierrehumbert and Wyman (1985), there exists a band of strong winds over the crest which is shifted downward behind the hill. This effect is strengthened with higher stability.

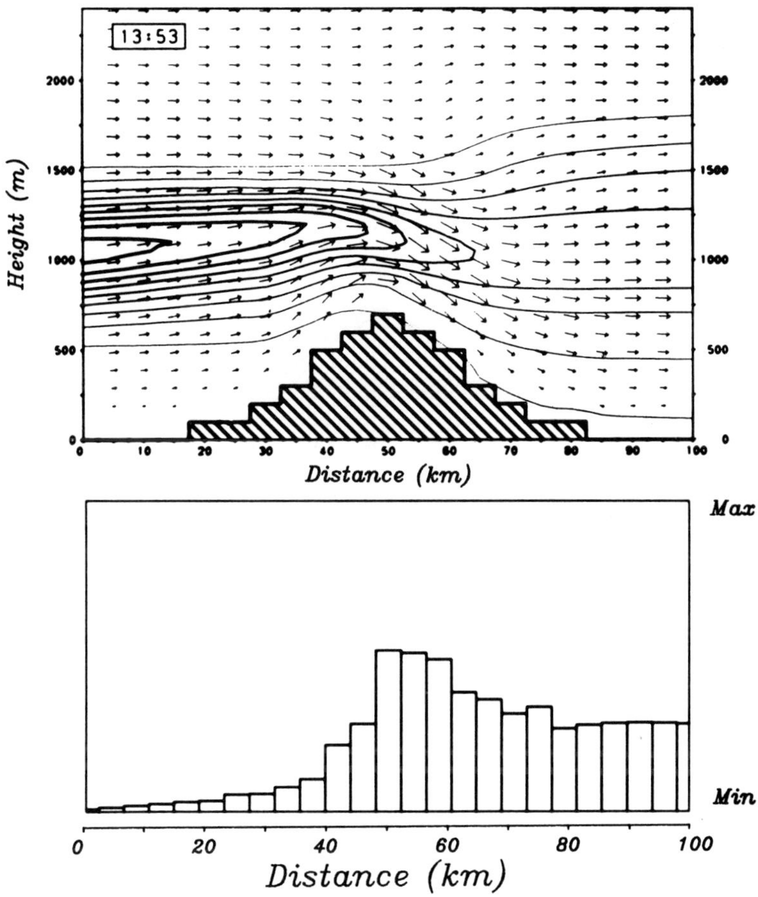

Fig. 2. Experiment T1. Bell-shaped mountain. The simulation time is 13 h 53 min (indicated in the *box left top*). The horizontal axis is labelled in km, the vertical axis in m. The windfield is indicated by *windvectors*. The isolines drawn are proportional to the levels of concentration. The maximum is normalized to 1.0 in all the following figures. Contour interval is 0.1. The *thickest line* corresponds to the highest values. The plume axis is 1000 m high. Transport takes place from *left* to *right*. *Lower picture*: near-ground concentration in relative units

Consequently, a plume transported in that band is also shifted downward, thereby increasing near-ground concentrations in the lee of the hill. This behavior is more pronounced in experiment T1 than in experiment T2.

3.2.2 Structured Topography

Now we turn to experiment T3 which is characterized by asymmetric topography. We selected a cross-section typical of a line from Forchheim in Bavaria to Eger (Cheb) in Czechoslovakia. An approximate average of a 20 km broad strip, recorded from a topographic map, represents a mean topography cross-section so

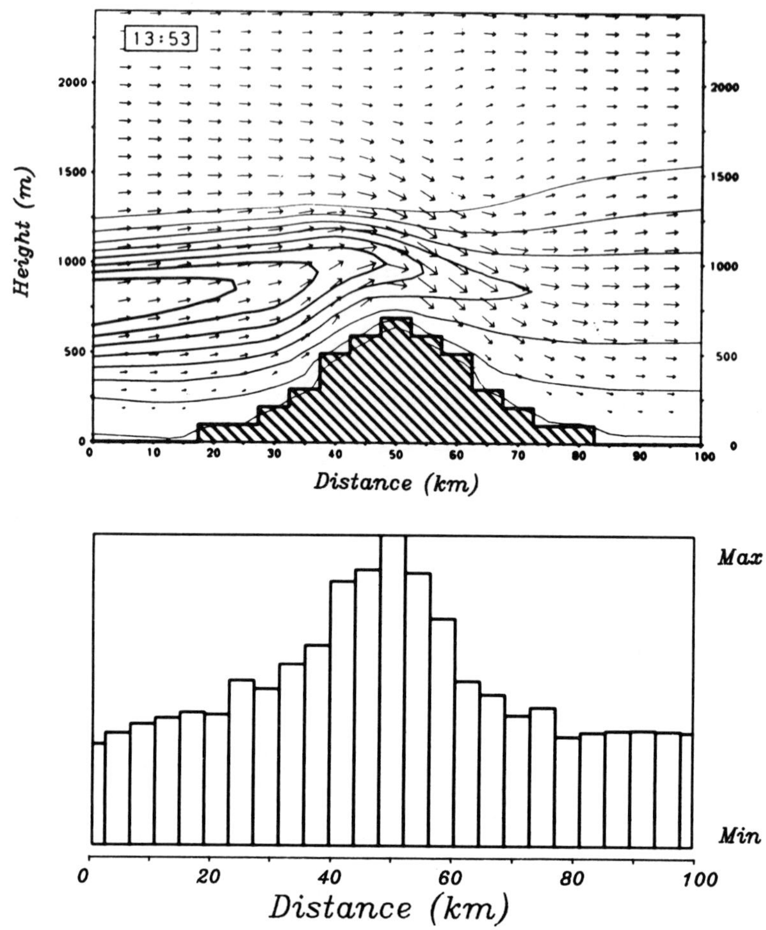

Fig. 3. Same windfield as before in T1. Now the plume enters at 750 m. The near-ground concentration has a maximum at the mountain top

that the Fichtelgebirge is flattened. If the standard deviation were added (so-called envelope-topography), we would expect the observed effects to be strengthened. In this theoretical study, we consider westerly winds only but, of course, easterly winds could be chosen as well. The main reason for selecting westerly winds is the prevalence of westerly winds in the Fichtelgebirge.

It is clear that the resolution used, poorly resolves all the characteristic features of the topography. The step-like topographic approximation appears more pronounced for asymmetric topography. In well-mixed, neutrally-stable cases, the flow close to the terraces is weak because the turbulent viscosity smoothes flow differences. While the model is not restricted to step-like topography, this simplification is maintained in the two-dimensional studies. The results agree with the previous calculations with the symmetric hill. In order to examine the effect of mountains, the difference in near-ground concentrations between the case with no topography (not shown) and the case with topography is plotted in Fig. 4 (*bottom frame*). This picture confirms earlier results found in T1 and T2 that the near-ground concentrations are low in the valleys but increased concentrations exist near the crests of the hills. Fig. 4 indicates that the near-ground concentrations are consistently increased at all the higher elevations. This can, of course, be strengthened in terms of the assumption of two-dimensionality, according to which the entire air mass is forced to climb over the hills instead of separating around them. This is not necessarily the case in the three-dimensional studies, but we have partly reduced this effect of the two-dimensionality by using a smoothed topography.

4 Three-Dimensional Predictions of Pollutant Transport

For the three-dimensional predictions, the topography of the Fichtelgebirge region is represented by area-averaged values of elevation in a 50 × 50 grid. On a horizontal plane the grid has a resolution of 1 km, whereas the vertical direction is resolved down to 100 m, as depicted in Fig. 5. Due to enormous computer time requirements, only two case studies are performed here.

The first case study involves a situation where there is a stable stratification of $N = 0.01\ s^{-1}$ as before, the second is a prescribed north-easterly flow in which the absolute value of the geostrophic wind is $7.5\ m\ s^{-1}$. This corresponds to a situation which often occurs especially in wintertime, when high air pollution is recorded in the Fichtelgebirge (e.g., at a recording station in Arzberg). Figure 6 shows wind vectors for a horizontal layer approximately 750 m a.s.l. Higher elevations of the Fichtelgebirge rise above this level as indicated by the blank areas in Fig. 6. The level of concentration is specified at the eastern and northern boundary of the computational domain (marked "E" and "N" in Fig. 7) with the same vertical profile as in T1 so that the maximum concentration occurs at approximately 1000 m a.s.l. This represents the advected concentration entering the domain under consideration. It is tacitly assumed that influx of the pollutants takes place chiefly at this height, since the stack heights and the thermal up-draught support this assumption. Of course, this is a very crude way to specify the source term in the transport equation.

At the ground we specify a no-flux condition (zero gradient). The effectiveness of the transport to the surface may depend on phenomena like combing-out of substances by leaves. In addition the effect of flow separation on the ground concentration is not clear. The result of the calculation is summarized in the halftone plot in Fig. 7. The values of the near-ground concentration have been normalized to a maximum value of 1. The scale on the right shows the corresponding values of ground concentration. A comparison between the topographic map and the windvector plot reveals that the highest concentrations are found close to the crests, indicated by a *circle* marked *a*, and to both sides of a gap, indicated by a *circle* marked *b*. An increase in concentration is evident also in the zone of blocked flow *c*. The result depends on the flow parameters chosen

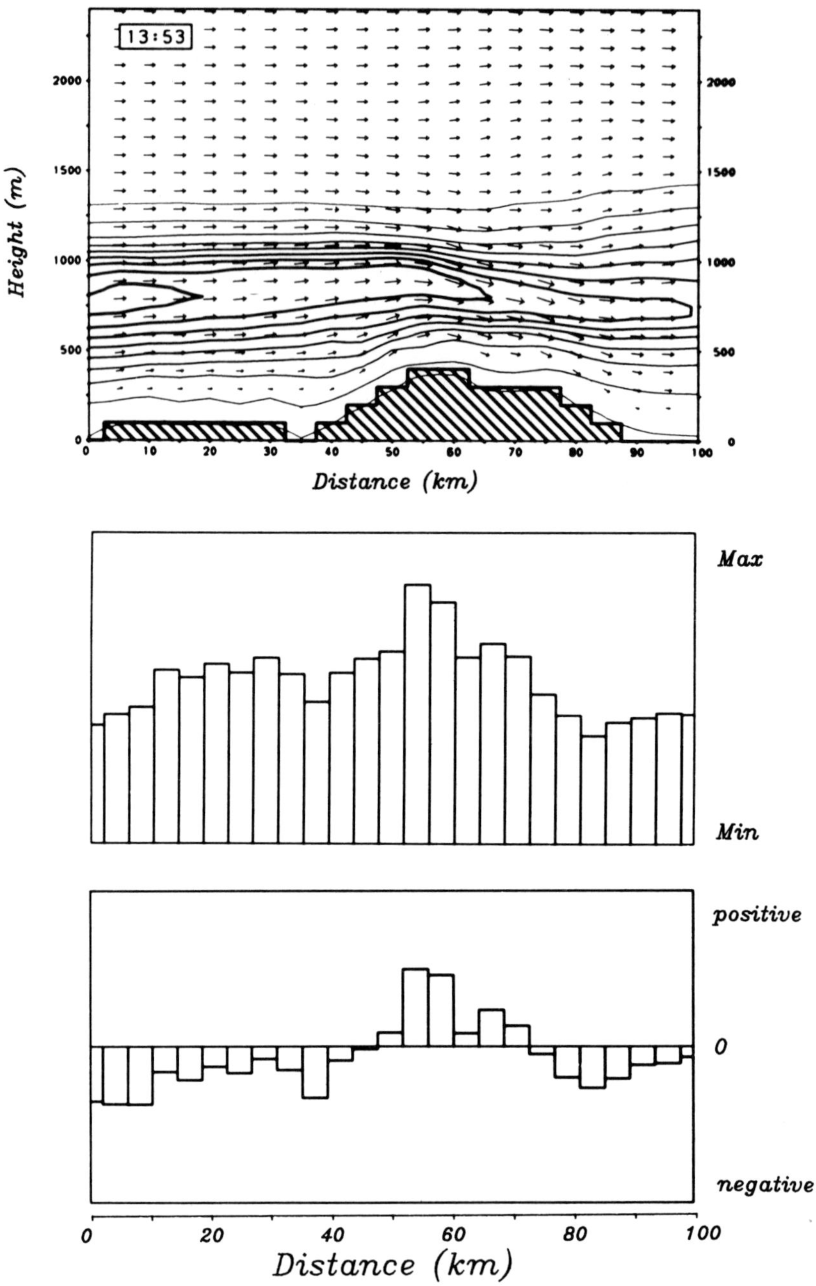

Fig. 4. Experiment T3. Simulation with a cross section from Forchheim (near Nürnberg) over the Fichtelgebirge to Czechoslovakia. Here the increased concentration near the top of the Fichtelgebirge becomes apparent. The reduced concentration in the valley (Main River near Bayreuth) is pronounced, too. The calculation is not completely stationary although the integration time corresponds to 13 h 53 min (as in all two-dimensional cases reported). *Bottom frame*: Representation of the difference between a run with and a run without topography. This displays the net mountain effect. The positive effect of the mountains and the negative effect of the valley on near ground concentrations is apparent

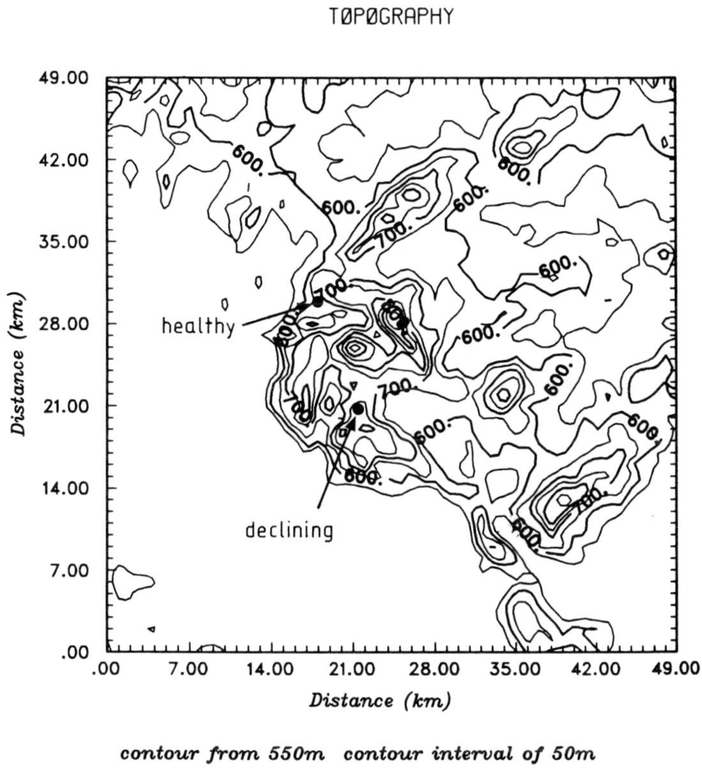

Fig. 5. Map of the topographic data of the Fichtelgebirge. As area averaged values (over one km^2) were used as input data, the topography is flattened. The horseshoe form of the Fichtelgebirge is apparent. The location of the healthy and declining site is indicated

and the concentration specified at the boundaries, but does not change dramatically for small changes in the parameters.

In a second case study, a very weak westerly flow with a geostrophic wind of 5.0 m s^{-1} is considered. The calculated flow pattern which corresponds to Fig. 5 is shown in Fig. 8. The flow around the main ridges of the Fichtelgebirge is more apparent in the first case study because the western inflow is at a lower elevation, so that the barrier effect of the mountains is even greater. The calculated, near-ground concentration is shown in Fig. 9. The entering plume is now specified on the western boundary marked *W*. The same profile is specified in the vertical along the inflow side as in the previously shown case. For a more realistic simulation, these inflow values need to be replaced by actual data, but no such data are presently available.

Although Fig. 9 gives the impression that the near-ground concentrations have smaller values everywhere, there is only one inflow side *W* in contrast to the two inflow sides in the first case. In any event, the significant parameter is the difference in concentrations. It is evident from Fig. 9 that the highest concen-

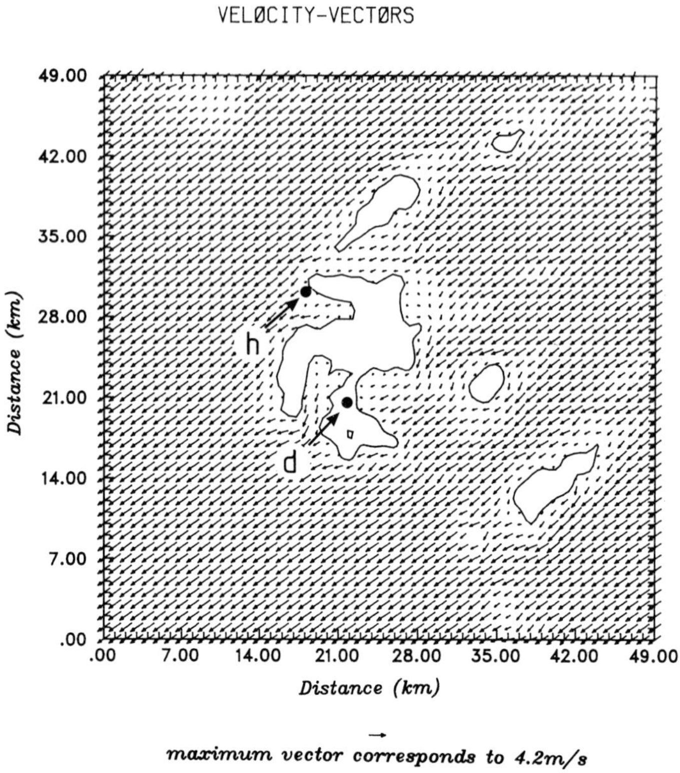

Fig. 6. Windvectors calculated in a horizontal plane 750 m a.s.l. Higher parts of the Fichtelgebirge rise out of the plane (indicated by *blank areas*). *h* and *d* indicate the location of the healthy and declining sites respectively

trations are found at the crests. This effect is more pronounced than in the first case, primarily because the western inflow is at a lower elevation and the plume reaches the ground later (if the transport distance is translated to travel time). At the mountain range *K* (Kösseine) the pollutant concentration is higher than that at *S* (Steinwald), although both mountain ridges have approximately the same peak height. This is caused by the lee effect of the central part of the Fichtelgebirge marked *O* (Ochsenkopf). This result is valid only for the particular wind direction considered.

5 Conclusion

The numerical simulations presented indicate that the crests of the Fichtelgebirge are likely to contain high air pollution due to the influx of pollutants from long distances (100–200 km). Stable stratification leads to accentuated differences in near-ground concentrations between valleys and crests. If the stable layer is thin,

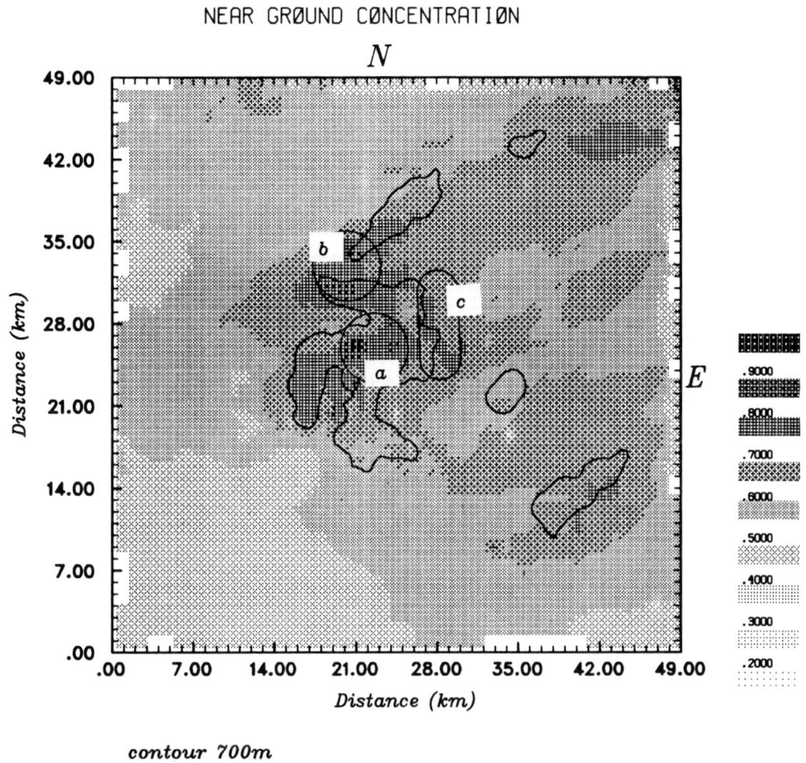

Fig. 7. Halftone plot of the calculated near-ground concentration. The highest elevations in the center of the Fichtelgebirge have the highest concentrations. Mountain ranges higher than 700 m are marked by a solid line. The inflow sides are *N* and *E*. See text for discussion of the areas indicated for *a* to *c*

the most burdened areas may be situated above some minimum level lower than the crests. Furthermore, we have shown that mesoscale modeling is, like wind-tunnel experiments, a useful tool to simulate transport processes.

It should be noted that some of the assumptions made in this paper are extremely crude. Furthermore, some of the important processes such as moisture formation and chemical reactions have been excluded. Finally, all this effort in modeling does not replace the task of taking measurements.

Appendix I. Integration Method of Meteorological Equations in the Two-Dimensional Case

In the two-dimensional case a streamfunction-vorticity system (ψ-ζ-System) is chosen (in contrast to the three dimensional case). The horizontal and vertical velocity-components u and w are related via streamfunction ψ and vorticity ζ according to:

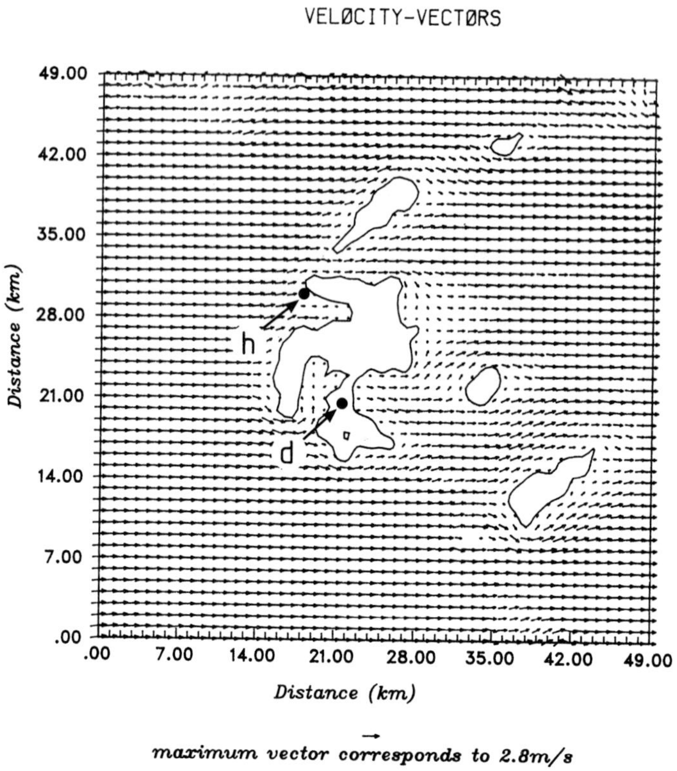

Fig. 8. Case 2. Windvectors calculated in a horizontal plane 750 m a.s.l. The higher elevations of the Fichtelgebirge rise out of the plane (indicated by *blank areas*). *h* and *d* indicate the location of the healthy and declining sites, respectively

$$u = -\frac{\partial \psi}{\partial z}, \; w = \frac{\partial \psi}{\partial x} \qquad \zeta = \left(\frac{\partial w}{\partial w} - \frac{\partial u}{\partial \; z}\right) \tag{AI.1}$$

We get the prognostic vorticity equation as:

$$\frac{\partial \zeta}{\partial t} + \frac{\delta}{\partial x}(u\zeta) + \frac{\delta}{\partial z}(w\zeta) = g\frac{\partial \sigma}{\partial x} - f\frac{\delta}{\partial z}(v - v_g) + F_\zeta \tag{AI.2}$$

in which v is the meridional component of $\underline{v}$, v_g the meridional component of the geostrophic wind $\underline{v}_g$. The above definition of u and w recommends a position of u and w staggered to the grid points of the streamfunction. The topography coincides with the streamline $\psi = 0$. This is also true if the topography is no longer approximated by terraces. Formally, the equations are written in difference formulation with the time levels n, n + 1, n + ½ as superscripts:

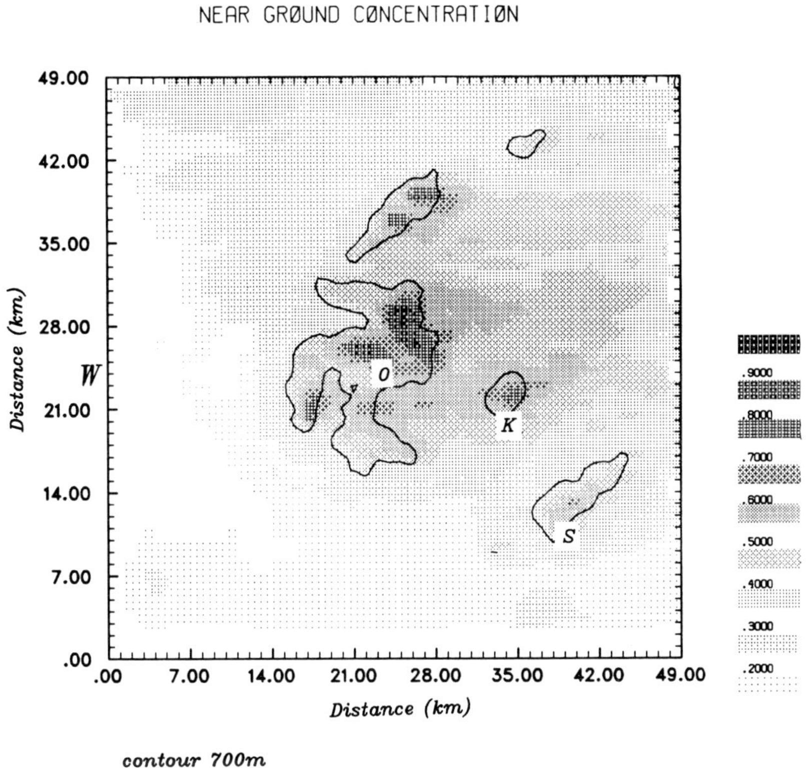

Fig. 9. Case 2. Halftone plot of the calculated near-ground concentration. *O*, *K*, and *S* mark mountain ranges Ochsenkopf, Kösseine and Steinwald respectively. *W* indicates the inflow boundary

$$\frac{\zeta^{n+1} - \zeta^{n}}{\Delta t} = -\mathrm{ADV}(\zeta^{n}) + g\frac{\delta\sigma^{n}}{\delta x} + f\frac{\delta}{\delta x}(v^{n+1} - v_g) + F_{\zeta}^{n+1/2} \qquad \text{(AI.3)}$$

$$\frac{v^{n+1} - v^{n}}{\Delta t} = -\mathrm{ADV}(v^{n}) - f(u^{n} - u_g) + F_{v}^{n+1/2} \qquad \text{(AI.4)}$$

Correspondingly the thermodynamic equation is written:

$$\frac{\sigma^{n+1} - \sigma^{n}}{\Delta t} = -\mathrm{ADV}(\sigma^{n}) - \frac{N^2 w}{g} + F_{\sigma}^{n+1/2} \qquad \text{(AI.5)}$$

F_ζ, F_v, F_σ contain the subgrid terms. $\frac{\delta}{\delta x}$ stands for the discrete counterpart of $\frac{\partial}{\partial x}$. The advection terms abbreviated by ADV in the equations of motion are evaluated by a scheme proposed by Arakawa (1966), which conserves energy and enstrophy. In the thermodynamic equation, a donor-cell scheme (e.g., Roache 1976) is applied to damp wiggles. For nonregular difference stars (when topography is no longer restricted to steps), Taylor approximations of first order are used. After every timestep, an elliptic equation for the streamfunction has to be inverted in the irregular domain of computation:

$$\nabla^2\psi = \zeta \tag{AI.6}$$

The bottom (topography) and top boundaries as well as the inflow side are of the Dirichlet type, whereas the outflow side is of Neumann type. This inversion is achieved very efficiently by the error-vector-propagation method (EVP-method, Roache 1971), a direct solution method. The inflow boundary values are calculated from a one-dimensional equivalent of the equations of motion (in z-direction). In the case of constant exchange, this would reduce to the Ekman solution. This solution is frozen throughout the calculation at the inflow side. At the outflow side a radiation condition (Orlanski 1976) is applied.

Appendix II. The Solution Technique for the Meteorological Equations in the Three-Dimensional Case

In the three-dimensional case, a pressure solution technique is used, which is similar to that of Hirt and Cook (1972):

The momentum equations:

$$\frac{d}{dt}\underline{v} = -\alpha_0\nabla p + g\sigma\underline{k} - f\underline{k} \times (\underline{v} - \underline{v}_g) + \underline{F}_m \tag{AII.1}$$

are discretized as follows:

$$\frac{\delta}{\delta t}\underline{v} = -\hat{\nabla}\cdot(\underline{vv}) - \hat{\nabla}p + g\sigma\underline{k} - f\underline{k} \times (\underline{v} - \underline{v}_g) + \hat{\nabla}\cdot(\underline{K}^m\cdot\hat{\nabla}\underline{v}) \tag{AII.2}$$

where we have written $\hat{\nabla}$ for the discretized del-operator. This can be written more simply as:

$$\frac{\delta}{\delta t}\underline{v} = -\alpha_0\hat{\nabla}p + \underline{\chi} \tag{AII.3}$$

where $\underline{\chi}$ contains the remaining terms.

The thermodynamic equations:

$$\frac{d\sigma}{dt} + \frac{N^2 w}{g} = Q_{dia} + F_T \qquad Q_{dia} = \frac{1}{c_p}\frac{ds}{dt} \tag{AII.4}$$

s : specific entropy

is discretized as follows:

$$\frac{\delta\sigma}{\delta t} + \hat{\nabla}\cdot(\underline{v}\sigma) + N^2 w/g = Q_{dia} + \hat{\nabla}\cdot(\underline{K}^h\cdot\hat{\nabla}\sigma) \tag{AII.5}$$

It is clear that in the discrete case we have to solve a prognostic rather than a diagnostic equation for the pressure:

$$-\alpha_0 \hat{\nabla}^2 p = \frac{\delta}{\delta t}(\hat{\nabla} \cdot \underline{v}) + \hat{\nabla} \cdot \underline{\chi} \qquad \text{(AII.6)}$$

This can be substituted by an adjustment process for the pressure changes δp. The iteration process is defined as follows. In the first step, preliminary velocity-values $\underline{v}^*$ are predicted by an Euler timestep:

$$\underline{v}^* = \underline{v} + (\delta \underline{v}/\delta t)^n \Delta t \qquad \text{(AII.7)}$$

or by an Adams-Bashforth time-step:

$$\underline{v}^* = \underline{v}^n + \frac{3}{2}(\delta \underline{v}/\delta t)^n \Delta t + \frac{1}{2}(\delta \underline{v}/\delta t)^{n-1} \Delta t \qquad \text{(AII.8)}$$

The superscripts n, n–1 indicate the time level. These velocities do not satisfy the continuity equation. The fault is:

$$D^* = \hat{\nabla} \cdot \underline{v}^* \qquad \text{(AII.9)}$$

Now in the second step, the pressure and the velocities in every cell are adjusted by an iteration process, as described below

a) a pressure correction per cell ($\Delta x, \Delta y, \Delta z$) proportional to D^*:

$$\delta p = -\beta D^* \qquad \text{(AII.10)}$$

where:

$$\beta = \beta_0 \left(\frac{1}{\Delta x^2} + \frac{1}{\Delta y^2} + \frac{1}{\Delta z^2} \right) \qquad 1 \leq \beta_0 \leq 2 \qquad \text{(AII.11)}$$

β_0 has the meaning of a relaxation factor.

b) a correction to satisfy continuity:

$$\underline{v}^{**} = \underline{v}^* - \frac{\delta}{\delta t}(\alpha_0 \hat{\nabla} \delta p) \qquad \text{(AII.12)}$$

This takes a simple form in the Arakawa C-grid:

$$u^{**}_{i+1/2} = u^{*}_{i+1/2} + \Delta t\, \alpha_0 \frac{\delta p}{\delta x}$$

$$u^{**}_{i-1/2} = u^{*}_{i-1/2} - \Delta t\, \alpha_0 \frac{\delta p}{\delta x}$$

$$v^{**}_{j-1/2} = v^{*}_{j-1/2} + \Delta t\, \alpha_0 \frac{\delta p}{\delta y}$$

$$v^{**}_{j-1/2} = v^{*}_{j-1/2} - \Delta t\, \alpha_0 \frac{\delta p}{\delta y}$$

$$w^{**}_{k+1/2} = w^{*}_{k+1/2} + \Delta t\, \alpha_0 \frac{\delta p}{\delta z}$$

$$w^{**}_{k-1/2} = w^{*}_{k-1/2} - \Delta t\, \alpha_0 \frac{\delta p}{\delta z}$$

The subscripts indicate the position of the variables in the grid. The iteration process has converged when $D^* < \varepsilon$, where ε is a limiting tolerable fault. Typical values lie in the range $\varepsilon \approx 10^{-3} \times \frac{U}{L}$, with the velocity scale U and the grid-spacing d.

Acknowledgement. The input data for the three-dimensional simulation were made available by J. Graf who is involved in a research project (J. Graf 1987) on air pollution in northern Bavaria sponsored by the Umwelt-Bundesamt in Berlin.

References

Arakawa A (1966) Computational design for long-term numerical integration of the equations of fluid motion: two-dimensional incompressible flow. J Comput Phys 1:119–143

Businger JA, Wyngaard JC, Izumi Y, Bradley EF (1971) Flux profile relationships in the atmospheric surface layers. J Atmos Sci 27:1015–1020

Crowley WP (1968) Numerical advection experiments. Mon Wea Rev 96:1–11

Dutton JA, Fichtel GH (1969) Approximate equations of motion for gases and liquids. JAS, 24:241–254

Graf J (1987) Simulation typischer Strömungsstrukturen in Nordost-Bayern und Berechnung des Schadstoffeintrages. Manuscript available from the Umwelt-Bundesamt Berlin

Hirt CW, Cook JL (1972) Calculating three-dimensional flows around structures and over rough terrain. J Comput Phys 10:324–340

Hanna SR (1982) Applications in air pollution modeling. In: Nieuwstadd FTM, Von Dop H (eds) Atmospheric turbulence and air pollution modeling. Atmos Sci Lib, Dordrecht

Orlanski I (1976) A simple boundary condition for unbounded hyperbolic flows. J Comput Phys 21:251–269

Pierrehumbert RT, Wyman B (1985) Upstream effects of mesoscale mountains. J Atmos Sci, 42(10):977–1003

Roache PJ (1971) A new direct method for the discretized Poisson equation. In: Holt M (ed) Proc 2nd Int Conf numerical methods in fluid dynamics. Lect Notes Phys vol 8, Springer, Berlin Heidelberg New York

Roache PJ (1976) Computational Fluid Dynamics. Hermosa, Albuquerque

Smolarciewicz P (1983) A simple positive definite advection scheme with small implicit diffusion. Mon Wea Rev 111:479ff

Ulrich W (1987) Simulationen von thermisch induzierten Winden und Überströmungssituationen. Münch Universitätsschr, Fakultät Phys Univ Münch – Meteorol Inst Wiss Mitt Nr 57

Ulrich W, Nemmaier C (1985) Schadstoffausbreitung über dem Fichtelgebirge – eine numerische Simulation. Beitr Landespflege Rheinland-Pfalz 10:180–199

1-B Air Pollution and Deposition

R. EIDEN (Coordinator)
with contributors by J. FÖRSTER (Sects. 3.2 and 4.1);
K. PETERS (Sect. 4.2);
F. TRAUTNER (Sects. 4.3 and 4.4);
R. HERTERICH (Sect. 5);
G. GIETL (Sect. 6.1)

1 Introduction

The meteorological conditions of the atmosphere and its earth surface boundary determine the transport of natural and anthropogenic trace gases and particulate matter, as well as these trace substances' residence in, and their removal from, the atmosphere. The long-range transport of particles and their local deposition and distribution on the surface of branches, leaves and needles depend on macro- and micrometeorological conditions in relation to the structure, or "roughness", of the particular local surface. Such mesoscale atmospheric phenomena caused by orographic structures as mountain and valley wind systems, temperature inversions, radiation fog, and cold-air lakes also have to be considered. The modeling of the atmospheric transfer processes are described in Chapter 1-A.

The research described here analyzed and monitored as many substances as possible whose immission have been assumed to destructively interfere with plant function. Studying a large number of substances was necessary, since any competing hypotheses require the simultaneous occurrence and the combined reactions of several atmospheric trace constituents. As immission, the concentration of a trace substance in the near-surface air layer, is only an implicit measure of the impact on vegetation, the efficacy, or at least the flux on or through the surface of the plants or soil, can only be calculated on the basis of indirectly derived transfer coefficients. This method has been applied by Hantschel (1987). However, a direct measurement of such fluxes is, if at all possible, very cumbersome, time consuming and expensive.

Such difficulties in measuring fluxes apply essentially to dry deposition, i.e., the "transfer" of gas molecules and aerosol particles by sedimentation, molecular and turbulent fluxes. At present, a routine measurement of fluxes of this kind is almost impossible. While wet deposition by precipitation and fog can be measured at least approximately, with less effort, a detailed study of the depositional distribution of fog droplets within a stand involves rather complex processes. Such processes include the roughness structure of the vegetation, the droplet size-distribution and the wind speed within the stand.

The chemical composition of a deposited droplet also is the result of a history which cannot always be clarified. Both precipitation elements and fog droplets acquire their chemical load through several different physical and chemical processes. They grow through condensation or sublimation by incorporating the

E.-D. Schulze, O.L. Lange, R. Oren (Eds.)
Ecological Studies Vol. 77

condensation or freezing nuclei ("rain out"), by capturing additional liquid and solid particles by diffusion and collision, and by diffusive solving of gaseous atmospheric compounds ("rain and wash out"). These chemical selective processes depend on the aerosol particle size-distribution and the transfer coefficient of the different gases (Twomey 1979).

Recently Woodman and Cowling (1987) stressed that there is still a lack of chemical monitoring data from remote areas and high elevation forests. The data given in the following sections may alleviate this deficiency. The atmospheric situation experienced in the Fichtelgebirge is not so unique that it may not be characteristic of other areas, too.

2 General Geographical and Climatological Description of the Observation Areas

The main research area to be considered is the Fichtelgebirge, although some comparative measurements have also been taken in the "Große Ohe". Both areas are located in the German highlands. They are situated within the west-wind belt of the temperate mid-latitudes and to the eastern side of Central Europe. During summer, rather mild Atlantic air masses determine the maritime weather conditions, which include westerly winds with moderate air temperatures, high cloudiness and sufficient precipitation. During winter, these regions experience a more continental influence, with a cloudless sky, a prevailing easterly wind flow and the resulting advection of dry but very cold air from Russia. The areas of high elevation have a continuous snow cover of long duration from which the terrestrial (infrared) radiation greatly intensifies the anticyclonic high pressure system over the region.

The Fichtelgebirge is located at approximately 49°59′ N and 11°45′ E in the northeastern part of Bavaria (Fig. 1). The middle range mountains covered with extended forests ascend from Bayreuth (330 m a.s.l.) in the southwest to the summit of the Ochsenkopf mountain (1024 m a.s.l.) in the northeast over a distance of 18 km. There are some small glass processing plants around Warmensteinach, a village between the Ochsenkopf mountain and Bayreuth, and a coal-fired power plant as well as porcelain processing plants ca. 25 km east of the area. Sixty km southwest is the industrial area of Nürnberg. Finally, there are the vast chemical plants beyond the Czechoslovakian border (CSSR), only 50 km east, and in the German Democratic Republic (GDR), ca. 100 km north. The prevailing wind directions may be seen from the windroses in Fig. 2a-c. Whereas the official Weather Bureau station at the Ochsenkopf summit indicates predominantly westerly winds unimpaired by nearby obstacles, the windroses at Wülfersreuth, another observation station in the Fichtelgebirgs area, and those at the Ochsenkopf fog station are both characterized by the local secondary wind guiding effects of a forest clearing. The local directional dependence of the horizontal flux of trace substance with air masses is indicated by the relative frequency of the direction of the wind velocity (Fig. 2b, c). Representative climatic data of the area are given in Table 1[1]. The climate varies with altitude and

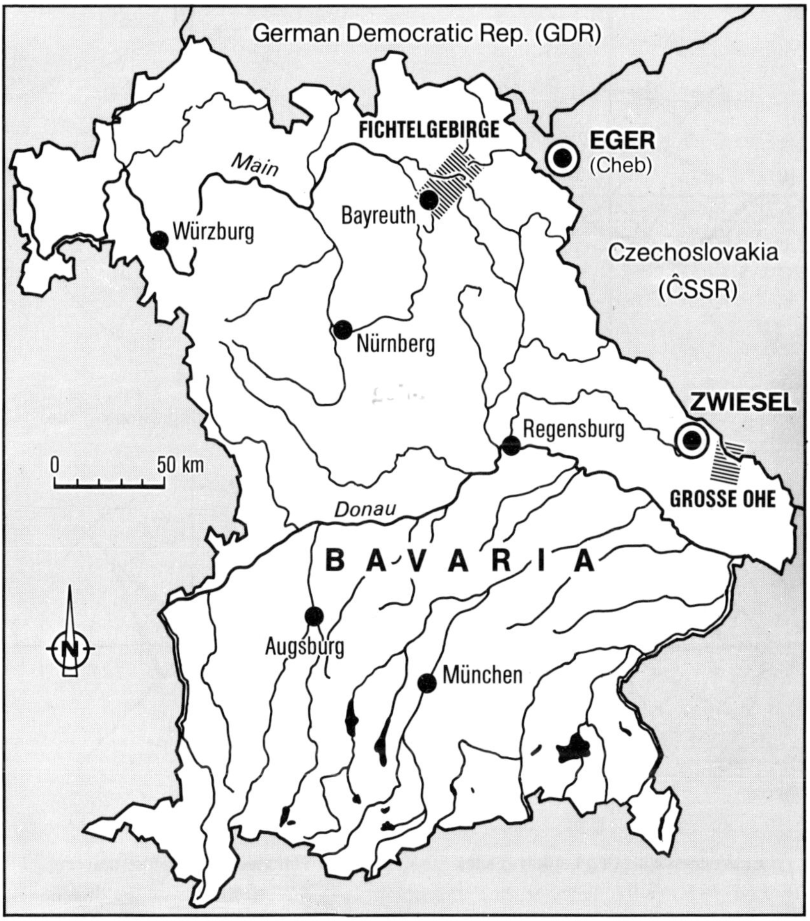

Fig. 1. Bavaria (FRG): location of the research areas Fichtelgebirge and Große Ohe

may be classified as "humid continental" with short, cool summers and long cold winters. There are no fundamental climatic differences between the observation sites on the south and west slope of the Ochsenkopf massif, Wülfersreuth (680 m a.s.l.) and Oberwarmensteinach (Wagenthal, 760 m a.s.l.) respectively. The north and east slopes exhibit a more continental climate with less precipitation (~ 600 mm/yr) and a colder winter.

[1]The precipitation data listed in Table 1 have been measured according to the standards of the German Weather Bureau, with rain gauges exposed to the free wind flux. The rain gauges of the research group, however, have been installed in the wind shade of small clearings (Oberwarmensteinach) or in the stagnation zone close to a forest edge (Wülfersreuth). Since the wind velocity within these regions is reduced, the loss of the smaller drops is also reduced (Rodda 1971). Thus, the amount of precipitation reported by the research group may be 20–30% higher than the official data of the Weather Bureau.

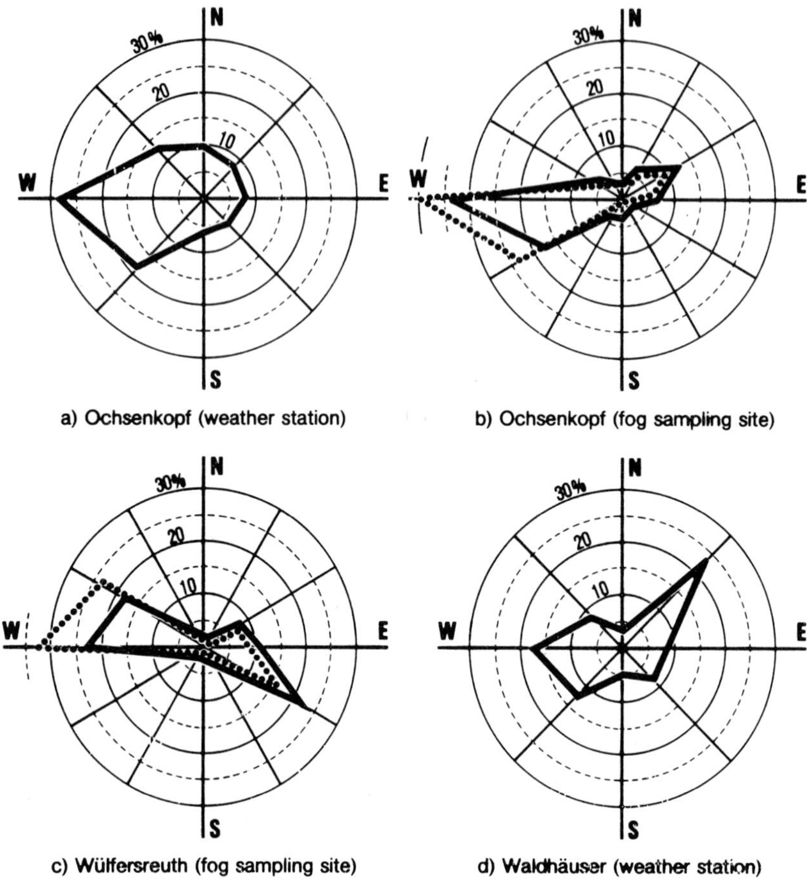

Fig. 2a-d. Relative frequency of wind direction (———); rel. frequency of the direction of wind velocity (-----) rel. windflux. **a** Ochsenkopf – Fichtelgebirge, German Weather Bureau station, 1019 m a.s.l., 1961–1980; **b** Ochsenkopf – Fichtelgebirge, fogwater collection station ~ 1000 m a.s.l., F. Trautner, June 20, 1986 – May 24, 1987; **c** Wülfersreuth – Fichtelgebirge, fogwater collection station, 680 m a.s.l., F. Trautner, July 23, 1984 – June 24, 1987; **d** Waldhäuser – Große Ohe, station controlled by the Bavarian Forest Research Institution's, Institute for Meteorology, Munich (Institut für Meteorologie der Bayerischen Forstlichen Forschungsanstalt, München), 945 m a.s.l., 1974–1976

The forest research and representative hydrological catchment area "Große Ohe" is at 48°57′ N, 13°26′ E, located in the southeastern part of Bavaria (Fig. 1) in the national park "Bayerischer Wald". A basin-shaped mountainous region with a diameter of ca. 5 km ascending from south (base station, 769 m a.s.l.) to north ("Großer Rachel", 1453 m a.s.l.), the area is situated amidst large, extended forests far from industrial installations. The only industrial pollution sources worth mentioning are glass producing and processing plants ca. 15 km northwest of the site. The Bavarian capital, Munich, is ca. 160 km to the southwest and the Fichtelgebirge ca. 130 km to the northwest. Some representative climatic data are

Table 1. Climatic data of the observation sites Fichtelgebirge and Große Ohe. Location and altitude of weather stations[a]

	Fichtelgebirge			Große Ohe
	Bayreuth 330 m a.s.l.	Warmensteinach 575 m a.s.l.	Ochsenkopf 1019 m a.s.l.	Waldhäuser 945 m a.s.l.
Mean annual air temperature	7.7°C	5.8°C	4.2°C	5.2°C
Mean monthly air temperature: January	-1.7°C	-3.6°C	-4.3°C	-4.2°C
Mean monthly air temperature: July	16.9°C	14.7°C	12.8°C	14.0°C
Mean annual depth of precipitation	676.5 mm*	1103 mm	1253°C	1276 mm
Ratio of mean depth of summer to winter precipitation	1.25*	1.00	0.96	1.15
Mean annual sunshine duration	1548 h	1367 h	1521 h	1535 h**

[a] Sources: Fichtelgebirge – Thommes and Schultz (1983), period: 1961–80 (*1951–80); Große Ohe – Noack (1979), period: 1951–70 (**1974–76).

listed in Table 1. The prevailing wind directions are shown in Fig. 2d. Besides the westerly winds, northeasterly winds are also common. The basic climates of the Große Ohe and the Fichtelgebirge are almost identical.

3 Gaseous and Particulate Atmospheric Pollutants and Atmospheric Conditions

3.1 Basic Immission Data

Atmospheric trace gases in the observation area were monitored by the Bavarian Environmental Protection Authority (Bayerisches Landesamt für Umweltschutz–BLfU). Sensors and measuring devices had been installed within the spruce stand of the observation site Oberwarmensteinach. Besides the general meteorological data (temperature, relative humidity, pressure, wind) the trace gases sulfur dioxide (SO_2), nitric oxide (NO) and nitrogen dioxide (NO_2) as well as ozone (O_3) were continuously monitored.

The mean seasonal distribution of these gases for two annual periods is shown in Fig. 3a. The main source of SO_2 is the combustion of fossil fuel. Thus, the highest concentrations of this atmospheric constituent were expected during the winter, especially late winter. In general, SO_2 concentration is modulated by wind direction, so that lower values can be attributed to westerly and southwesterly winds and higher values to winds from the northeast or southeast. Short-term peak values are mainly correlated with easterly winds and temperature inversions

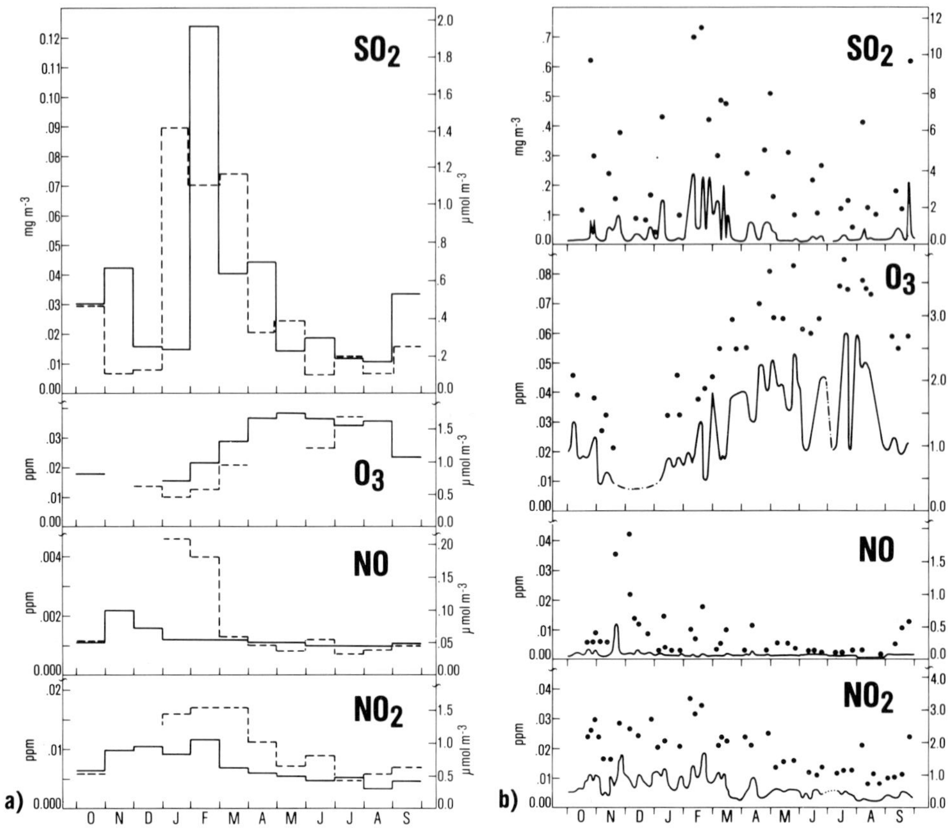

Fig. 3a,b. Near surface atmospheric trace gases SO_2, O_3, NO and NO_2; observation site Oberwarmensteinach, 770 m a.s.l.; data measured by BLfU (1985, 1986, 1987). **a** Mean monthly concentrations: —— Oct. 1985 – Sept. 1986; ----- Oct. 1986 – Sept. 1987. **b** Mean daily concentrations (——), smoothed; *dots* designate the three maximum half hourly peak concentrations of the month, Oct. 1985 – Sept. 1986

during high pressure conditions, when the peak value (half-hour mean value) may exceed the given mean value (Fig. 3a, b) by a factor of 20 or even more. Peak values can be observed throughout the year.

The concentrations of the nitrogen oxides, which can mainly be attributed to automobile exhaust, also show an increase in winter. The weak increase usually observed during this season can be attributed to the reduced atmospheric exchange during the more stable winter temperature inversions. The enhanced increase during the period 1986–87 was probably caused by the intensified local winter sport activities close to the observation site. Short-term concentration peaks were also noticed throughout the year. The peak values exceeded the corresponding mean values by a factor of ca. 10 in the case of NO and a factor of ca. 3.5 in the case of NO_2.

The ozone concentration was highest during the summer. Peak concentrations were only up to 2.5 times as large as the mean values given in Fig. 3a. Ozone is produced primarily in the stratosphere as a result of photolysis of molecular oxygen, O_2, by solar radiation ($\lambda < 320$ nm). Tropospheric O_3 is derived in part from transport of O_3 from the stratosphere and in part from in situ photochemical reactions ($\lambda < 385$ nm) in the troposphere with nitrogen oxides. Downward transport after an enhanced stratospheric production during spring and early summer explains not only the summer increase of O_3 near the surface but also O_3 fluctuations as a result of subsidence motion in high pressure systems. The production of O_3 near the surface plays a secondary role, mainly in urban areas with high traffic density (Liu et al. 1980).

3.2 Some Selected Gaseous and Particulate Pollutants

Since both specific surface area and aerodynamic resistance of forests are comparatively high, forested ecosystems are likely to constitute effective receptors for dry deposition of pollutants from the gas and particulate phases (Hicks 1980, Höfken and Gravenhorst 1983, Wiman and Agren 1985). The role of atmospheric trace gases (SO_2, O_3) as potential forest damaging agents has been intensively studied (e.g., Guderian 1985, Scholz and Lorenz 1984), whereas data concerning immission and deposition of particulate pollutants in forests are still sparse (Wiman and Lannefors 1985).

Precise measurement of dry deposition rates in forest ecosystems is almost impossible because of the complex geometrical and aerodynamical structure (Pruppacher et al. 1983). However, immission values can be used as a measure of pollutant input to forest stands. Since one research objective was to assess the influence of individual weather conditions on atmospheric pollution and to detect peak concentrations, our equipment allowed for as short as possible (~ 1 h) successive single sampling intervals. Discontinuous sampling was conducted July 1985–April 1986. The sampling periods varied in length according to the actual weather conditions. The observation sites chosen were Wülfersreuth and Oberwarmensteinach. At both sites the measurements were taken in a forest clearing to guarantee an unimpeded influx of air. The results of both sites are generally treated and discussed together, because neither site significantly differed in atmospheric pollution conditions.

Aerosol particles were analyzed for the most important inorganic ions (SO_4^{2-}, NO_3^-, Cl^-, H_3O^+, NH_4^+, Mg^{2+}, Ca^{2+}, Na^+, K^+) as well as for polycyclic aromatic hydrocarbons (PAHs). For comparison, atmospheric gas was analyzed sporadically for PAHs and some chlorinated hydrocarbons. PAHs serve as tracers for pollutants from anthropogenic combustion processes, since their main sources in Central Europe are traffic, domestic heating and the steel industry. Thus, the organic micropollutants' wide range of physico-chemical properties allows the estimation of the environmental behaviour of other organic compounds of interest.

Atmospheric particulate matter was sampled on quartz fiber filters using a high-volume-sampler. One half of each filter sample was used for inorganic analyses and the other half for extraction with an organic solvent. Anions were analyzed by ion chromatography (IC) and cations by atomic absorption spectrometry (AAS), flame photometry and potentiometry (NH_4^+). H_3O^+ was calculated from measured pH values. Chlorinated hydrocarbons were quantified by gas chromatography, while PAHs were analyzed by high performance liquid chromatography (HPLC) and high performance thin layer chromatography (HPTLC). Gaseous pollutants were trapped by pumping filtered air past an adsorption column filled with XAD-2.

The ranges of gas phase concentrations detected for chlorinated hydrocarbons and some volatile PAHs are given in Table 2a. PAHs could be detected by their vapor pressure in both the gas and particulate (adsorbed) phases. The distribution between these physical states is also influenced by air temperature (Yamasaki et al. 1982). Hexachlorocyclohexane (HCH) concentrations show a distinct seasonal trend, with absolute maximum values in early summer showing 90–100% in the gas phase. The highest gas phase values of PAHs, however, were found in a winter sample (February 1986). The gas/particle distribution strongly depends on molecular weight (Table 3).

Table 2. Concentration ranges of organic micropollutants in the atmosphere, Wülfersreuth, Oberwarmensteinach

a) Concentrations in the gas phase (May 1984–April 1986)	
Chlorinated hydrocarbons	
α-HCH	0.04–0.80 ng m^{-3}
γ-HCH	0.28–2.84 ng m^{-3}
Hexachlorobenzene	0.18–1.72 ng m^{-3}
PAH	
Naphtalene	3.5 –159 ng m^{-3}
Fluoranthene	2.2 –336 ng m^{-3}
Benzo(a)anthracene + chrysene	0.5 – 18 ng m^{-3}
b) Concentrations in atmospheric particulate matter (July 1985–April 1986)	
Fluoranthene	0.07–92.5 ng m^{-3}
Pyrene	0.04–54.6 ng m^{-3}
Chrysen + benzo(a)anthracene	0.08–52.8 ng m^{-3}
Benzo(b) + (k)fluoranthene	0.04–35.2 ng m^{-3}
Perylene	< 0.02–14.0 ng m^{-3}
Benzo(a)pyrene	0.02–10.0 ng m^{-3}
Indeno(cd)pyrene	0.03–10.9 ng m^{-3}
Benzo(ghi)perylene	0.04–20.0 ng m^{-3}

The extreme concentrations of the different atmospheric particulate inorganic ions were typically associated with different weather conditions (for concentration ranges see Table 4). For example, the sulfate maximum was measured on February 10, 1986, after a snowfall, when extremely cold air masses moved in from the northeast. The highest nitrate value was detected with westerly winds from October 31 to November 1, 1985, at the end of a long period with no precipitation. The analyzed ions formed about 41% of this mass.

Table 3. Averaged distribution of PAH on the gas and particulate phases, Oberwarmensteinach (1985–1986); number of samples n = 8[a]

	Naphtalene	Fluoranthene	Pyrene	Chrysene + Benzo-(a)anthracene	PAH with 5 and more rings
Gas phase (%)	≈ 99	80 ± 8	59 ± 16	51 ± 19	0
Molecular weight	128	202	202	228	≧ 252
Vapor pressure	10.4	$1.23 + 10^{-3}$	$6 \cdot 10^{-4}$	$\approx 2 \times 10^{-5}$	$< 10^{-6}$

[a] For comparison the molecular weight and the vapor pressure (Pa, 25 °C) of the substances analyzed are added.

Table 4. Range of concentrations of inorganic ions in atmospheric particulate matter per volume of atmospheric air, Wülfersreuth and Oberwarmensteinach (1985–1986)

SO_4^{2-}	14–531 nmol m^{-3}
NO_3^-	< 5–281 nmol m^{-3}
Cl^-	0.2– 42 nmol m^{-3}
NH_4^+	9–861 nmol m^{-3}
Na^+	< 3–176 nmol m^{-3}
Mg^{2+}	0.1– 25 nmol m^{-3}
K^+	< 1– 36 nmol m^{-3}
Ca^{2+}	0.3– 67 nmol m^{-3}
H_3O^+	0.4–182 nmol m^{-3}
Particle Mass	14–167 $\mu g\ m^{-3}$

The ionic charge balance (Fig. 4) shows that the acidic anions are counterbalanced mainly by ammonium (62%) with minor contributions from the soil-borne cations, while the proportion of free acid (H_3O^+) in particle leachate is small. PAHs in the particulate aerosol have a summer minimum and winter maximum with a wide range of concentrations (Table 2b). Unlike the inorganic ions, mutual fractions of PAHs are almost constant over time. The maximum concentrations of PAH were all recorded February 9–10, 1986, when cold air was advected from the northeast and an inversion layer reduced vertical atmospheric mixing.

Figures 5a and 5b represent results from aerosol monitoring periods in which concentrations of trace substances in the particle phase were compared with meteorological parameters and atmospheric trace gas concentrations (O_3, SO_2, NO_x), measured at the site by the BLfU.

The influence of the air mass advection direction on the atmospheric pollutant concentrations in the Fichtelgebirge is illustrated by Fig. 6. To determine the directional derivative of the concentration means, each sample was assigned to a 90°-directional class. To allow a better directional interpretation of the data, the concentration windrose has been turned 10° from normal. To coordinate a sampling interval with a directional class, three independent information sources were used: weather charts (geostrophic winds), wind direction measured at the

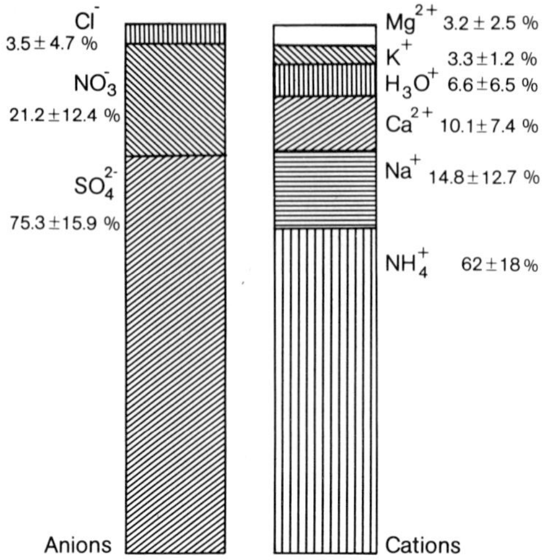

Fig. 4. Mean fractional ionic charge composition of atmospheric aerosol particles in samples from the Fichtelgebirge (Wülfersreuth and Oberwarmensteinach), July 1985 – April 1986

Wülfersreuth site, and radio sonde readings. The classification "Winter" almost always indicates a snow layer on the ground (excluding soil dust sources). The concentration windroses clearly show an increased immission of substances originating from anthropogenic combustion processes during winter, and especially high concentrations when air masses have passed over the GDR (north-northeast) or Czechoslovakia (east-northeast). Multivariate statistical analyses of the aerosol data together with meteorological variables confirmed the significantly differing characteristics of sulphate and nitrate immission. Principal component analysis (PCA) grouped SO_4^{2-}, NH_4^+, H_3O^+, the soil-borne cations and the PAHs together, highly loading low temperature, high pressure and wind direction parameters (north and east), but no west winds on the first principal component (PC) while NO_3^- was grouped with Cl^- on a second PC.

4 Chemical Composition of Precipitation and Fogwater

Water in some form is a conspicuous feature of the humid forested landscape. Rain and snow, as well as fog, supply the water that replenishes groundwater reserves. However, the water that reaches the surface of the soil or is intercepted by the leaves or branches of the trees also transports chemicals from anthropogenic activities. Remote sources, as well as local ones, contribute to this chemical load, which is heterogeneous in its input and distribution to the soil and vegetation. This heterogeneity results from different physical and chemical atmospheric processes and the roughness structure of the trees and the herbaceous layer. Variable distribution has to be anticipated, especially for the fog droplet input. This may be significant if accumulating or dosage effects exceeding a threshold value lead to biological injury or malfunction.

4.1 Properties of Selected Precipitation Samples

Current published data on the chemical composition of precipitation water refers mostly to bulk precipitation averaged over days or weeks. In addition to hydrometeors (rain droplets, snow flakes, etc.), bulk water samples contain atmospheric dry particulate matter deposited in the precipitation collector over the sampling period. However, to assess the potentially toxic impact of precipitation water constituents on vegetation, separating wet-only deposition from dry deposition is necessary. Wet and dry components of the deposits have a different chemical and physical prehistory (compare also Sect. 3). The two deposition processes themselves are physically and chemically different and influence vegetation differently. Short sampling intervals not only detect peak concentrations and short-term fluctuations, but provide a better understanding of the washout process.

In this section, determinations of the concentrations of trace (organic and inorganic) pollutants in samples of precipitation water collected from the observation sites in the Fichtelgebirge are described. The analyzed substances were from the groups of PAHs and HCHs (see Sect. 3 this Chap.) and inorganic ions (H_3O^+, Na^+, K^+, Ca^{2+}, Mg^{2+}, NH_4^+, Cl^-, NO_3^-, SO_4^{2-}). Sampling consisted of discontinuously selecting and studying individual precipitation events.

The sampling equipment allowed for such small increments of precipitation that enough sample volume was obtained for each of the analyses. A set of up to eight parallel mounted glass funnels (total area = 5656 cm^2) was used to sample water for organic trace pollutants. Four parallel polyethylene funnels (total area = 3020 cm^2) were used for the inorganic analyses. All funnels and bottles were cleaned with distilled water just before the onset of a precipitation event to prevent dry deposition contamination. Samples were stored at 4° C and analyzed within 2 weeks. For analytical methods see Section 3 of this Chapter and Thomas (1981).

The results from the two observation sites, Wülfersreuth and Oberwarmensteinach, are not presented separately, as the precipitation water samples from these sites did not differ significantly from each other. For comparison, precipitation water was also sampled at the summit of the Ochsenkopf (1000 m a.s.l.) and in the lee of the Fichtelgebirge's main ridge at Meilerhütte (695 m a.s.l.). Deposition values were higher at Mt. Ochsenkopf because the greater elevation leads to a higher amount of precipitation. At both stations, however, the measured concentrations neither differed greatly from the values obtained from the Wülfersreuth and Oberwarmensteinach sites, nor showed any consistant deviation.

Table 5 presents a summary of the concentration ranges and means of all the analyzed substances. Since sampling was not performed continuously and only exemplary events were selected, the calculated mean values cannot be taken as true annual means but indicate approximate value. They are based on a total amount of precipitation of N = 483.7 mm. Almost all concentration means of the pure wet precipitation water are lower than the mean ionic concentrations in bulk precipitation water collected at the same sites (Hantschel 1987). An exception is the pH value which is almost identical in bulk and wet precipitation (pH = 4.2).

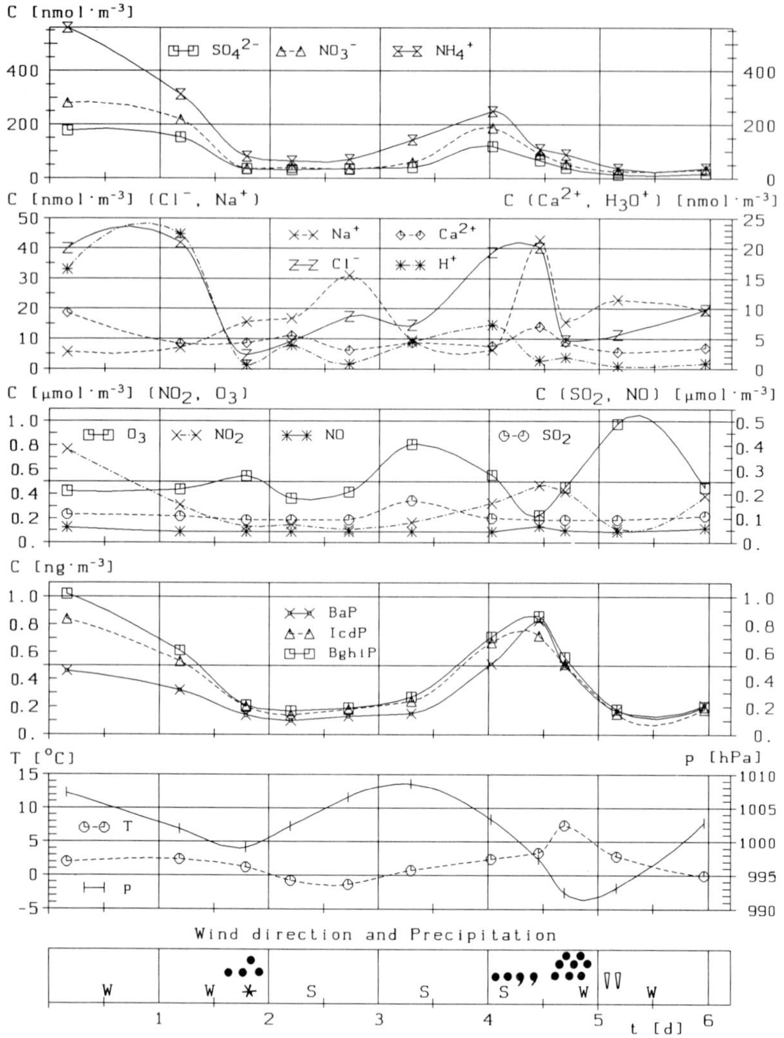

Fig. 5a

Fig. 5a,b. Substance concentrations in atmospheric aerosol in comparison with meteorological conditions; observation site Oberwarmensteinach, Fichtelgebirge. The *marker points* are centered in the middle of the sampling intervals. **a** Time period Oct. 31 – Nov. 07, 1985. Passage of a cold front from NNW (Nov. 02, 1985) and a warm (Nov. 05, 1985, morning) and cold front (Nov. 05 and 06, 1985) of a Biscay-low leading to a decrease in acid ion and PAH concentrations. [*BaP* = Benzo(a)pyrene; *IcdP* = Indeno(cd) pyrene; BghiP = Benzo (ghi) perylene.] The data on atmospheric O_3, SO_2, NO, and NO_2 are from BLfU (1985). **b** Time period Feb. 04–16, 1986. Mid-winter period of high emission level. The data on atmospheric O_3, SO_2, NO, and NO_2 are from BLfU (1986)

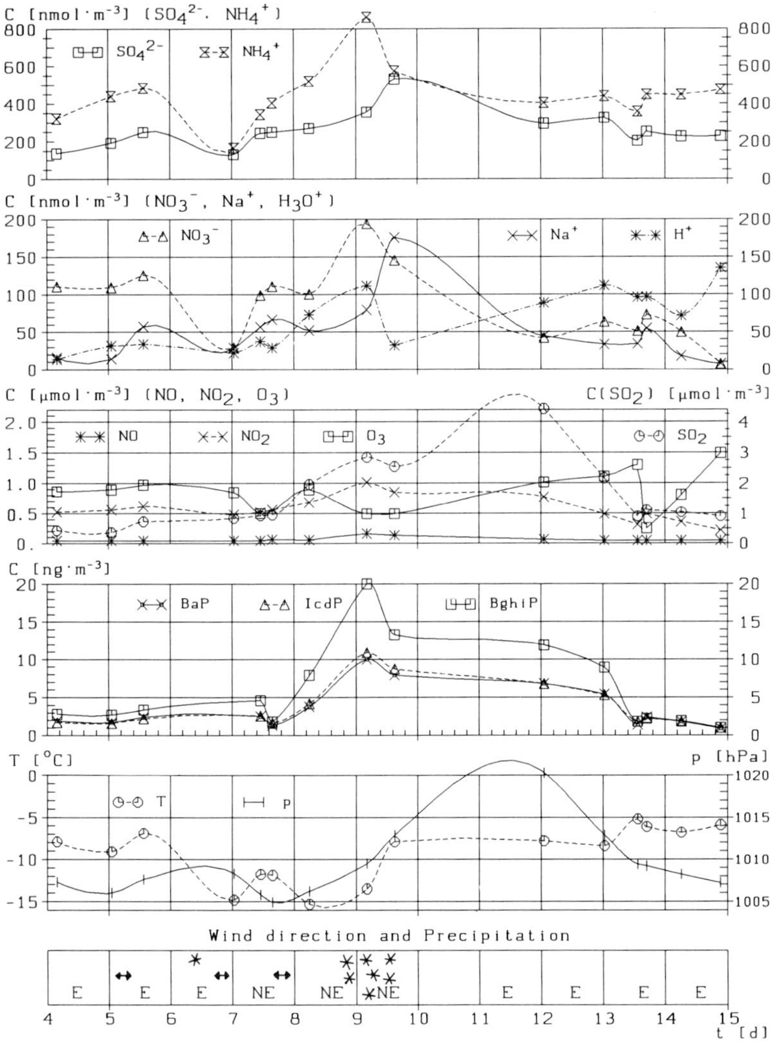
C [nmol·m⁻³] (SO4²⁻, NH4⁺)
SO4²⁻
NH4⁺
C [nmol·m⁻³] (NO3⁻, Na⁺, H3O⁺)
NO3⁻
Na⁺
H⁺
C [μmol·m⁻³] (NO, NO2, O3)
C(SO2) [μmol·m⁻³]
NO
NO2
O3
SO2
C [ng·m⁻³]
BaP
IcdP
BghiP
T [°C]
p [hPa]
T
p
Wind direction and Precipitation
E
NE
t [d]

Fig. 5b

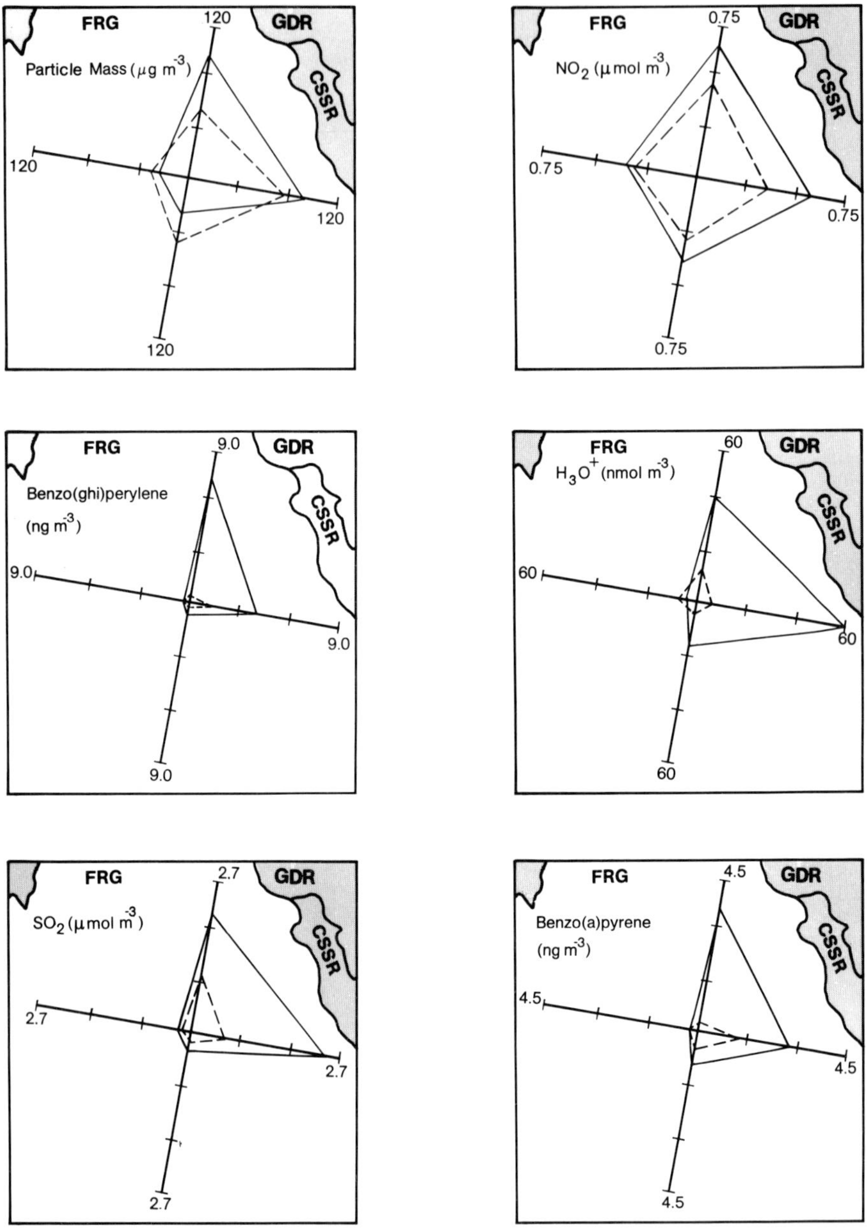

Fig. 6. Directionally differentiated concentration means of atmospheric pollutants; observation sites Wülfersreuth and Oberwarmensteinach, Fichtelgebirge; July 1985 – April 1986. To allow a better directional interpretation of data, the concentration windrose has been turned 10° C from the north using the normal to the borderline between GDR and CSSR as a starting line. A mean air temperature of O° C during the sampling intervals was taken as the limit for classification of "winter" (T ≦ O° C) or "summer" (T > O° C). (----- summer; —— winter)

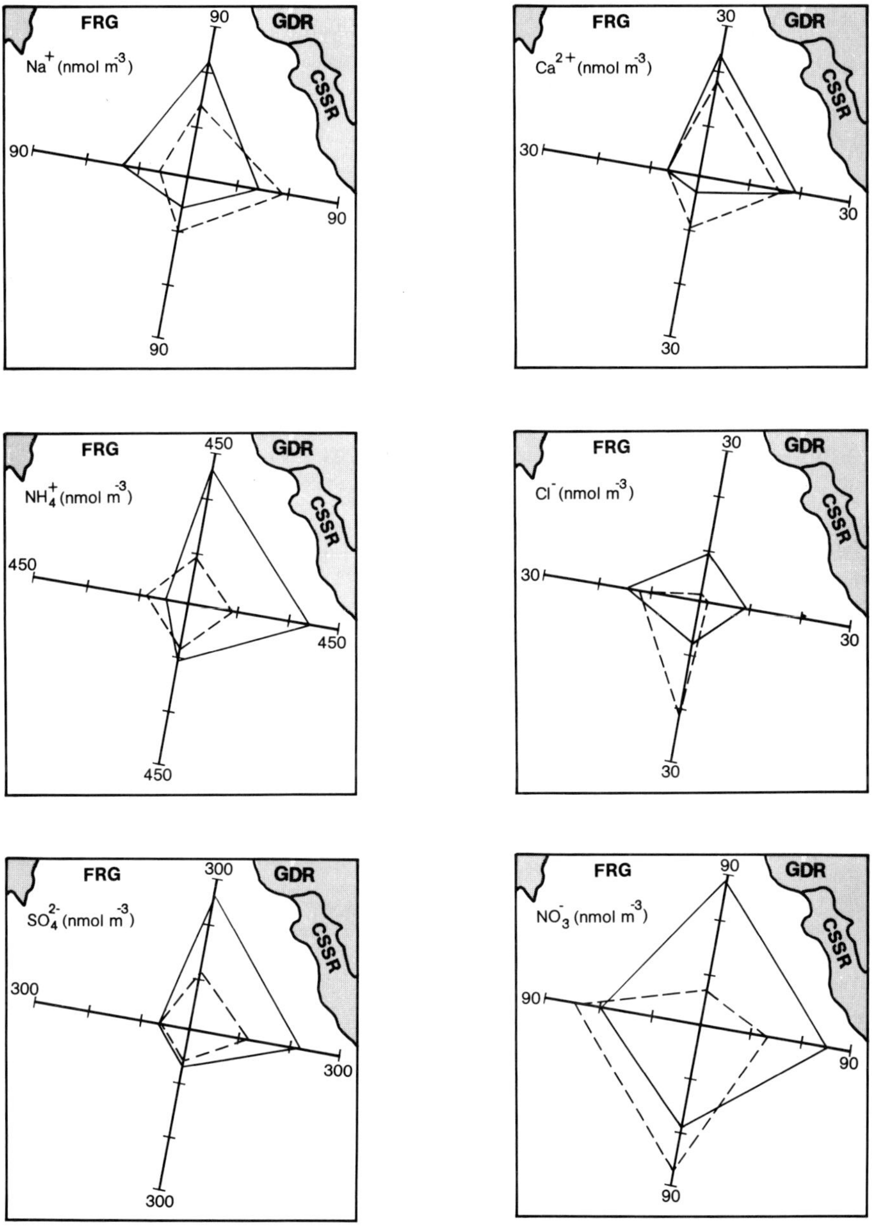

Fig. 6. continued

Table 5. Concentration ranges and means of ions and organic compounds in precipitation water

Parameter[a]	Range	Weighted mean[b]	
pH	3.4–6.0	4.2	
LF	6.0–153 μS cm^{-1}	34 μS cm^{-1}	
K^+	0.1–109 μmol l^{-1}	4.1 μmol l^{-1}	
Na^+	0.4– 97 μmol l^{-1}	5.1 μmol l^{-1}	
Ca^{2+}	0.5–109 μmol l^{-1}	5.4 μmol l^{-1}	
Mg^{2+}	0.1– 26 μmol l^{-1}	1.55 μmol l^{-1}	
NH_4^+	6.0–134 μmol l^{-1}	33.2 μmol l^{-1}	
Cl^-	2.5–101 μmol l^{-1}	11.1 μmol l^{-1}	
NO_3^-	4.8–187 μmol l^{-1}	24.6 μmol l^{-1}	
SO_4^{2-}	5.2–195 μmol l^{-1}	30.6 μmol l^{-1}	
α-HCH	2.7– 99 ng l^{-1}	12.7 ng l^{-1}	Oct.–March
		23.6 ng l^{-1}	April–Sept.
γ-HCH	4.4–434 ng l^{-1}	28.1 ng l^{-1}	Oct.–March
		95.5 ng l^{-1}	April–Sept.
Fluoranthene	3.0–690 ng l^{-1}	48.0 ng l^{-1}	
Benzo(a)pyrene	0.7–194 ng l^{-1}	8.6 ng l^{-1}	
Indeno(od)pyrene	0.8–159 ng l^{-1}	11.2 ng l^{-1}	
Benzo(ghi)perylene	1.3–139 ng l^{-1}	12.9 ng l^{-1}	
Cu	1 –215 μg l^{-1}	16 ng l^{-1}	
Zn	28 –3250 μg l^{-1}	211 μg l^{-1}	
Pb	5 –132 μg l^{-1}	15 μg l^{-1}	
Cd	<.1– 13 μg l^{-1}	0.6 μg l^{-1}	

[a]Selected samples from the observation sites Wülfersreuth, Oberwarmensteinach, Ochsenkopf and Meilerhütte; time period May 1984 through August 1986.
[b]The values listed are based on different numbers of individual samples, since not all parameters could be analyzed for all samples. The means are corrected for precipitation volume.

Figure 7 shows the mean composition of ionic charges in precipitation water. In comparison to the ionic content of particles (Fig. 4, see Sect. 3 this Chap.), a relatively higher proportion of Cl^- and NO_3^- among the anions and a much higher proportion of H_3O^+ among the cations in precipitation water is evident. These differences in ionic charge fractions can be at least partly understood when the directional differences in air mass advection are taken into account. The average of the particulate matter fractional ionic content (Fig. 4, see Sect. 3 this Chap.) includes samples associated with air mass advection from all directions, with maximum concentrations from northeasterly directions. However, precipitation most frequently occurs when air masses move from southwest to northwest, so that traffic emissions and maritime air masses with seaspray constituents predominate. The comparatively high proportion of H_3O^+ in precipitation ($\approx$ 50% of cationic charge), which has also been observed at other locations (Sartorius 1985), implies that gas-phase reactions, rather than the scavenging of almost neutralized aerosol particles, account for the acidity of the rain. The organic micropollutants investigated (HCHs, PAHs) were detected in both the adsorbed

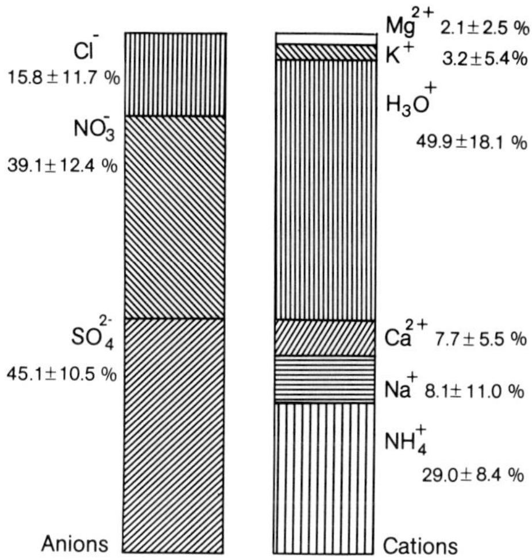

Fig. 7. Mean fractional ionic charge composition of precipitation water samples from the Fichtelgebirge (Wülfersreuth, Oberwarmensteinach, Meilerhütte, Ochsenkopf); May 1984 – August 1986

and soluted phases; their phase distribution (Table 6) is a result of gas and particulate phase scavenging and reflects their aqueous solubilities.

Plotted as a function of time, the concentrations of the measured ions and PAHs typically form a U- or V-shaped curve (Fig. 8a, b) almost independently of the storm's type and duration. Starting with high pollutant concentrations, the values usually decrease until they reach a sometimes expanded minimum. Towards the end of the rainfall event, the concentrations again increase. This final increase is observed less frequently than the decrease at the beginning, which is occasionally preceded by an increase. The general type of pollutograph described is almost independent of the season and the duration of a single precipitation event. While high peak concentrations have been recorded during all seasons, the centre value of the U-shaped curve tends to be lower in summer than in winter. Similar types of concentration graphs have been described earlier, e.g., by Asman (1980, Den Helder/NL) and Castillo et al. (1985, Albany/USA), while Georgii et al. (1983) associated U-shaped curves with convective showers without a final increase in concentrations for frontal precipitation.

HCH concentrations demonstrated a different pattern from all other substances investigated (Fig. 8b): They remain fairly constant during a single precipitation event but exhibit a clear seasonal variation with a maximum in early summer (after application of lindane in agriculture) and a minimum in the cold season.

4.2 Results of Time-Resolved, On-Line Monitoring of the pH Value and the Electrical Conductivity of Precipitation

From December 1984 to September 1986, with an interruption from January through April 1986, a precipitation monitor was installed at the observation site

Table 6. Percentage of pollutants in rain absorbed onto particles[a]

α-HCH	γ-HCH	Fluor-anthene	Benzo(a)-pyrene	Indeno(cd)-pyrene	Bezo(ghi)-perylene
11%	7%	81%	66%	86%	88%
(1.4)	(6.0)	(0.265)	($3 \cdot 10^{-3}$)	–	($2.6 \cdot 10^{-4}$)

[a]For comparison the water solubility of the substances are added (number in parentheses: mg l^{-1}).

Oberwarmensteinach in the Fichtelgebirge. The monitor (Winkler 1977) analyzes and records the electrical conductivity X_{25} (25 °C), the pH value, and the precipitation rate in situ with a high time resolution. The inlet funnel of the monitor is protected against dry deposition by a cover which is removed automatically during periods of precipitation. Thus, not the bulk but the direct precipitation water, the wet deposition only, is analyzed.

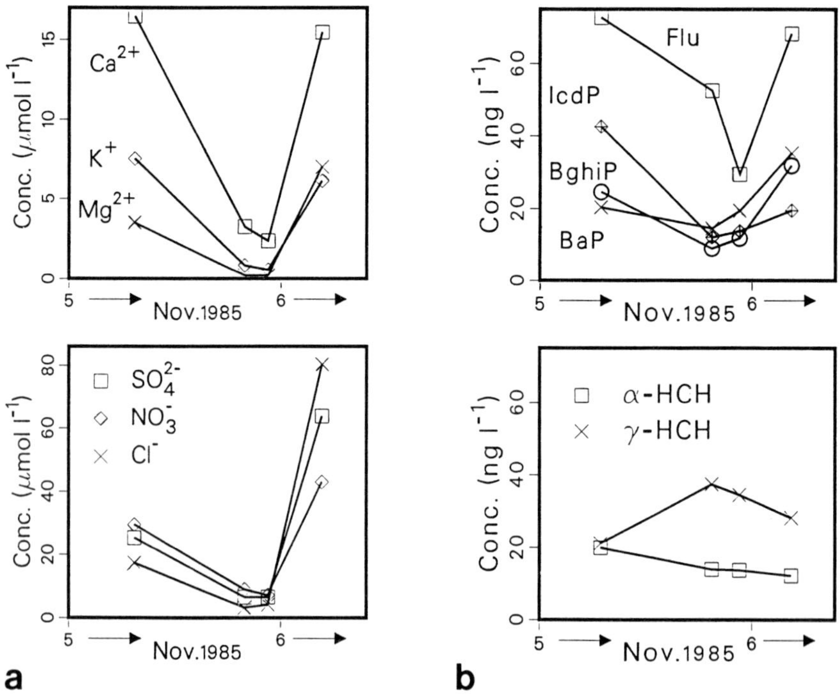

Fig. 8a,b. Concentration of trace substances in precipitation water versus time. During the passage of a frontal system of a Biscay-low (Nov. 05, 1985, 6 a.m. – Nov. 06, 1985, 6 a.m.) a total amount of 24,2 mm was collected. Atmospheric particulate and gas phase concentrations are given in Fig. 5a. The *marker* representing a single sample is set to the point of the time interval where 50% of the water sample analyzed has been collected. **a** Ion concentrations, Wülfersreuth. **b** Concentration of organic trace substances, Oberwarmensteinach. *Flu* = Fluoranthene; *BaP* = Benzo(a)pyrene; *IcdP* = Indeno(cd)pyrene; *BghiP* = Benzo(ghi) perylene; *α-HCH* = α-Hexachlorocyclohexane; *γ-HCH* = γ-Hexachlorocyclohexane

Of the total separate precipitation events, 91, or ca. 30%, could be monitored correctly, as lightning disturbances and heavy snow storms caused fragment instrument failure. Of these 91 reliably recorded events, 61 occurred during the summer and 30 during winter. The mean pH value and mean conductivity value, deduced from these measurements and based on single short-term readings, each corresponding to a constant amount of 0.05 mm rainwater, were for all the summer events pH = 4.42, and χ_{25} = 42.8 μS cm^{-1}, and for all winter events pH = 4.43, and χ_{25} = 40.6 μS cm^{-1}.

These data do not show significant seasonal differences. Sporadic results yielding lower pH values for winter precipitation compared to summer precipitation could not be confirmed with these measurements (Kayser et al. 1974, Verhoeven 1984). Frontal precipitation and showers in general yield lower pH values than continuous precipitation. The mean values given above are based on all types of precipitation events. The pH value reported by Förster (pH = 4.2, Table 5) represents the frontal and shower-like precipitation.

To ascertain to what degree the pH value of the precipitation depends on the origin of the wind direction, the events were subdivided into those occurring during westerly winds and those during easterly winds. In Fig. 9 mean values for all individual events are plotted in a χ_{25}-pH diagram with wind direction. These show that no significant relationship between χ_{25} and wind direction, nor between pH and wind direction, exists. Winkler (1984) also failed to find a unique dependency for similar measurements performed in Hamburg, even after computing the trajectories of the air masses. However, it should be noted that the extreme pH values, the lowest as well as the highest, are generally associated with westerly winds. This observation can be explained by the fact that cloud and precipitation elements are formed in humid air masses originating from regions to the west of the observation area. Precipitation from a dry continental air mass is unlikely. Thus, the chemistry of precipitation water is essentially determined by pollution sources located west of the observation area. This is not true for the gaseous and particulate atmospheric trace substances and for fog of local origin.

Several studies have indicated that in the range of pH = 3–5, almost 100% of the dissolved matter in precipitation water consists of nine ions: SO_4^{2-}, NO_3^-, Cl^-, H_3O^+, NH_4^+, K^+, Na^+, Ca^{2+}, and Mg^{2+} (Granat 1972). Thus, the value of electrical conductivity represents all the dissolved material because of the ionic nature of these constituents. The pH value is determined by the strong acids H_2SO_4, HNO_3 and HCl. Therefore, for the wide range of pH = 3–5, adding or removing pure water leads to direct changes in pH and χ_{25} in the ratio of 1:1 as long as the limits of the indicated pH range are not exceeded.

The acid fraction AF_{eq}, however, defined as

$$AF_{eq} = \frac{\text{concentration of the hydronium ion in equivalents}}{\text{sum of concentrations of all cations in equivalents}}$$

remains constant as long as only condensation and evaporation processes occur. Thus AF_{eq} is a characteristic quantity of the chemical composition of precipitation water. AF_{eq} can be calculated from pH and χ_{25} as:

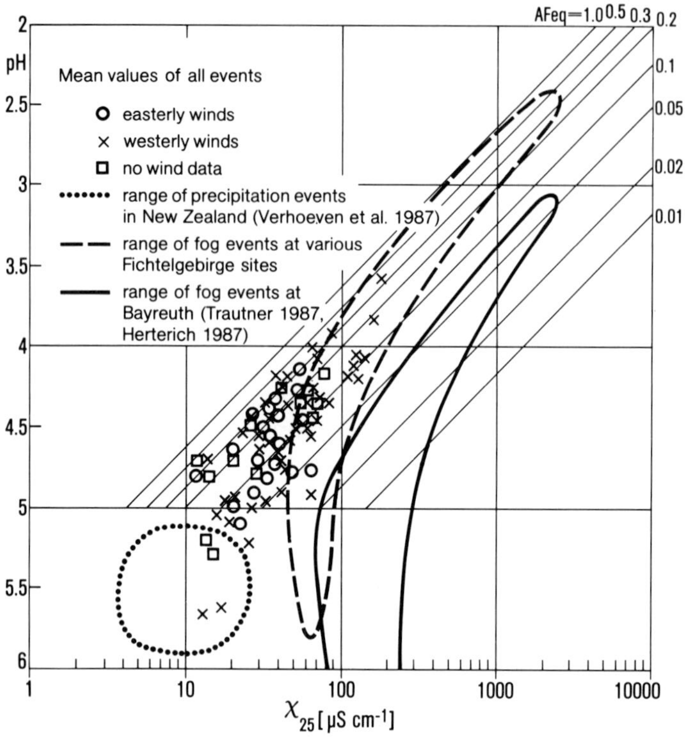

Fig. 9. Mean pH value of samples of precipitation water plotted versus electrical conductivity χ_{25} of the same single precipitation event (□, ○, x), observation site Oberwarmensteinach, Dec. 1984 – Dec. 1985, April 1986 – Sept. 1986. The *inclined straight lines* represent constant acid fractions AF_{eq}. The *encircled areas* represent the range of comparison data as indicated; single events are not marked

$$AF_{eq} = \frac{\hat{\lambda}_{Cat} + \hat{\lambda}_{An}}{\frac{\chi_{25}}{10^{-pH}} - \lambda_{H_3O^+} + \hat{\lambda}_{Cat}}$$

with

$\lambda_{H_3O^+}$ = equivalent conductivity of $H_3O^+ = 0.350 \frac{\mu S \cdot l}{cm \cdot \mu eq}$;

$\hat{\lambda}_{Cat}$ = mean equivalent conductivity of all cations except H_3O^+ weighted with the portion of each cation $= 0.0656 \frac{\mu S \cdot l}{cm \cdot \mu eq}$;

$\hat{\lambda}_{An}$ = mean equivalent conductivity of all anions weighted with the portion of each anion $= 0.0762 \frac{\mu S \cdot l}{cm \cdot \mu eq}$.

As a change of AF_{eq} during a precipitation event implies a shift in the chemical composition of the precipitation water, evaporation or condensation processes are characterized in spite of changing values of pH and χ_{25} by a constant AF_{eq}. The precipitation monitor can measure the development of pH and χ_{25} in the course of a precipitation event together with the precipitation rate, and thus AF_{eq}.

Figure 10 shows the cumulative precipitation curve of a shower recorded during April 13, 1985, as well as the simultaneously measured χ_{25}-, pH-, and AF_{eq}-values. A comparison of the different curves reveals that periods of enhanced precipitation rate are accompanied by minima of χ_{25}. This phenomenon has been observed previously (Asman 1980, Castillo et al. 1985) as noted earlier. The decreasing first part of the U-shaped course of χ_{25} reflects the dilution of rainwater concentration by the enhanced precipitation rate when particles are generally not yet removed very effectively. The subsequent increase of χ_{25} reflects the increasing scavenging efficiency of the smaller rain drops which characterize a ceasing shower. Whether or not these variations of χ_{25} are mainly due to subcloud scavenging or to dilution processes within the cloud is determined by the acid fraction AF_{eq}. Measurements compiled by Winkler (1980) show that for Central Europe, AF_{eq} of the water soluble portion of the aerosol particles is lower than that for precipitating raindrops leaving the cloud base. Thus a reduction of subcloud particulate matter in precipitation water, i.e., a reduced scavenging efficiency, should be reflected in a simultaneous increase of AF_{eq}. In fact during the shower-like precipitation represented in Fig. 10, minima of χ_{25} are accompanied by maxima of AF_{eq}. Therefore, it can be concluded that subcloud scavenging plays a major role in shower precipitation chemistry.

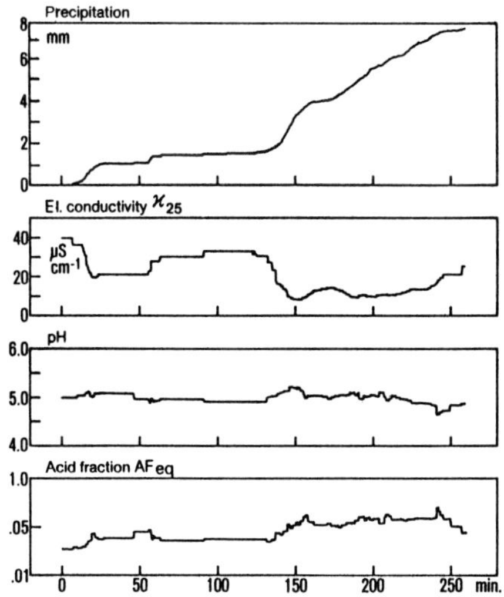

Fig. 10. Simultaneously measured representative quantities of precipitation water as a function of time, Oberwarmensteinach, Fichtelgebirge. Shower, single event, April 13, 1985

During continuous precipitation, scavenging processes also cause the removal of particles and gases from the atmosphere, and with them a gradual reduction of the dissolved matter in precipitation water. The χ_{25}-value steadily decreases, whereas the pH value steadily increases. However, the acid fraction AF_{eq} is not very sensitive to these changes. Even during extreme changes in the precipitation rate, no shifts of AF_{eq} have been observed. Because the airborne particle concentration reaches a very low level, the chemical composition of rain drops can hardly be influenced by scavenging or dilution effects.

4.3 Collection and Properties of Fog

Fog studies are aimed at quantitatively determining fogwater properties at the fringe of and inside a spruce stand, and quantifying the amount of water retained on the surface of a spruce. The ion concentrations of fog droplets are much higher than the comparable concentrations of rain drops, e.g., SO_4^{2-}, NO_3^-, Cl^-, H^+ (Prinz et al. 1982, Schrimpff et al. 1984, Georgii and Schmitt 1985, Forschungsbeirat Waldschäden und Luftverunreinigung 1986). Furthermore, spruce trees are able to collect fog droplets very effectively through impaction processes. The droplets, which partly adhere for a long time to the leaf and needle surface, incorporate solute particulate matter previously dry-deposited. Subsequent evaporation additionally enhances the concentration of ions contained in the fogwater and may lead to an increased leaching of nutrients out of the needles (Wood and Boorman 1975, Frevert and Klemm 1984). In addition, fogwater dripping off leaves and branches affects the forest soil.

The areas of conspicuous severe forest injury coincide with those areas of high fog frequencies, namely, the higher parts of the German subalpine mountains, e.g., the Fichtelgebirge, and mountain summits shrouded with incumbent clouds.

Fog is characterized by a reduction of visibility, droplet distribution and liquid water content. According to the international definition (World Meteorological Organization – WMO), fog reduces visibility below 1000 m. During the study reported here, fogwater was sampled at a visual range of ca. 200 m and less. Where clouds are elevated dispersed droplet systems, fog differs from clouds only in that the base of fog touches the earth surface. Fog is composed of dispersed droplets with radii from $r \approx 0.5\ \mu m$ up to $r \approx 50\ \mu m$. The size distribution within this interval depends on the aerosol particle and water supply and changes during the life time of a fog event (Bachert 1981). The average radius of a fog droplet is ca. $r = 5\ \mu m$. The water condensed within a volume of air may be calculated from the drop size distribution (Fuzzi 1986)

$$LWC = \frac{4}{3} \pi \rho_w \Sigma n_i r_i^3 \qquad (g\ m^{-3})$$

with r_i = radius of droplet, n_i = number of droplets per radius interval and air volume, ρ_w = density of water. The liquid water content of fog may vary from $LWC \approx 0.01\ g\ m^{-3}$ up to $0.3\ g\ m^{-3}$.

Fog may be distinguished according to whether it forms through radiation or advection. Radiation fog is produced by radiational cooling below the dew point of soil-vegetation surfaces and the adjacent air layer. In general, it starts to form during cloudless nights and is favored by humid surfaces and light surface winds. It is a more or less local process. In hilly country it fills the valleys (valley fog) and is accompanied by a temperature inversion. Radiation fog is most frequently observed after the growing season.

Advection or mountain fog (orographic fog, mixing fog, evaporation fog) is formed when air flows over mountains, adiabatically cools and mixes with water vapor from a very humid surface layer, especially one following a precipitation event. An extended cloud layer shrouding a mountain summit may be considered advection fog. As a characteristic feature of advection fog is high windspeed, the chemical properties of the air masses involved are essentially influenced by remote sources. Advection fog is observed throughout the year.

Both types of fog may be modified depending on special atmospheric or orographic conditions (see Sect. 5 of this Chap.). The frequency of fog days increases with altitude (Fig. 11) and is highest for spring and autumn.

Fog droplet water samples first taken with passive fogwater collectors during sampling periods of one week proved difficult to interpret, as aerosol particles and rain drops were also collected. The system was replaced with active collectors previously developed by Andre (1976) and improved by the author. These consist of a plexiglas tube (Fig. 12a) through which air is aspirated by a fan (240 m^3 h^{-1}). The shape of the inlet impedes the collection of raindrops and water dripping down from trees.

The fog droplets are deposited on 3 nets of polyethylene. Different string diameters (0.1 and 0.3 mm) provide a crude differentiation of droplet size classes. Fig. 12b shows the collection efficiency of the system. The ionic concentrations of

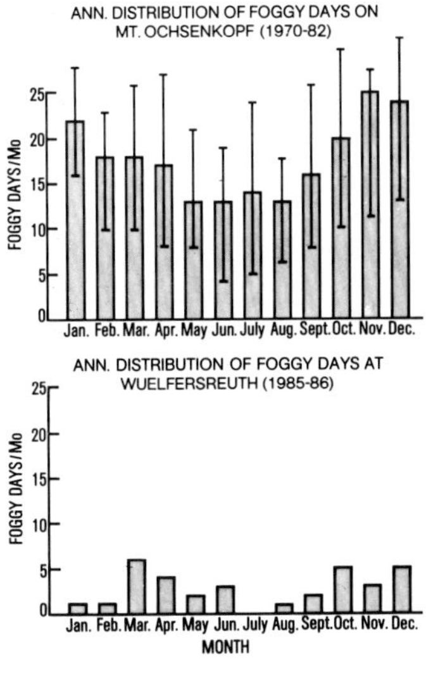

Fig. 11. Annual variation of monthly sums of days with fog, Mount Ochsenkopf (alt. 1020 m a.s.l.) 1970–1982 Wülfersreuth (alt. 680 m a.s.l.) 1985

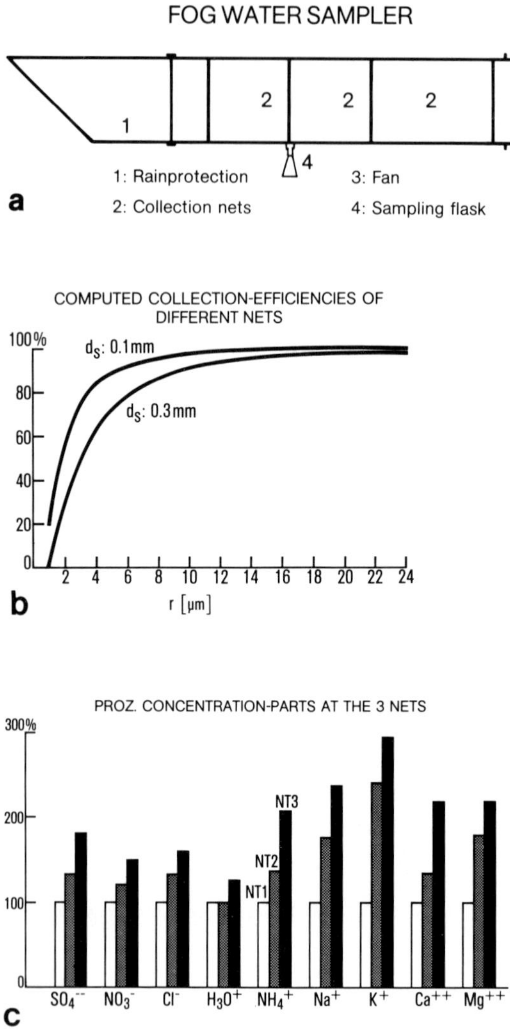

Fig. 12a-c. Vacuum (or active) fogwater sampler. **a** Schematic representation. **b** Computed collection efficiencies of collecting nets versus droplet radius r (Langmuir & Blodgett 1946). Net No. 1 and 2, string diameter d_s = 0.3 cm; net No. 3, string diameter d_s = 0.1 mm (dry nets). **c** Relative increase of ion-concentration of fogwater samples with decreasing diameter of string (and sequence of net), December 1985. Data normalized: net No. 1 $\hat{=}$ 100%

the fractionated fog droplet sample are generally highest in the third fraction. This may be the result of an increase in concentration with decreasing drop-size (Fig. 12c).

Measurements were conducted 70 m inside and 25 m in front of the spruce stand of Wülfersreuth where trees averaged 18 m high. To collect fogwater at two different levels, 6 m and 16 m altitude, platforms were installed, which could also be used to determine the vertical profile of horizontal wind.

At Wülfersreuth, fogwater was collected during the winter of 1985–1986. Mean, maximum, and minimum values of fog water samples analyzed from this study are given in Table 12. Exemplary measurements of ionic concentrations are given in Fig. 13a and b. No significant differences were observed at the different

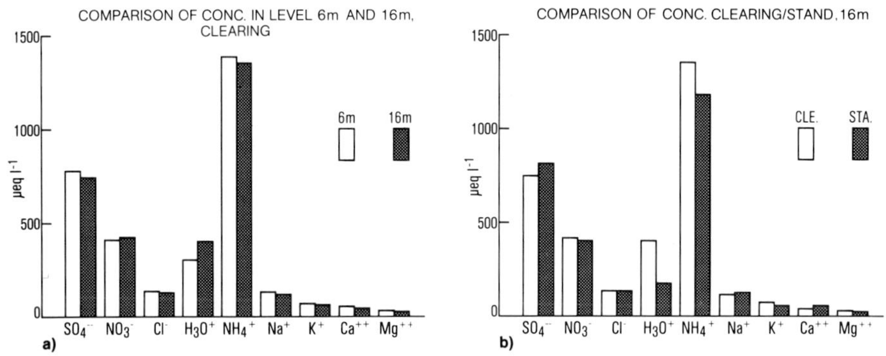

Fig. 13a,b. Illustrative example of ion concentrations of fogwater samples, collected simultaneously, Wülfersreuth, December 1985, cyclonic weather conditions (westerly winds), **a** 6 and 16 m above ground close to the edge of a spruce stand; **b** inside and outside of spruce stand, 16 m above ground

levels or inside and outside the spruce stand. (The differences in H_3O^+ shown in Fig. 13 were not observed for other fog events.)

The degree to which the ion concentrations vary, however, depends on the general meteorological situation. The highest concentrations were observed during stable anticyclonic high pressure conditions with easterly-northeasterly winds (Fig. 14). The concentrations were enhanced by a factor of almost 10 compared to those measured during cyclonic conditions with westerly-southwesterly winds (compare also Fig. 6). Fogwater conductivity reached values around 3000 μS cm^{-1} while the pH values dropped to 2.4. The liquid water content (LWC) of the air was always higher in front of the spruce stand than inside it. This is demonstrated in Fig. 15 which shows the empirical LWCs (LWC extracted by the collecting device) of different fog events inside and in front of a spruce stand.

The mass C_e of the individual element (or ion) contained in fog droplets per unit volume of fog air is called "fogwater element concentration of the atmosphere" and is given by (Georgii et al. 1986)

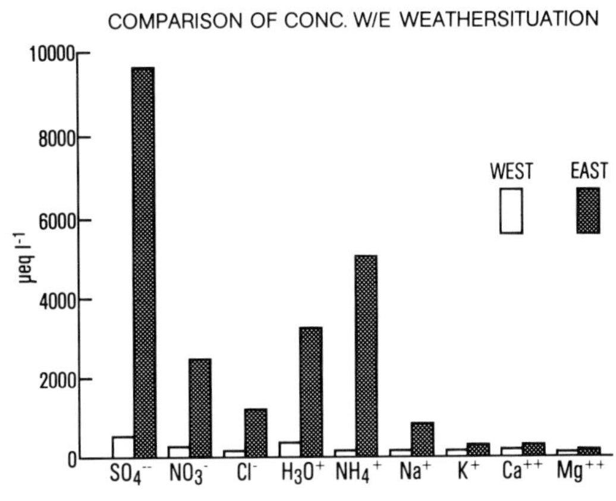

Fig. 14. Comparison of two fogwater samples taken during cyclonic (westerly winds, December 1985) and anticyclonic weather conditions (easterly winds, March 1986)

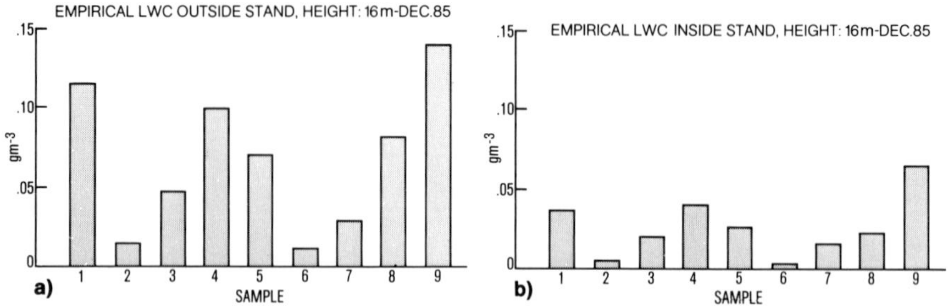

Fig. 15a,b. Liquid water content (LWC) of different fog events during which fogwater samples were taken, Wülfersreuth, December 1985, 16 m above ground, **a** outside a spruce stand and **b** inside

$$C_e = \frac{c \cdot V_{fw}}{V_a} \qquad (\mu eq\ m^{-3})$$

with c = element (ion) concentration of fogwater (μeq/ml);
V_a = volume of air aspirated (m^3);
V_{fw} = volume of fogwater collected (ml),

The calculation of C_e reveals that the mass of trace substances included in fog droplets and transported within a volume of atmospheric air is higher at the 16 m than at the 6 m level (Fig. 16a). The difference is even more dramatic if we compare C_e inside and outside the stand. Figure 16b indicates a decrease in element concentration C_e of more than 50%, moving from the fringe into the spruce stand. If we multiply the mass concentration C_e by the horizontal wind speed v_h (m s^{-1})

$$F_e = C_e \cdot v_h \qquad (\mu eq\ m^{-2}\ s^{-1})$$

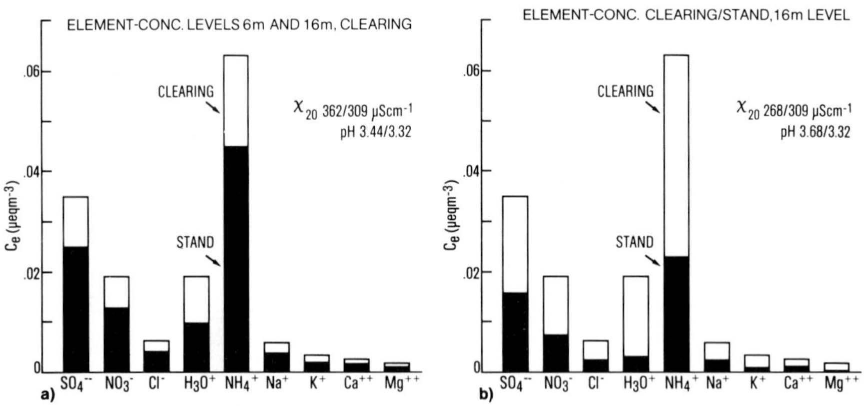

Fig. 16a,b. Exemplary data set of fogwater element concentration C_e, Wülfersreuth, December 1985, cyclonic weather conditions (westerly winds). **a** At two levels, 6 and 16 m above ground, outside a spruce stand close to the edge and **b** inside and outside a spruce stand 16 m above ground; χ_{20} = electrical conductivity (20°C) and pH = exp. of hydrogen ion concentration of samples

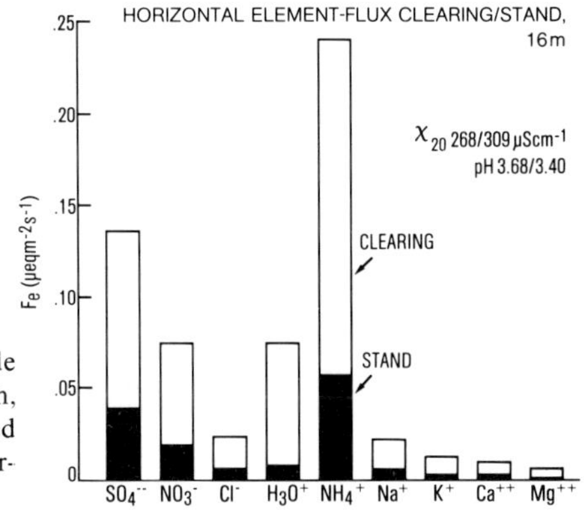

Fig. 17. Horizontal fogwater element flux inside and outside a spruce stand at Wülfersreuth, December 1985, 16 m above ground, calculated using measured element concentration and horizontal wind speeds (χ_{20}, pH cf. Fig. 16)

the product represents the horizontal element (or ion) flux. Since the horizontal wind speed and also the element concentration C_e increases with height, the horizontal flux will increase with height and will be more pronounced outside than inside a stand. If the same impaction efficiencies of fog droplets are assumed for all trees independent of their position, the horizontal flux F_e is proportional to the actual deposition of trace elements on the surface of the trees. Thus, the element input to trees by fog at Wülfersreuth has been at least three times higher at the edge of the spruce stand than inside it (Fig. 17).

4.4 Deposition of Fogwater upon a Spruce

Fogwater input to spruce trees depends on their capacity to retain water on the branches and needles. This property also determines the input of trace substances which accompany the droplets. In order to measure the quantity of fogwater retained on the surface of a tree, initial investigations were begun in connection with the fog project reported in the previous section. A preliminary result is reported here.

Fig. 18 shows the schematic representation of the measuring device which consists of a sensitive balance (Sauter) and a small tree planted in a pot. Balance and pot are protected against water dripping from the branches and needles of the spruce. The changing water load of the tree is measured by the balance and continuously recorded. The leaf area was calculated from a correlation between this surface quantity and the stem diameter (Oren et al. 1986). Measurements were carried out with radiation fog at the university campus on different nights in the autumn of 1986.

The results are shown in Fig. 19, plotted as a function of time. The maximum water deposition on the surface of the tree was about 100 ml m^{-2}. All measurements show a similar time dependence. The relative maximum around 1 a.m. was probably caused by ceasing dew fall followed by a late-starting fog formation. Between 8 and 9 a.m., water deposit again decreased in spite of a continuing fog.

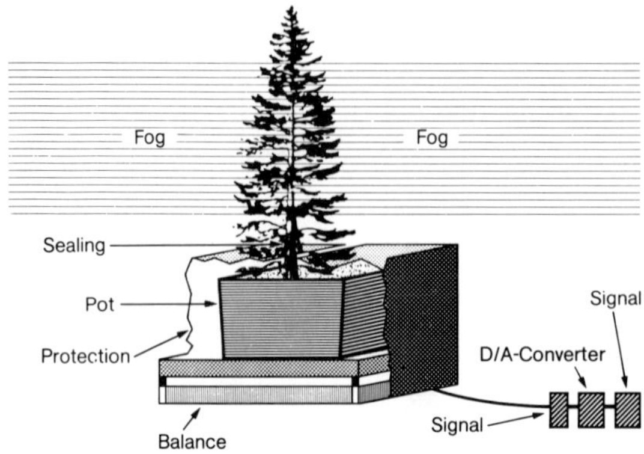

Fig. 18. Deposition of fogwater droplets on a spruce tree. Schematic representation of experimental measuring device

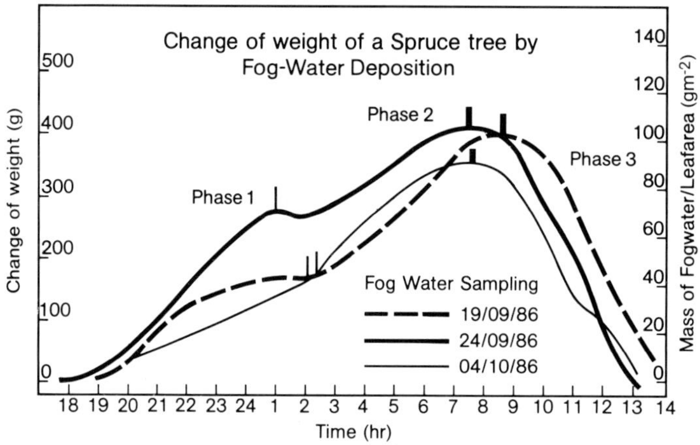

Fig. 19. Fogwater deposition on a spruce tree versus time. *Phase 1:* dew formation on the needle surface; *Phase 2:* fog droplet deposition; *Phase 3:* dripping off and evaporation from needle surface

This increase in water loss may have been caused by the decreasing LWC of the fog or by the coalescence of droplets on the surface of the needles, thereafter dropping off. This effect may have also enhanced the drying of the tree surface as the fog abated.

5 Concentration of Hydrogen Peroxide in Precipitation and Fogwater

Masuch et al. (1986) studied acid fog containing H_2O_2 at concentrations up to 5 ppm, levels which corresponded to H_2O_2-measurements in cloudwater at Whiteface Mountain (Kadlecek et al. 1983). Temporary exposure of young Norway spruce trees to H_2O_2 in artificial fogwater affected the capability to transport and assimilate water because of major changes in the vascular bundle. H_2O_2 contained in cloud- and fogwater intercepted by spruce should therefore be included in studying the role of photooxidants in forest die-back.

Recent analysis of H_2O_2 rain, cloud and fogwater sampled under various meteorological conditions in the Fichtelgebirge area will be presented and discussed in this section. In previous studies of aqueous phase H_2O_2, Bufalini et al. (1979), Kok (1980, 1982), Zika et al. (1982), Klockow and Jacob (1986) observed H_2O_2-concentrations in rainwater ranging from 0.001 ppm–3 ppm. Airborne cloudwater data at the same order of magnitude have been reported by Kok (1982), Richards et al. (1983), Römer et al. (1984), and Kelly et al. (1985). In Los Angeles fogwater, the H_2O_2-concentrations were found to be only 0.001 ppm (Hoffmann 1983).

Low concentration levels in polluted areas and during the cold season follow from the possible generation and decomposition reactions of H_2O_2 and precursors. Sources of aqueous phase H_2O_2 (aq) are scavenging of gaseous H_2O_2 (g) and O_2H (g). The imbalance of O_2H (g), the hydrated form ($H_2O{\cdot}HO_2$) (g) and O_2H (aq) yield H_2O_2 (g) and H_2O_2 (aq), respectively. O_2H (g) is produced from the photochemical oxidation of CO and to a lesser extent from hydrocarbons and aldehydes initiated by the attack of OH-radicals, which originate mainly from ozone photolysis. The primary trace gases NO and SO_2 also play an important role in the occurrence of atmospheric H_2O_2. O_2H (g) reacts with NO to form NO_2 and H_2O_2 (aq), which is quickly removed by HSO_3^- to reach equilibrium with gaseous SO_2. More detailed information is given by Chameides and Davis (1982) and McElroy (1986).

In our analysis of H_2O_2, rainwater was collected with a polyethylene funnel (diam. = 30 cm) and volume-calibrated, dark-colored glass bottles. The clean sampling device was exposed to precipitation just at the outset of the precipitation. Sampling intervals were generally within 30 min. Fog respectively cloud water was collected with a teflon-coated, rotating-arm collector developed and supplied by the Deutsche Forschungs- und Versuchsanstalt für Luft- und Raumfahrt (DFVLR). Sampling intervals of fogwater did not exceed 30 min and were instantly followed by analysis preparation. The concentration of H_2O_2 was determined fluorimetrically by an enzymatic method based on the peroxidase-catalysed dimerisation of p-hydroxyphenyl acetic acid in the presence of H_2O_2 (Guilbault et al. 1968, Lazrus et al. 1985). Catalase was used to provide analytical blanks. If organic peroxides are present, both H_2O_2 and R-OOH produce a fluorescence signal. Experiments with shorter reaction time, however, did not show any change of signal, indicating that the quantity of organic peroxy compounds was insignificant and negligible.

Slight modifications in the composition of reagents and reaction time for a nonautomated procedure have been specified by Herterich (1987). To simplify analysis, the fluorometer was operated at reduced sensitivity enabling determination within a range of 0.3–60 μmol l^{-1} (0.01–2 ppm) and a reproduceability of $\pm$ 10%. A Dionex Ion Chromatograph was used to analyze anions. The detection limit of S(IV) was dependent on the dilution of the sample, which did not exceed 5 μmol l^{-1}. Freezing was the sole means of preservation used.

The concentration of H_2O_2 in rainwater varied from less than 0.3 to up to 56 μmol l^{-1}. As demonstrated in Fig. 20, these variations could be correlated to such meteorological conditions as UV-radiation intensity and wind direction, as well

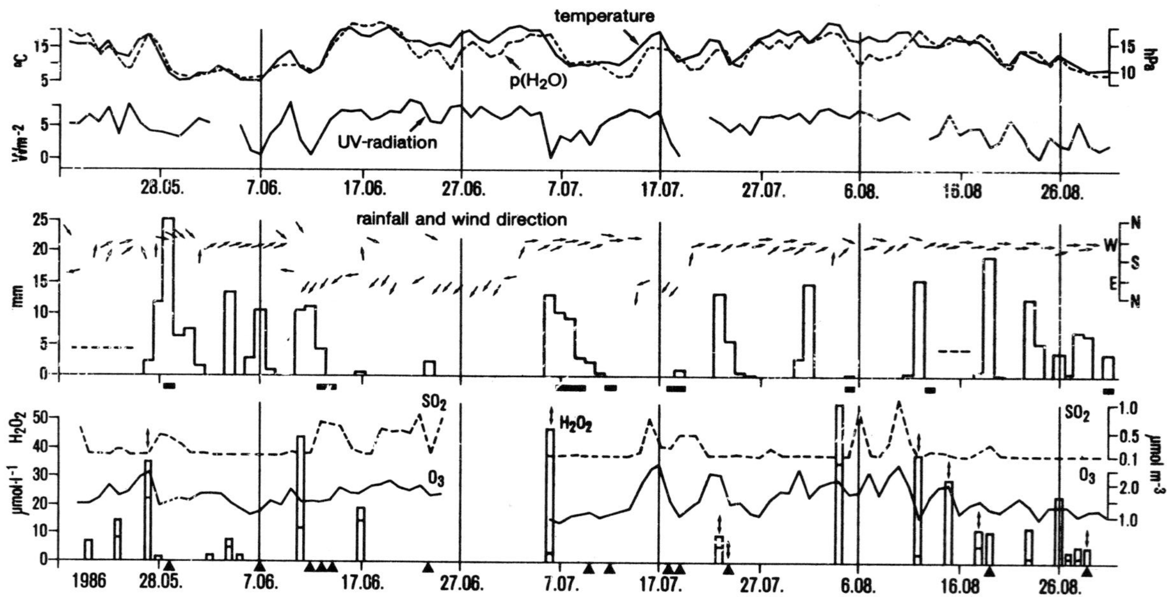

Fig. 20. Daily means of meteorological parameters recorded near Bayreuth and at the Mt. Ochsenkopf summit (alt. 1020 m a.s.l.), SO_2 and O_3 concentrations at Oberwarmensteinach (alt. 760 m a.s.l.). The daily amount of rainfall and the maximum and minimum H_2O_2 concentrations of one day are represented by *columns (scale on the left)*. The rainwater samples have been taken in a rural residential area 5 km southwest of Bayreuth. $p(H_2O)$: partial pressure of water vapor; ▲ : H_2O_2 concentration < 10 ppb; ≡ : fogwater collected and analyzed; ↕: thunderstorm discharge; –: no data; H_2O_2: 10 ppb ≙ 0.3 µmol l^{-1}. The data on atmospheric O_3 and SO_2 are from BLfU (1986)

as to physical and chemical properties of the air masses. At least 3 different weather patterns were distinguished:

1. Early precipitation, which followed advection of warm and moist southerly and southwesterly air, contained high and medium scale H_2O_2 concentrations. If precipitation preceded sunny weather, depth and rate of rainfall seem to be important.
2. During periods of cold and rainy weather, H_2O_2-content was near or below 0.3 μmol l^{-1}.
3. Rainfall associated with increased atmospheric SO_2 concentrations did not show detectable H_2O_2 concentrations. This observation agrees with cloud-water measurements by Kelly et al. (1985).

In northeastern Bavaria, an increase in SO_2 concentration can frequently be observed with northerly and easterly winds. This was the case on June 23–24 and July 19 when precipitation fell from a stratiform cloud layer. An increase in SO_2 concentration was also recorded in the early morning of August 19. A first heavy rainfall from a warm front cloud system contained less than 0.3 μmol l^{-1} H_2O_2, whereas in a single independent shower in the afternoon 12 μmol l^{-1} H_2O_2 was detected. In Table 7 analysis of sulphate and S(IV) and meteorological observations indicate that primarily near-surface SO_2 and low nocturnal mixing rates are responsible for the lack of H_2O_2 in rain from stratus clouds in the early morning. Low concentrations particularly in rainwater of large postfrontal

Table 7. Concentration of some trace constituents of rainwater near Bayreuth and fogwater at Ochsenkopf Mountain (alt. 1020 m a.s.l.) Fichtelgebirge, Bavaria

Date	MEZ[a]	Sample[b]	AP or LWC		H_2O_2	SO_4^{2-}	NO_3^-	S (IV)[d]
					μmol·l^{-1}			
24/06/86	07:37–07:46	R	0.18	l m^{-2}	< 0.3	55	71	8
07/07/86	06:10–06:15	OR	0.25	l m^{-2}	0.9	13	16	–
07/07/86	06:19–06:33	F	0.06	g m^{-3}	0.6	42	84	–
18/07/86	09:55–09:28	F	0.02	g m^{-3}	< 0.3	380	210	14
19/07/86	06:10–06:30	R	0.09	l m^{-2}	< 0.3	510	240	20
19/07/86	09:38–09:58	F	0.07	g m^{-3}	< 0.3	370	150	22
04/08/86	13:10–13:40	R	0.20	l m^{-2}	56	80	81	–
05/08/86	08:48–09:06	F	0.04	g m^{-3}	1.0	180	510	–
12/08/86	00:10–01:05	R1	0.75	l m^{-2}	6.8	110	130	–
12/08/86	01:05–01:11	R2	1.25	l m^{-2}	5.3	38	45	–
12/08/86	02:02–02:30	R6	0.19	l m^{-2}	30	6	21	–
12/08/86	18:05–18:31	OR	1.1	l m^{-2}	0.7	70	85	–
12/08/86	18:32–18:50	F	0.10	g m^{-3}	0.6	120	150	–
12/08/86	19:10–19:20	F	0.08	g m^{-3}	< 0.3	140	180	–
19/08/86	02:00–06:30	R	11.5	l m^{-2}	< 0.3	39	60	8
19/08/86	15:20–17:10	R	3.8	l m^{-2}	12	27	49	–
03/09/86	14:30–18:10	R	5.3	l m^{-2}	22	21	29	–

[a]MEZ: mean European time.
[b]F: fog; R: rain; OR: wet fog.
[c]AP: amount of precipitation (l m^{-2}) or LWC: liquid water content of fog (g m^{-3}). LWC has been calculated on the basis of sampled air volume and weighed liquid water.
[d]–: below detection limit.

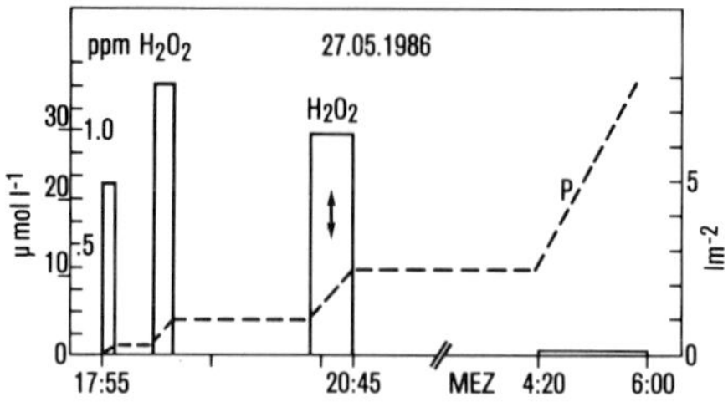

Fig. 21. H_2O_2 concentration in bulk rainwater samples and cumulative amount of rain (P) during the passage of a strong cold front on May 27 and 28, 1986. The width of the column represents the sample interval. *MEZ*: mean European time; ↕: thunderstorm activity

precipitation systems, on the other hand, can be attributed to preceding rainout of H_2O_2 (g) and a change in air mass (Fig. 20, June 7 and 12, July 7, 1986).

A case study supports this assumption (Fig. 21). During May 26 and 27 subtropical, maritime air was transported into Central Europe in front of an eastward-moving cyclonic system. In the evening of May 27, the approaching cold front was announced by three individual prefrontal showers with significant H_2O_2 concentrations. After the passage of the front around 4 a.m. the following morning, only the fraction sampled through 6 a.m. contained more than 0.3 μmol l^{-1}. The concentration of samples collected from 6 a.m.–11 p.m. (a total of 4.5 l m^{-2}) was less than 0.3 μmol l^{-1}.

Another typical characteristic of H_2O_2 in the course of individual rainfall has already been reported by Zika et al. (1982). Unlike ionic compounds, the concentration of H_2O_2 does not generally decrease with lasting rainfall. However, it very often irregularly fluctuates, especially during the early rain, when it is only slightly dependent on precipitation rate (Fig. 22). As Zika et al. (1982) found no indication that the amount of H_2O_2 is correlated with any other trace substance, including NO_3^-, which is mainly incorporated by gas phase scavenging, they suggested that diurnal H_2O_2 might be generated within cloud droplets after scavenging of OH- and O_2 H-radicals. Since H_2O_2 concentration patterns similar to those of daylight hours can be found for nighttime showers as well, an in-cloud production of free radicals does not satisfactorily explain erratic concentration changes. This is also true for thunderstorm activity, which has previously been assumed to be important in H_2O_2-formation (Kok 1980).

The dynamics of a moving convective cloud, i.e., the entrainment, partial evaporation and downdraft movement of cloud and rain elements, may account for the irregular concentration patterns observed during the lifetime of a rain

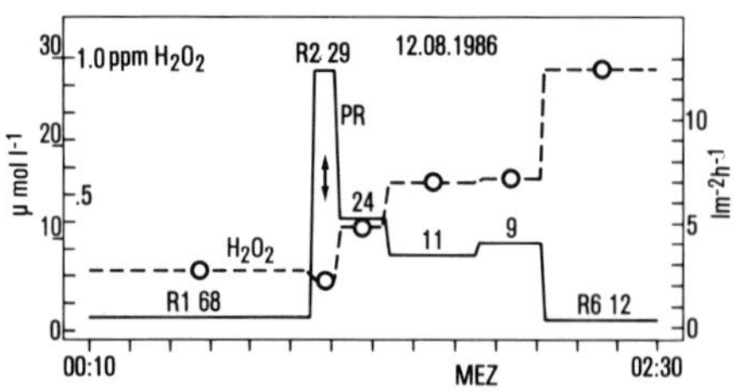

Fig. 22. H_2O_2 concentration, electrical conductivity (μS cm^{-1}) noted above each of the sample intervals and precipitation rate (PR) of a thunderstorm shower. The *symbols R1–R6* refer to the Table 7. Decomposition of H_2O_2 is to be taken into account for the first sample. Electrical discharges were noticed during the second sample interval only. *MEZ:* mean European time

shower. In addition to nucleation, the more effective flux of a highly soluble gas like H_2O_2, compared with the flux of particles to the droplet (Georgii 1965), may contribute to the weak correlation of H_2O_2 with rain intensity and concentration of ionic trace substances. Moreover, the decomposition of H_2O_2 by dissolved SO_2 and organic particulates in cloud and falling rain droplets (Kok 1982, Bufalini et al. 1979) may even prevent an enrichment of H_2O_2 (aq) during the onset of rain. Thus, the instability of H_2O_2 (aq) permits an increase in H_2O_2-content towards the end of the rain when reaction partners are removed. It should be pointed out, however, that all attempts to fully explain the variability of H_2O_2-content in rainfall have been more or less speculative. To address this question, supporting gas-phase measurements at different tropospheric heights have to be performed. Since our findings give no indication of a diurnal variation of H_2O_2 (aq) in rain from convective clouds, we concluded that a rather uniform daily pattern of gaseous H_2O_2 existed in the clean upper troposphere.

Fogwater was collected at the summit of the Ochsenkopf (alt. 1020 m a.s.l.) where three different types of fog formation can be distinguished:

1. Advection, or mountain fog, forming below precipitation postfrontal convective clouds due to drops in temperature and abundance of moisture. Following precipitation, advection persists as a cloud cap with continuous transition to the cloud layer above.
2. Nonprecipitating stratiform clouds touching the surface.
3. Radiation (high inversion) fog, which develops during increasing anticyclonic influence as a result of nocturnal radiative cooling from and below a dissipating cloud cover and often preceded by precipitation. During the warm season, precipitation occurs exclusively at night and in the early morning.

Types 1 and 3 are special cases of fog described earlier in Section 4 of this chapter.

In Fig. 20 (cf. fog symbols), the amount of H_2O_2 in rainwater was very low when weather conditions favored fog formation. In cloud- and fogwater, even lower concentrations could be detected. In most fog samples, the concentration was less than 0.3 μmol l^{-1}. A maximum value from August 5 of 1 μmol l^{-1} in dissipating stratus corresponds to the highest rainwater values measured the day before. Simultaneous and sequential sampling of rain and fogwater (Type 1) showed equal or higher values for rainwater (Table 7). This may indicate that H_2O_2 contained in fog droplets originates from upper cloud cells and is transported downward during episodes of shower activity. This implies, of course, a reverse relationship between H_2O_2 and the concentration of ionic trace substances already reported elsewhere (Kadlecek et al. 1983, Römer et al. 1984).

While the cloud- and fogwater data from Ochsenkopf contrast with the previously cited data of airborne collection, as well as with ground-based samples taken at Whiteface Mountain (Kadlecek et al. 1983), it should be realized that because of the low elevation of Ochsenkopf, intercepting cloud water is restricted to periods with very low photochemical activity and preceding excessive rain-out of H_2O_2 (g) and precursors. With radiation fog, the nighttime depression of H_2O_2 (g) in the atmospheric boundary layer (Jacob et al. 1987) is a further reason for very low H_2O_2 concentrations in fog droplets. These low concentrations of a few ppb may also be produced in homogeneous and heterogeneous aquatic redox processes (Graedel et al. 1986, Klokow and Jacob 1986) that begin with the hydration of condensation nuclei. In opposition to postfrontal fog events, low-level stratus clouds formed by an upwards sliding motion (Type 2) can also be observed after dry and sunny weather. However, these always coincide with periods of reduced vertical mixing and very often with northeasterly winds. Under this meteorological condition in rural northeastern Bavaria, even in summertime, daily means of SO_2 higher than 0.8 μmol m^{-3} occur. These give a mixing ratio of SO_2:H_2O_2 equal to 17:1 if a H_2O_2 (g) concentration of 10^{-9} atm is assumed (Radojevic 1983). Considering its high solubility and the very fast decomposition of H_2O_2 via aqueous phase oxidation of SO_2, the absence of liquid and gas phase H_2O_2 can be expected at Ochsenkopf.

6 Comparison of Bulk Precipitation Water and Fogwater Data from Surrounding and Remote Areas

6.1 Bulk Deposition Studies in the Catchment "Große Ohe"

Since 1978, forest hydrological studies have been carried out in the catchment "Große Ohe" in the national park "Bayerischer Wald". Eight observation sites within the catchment area were chosen for a survey of the regional deposition including its distribution by altitude and differing stands (Gietl 1985).

Because of the high and persistent snow cover in the upper parts of the catchment, only those sites below 800 m a.s.l. allow regular measurements throughout the year. Thus, Taferlruck was used as base station 769 m a.s.l. in the open field and Grüben at 790 m a.s.l. as a station in a spruce stand.

Instrumentation followed the recommendation of the German Union for Hydroeconomy and Cultivation Technique (Deutscher Verband für Wasserwirtschaft und Kulturbau: DVWK 1984). For precipitation collection, all sites were equipped with LÖLF-bulk-collectors which do not separate dry- and wet-deposited substances. These were emptied weekly, provided there was enough precipitation water for analysis ($\geq$ 5 mm/week). Immediately following the weighing of the precipitation samples, pH value and electrical conductivity were determined on site. Further sample analysis was carried out in the laboratory (Bosse 1987). Measurements were taken from July 1983 (open air and throughfall in stands) until December 1986.

Bulk precipitation data did not show any significant seasonal variation of element concentration or deposition rates. Data from the base station Taferlruck are listed in Table 8.

Temporal deposition variation depends mainly on the weather conditions as air masses flowing in from west to north are responsible for the highest acid concentrations and deposition rates. This can be explained by the long-range transport from industrial or urban areas west of the Große Ohe (Bosse 1987), a finding which corresponds to the situation of the Fichtelgebirge. As noted earlier,

Table 8. Observation site Taferlruck (769 m a.s.l., Große Ohe)

Year	Precipitation (mm)	H^+ (pH)	SO_4-S	NO_3-N	Cl	Elements[a] (10^{-3} mol m^{-3}) NH_4-N	Na	K	Ca	Mg	P-total
[a]1984	981	4.3	93	34	27	31	17	4.3	9.5	3.8	1.6
1985	1050	4.25	91	48	—	52	16	6.4	8.5	—	1.0
1986	1262	4.25	97	37	35	36	17	3.6	8.0	—	1.9
						Elements (10^{-3} mol m^{-3})					
			Al	Cd	Cu	F	Fe	Hg	Mn	Pb	
[b]1984	981		1.3	0.0017	0.044	6.3	0.38	0.002	0.10	0.29	
						Detection Limit (10^{-6} mol m^{-3})					
			40	1.0	15	5000	20	0.5	20	5	

[a]Mean weighted element concentration in the bulk deposition in the open air (not enough data available).
[b]Mean concentration of trace metals in the bulk deposition in the open air.

cloud elements are formed in humid, marine air advected from the west. Abundant snow masses, which are brought into the area with southeastly winds around the Alps by Adriatic cyclones in fall and winter, also originate from western regions. In contrast to the Fichtelgebirge, the highest concentrations of gaseous and particulate atmospheric trace substances in Große Ohe are usually advected with westerly winds as well.

Spatial variation of the bulk deposition is, if at all, mainly caused by the precipitation distribution, primarily by the increase of precipitation with increasing altitude. Because of the location of the collectors underneath the canopies, bulk deposition in the stands contains those substances rained and washed out from the atmosphere, as well as those filtered out by the trees, deposited on the leaves and branches, and washed down again. These along with interception losses explain higher element concentration in precipitation samples in the stands. Furthermore, samples contain material from the internal element cycle of the stands, i.e., from leaching (Ulrich 1979). Therefore, the bulk deposition not only yields integral values of various processes but depends on the stand, too.

Table 9 shows typical element variation in precipitation in the stands as compared to that in the open. The enrichment of nutrients (Ca, K, Mg) in the stands can be deduced from the internal element cycle. These correspond to the deposition rates of protons (Gietl and Rall 1986). As a comparison, the same data from the Fichtelgebirge show different quantities but the same trend. The significantly higher deposition sums of SO_4–S in the Große Ohe than in the Fichtelgebirge can be attributed to the combustion processes of the densely populated and industrialized region of Munich and the air guiding properties of the Alps. On the other hand, the NO_3–N and NH_4–N deposition sums are lower because of a reduced source strength of these atmospheric trace constituents in the sources, namely, stock farming and car traffic, close to the Große Ohe and the surrounding national park area. The decrease in the NH_4–N deposition rate in the stand in relation to the open air, especially during the summer, is probably caused by microbial reduction processes in the canopies (Müller et al. 1982).

6.2 Comparison of Representative Data of Different Regions

The Fichtelgebirge is a forested, nonindustrialized area with only a few small industrial installations. The area may thus be classified as a primarily clean air region with a low flux of pollutants from the atmosphere into the biosphere. However, the immission data described in the preceding sections belie this cursory supposition. Comparisons of the chemical composition of rain- and fogwater represented by selected characteristic ions with concentrations of the same ions from other remote areas indicate the greater-than-average stress of wet-deposited airborne pollutants on the forested Fichtelgebirge ecosystem.

Comparisons have generally involved bulk precipitation water-data. The average pH value of precipitation water shows (Table 10) a slight increase from

Table 9. Mean bulk deposition sums (10^{-3} mol m^{-2}) for summer (May-Oct., 1984–1986) and winter (Nov.-April, 1983/84–1985/86) for the sites Taferlruck, 769 m a.s.l., and Grüben, 790 m a.s.l., Große Ohe

Site	Precipitation (mm)	Elements									Remarks
		SO_4-S	No_3-N	Cl	H^+	NH_4-N	Na	K	Ca	Mg	
Taferlruck	602	59	19	20.1	35	22	9.2	2.2	3.8	0.7	Open air
Oberwarmensteinach	652.5	40	44	18	–	34	–	8.5	9.0	3.8	Open air
Wülfersreuth	578.2	29	41	19	–	35	–	6.0	11	4.0	Open air
Grüben[a]	436	108	29	–	51	15	6.8	23	13	3.5	Throughfall
Oberwarmensteinach	495.9	56	40	20.7	–	20	–	48	17.8	5.1	Throughfall
Wülfersreuth	438.1	37	44	17.6	–	28	–	25	14	4.6	Throughfall
Taferlruck	495	44	25	15.4	26	22	8.8	2.9	5.6	3.1	Open air
Oberwarmensteinach	450.5	19	15	16	–	16	–	6.3	8.0	3.9	Open air
Wülfersreuth	379.5	25	25	13	–	19	–	5.0	10	5.0	Open air
Grüben[a]	376	99	34	–	59	19	11.2	16	16	6.0	Throughfall
Oberwarmensteinach	346.5	56	25	21	–	18	–	24	15	5.4	Throughfall
Wülfersreuth	288.7	34	25	12	–	18	–	10	9.0	4.0	Throughfall

[a] For comparison the corresponding data of Oberwarmensteinach, 770 m a.s.l., and Wülfersreuth, 680 m a.s.l., Fichtelgebirge, for winter 1984/85 and summer 1985 are added (Hantschel 1987).

Table 10. Ion concentration of bulk precipitation water ($\mu eq\ l^{-1}$) at the Fichtelgebirge (Bavaria, FRG) compared with observations reported from some other forested areas[a]

Location/Remarks		pH	SO_4^{-2}	NO_3^-	Cl^-	H_3O^+	NH_4^+	Na^+	K^+	Ca^{2+}	Mg^{2+}	P (mm/a)	Reference
Oberwarmensteinach and Wülfersreuth (Fichtelgebirge) 680/720 m a.s.l.	c_v	4.2	164	82	37	64	73	34	17	48	18	1188	Hantschel (1987)
Solling northern FRG 500 m a.s.l.	c_v	4.1	169	58	51	83	88	34	9.5	60	17	950	Ulrich et al. (1979)
Bodenmais south-eastern FRG 950 m a.s.l.	c_v	4.5	119	43	34	32	57	22	13	30	8	1489	Dunkl and Rehfuess (1988)
Taubenberg southern FRG 770/780 m a.s.l.	c_v	4.5	94	36	23	32	50	22	5	30	8	1438	Dunkl and Rehfuess (1988)
Hubbard Brook New Hampshire USA 500/800 m a.s.l.	c_v	4.1	60	24	13	72	12	5.2	1.8	8.0	3.3	1322	Likens et al. (1977)
Maimai New Zealand (investigation period) from 30.10.85–12.12.85) 300 m a.s.l.	c_v	5.6	< 3	0.4	28	3	< 3	21	2.0	1.2	2.4	2600	Verhoeven et al. (1987)

[a] Volume-weighted annual means (c_v), annual depth of precipitation (P).

the northern (pH = 4.1, Solling) to the southern German highlands (pH = 4.5, Taubenberg). The Große Ohe reflects this trend with pH = 4.25–4.30 (Table 8) with the Fichtelgebirge (pH = 4.2) at the lower end of the pH-interval. This reflects the generally decreasing ion load of rainwater with the decreasing concentration of primary air pollutants from north to south along the eastern border of the FRG. The higher precipitation rates at the southern stations tend to compensate the concentration decrease, since the meañ annual bulk precipitation input shows a reduced north-south inclination (Table 11). While the Fichtelgebirge distribution pattern of analyzed ions is similar to that of continental European and comparable northamerican regions (e.g., the forest research station Hubbard Brook), the concentrations are significantly higher in the Fichtelgebirge. Except for SO_4-S, this is also true for the Große Ohe (see Table 8). Although a portion of the sulphate analysed may have originated from dry-deposited ammonium sulphate as a result of the rainwater sampling method, the precipitation acidity is nevertheless mainly determined by sulphuric and nitric acid.

New Zealand rainwater presents a completely different ion pattern. Its maritime character is clearly demonstrated by the dominating chloride and sodium and the low ammonium and sulphate components (pH = 5.6). The low rainwater concentration indicates the origin of the rain to be a "clean atmosphere".

Fogwater has higher ion concentrations than rainwater by a factor of 5–10 for continental regions which are influenced by even remote anthropogenious sources. Maritime, clean air fog shows a reduced increase of ion concentrations. However, the liquid water content of fog is low and fogwater input is considerably smaller compared to rainwater input (Sect. 3 this chapter).

From the areas listed in Table 12, the Fichtelgebirge shows almost the highest concentrations of trace substances in fogwater. Even at Bakersfield, California, a site of agricultural activity and oil recovery operations, concentrations of fogwater constituants only approach these values. In addition, the fogwater measurements of the Fichtelgebirge reveal significant vertical differences. Radiation fog samples yield the highest ion load mainly at 700–800 m a.s.l., just below the temperature inversion.

The chemistry of fog formed within and below the inversion layer and associated with low wind speeds is essentially determined by pollution sources located in the valleys and basins of the region. Temperature inversions are primarily observed with anticyclonic weather and slight easterly winds, as the pollution sources are predominantly located east of the observation site. The radius of influence may be ca. 50 km. Because model studies by Ulrich (Chap. 1–A) of the atmospheric transport of gaseous pollutants do not take into account short range sources, the conclusion that highest concentrations exist close to the mountain summit does not disagree with this finding. In the regime of mountain fogs at the summit of the Ochsenkopf massif, the fogwater ion concentrations are lower. In this case, atmospheric background conditions and remote sources play the more dominant but less influential role.

Table 11. Mean annual input with bulk precipitation (10^{-3} mol m^{-2} a^{-1}) at the Fichtelgebirge (Bavaria, FRG) compared with observations reported from some other forested areas

Location/remarks	SO_4-S	NO_3-N	Cl	H^+	NH_4-N	Na	K	Ca	Mg	Period[a] (years)	Reference
Oberwarmensteinach and Wülfersreuth (Fichtelgebirge) 680/720 m a.s.l.	73.4	72.1	35	60	63	–	17.4	20	7.8	2	Hantschel (1987)
Solling northern FRG 500 m a.s.l.	75	50	48	80	79	33	8.5	28	8.3	10	Ulrich et al. (1979)
Bodenmais south-eastern FRG 950 m a.s.l.	88	64	51	50	86	31	18	22	6.6	4	Dunkl and Rehfuess (1988)
Taubenberg southern FRG 770/780 m a.s.l.	69	50	34	40	71	34	67	21	6.6	4	Dunkl and Rehfuess (1988)
Hubbard Brook New Hampshire USA 500/800 m a.s.l.	40	32	20	97	16	6.9	2.3	5.4	2.4	11	Likens et al. (1977)
Maimai New Zealand 300 m a.s.l.	< 2	1.1	71	0.8	< 3	5.7	5.1	1.5	3.1	1*	Verhoeven et al. (1987)

[a] Measuring period *extrapolated.

Table 12. Fogwater ion concentration ($\mu eq\ l^{-1}$) at the Fichtelgebirge (Bavaria, FRG) compared with observations reported from some other forested areas (except Bakersfield)

Location/remarks[a]			pH	SO_4^{2-}	NO_3^-	Cl^-	H_3O^+	NH_4^+	Na^+	K^+	Ca^{2+}	Mg^{2+}	Reference
Ochsenkopf Mt.		c	3.3	840	640	100	490	860	96	38	90	24	
(Fichtelgebirge)	Cl	c_{min}	6.2	48	36	9	0.6	28	5.2	4.1	9.0	3.3	Herterich & Paffrath (1986)
1020 m a.s.l.		c_{max}	2.6	3960	2600	380	2630	2860	1090	160	380	110	
Fichtelgebirge		c	3.0	1600	730	310	1020	1600	160	75	90	31	
different sites,	Ra	c_{min}	4.9	230	140	43	12	11	12	9.0	9.5	3.9	Herterich (1987)
650–800 m a.s.l.		c_{max}	2.2	9500	4900	1310	6600	19500	770	250	390	90	Trautner (1987)
Bayreuth and sur-		c	4.0	1220	630	160	90	1380	190	65	290	58	
roundings	Ra	c_{min}	6.6	160	130	23	0.2	390	5.7	5.4	12	3.3	
370–550 m a.s.l.		c_{max}	3.0	9040	5600	920	950	7330	830	440	3350	580	
Taunus, FRG		c	3.8	830	540	190	160	470	120	28	120	25	
	Cl	c_{min}	7.9	250	–	–	< 0.1	170	7	5.1	40	8.3	Georgii et al. (1986)
800 m a.s.l.		c_{max}	2.4	5600	2900	1100	4200	470	1800	130	440	270	
Bakersfield, Cal.		m	4.2	1160	850	47	60	1440	20	9.3	47	6.3	
USA	Ra	c_{min}	–	194	200	1	0.1	490	1.4	1.6	7	1.1	Jacob et al. (1984)
180 m a.s.l.		c_{max}	–	9400	6800	980	2750	13300	325	368	3500	430	
Whitface Mt., N.Y.		c_v	4.2	140	110	3.1	280	89	11	20	17	6.1	
USA	Cl												Castillo et al. (1983)
1483 m a.s.l.		σ	0.2	150	52	4.2	220	61	20	25	14	4	
Craigieburn													
New Zealand	Cl	c_v	6.1	–	5.1	34	0.4	18	44	2.7	1.8	2.5	Verhoeven et al. (1987)
1600 m a.s.l.		σ	–	–	7.8	61	0.7	6	64	2.3	1.8	2.4	

[a] Arithmetic means (c), volume-weighted means (c_v), maximum/minimum values ($c_{max}/_{min}$), median (m), standard deviation (σ); mountain (orographic) fog (Cl), respectively, radiation (valley) fog (Ra) predominant.

7 Comments and Conclusions

The "classic" forest damage caused by flue gases may be considered a strong interaction between airborne chemicals and trees. On the other hand, the possible contribution of the airborne chemicals to modern forest decline can be labeled at best a weak interaction. Air pollution stress is within the range of natural biological and climatological stress, than most symptoms attributed to air pollution are difficult to distinguish clearly from those caused by natural stress (Huttunen 1984, Smith 1981). At the beginning of the research reported in this chapter, no strong pattern of consistency or responsiveness nor proven biological mechanism was known to explain pathological tree symptoms by suspected causal factors. Thus, field studies have had to be conducted to provide those data from which cause-and-effect and dose-response relationships could be deduced appropriately.

As a result of these studies we emphasize that:

- In the Fichtelgebirge, the near-surface atmospheric gaseous and particulate pollution load (the potential for dry deposition) is highest in winter and for air masses advected from northerly to easterly directions.
- The main air pollution sources are steel and chemical industries and coalfired power plants presumably located in the GDR and Czechoslovakia. Although a power plant operates near Arzberg, there are no major emittants in FRG territory north and east of the Fichtelgebirge, where the population density is low.
- Air masses advected from west to south have a relatively higher proportion of traffic-born pollutants (e.g., No_3^-) than those from east to north.
- Stable winter temperature inversions prevent vertical exchange and enhance pollution. These may lead to severe smog episodes, thereby increasing the potential phytotoxicity of atmospheric pollutants. Severe winter conditions should be considered when planning future botanical research on forest damage.
- The surprisingly high concentrations of ammonium in atmospheric particles represent a further source of nitrogen, which may also trigger vegetative injuries (Nihlgård 1985). The principal source of ammonium is probably stock farming in the neighbouring regions.
- The nongaseous atmospheric input of acidity (H_3O^+) to the forest ecosystems of the Fichtelgebirge comes almost exclusively from wet deposition (see Sect. 4 this Chap. for dry deposition).
- The monitoring of pH values and conductivity of precipitation water shows no significant wind direction differences. Extreme (maximum and minimum) values, however, are always observed during westerly winds. In the beginning of shower-like precipitation, an occasional decrease of acid fraction occurs with increasing precipitation rate. The acid fraction, however, increases again during the final intensive precipitation period of the same event. During continuous precipitation, the acid fraction remains constant.
- Airborne pollutants dissolved in water droplets (ions, most polycylic aromatic

hydrocarbons) and transported with precipitation often show U-shaped pollutographs. Thus, surfaces of forest trees and other plants will experience higher concentrations at the beginning and at the end of individual precipitation events.

- Compared with other remote forested areas in the United States and in New Zealand, the chemical analysis of rain- as well as fogwater indicate an ion concentration and ion input flux clearly above the average.
- During precipitation events, volatile pollutants like hexachlorocyclohexane show almost constant concentrations, which obviously depend on their gas phase concentrations and aqueous solubility.
- There is no significant regional variation in precipitation chemistry in the Fichtelgebirge.
- The ion concentrations of fogwater on the Fichtelgebirge mountain depend on the general weather conditions. High concentrations and low pH values were found during anticyclonic atmospheric circulation characterized by east to northeast surface winds, whereas cyclonic atmospheric circulation with west to southwest winds show lower ion concentrations and higher pH values.
- The ion concentration of fogwater is the same inside and outside a spruce stand. That is not true for the fogwater element (or ion) concentration per unit volume of fog air. This "atmospheric" element (or ion) concentration is highest outside the spruce stand and near the tree top.
- The horizontal element (or ion) flux with fog droplets is highest in the vicinity of the upper canopy range of the trees growing at the edge of a stand. Assuming the same depositional efficiencies of fog droplets for trees in- and outside a stand, trees at the edge of a forest, and especially at the upper ranges of their canopies, are exposed to substantially higher ion and acid loads.
- Spruce trees weighed during a fog event indicated an average surface wetting of at least 100 ml m^{-2} leaf area. Preliminary investigations showed that a fraction of the water deposited during beginning fog formation may be attributed to dew. The droplet deposition itself seems to be limited by coalescense processes on the waxy needle surface. Large drops produced by this process bend the tip of the needles down and drip off.
- As found in other studies, the H_2O_2 concentration of rain varied widely from less than 0.3 μmol l^{-1} up to as much as 60 μmol l^{-1}. Meteorological conditions and air quality determine the amount of H_2O_2 in rainwater. However, previous ground-based cloudwater studies (Kadlecek et al. 1983) could not be confirmed. Fog- and near surface cloudwater sampled in the Fichtelgebirge contained H_2O_2 only at the very low level of 0.5 μmol l^{-1} or less. Therefore, biological hazards eventually arising from H_2O_2 in fogwater (Masuch et al. 1986) must be eliminated as a factor contributing to forest decline in this area. Although increased H_2O_2 input can be expected in the Black Forest and the Alpine Mountains because of the higher ozone and lower SO_2 values, the formation and decomposition pathways of aqueous H_2O_2 do not basically favor H_2O_2 concentrations above the level of 0.1 μmol l^{-1} in fogs and clouds covering highland mountain areas. As acid deposition is in fact a more severe problem in the Fichtelgebirge, we argue that even in summer, SO_2 emissions

in the GDR and CSSR can exceed oxidation capability of the atmosphere. In addition, a considerable reduction in SO_2 emission would therefore not yet essentially diminish acid-deposition.

References

Andre K (1976) Bestimmung des Massenabsorptionsindex an Niederschlags- und Nebelwasserrückständen sowie nicht aktivierten Aerosolpartikeln für den Spektralbereich 380–920 nm. Masters thesis, Univ Mainz

Asman WAH (1980) Draft, construction and operation of a sequential rain sampler. Water Air Soil Pollut 13:235–245

Bachert R (1981) Zeitliches und kleinräumiges Verhalten von Tropfengrößenverteilungen in Nebeln und Wolken. Aufbau und Eichung der Meßapparatur. Erste Messungen. Masters thesis, Univ Mainz

BLf U – Bayerisches Landesamt für Umweltschutz (1985) Lufthygienische Monatsberichte ISSN 0178-9074

BLf U – Bayerisches Landesamt für Umweltschutz (1986) Lufthygienische Monatsberichte ISSN 0178-9074

BLf U – Bayerisches Landesamt für Umweltschutz (1987) Lufthygienische Monatsberichte ISSN 0178-9074

Bosse M (1987) Säure- und Stoffeintrag mit dem Niederschlag. Wasserhaushalt und Stoffbilanzen im naturnahen Einzugsgebiet Große Ohe, Vol 2. Ber Nationalpark Bayer Wald, Grafenau, pp 1–167

Bufalini JJ, Lancaster HT, Namie GR, Gay BW (1979) Hydrogen peroxide formation of the photooxidation of formaldehyde and its presence in rainwater. J Environ Sci Health A 14:135–141

Chameides WL, Davis DD (1982) The free radical chemistry of cloud droplets and its impact upon the composition of rain. J Geophys Res 87:4863–4877

Castillo RA, Jiusto JE (1983) The pH and ionic composition of stratiform cloud water. Atmos Environ 17:1497–1505

Castillo R, Lala G, Jiusto JE, Fuzzi S (1985) The chemistry and microphysics of intrastorm sequential precipitation samples. Tellus 37 B:160–165

Dunkl I, Rehfuess KE (1988) Stoffdeposition mit dem Freiland- und dem Bestandsniederschlag an fünf Waldstandorten in Südbayern. Forstl Forschungsber 86, Lehrstuhl Bodenkunde, Univ Munich, pp 27–81

Durham JL (ed) (1984) Acid precipitation series. Vol 2: Chemistry of particles, fogs and rain. Butterworth Publ, Boston

DVWK – Deutscher Verband für Wasserwirtschaft und Kulturbau (1984) Ermittlung der Stoffdeposition in Waldökosysteme. DVWK Regeln 122, Paul Parey-Verlag, Hamburg

Förster J (1986) Schadstoffzusammensetzung der atmosphärischen Partikelphase im Waldschadensgebiet Fichtelgebirge in Abhängigkeit von der Wetterlage. Masters thesis, Univ Bayreuth

Forschungsbeirat Waldschäden und Luftverunreinigungen der Bundesregierung und der Länder (1986). 2. Ber, pp 5–220

Frevert T, Klemm O (1984) Wie ändern sich pH-Werte im Regen- und Nebelwasser beim Abtrocknen auf Pflanzenoberflächen? Arch Meteorol Geophys Bioclimatol Ser B, 34:75–81

Fuzzi S (1986) Radiation fog chemistry and microphysics. In: Jaeschke W (ed) Chemistry of multiphase atmospheric systems. NATO ASI Ser, Springer, Berlin Heidelberg New York Tokyo pp 213–226

Georgii HW (1965) Untersuchungen über Ausregnen und Auswaschen atmosphärischer Spurenstoffe durch Wolken und Niederschlag. Ber Dtsch Wetterdienst, Offenbach, 100:23

Georgii HW, Schmitt G (1985) Methoden und Ergebnisse der Nebelanalyse. Staub – Reinh Luft 45:260–263

Georgii HW, Perseke C, Rohbock E (1983) Feststellung der Deposition von sauren und langzeitwirksamen Luftverunreinigungen aus Belastungsgebieten. Umweltbundesamt-Ber 6/83, Schmidt, Berlin

Georgii HW, Grosch S, Schmitt G (1986) Feststellung der Schadstoffbelastung von Waldgebieten in der Bundesrepublik Deutschland durch trockene und nasse Deposition. Inst Meteorol Geophys Final Report Part A, Univ Frankfurt, pp 1–247

Gietl G (1985) Der Stoffeintrag mit dem Niederschlag im Einzugsgebiet Große Ohe. In: Nationalpark Bayerischer Wald, Tagungsber Vol 5/2, Grafenau, pp 431–444

Gietl G, Rall A (1986) Bulk Deposition into the catchment Große Ohe. In: Georgii HW (ed) Atmospheric pollutants in forest areas. Reidel, Dordrecht

Graedel TE, Mandich ML, Weschler CJ (1986) Kinetic model studies of atmospheric droplet chemistry. 2. Homogeneous transition metal chemistry in raindrops. J Geophys Res 91:5205–5221

Granat L (1972) On the relation between pH and the chemical composition in atmospheric precipitation. Tellus 24:550–560

Guderian R (ed) (1985) Air pollution by photochemical oxidants. Springer, Berlin Heidelberg New York Tokyo

Guilbault GG, Brignac PJ, Juneau M (1968) New substrates for the fluorometric determination of oxidation enzymes. Anal Chem 40:1256–1263

Hantschel R (1987) Wasser- und Elementbilanz von geschädigten, gedüngten Fichtenökosystemen im Fichtelgebirge unter Berücksichtigung von physikalischer und chemischer Bodenheterogenität. Thesis, Univ Bayreuth

Herterich R (1987) Eignung und Anwendung von Rotating Arm Collectoren zur Bestimmung von Nebeleigenschaften – Flüssigwassergehalt, ionische Spurenstoffe, Wasserstoffperoxid. Masters thesis, Univ Bayreuth

Herterich R, Paffrath D (1986) Untersuchung großräumiger Schadstoffbelastung im Zusammenhang mit den Waldschäden in Bayern – Meßergebnisse des Schadstoffgehalts in Nebelwasserproben aus dem Fichtelgebirge. Dtsch Forschungs- und Versuchsanstalt Luft- und Raumfahrt, IB 553 86/23, Oberpfaffenhofen

Hicks BB (1980) On the dry deposition of acid particles to natural surfaces. In: Toribara TY, Morton WM, Paul EM (eds): Polluted rain. Plenum Press, NY, pp 327–339

Hinds WC (1982) Aerosol technology. Wiley, NY, pp 1–424

Höfken KD, Gravenhorst G (1983) Untersuchung über die Deposition atmosphärischer Spurenstoffe an Buchen- und Fichtenwald. Umweltbundesamt-Ber 6/83, Schmidt, Berlin, pp 1–208

Hoffmann MR (1983) The chemistry of bisulfite aldehyde addition products in fog and cloudwater. Pap presented at NATO Advanced Study Inst on Chemistry of Multiphase Atmospheric Systems, Corfu, Greece

Huttunen S (1984) Interactions of disease and other stress factors with atmospheric pollution. In: Treshow M (ed) Air pollution and plant life. Wiley, NY, pp 321–356

Jacob DJ, Waldmann JM, Munger W, Hoffmann MR, (1984) A field investigation of physical and chemical mechanisms affecting pollutant concentrations in fog droplets. Tellus 36 B:272–285

Jacob P, Bambauer A, Barth S, Klockow D (1987) Zur Rolle des Wasserstoffperoxids in der atmosphärischen Mehrphasenchemie. In: 7. Seminarbericht "Waldschäden Luftverunreinigungen", 8. Statussemin Fraunhofer Inst Unweltchem Ökotoxikol, Schmallenberg, s 62–80

Kadlecek J, McClaren S, Camarota N, Mohnen V, Wilson J (1983) Cloud water chemistry at Whiteface Mountain. In: Pruppacher HR, Semonin RG, Slinn WG (eds) Precipitation scavenging, dry deposition, and resuspension. Elsevier, NY 1:103–112

Kayser K, Jessel U, Köhler A, Rönicke G (1974) Die pH-Werte des Niederschlags in der Bundesrepublik Deutschland 1967–1972. DFG-Komm Erforschung der Luftverunreinigungen. Mitt IX, Boldt, Boppard

Kelly TJ, Daum, PH, Schwartz SE (1985) Measurements of peroxides in cloudwater and rain. J Geophys Res 90:7861–7871

Klemm O, Frevert T (1986) Theorie und Anwendung einer Redox-Meßtechnik zur Bestimmung von SO_2 und schwefeliger Säure in Benetzungswässern. Staub – Reinh Luft 46:23–26

Klockow D, Jacob P (1986) The peroxyoxalate chemiluminescence and its application to the determination of hydrogen peroxide in precipitation. In: Jaeschke W (ed) Chemistry of multiphase atmospheric systems. NATO ASI Ser, Springer, Berlin Heidelberg New York Tokyo, pp 117–132

Kok GL (1980) Measurements of hydrogen peroxide in rainwater. Atmos Environ 14:653–656

Kok GL (1982) Measurements of hydrogen peroxide in the OSCAR program. Final rep Coordinating Res Counc, CAPA-21-80, Harvey Mudd College, Claremont, Calif
Langmuir T, Blodgett KB (1946) A mathematical investigation of water droplet trajectories. US Army Forces Techn Rep No 5418.40, pp 1-47
Lazrus AL, Kok GL, Gitlin NS, Lind JA, McLaren SE (1985) Automated fluorimetric method for hydrogen peroxide in atmospheric precipitation. Anal Chem 57:917-922
Likens GE, Bormann FH, Pierce RS, Eaton JS, Johnson NM (1977) Biogeochemistry of a Forested Ecosystem. Springer, Berlin Heidelberg New York
Liu SC, Kley D, McFarland M, Mahlman JD, Levy H (1980) On the origin of tropospheric ozone. J Geophys Res 85:7546-7552
Masuch G, Kettrupp A, Mallant RKAM, Slanina J (1986) Effects of H_2O_2-containing acid fog on young trees. Int J Environ Anal Chem 27:183-213
McElroy WJ (1986) Sources of hydrogen peroxide in cloudwater. Atmos Environ 20:427-438
Müller KP, Aheimer G, Gravenhorst G (1982) The influence of immediate freezing on the chemical composition of rain samples. In: Georgii HW, Pankrath J (eds) Deposition of atmospheric pollutants. Reidel, Dordrecht
Nihlgård B (1985) The ammonium hypothesis – an additional explanation to the forest dieback in Europe. Ambio 14:2-8
Noack EM (1979) Witterung und Klima im Nationalpark Bayerischer Wald. Schriftenr Nationalpark Bayer Wald 5, Passau, pp 11-132
Oren R, Werk KS, Schulze ED (1986) Relationship between foliage and conducting xylem in *Picea abies* (L.). Karst Trees 1/1:61-69
Prinz G, Krause GHM, Stratman H (1982) Waldschäden in der Bundesrepublik Deutschland. LIS-Bericht Nr 28, Essen, pp 1-154
Pruppacher HR, Semonin RG, Slinn WGN (eds) (1983) Precipitation scavenging, dry deposition, and resuspension. Elsevier, NY, 2 vols
Radojevic M (1983) H_2O_2 in the atmosphere and its role in the aqueous oxidation of SO_2. Environ Technol Lett 4:499-508
Richards LW, Anderson JA, Blumenthal DL, McDonald JA, Kok GL, Lazrus AL (1983) Hydrogen peroxide and sulfur (IV) in Los Angeles cloudwater. Atmos Environ 17:911-914
Rodda JC (1971) The precipitation measurement paradox – the instrument accuracy problem. World Meteorol Organ, Geneva, WMO-No 316
Römer FG, Veldkamp AA, Van Galen P (1984) Determination of H_2O_2 in cloud and rainwater. In: CEC (Rep) EUR 9436. Phys-Chem Behav Atmos Pollut, Reidel, Dordrecht, pp 74-82
Sartorius R (1985) Räumliche Verteilung und zeitliche Entwicklung der Immissionsbelastung. In: Bauer F (ed) Die Sache mit dem Wald. BLV, München, s 167-180
Scholz F, Lorenz M (1984) Schadensursachen und Wirkungsmechanismen bei den Waldschäden. Allg Forstz 39:1275-1278
Schrimpff E, Klemm O, Eiden R, Frevert T, Herrmann R (1984) Anwendung eines Grunow-Nebelfängers zur Bestimmung von Schadstoffgehalten in Nebelniederschlägen. Staub – Reinh Luft 44:72-75
Smith WH (1981) Air pollution and forests. Springer, Berlin Heidelberg New York
Thomas W (1981) Entwicklung eines Immissionsmeßsystems für PCA, Chlorkohlenwasserstoffe und Spurenmetalle mittels epiphytischer Moose – angewandt auf den Raum Bayern. Bayreuther Geowiss Arb 3:1-143
Thommes W, Schultz E (1983) Klimagutachten, Klimaanalyse für die Gemeinde Warmensteinach, Landkreis Bayreuth, zur Anerkennung als heilklimatischer Kurort. Dtsch Wetterdienst, Nürnberg
Trautner F (1987) Bavarian Study Group Foresttoxicology-Meteorology. Univ Bayreuth, FRG, private communication (measurements 1985-1987)
Twomey S (1979) Atmospheric Aerosols. Elsevier, Amst NY
Ulrich B (1979) Stoffhaushalt von Waldökosystemen. Scriptum 2 (course notes, 2 edn), Inst Bodenkd Waldernähr, Univ Göttingen
Ulrich B, Mayer R, Khanna PH (1979) Fracht an chemischen Elementen in den Niederschlägen im Solling. Z Pflanzenernähr Bodenkd 142:601-615

Verhoeven W (1984) Kationische und anionische Inhaltsstoffe in Niederschlagswasser im Bereich einer Gipfellage sowie entlang eines Höhenprofils im Fichtelgebirge in Abhängigkeit von der Witterung. Masters thesis, Univ Bayreuth

Verhoeven W, Herrmann R, Eiden R, Klemm O (1987) A comparison of the chemical composition of fog and rainwater collected in the Fichtelgebirge, FRG and from the South Island of New Zealand. Theor Appl Climatol 38:210–221

Wiman BLB, Ågren GT (1985) Aerosol depletion and deposition in forests – a model analysis. Atmos Environ 19:335–347

Wiman BLB, Lannefors HO (1985) Aerosol characteristics in a nature coniferous forest – methodology, composition, sources and spatial concentration variations. Atmos Environ 19:349–362

Winkler P (1977) Automatic analyser for pH and electrical conductivity of precipitation. In: Pap presented at the WMO Tech Conf on Instruments and Methods of Observation (TECIMO). Hamburg Genf, WMO-No 480:191–196

Winkler P (1980) Observations on acidity in continental and in marine atmospheric aerosols and in precipitation. J Geophys Res 85:4481–4486

Winkler P (1984) Observations on acid particles and droplets in the atmosphere. In: Durham JL (ed) Chemistry of particles, fogs and rain. Butterworth Publ, Boston Lond, pp 161–196

Wood T, Boormann FH (1975) Increase in foliar leaching caused by acidification of artificial mist. Ambio 4:169–171

Woodman JN, Cowling EB (1987) Airborne chemicals and forest health. Environ Sci Technol 21:120–126

Yamasaki H, Kuwata T, Miyamoto H (1982) Effect of ambient temperature on aspects of airborne polycyclic aromatic hydrocarbons. Environ Sci Technol 16:189–194

Zika R, Saltzmann E, Chameides WL, Davis DD (1982) H_2O_2 levels in rainwater collected in South Florida and the Bahama Islands. J Geophys Res. 87:5015–5017

Part 2. The Role of Fungi, Microorganisms, and Soil Animals

Preface

In the wake of the initial hypothesis of Ulrich (1980) that atmospheric pollutants and acid rain have caused observed large-scale forest decline, several alternative explanations have been suggested. Among these, decline sypmtoms have not been related to air pollution, but to some unknown epidemic of microorganisms (Kandler 1983) or needle fungi (Rehfuess 1983). The epidemic hypothesis has been mainly based on historic records in which similar symptoms on a smaller scale have been described in the last century. The large spatial and temporal variation in the spread of the symptoms also seemed to indicate the evolvement of epidemic organisms (see also Kandler 1985). Co-occurrence of fungal disease with areas of forest decline has lent credibility to the epidemic hypothesis (Rehfuess 1983).

Following the initial controversy over whether the primary cause for forest decline was epidemics or emissions, it became essential to test both hypotheses in the stands of this study. Experiments testing the epidemic hypothesis were carried out by two institutes of plant pathology, the Institute of Plant Pathology at the University of Bonn (Prof. Dr. F. Nienhaus) and the Institute of Microbiology of a German federal agency the "Biologische Bundesanstalt Berlin" (Dr. H. Petzold). The results of these studies are summarized in the introduction of Chapter 2-A. No microorganism found in, or isolated or transferred from, spruce could explain the observed pattern of decline. Nienhaus (1986) has summarized the results of intensive pathology studies in West Germany. Virus and virus-like structures are widely distributed in trees, but there is no indication that they are involved in forest decline. No viroids nor mycoplasma-like organisms have been found in forest trees. Bacteria have occasionally been seen in EM-preparations, but these findings have not been related to forest decline. From these studies it is concluded that forest decline is not the result of an epidemic caused by microorganisms.

In addition to microorganisms, needle-reddening disease was suggested to be a major factor in forest decline. It is mainly caused by two needle fungi, *Lophodermium piceae* and *Rhizosphaera kalkhoffii*, which caused spruce to lose needles (Rehfuess 1983). This hypothesis has been supported by the fact that needle-reddening disease has spread in successive years from southeast Germany over large areas of southern Germany, Switzerland, and France and coincided with observed decline (Forschungsbeirat Waldschäden 1986). In Chapter 2-B, the needle fungi responsible for needle-reddening are discussed. It is concluded that these needle fungi live in "commensalism" inside the needle for many years

E.-D. Schulze, O.L. Lange, R. Oren (Eds.)
Ecological Studies Vol. 77

without harmful effects. It is still unclear why these needle fungi became pathogenic in 1982 and led to major needle losses in southern Germany and in the Fichtelgebirge. The phenomenon is probably not directly related to concentrations of air pollutants. However, it has been proposed that either certain weather conditions have stimulated these fungi, or that nutritional imbalances (cation/nitrogen ratios) have disturbed this commensalism and activated the fungus (Forschungsbeirat Waldschäden 1986). In any case, the pathogen of the needle-reddening disease did not "spread", but was always present in spruce needles. The needle-reddening disease indicates that pathogens may be a secondary rather than a primary stress factor associated with forest decline. Primary factors are related to the impact of atmospheric pollutants (Chap. 3, 4).

In Chapter 2-A, results of studies on an important fungal component of forest ecosystems, mycorrhiza, are reported. It has been found when mycorrhizal fungi are grown in soil solution with aluminum concentrations similar to those observed at the experimental sites, that their growth and their magnesium and calcium contents decrease. Since spruce is obligatorily infected by mycorrhizal fungi, these declines should affect tree nutrition (see Chap. 4-E, 4-F). In Chapter 2-A, it is also shown that soil fungal flora has changed along with soil chemical conditions, which could potentially disturb the cation balance of the trees. This again indicates that soil chemical conditions, not epidemics, are involved in forest decline.

Additional evidence for the importance of soil parameters is given in Chapter 2-C. It is shown that the soil fauna changes dramatically when calcium concentrations in the soil is reduced. Calcium is an important building material for the body shells of many soil animals. A change in soil fauna may affect mineralization and humus chemistry (see Chapter 4-B).

Together, the Chapters in Part 2 of this book reject the theory that epidemics have caused the observed pattern of forest decline. Needle fungi, which cause needle loss, probably have become pathogenic as unknown factors have changed their behavior directly or through their host. Soil fungi, including mycorrhiza, and fauna are strongly affected by soil chemical parameters such as the ratio of calcium to aluminum. Uptake of important minerals by mycorrhiza, such as magnesium and calcium, have been reduced in the presence of aluminum in the cultural medium (see also Boxmann et al. 1986). Quantitative and qualitative changes in functional groups of soil organisms may affect tree resource availability and the proper function of the root system.

References

Boxman AW, Sinke RJ, Roelofs JGM (1986) Effects of ammonium on the growth and kalium (^{86}Rb) uptake of various ectomycorrhizal fungi in pure culture. Water Air Soil Pollut 32:517–522

Forschungsbeirat Waldschäden der Bundesregierung und der Länder (1986) 2. Ber Kernforschungszentrum Karlsruhe, 229 s

Kandler O (1983) Waldsterben: Emissions- oder Epidemie-Hypothese. Naturwiss Rundsch 11:488–490

Kandler O (1985) Imissions- versus Epidemie-Hypothese. In: von Kortzfleisch G (ed) Waldschäden. Oldenbourg, München Wien, s 20–50
Nienhaus F (1986) Zusammenfassung der Ergebnisse des IMA-Statusseminars "Biotische Schadfaktoren/Epidemiologie". Jül-Spez-364 ISSN 0343-7639:33–41
Rehfuess KE (1983) Walderkrankungen und Immissionen – eine Zwischenbilanz. Allg Forstz 38:601–610
Ulrich B (1980) Die Wälder in Mitteleuropa: Meßergebnisse ihrer Umweltbelastung, Theorie ihrer Gefährdung, Prognose ihrer Entwicklung. Allg Forstz 44:1198–1204

2-A Soil Fungi and Other Microorganisms

W. PAULUS and A. BRESINSKY

1 Introduction: Aluminum and the Possible Effects of Microorganisms

During the discussion about recent forest decline, it is often argued that biotic factors may be involved in several different ways:

1. An epidemic attack by highly virulent parasites such as viruses, bacteria or fungi. Such an attack could lead to reductions in growth and finally, extinction of formerly healthy stands.
2. An attack by less virulent parasites following previous damage (e.g., by pollutants);
3. Damage to symbiotic organisms, especially mycorrhizal fungi, which contribute to growth under undisturbed conditions.

Thus, biotic factors may be primary reasons for forest decline. In this case, anthropogenic pollutants in present concentrations may hardly influence the onset and course of disease symptoms. This would require evidence for highly virulent and infective organisms, e.g., in Dutch elm disease, caused by the ascomycete *Ceratocystis ulmi* (Buisman) C. Moreau, and spread by *Scolytus scolytus* Fr. On the other hand, where pollutants or mismanagement either increase predispositions to biotic damage, or create imbalanced conditions or symbiotic deficits, microorganisms can serve as secondary rather than as primary agents.

No viroids or bacteria have yet been found which cause epidemic primary infections responsible for forest decline in the Fichtelgebirge or other areas of Germany (Petzold 1987). Numerous viruses have been isolated from conifers and deciduous trees, but their role in explaining forest decline is not yet understood. It is, however, clear that different tree diseases can definitely be caused by microorganisms. For example, viroids are responsible for "cadang-cadang" of coconut trees in the Philippines (Rillo et al. 1972) and viruses for various diseases of oak and beech (Nienhaus 1975; Nienhaus et al. 1985). Mycoplasmlike organisms cause elm phloem necrosis (Sinclair et al. 1976) and the decline of coconut trees in the Caribbean Islands and Africa (Nienhaus et al. 1982). *Rickettsia*-like organisms are known to injure larch (Nienhaus et al. 1976; Nienhaus 1979) and other forms of bacteria cause blight in apple trees (Suhayada and Goodman 1981). These experiences justified the call for investigations about biotic primary reasons (Kandler 1985).

In addition, parasitic fungi have not been found to contribute to recent forest decline in the studied area as primary agents. However, several experiences (e.g., with Dutch elm disease) justified detailed studies on their role in forest decline (Suske and Acker 1987).

In this chapter we focus on the role of symbiotic mycorrhizal fungi in explaining forest decline. According to Meyer (1985), forest decline may relate to a degeneration of mycorrhiza; this may occur relatively early in a chain of other damaging factors or as a result of tree damage. Meyer suggests that mycorrhizal damage in forest trees occurs as a result of increased nitrogen input into soils (see Nihlgård 1985). In many studies, increased nitrogen caused a decrease in frequency and a change in morphology of mycorrhiza along with lowered production of fruiting bodies (Heinrich and Wojewoda 1976; Menge and Grand 1978).

Mycorrhiza serves a critical function in contributing to the supply of minerals and water to higher plants (Harley 1969; Marks and Kozlowsky 1973). In addition, the ectomycorrhizal fungal sheath serves as a mechanical and chemical barrier in protecting the roots of the host from infectious agents and other damaging effects (Zak 1964). Thus, while intact mycorrhizae are critically important for the stability of the forest ecosystem (Schlechte 1986), only a few studies have examined the influence of soil properties on the vitality and functioning of mycorrhiza and on the interrelationship between treetop damage and mycorrhiza (Blaschke et al. 1985). Little is known about the role of mycorrhiza in explaining forest decline, primarily because of experimental difficulties: e.g., many ectomycorrhizae-forming fungi cannot be cultivated or the amount which can be grown is extremely small. Thus suitable research objects have to be choosen. In addition, results on a specific organism cannot be generalized for all types of mycorrhiza formed by different fungi in the field.

The influence of aluminium and heavy metals on growth and nutrient uptake on the mycelium has been studied at the example of *Suillus variegatus* (Schwartz ex FR.) O. Kuntze. Aluminium as well as heavy metals may damage the mycelium of the mycorrhizal fungi, hinder nutrient uptake and thereby reduce the supply of mineral nutrients to the tree. This is important because the widely ramified reticulum of the hyphae is able to accumulate and transport metals.

Suillus variegatus occurs in the study area as the mycorrhizal partner of occasionally interspersed pines (*Pinus sylvestris* L.). In contrast to other mycorrhizal fungi, it grows rapidly with a high mycelium yield in liquid culture medium. Determining the nutrient uptake is thus relatively easy in contrast to that of the mycorrhizal fungi of the predominant spruce species (*Picea abies* L.), which are more relevant to the study of local forest decline, but more difficult to cultivate.

In another part of our study, we attempted to determine the interactions between tree damage and mycorrhizal fungi flora, and how different species might differentially be found in response to forest decline. In particular, we tried to determine whether different mycorrhizal species of fungi occur in damaged and undamaged stands, as well as in stands of different ages.

2 The Effect of Aluminum and Heavy Metals on the Mycorrhizal Fungus *Suillus variegatus*

The mycelium culture of the ectomycorrhizal fungus *Suillus variegatus* was isolated under sterile conditions from fruiting bodies collected from the healthy study site at Wülfersreuth.

Both influence on mycelium growth and on calcium, magnesium, zinc, sodium, lead, and aluminum contents at different lead and aluminum concentrations in the culture medium were studied.

A liquid, modified mycorrhiza medium (MMM) was used as the standard culture medium as follows: 10.0 g glucose; 20.0 g maltose; 1.0 g ammonium tartrate; 1.0 g malt extract (Difco); 0.2 g KH_2PO_4; 0.1 g $MgSO_4 \times 7\,H_2O$; 20.0 mg NaCl; 26.0 mg $CaCl_2 \times 2\,H_2O$; 0.88 mg $ZnSO_4 \times 7\,H_2O$; 0.81 mg $MnSO_4 \times 4H_2O$; 0.80 mg $FeCl_3 \times 6\,H_2O$; 0.11 mg thiamine; 0.14 mg pyridoxine; 0.15 mg riboflavine; 25 μg biotin; 0.5 μg cyanocobalamine; 7.5 mg L-ascorbate; 0.7 mg α-tocopherol acetate; 1.0 mg panthothenate; 0.1 mg p-NH_2-benzoate and 10.0 mg inositol filled to 1000 ml with double distilled water.

In the studies on growth and element uptake, two mycelium cultures were used for each concentration. Comminuted mycelium, which was preincubated for 2 weeks in 20 ml MMM, was used as the inoculation medium. The main cultures (200 ml MMM) were inoculated with this in Fernbach flasks. Aluminum (III), cadmium (II), copper (II), mercury (II), and lead (II) were added as chloride salts to the basic medium in specific concentrations, each corresponding to the concentrations of cations added. Autoclaving adjusted pH to a value of 4.75 $\pm$ 0.25 in the culture medium, except for the media with aluminum concentrations of 3.7 mmol l^{-1}, which had a pH of 3.1.

After inoculation, main cultures were incubated in a dark room for 4 weeks at 23 °C and at a relative humidity of 60%. Mycelia were harvested by washing twice with 200 ml double distilled water using a vacuum pump. Then dryweight was determined (Seeger and Groß 1981). After wet incineration with HNO_3/H_2O_2 in a quartz glass apparatus, determination of contents of the elements were made flamephotometrically using an absorption spectrometer (Heumann et al. 1981).

Mycelial contents of lead, aluminum, calcium, magnesium, zinc and sodium were plotted as means of the mycelial contents with the variation of the single values (two parallels) for each concentration of lead and aluminum in the medium (Fig. 1). Correlation coefficients were then calculated (r for $p = 0.001$). Figure 2 shows the mycelial contents of calcium, magnesium, aluminum as well as growth in g dry weight and change in pH of the medium versus the time of incubation after the addition of 370 μmol l^{-1} aluminum to the medium.

2.1 *Influence of Aluminum and Heavy Metals on Growth*

Growth of the mycelium of *Suillus variegatus* is subjected to large variations in liquid culture medium (Table 1). Calculated mean mycelium yield was 6.50 g_{dw}

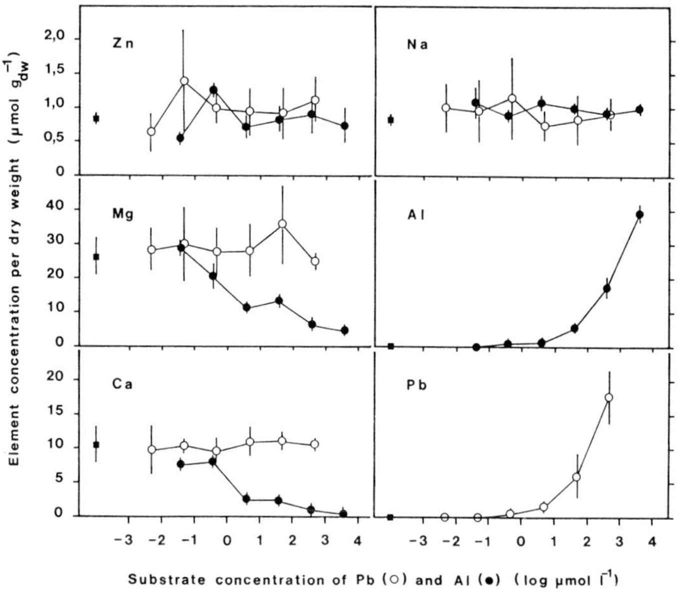

Fig. 1. Contents of aluminum, lead, calcium, magnesium, zinc, and sodium (μmol g_{dw}^{-1}) in the mycelium of *Suillus variegatus* at different lead- (○) and aluminum (●) concentrations in the culture medium (μmol l^{-1}). Each presented value is based on the results of two parallel experiments. Deviations of the single values are shown as *bars*

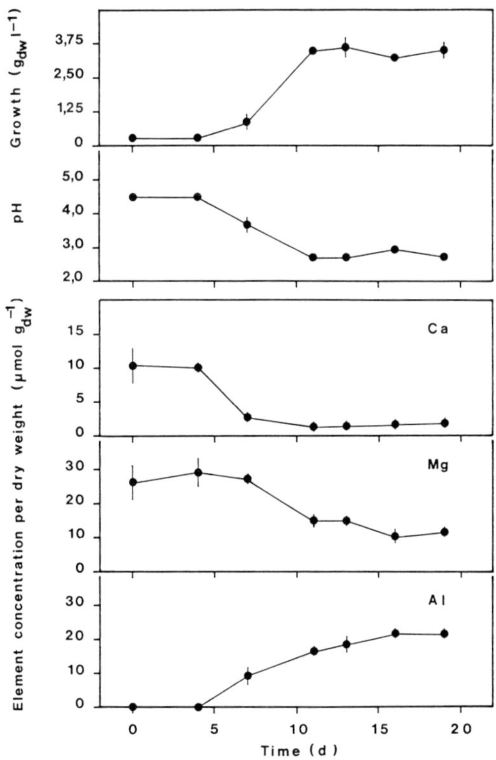

Fig. 2. Aluminum, calcium and magnesium (μmol g_{dw}^{-1}) contents in the mycelium of *Suillus variegatus*, pH values of the culture liquids and mycelium yield after different incubation time (days). At the time of inoculation, aluminum concentration in the culture medium was adjusted to 0.37 mmol l^{-1}

Table 1. Mean growth of mycelium of *Suillus variegatus* (g_{dw} l^{-1}) at different concentrations of aluminium, cadmium, copper, mercury, and lead

	Concentration in substrate (μmol l^{-1})					
Aluminum						
Concentration (μmol l^{-1})	$3.7 \cdot 10^{-2}$	$3.7 \cdot 10^{-1}$	$3.7 \cdot 10^{0}$	$3.7 \cdot 10^{1}$	$3.7 \cdot 10^{2}$	$3.7 \cdot 10^{3}$
Growth (g_{dw} l^{-1})	7.05 ± 0.115	3.75 ± 0.290	7.10 ± 0.090	5.95 ± 0.070	7.30 ± 0.380	7.25 ± 0.930
End pH	2.7 ± 0.00	2.8 ± 0.00	2.8 ± 0.00	2.75 ± 0.05	2.75 ± 0.05	2.7 ± 0.00
Cadmium						
Concentration (μmol l^{-1})	$8.9 \cdot 10^{-3}$	$8.9 \cdot 10^{-2}$	$8.9 \cdot 10^{-1}$	$8.9 \cdot 10^{0}$	$8.9 \cdot 10^{1}$	$8.9 \cdot 10^{2}$
Growth (g_{dw} l^{-1})	7.55 ± 0.115	6.40 ± 0.765	1.75 ± 0.135	0.55 ± 0.025	–	–
End pH	2.8 ± 0.00	2.8 ± 0.00	3.35 ± 0.05	4.35 ± 0.05	–	–
Copper						
Concentration (μmol l^{-1})	$1.6 \cdot 10^{-2}$	$1.6 \cdot 10^{-1}$	$1.6 \cdot 10^{0}$	$1.6 \cdot 10^{1}$	$1.6 \cdot 10^{2}$	$1.6 \cdot 10^{3}$
Growth (g_{dw} l^{-1})	8.55 ± 0.355	8.25 ± 0.075	6.75 ± 1.375	6.65 ± 0.070	7.70 ± 0.565	2.40 ± 0.450
End pH	2.8 ± 0.00	2.8 ± 0.00	2.8 ± 0.00	2.8 ± 0.00	2.8 ± 0.00	3.1 ± 0.05
Mercury						
Concentration (μmol l^{-1})	$5.0 \cdot 10^{-3}$	$5.0 \cdot 10^{-2}$	$5.0 \cdot 10^{-1}$	$5.0 \cdot 10^{0}$	$5.0 \cdot 10^{1}$	$5.0 \cdot 10^{2}$
Growth (g_{dw} l^{-1})	7.70 ± 0.600	7.35 ± 0.105	9.60 ± 0.295	9.10 ± 0.260	7.85 ± 1.515	–
End pH	2.8 ± 0.00	2.8 ± 0.00	2.8 ± 0.00	2.8 ± 0.00	2.8 ± 0.00	–
Lead						
Concentration (μmol l^{-1})	$4.8 \cdot 10^{-3}$	$4.8 \cdot 10^{-2}$	$4.8 \cdot 10^{-1}$	$4.8 \cdot 10^{0}$	$4.8 \cdot 10^{1}$	$4.8 \cdot 10^{2}$
Growth (g_{dw} l^{-1})	8.05 ± 0.325	7.70 ± 0.805	5.60 ± 0.080	7.50 ± 1.910	5.35 ± 2.155	4.25 ± 0.250
End pH	2.8 ± 0.00	2.8 ± 0.00	2.8 ± 0.00	2.8 ± 0.00	2.8 ± 0.00	2.8 ± 0.00
Control						
Growth (g_{dw} l^{-1})	6.85 ± 0.815					
End pH	2.7 ± 0.00					

[a]The culture medium (μmol l^{-1}) and pH values of the culture liquids after incubation for 4 weeks. Each presented value is based on the results of two parallel experiments.

per liter of culture medium (s = 2.325) after incubation for 4 weeks. In the medium, pH- values decreased to pH 2.8 ± 0.1, independent from pH at the beginning of the experiment and of the mycelium yield. Increases, however, occured with concentrations of copper of 1.6 mmol l^{-1} (final pH = 3.15 ± 0.05) and cadmium of 0.89 μmol l^{-1} (final pH = 3.35 ± 0.05) and 8.9 μmol l^{-1} (final pH = 4.35 ± 0.05). For copper and cadmium, applied concentrations were obviously close to tolerance levels of mycelium growth.

Of the metals studied, aluminum reduced growth least with no effect at 3.7 mmol l^{-1}. Heavy metals, on the other hand, inhibited mycelium growth at low concentrations with pronounced growth reduction at cadmium concentrations as low as 8.9 μmol l^{-1} and completely inhibited growth at 89 μmol l^{-1}. Mercury caused no detectable reductions in yield up to 50 μmol l^{-1}; lethal concentrations are reached at 500 μmol l^{-1}. Lead was tolerated with reduced growth at concentrations of 0.48 mmol l^{-1} and copper at concentrations of 1.6 mmol l^{-1}. The degree of toxicity is indicated in the following sequence for the metals studied: aluminum < copper < lead < mercury < cadmium.

2.2 Element Contents in the Presence of Lead and Aluminum

The concentration of lead and aluminum of the mycelium (Fig. 1) is positively correlated with the concentration of these metals in the medium (r_{Pb} = 0.9188, p = 0.001; r_{Al} = 0.9368, p = 0.001). Mycelial contents of zinc and sodium, on the other hand, were not influenced by lead and aluminum in the medium. For zinc, the mean content was 0.85 μmol/g_{dw} (s = 0.306) in the case of increasing aluminum concentrations, and 1.10 μmol/g_{dw} (s = 0.543) in the case of increasing concentrations of lead. Sodium contents were 99.4 μmol/g_{dw} (s = 13.81) for increasing aluminum concentrations and 93.8 μmol/g_{dw} (s = 44.54) for increasing lead concentrations respectively. While mycelial contents of calcium (10.5 μg/g_{dw}, s = 2.25) and magnesium (29.3 μmol/g_{dw}, s = 9.96) were not affected by increasing concentrations of lead in the medium, they were reduced by increasing aluminum concentrations. This resulted in a significantly negative correlation between the logarithm of the external concentration of aluminum and calcium contents (r_{Ca} = 0.9183, p = 0.001; r_{Mg} = 0.9201, p = 0.001). Under control culture conditions with no aluminum, 10.5 μmol/g_{dw} of calcium and 26.5 μmol/g_{dw} of magnesium were taken up into the mycelium. With 3.7 mmol l^{-1} of aluminum in the medium, mycelial contents of calcium and of magnesium were reduced to 2.4% and 18.5%, respectively.

Figure 2 shows the temporal course of aluminum uptake after the concentration of aluminum in the medium was adjusted to 370 μmol l^{-1} at the beginning of the experiment. Mycelial contents of calcium and magnesium, the pH-value of the medium, and mycelial yield are plotted versus time of incubation. Aluminum uptake begins after a lag time of 4 days and reaches a saturation value of about 21 μmol/g_{dw} after 16 days. Decreasing contents of calcium were evident by the fourth day after addition of aluminum to the medium. Contents of magnesium began to fall after about 7 days when calcium concentrations already reached their final content in the mycelium. Magnesium reaches a constant concentration in the mycelium at the same rate as aluminum 16 days after the experiment had begun.

At pH = 4.7, the pH of the medium begins to decline with the course of the experiment and reaches a constant value after 11 days (pH = 2.8). Time of optimal growth under the experimental conditions is from day 7 to day 13.

3 Mycorrhizal Flora of Two Forest Locations

The comparative studies on fungal sociology and production of mycorrhizal fruit bodies were conducted in damaged (location: Oberwarmensteinach) and undamaged stands (location: Wülfersreuth) of spruce. The experimental areas were the same in terms of stand age and companional flora as well as size (900 m^2). In addition to *Picea abies* (L.) H. Karsten, the dominant tree species, trees among or in adjacent stands included *Pinus sylvestris* L., *Larix decidua* Mill., *Betula pendula* Roth, *Fagus sylvatica* L. and *Sorbus aucuparia* L. . At both locations, young stands (30–50 years old) and old stands (80–120 years old) were studied (Table 2). These studies were carried out to determine whether differences in the species spectra and productivity in terms of biomass of fruiting bodies of the mycorrhiza species differed with the degree of damage and the age of the stand. The study sites were visited in weekly intervals from August to October during three sucessive years (1985–1987) and the present species were noted. To estimate productivity, all differentiated carpophores of the mycorrhizal fungi (meaning all able to sporulate) at the two study sites were harvested, dried to constant weight, and their dry weights measured (Hering 1966; Fogel 1975).

The total annual production of fruit bodies by mycorrhizal fungi at the two study sites differed in the three successive years. Obviously, in damaged stands, the diversity and frequency of fungal species were reduced for both young and old stands. In young and old damaged stands and in old undamaged stands, total annual production fluctuated about twice as much (%s = 89–110) as in the young undamaged stands (%s = 50). These fluctuations can be explained by the greatly

Table 2. Mean production of fruiting bodies (x ± s) in g_{dw} m^{-2} a^{-1} and numbers of mycorrhizal fungi species (n) in damaged and intact spruce stands

	$\bar{x} \pm s$	$\bar{n} \pm s$
Young stands, undamaged	1.839 ± 0.927	22.3 ± 9.3
Young stands, damaged	0.200 ± 0.179	9.3 ± 6.3
Young stands, damaged	0.309 ± 0.288	11.6 ± 1.3
Old stands, undamaged	0.279 ± 0.307	6.0 ± 1.0
Old stands, damaged	0.028 ± 0.027	4.0 ± 0.0

[a]At both locations, young (30–50 years) and old (80–120 years) stands were sampled.

reduced production of fruit bodies in the second year of the examination due to low precipitation during that late summer and early fall.

Despite the generally reduced production of fruit bodies in that year, no remarkable changes in the spectrum of species occurred. In comparison to young stands, old stands were poorer in species in damaged as well as undamaged stands with a simultaneous reduction in productivity of mycorrhizal fungi. In comparison to old undamaged stands, young damaged stands were similar in productivity and fungal species spectrum. In both cases the following occur: *Lactarius necator* (Bull. em Pers. ex FR.) Karst., *Hygrophorus olivaceoalbus* (Fr. ex Fr.) Fr., *Paxillus involutus* (Batsch) Fr., *Laccaria laccata* (Scop. ex Fr.) Bk. & Br., and above all, *Russula ochroleuca* (Pers.) Fr. . Young undamaged stands differed from the other stands in having more diverse species. Here, *Lactarius rufus* (Scop.) Fr., *Lactarius necator* (Bull. em Pers. ex Fr.) Karst., *Amanita rubescens* (Pers. ex Fr.) Gray, *Xerocomus badius* (Fr.) Kühn ex Gilb., *Russula emetica* (Fr.) and *Hygrophorus olivaceoalbus* (Fr. ex Fr.) Fr. were the dominant species.

4 Conclusion

Several authors regard soil pollutants, in addition to direct effects of emissions from the air, as possible causes of roots and mycorrhiza damage (Mayer 1981; Mayer and Heinrichs 1981). Runge (1981) considers the aluminum tolerance of higher plants to be a measure for their natural abundance. Damaging effects of aluminum on the root system have been experimentally demonstrated (Ulrich et al. 1979; Metzler and Oberwinkler 1986; Rost-Siebert 1983; Simon et Rothe 1985).

In addition to aluminum, the release of heavy metals from natural and anthropogenic sources accompanying increasing acidification could also be a cause for root damage. In heavily stressed habitats, e.g. mine dumps, the filtration effect of mycorrhiza is obvious, as mycorrhized plants exhibit a higher resistance to heavy metals than non-mycorrhized plants (Bradley et al. 1981; Brown and Wilkins 1985). In vitro studies show that different heavy metals impact the growth of mycorrhiza, since the different fungi and lines from different proveniences of one species react in distinct ways to stress by heavy metal stress (Thompson and Medve 1984; Willenborg 1987).

In addition to an often insufficient supply of calcium and magnesium in the soil, nutrient uptake by plant roots may be hindered due to increased concentration of aluminum in the nutrient solution (Foy et al. 1987; Jorns and Hecht-Buchholz 1985). From our study, the amount of calcium and magnesium supplied by the mycelium of the ectomycorrhizal fungi was found to decrease because of the negative impact of aluminum concentration. This could explain why needles from damaged spruce contain less calcium and magnesium than the ones from undamaged trees at the experimental sites in the Fichtelgebirge (see Chap. 4-F) because aluminum in the soil solution can occur in concentrations in the order of magnitude from about 0.1 to 0.5 mmol l^{-1} while other trace elements, e.g., heavy metals, occur only in much lower concentrations (Fiedler and Rösler 1988). Listl (1985) also found that magnesium in the fungal fruiting bodies from undamaged stands was higher on the average than those from damaged ones.

While the detrimental effect of aluminum on plant growth has long been known (Johnson and Jackson 1964; Lee 1971), it was not clear until now how aluminum hinders the uptake of other nutrients. Studies by Roth et al. (1987) have shown that aluminum as well as calcium and magnesium exist in polyphosphate granules (PPGs) from bacteroids symbiontically associated with higher plants. Incubation with higher aluminum concentrations in the substrat results in an increase in the aluminum content of the PPGs. Therefore, it is likely that they are also the location where antagonistic interactions between aluminum and calcium and magnesium occur.

In contrast to aluminum, increased concentrations of lead showed no effect on the absorption of nutrients, even through lead is similar to aluminum in the way it accumulates in the mycelium.

The stability of forest ecosystems depends on intact root systems and functional mycorrhiza (Schlechte 1986). In areas with damaged forests, mycorrhizal impairment has been found to accompany observed damage in the crown (Murach 1984). In addition to progressing acidification of the soil and the concommitant increase of aluminum and heavy metals, the increased input of nitrogen may be a trigger for these disturbances. Meyer (1985) and Nihlgård (1985) have suggested that nitrogen may reduce production of growth substances by the mycorrhizal fungi, and thus hinder the formation of new mycorrhiza (Moser 1959; Slankis 1971). Finally, when roots are insufficiently branched, reduced mycorrhizal trimming occurs, resulting in the hindrance of water and nutrient supply to the host (Meyer 1985).

Large differences in the content of heavy metals (Seeger and Beckert 1979; Seeger 1982) and in mineral nutrients (Mutsch et al. 1979; Ohtonen 1982) are found in fruiting bodies of different mycorrhizal fungi species. Some species may thus supply nutrients to their hosts more efficiently than other species, and these differences may vary with the external conditions. Such changes in soil conditions, e.g., higher nitrogen input, increasing acidification, increasing aluminum and heavy metals contents, and an insufficient nutrient supply may thus displace some mycorrhizal fungal species for others better adapted to the new environmental conditions. Our studies show that depending on the age of the trees, the number of fruiting bodies as well as the diversity of species of mycorrhizal fungus species in undamaged stands are greater than in damaged stands. These results are similar to other findings (Benkert 1982; Derbsch and Schmitt 1987; Fellner 1983; Schlechte 1986) and further evidence that the decline of mycorrhizal fungi observed in the last centuries is related to the latent tree damages (Keller 1977).

References

Benkert D (1982) Vorläufige Liste der verschollenen und gefährdeten Großpilzarten der DDR. Boletus 6:21–32

Blaschke H, Bremer K, Schwarz H (1985) Wurzelschäden und Waldsterben: Zur Bestimmung morphometrischer Kenngrößen von Feinwurzelsystemen mit dem IBAS- erste Ergebnisse. Forstwiss Centralbl 104:199–205

Bradley R, Burt AJ, Read DJ (1981) Mycorrhizal infection and resistance to heavy metal toxicity in *Calluna vulgaris*. Nature (Lond) 292:335–337

Brown MT, Wilkins DA (1985) Zink tolerance of mycorrhizal *Betula*. New Phytol 99:101–106

Derbsch H, Schmitt JA (1987) Atlas der Pilze des Saarlandes. Teil 2: Nachweise, Ökologie, Vorkommen und Beschreibungen. Wiss Schriftenr der Obersten Naturschutzbehörde, Saarbrücken, Sonderband 3:1–816

Fellner R (1983) Mycorrhizae-forming fungi in climax forest communities at a timberline in Giant Mountains. Ceska Mykol 37:109

Fiedler HJ, Rösler HJ (1988) Spurenelemente in der Umwelt. Enke, Stuttgart

Fogel R (1975) Ecological studies of hypogaeous fungi. II. Sporocarp phenology in a western Oregon douglas fir stand. Can J Bot 54:1152–1162

Foy CD, Chaney RL, White MC (1978) The physiology of metal toxicity in plants. Annu Rev Plant Physiol 29:511–566

Frank AB (1885) Über die auf Wurzelsymbiose beruhende Ernährung gewisser Bäume durch unterirdische Pilze. Ber Dtsch Bot Ges 3:128–145

Harley JL (1969) The Biology of Mycorrhiza. Leonard Hill, Lond

Heinrich Z, Wojewoda W (1976) The effect of fertilization on a pine forest ecosystem in an industrial region. IV. Macromycetes. Ekol Pol 24:319–330

Hering TF (1966) The terricolous higher fungi of four Lake District woodlands. Trans Br Mycol Soc 49:369–383

Heumann KG, Kastenmayer P, Zeininger H (1981) Pb- und Tl-Spurenbestimmung im ppm- bzw. ppb-Bereich in biologischem Material durch massenspektrometrische Isotopenverdünnungsanalyse. Fresenius Z Anal Chem 306:173–177

Johnson RE, Jackson WA (1964) Calcium uptake and transport by wheat seedlings as affected by aluminum. Soil Sci Am Proc 28:381–386

Jorns A, Hecht-Buchholz C (1985) Aluminiuminduzierter Magnesium- und Calciummangel im Laborversuch bei Fichtensämlingen. Allg Forstz 46:1248–1252

Kandler O (1985) Immissions- versus Epidemie-Hypothesen. In: Kortzfleisch G (ed) Waldschäden. Oldenbourg, München Wien, pp 20–50

Keller T (1977) Begriff und Bedeutung der "latenten Immissionsschädigung". Allg Forst-Jagdztg 148:115–120

Lee CR (1971) Influence of aluminum on plant growth and mineral nutrition of potatoes. Agron J 63:604–608

Listl M (1985) Pilze einer geschädigten und nicht geschädigten Waldfläche in Fichtelgebirge unter besonderer Berücksichtigung ihrer Metallgehalte und Versuche zum Metallaufnahmeverhalten von Pilzen. Zulassungsarbeit, Univ Regensburg

Marks GC, Kozlowski TT (1973) Ectomycorrhizae. Their ecology and physiology. Academic Press, Lond NY

Mayer R (1981) Natürliche und anthropogene Komponenten des Schwermetallhaushalts von Waldökosystemen. Gött Bodenkdl Ber 70:1–22

Mayer R, Heinrichs H (1981) Gehalte von Baumwurzeln an chemischen Elementen einschließlich Schwermetallen aus Luftverunreinigungen. Z Pflanzenernähr Bodenkd 144:637–646

Menge JA, Grand GL (1978) Effect of fertilization on production of epigeous basidiocarps by mycorrhizal fungi in lobolly pine plantations. Can J Bot 56:2357–2362

Metzler B, Oberwinkler F (1986) Charakteristische Meristemschäden in Fichtenwurzeln durch niedrigen pH-Wert und Aluminiumionen. Allg Forstz 25:649–651

Meyer FH (1985) Einfluß des Stickstoff-Faktors auf den Mykorrhizabesatz von Fichtensämlingen im Humus einer Waldschadensfläche. Allg Forstz 9/10:208–219

Moser M (1959) Beiträge zur Kenntnis der Wuchsstoffbeziehungen im Bereich ektotropher Mykorrhizen. Arch Mikrobiol 34:251–269
Murach D (1984) Die Reaktion der Feinwurzeln von Fichten (Picea abies Karst.) auf zunehmende Bodenversauerung. Gött Bodenkdl Ber 77:1–126
Mutsch F, Horak O, Kinzel H (1979) Spurenelemente in höheren Pilzen. Z Pflanzenphysiol 94:1–10
Nienhaus F (1975) Viren und virusverdächtige Erkrankungen in Eichen (Quercus robur und Quercus sessiflora). Z Pflanzenkr Pflanzenschutz 82:739–749
Nienhaus F (1979) Lärchen-Degeneration durch *Rickettsien*-ähnliche Bakterien. Allg Forstz 6:130–132
Nienhaus F, Brüssel H, Schinzer U (1976) Soil-borne transmission of *Rickettsia*-like organisms found in stunted and witches' broom diseased larch tree (*Larix decidua*). Z Pflanzenkr Pflanzenschutz 83:309–316
Nienhaus F, Schuiling M, Gliem G, Schinzer, Spittel A (1982) Investigations on the etiology of the lethal disease of coconut palm in Tanzania. Z Pflanzenkr Pflanzenschutz 89:185–193
Nienhaus F, Ebrahim-Nesbat F, Fricke M, Büttner C, Welter K (1985) Investigations on viruses from declining beech trees (*Fagus sylvatica* L.) in Rhineland and Westfalia, Federal Republic of Germany. Eur J For Pathol 15:402–411
Nihlgård B (1985) The ammonium hypothesis.-An additional explanation to the forest dieback in Europe. Ambio 14:2–8
Ohtonen R (1982) Mineral concentrations in some edible fungi and their relation to fruit-body size and mineral status of substrate. Ann Bot Fenn 19:203–209
Petzold H (1987) Übertragungsversuche, licht- und elektronenmikroskopische Untersuchungen zur Prüfung der ursächlichen Beteiligung von Prokaryonten (MLO, RLO) an den neuartigen Waldschäden. Statussemin BMFT-Förderschwerpunkt Ursachenforschung Waldschäden 30.3.–3.4.1987. Spez Ber Kernforschungsanlage Jülich 413:37–38
Rillo EP, Pableo GO, Price WC (1972) An anatomical study of coconut leaves from healthy trees and those affected by cadang-cadang. Bull Torrey Bot Club 99:271–277
Rost-Siebert K (1983) Al-Toxizität und -Toleranz an Keimpflanzen von Fichte (*Picea abies* Karst.) und Buche (*Fagus silvatica* L.). Allg Forstz 26/27:686–689
Roth LE, Dunlap JR, Stacey G (1987) Localizations of aluminium in soybean bacteroids and seeds. Appl Environ Microbiol 53:2548–2553
Runge M (1981) Zur Bedeutung des Aluminiums für die Ausbildung der natürlichen und naturnahen Vegetation. Mitt Erg Stud Ökol Umweltschutz 7:16–38
Schlechte G (1986) Zur Mykorrhizapilzflora in geschädigten Forstbeständen. Z Mykol 52:225–232
Sinclair WA, Braun EJ, Larsen AO (1976) Update on phloem necrosis of elms. J Arboric 2:106–113
Seeger R (1982) Toxische Schwermetalle in Pilzen. Dtsch Apothztg 37:1835–1844
Seeger R, Beckert M (1979) Magnesium in höheren Pilzen. Z Lebensm Unters Forsch 168:264–281
Seeger R, Groß M (1981) Thallium in höheren Pilzen. Z Lebensm Unters Forsch 173:9–15
Simon B, Rothe GM (1985) Aluminium-bedingte Stoffwechseländerungen in Fichtenkeimlingen. Allg Forstz 36:931–936
Slankis V (1971) Formation of ectendomycorrhizae of forest trees in relation to light, carbohydrates, and auxin. In: Hacskaylo E (ed) Mycorrhizae. US Dep Agric Misc Publ 1189:151–167
Suhayda CS, Goodman RN (1981) Early proliferation and migration and subsequent xylem occlusion by *Erwinia amylovora* and the fate of its extracellular polysacharides (EPS) in apple shoots. Phytopathology 71:697–707
Suske J, Acker G (1987) Internal hyphae in young, syptomless needles of *Picea abies*: electron microscopic and cultural investigation. Can J Bot 65:2098–2103
Thompson GW, Medve RJ (1984) Effects of aluminium and manganese on the growth of ectomycorrhizal fungi. Appl Environ Microbiol 48:556–560
Ulrich B, Mayer R, Khanna PK (1979) Depositionen von Luftverunreinigungen und ihre Auswirkungen in Waldökosystemen im Solling. Schr aus der Forstl Fakultät Univ Göttingen und der Niedersächsischen Forstl Versuchsanstalt. Bd 58:1–291
Willenborg A (1987) In-vitro-Untersuchungen zum Verhalten verschiedener in Kultur genommener Mykorrhizapilze gegenüber biotischen und abiotischen Faktoren unter besonderer Berücksichtigung der neuartigen Waldschäden. Mitt Versuchskammer Pilzanbau Landwirtschaftskammer Rheinland, Krefeld-Großhüttenhof. Sonderheft 5:1–259
Zak B (1964) Role of mycorrhizae in root disease. Annu Rev Phytopathol 2:377–391

2-B Endophytic Needle Fungi: Culture, Ultrastructural and Immunocytochemical Studies

J. SUSKE and G. ACKER

1 Introduction

Pathogenic fungi can cause various needle diseases in Norway spruce (Butin 1983). At the same time, many deciduous and evergreen plants have been reported to host fungal endophytes, i.e., fungi that cause unapparent, asymptomatic infections entirely within the plant tissues (e.g., Butin 1986; Carroll 1986; Petrini 1986). However, endophytes may cause pathogenic symptoms if the host has been previously stressed by different factors (for literature review and discussion, Carroll 1986).

In Norway spruce needles collected in the area investigated in this study, *Lophodermium piceae* (Fckl.) Höhn. and *Rhizosphaera kalkhoffii* Bubak were the most frequent needle colonizers (Suske and Acker 1987a,b). *L. piceae* has been described alternatively as a potentially aggressive pathogen (Rostrup 1891), a weak parasite, or a saprophyte (Darker 1932). Rehfuess and Rodenkirchen (1984, 1985) considered both *L. piceae* and *R. kalkhoffii* as pathogens of Norway spruce, which together cause needle-reddening disease (*Nadelröte*). This disease is characterized by a short period of chlorosis of mainly older spruce needles in autumn followed by reddish-brown discoloring of needles and needle fall. Butin and Wagner (1985) reported isolation experiments with brown needles collected from different German locations in which the infection frequency varied strongly from one spruce stand to another. They suggested that both *L. piceae* and *R. kalkhoffii* were only early colonizers of dying or dead needles, and that these fungi were not the cause of needle-reddening disease. This conflicting view of the pathogenicity of *L. piceae* and *R. kalkhoffii* (Butin 1986; Rehfuess and Rodenkirchen 1985; Schütt 1985) is a conclusion mainly drawn from isolation experiments on diseased or dead needles, and has not been directly confirmed (Butin 1986). Understanding the life cycles of these fungi is important for interpreting forest-decline phenomena, especially the symptom of early needle loss. If *L. piceae* and *R. kalkhoffii* are saprophytes, then their presence in declining forests has nothing to do with the primary cause of needle loss. If they are aggressive pathogens which are responsible for the tree decline, these fungi would not occur in green, symptomless needles but in diseased foliage. Such an occurrence would then support the "epidemic hypothesis" of forest decline (Kandler 1984). Finally, if these fungi are endophytes, which normally grow in green needles and which become aggressive only in certain stress situations, then emissions or climatic extremes could trigger fungal needle loss as a consequence of pollutant impact.

E.-D. Schulze, O.L. Lange, R. Oren (Eds.)
Ecological Studies Vol. 77

In the present study, internal fungi were isolated from both green, symptomless and symptomatic Norway spruce needles. Hyphae of *L. piceae* were detected in green, symptomless needles, and were studied in detail by scanning and transmission electron microscopy.

2 Electron Microscopy of Needle Fungi

2.1 Isolation of Internal Fungi

Needles of Norway spruce were collected from 4–6 m high branches of roughly 30- to 80-year-old trees. For each of the needle samples the condition of the needle, its age, and the tree from which it was taken were recorded. The following needle categories were considered:

1. green, symptomless foliage taken from the internodes of apparently healthy trees in a sparse stand,
2. green, symptomless foliage taken from declining trees in a sparse stand,
3. green, symptomless, but permanently shaded foliage taken from healthy and declining trees in a dense stand,
4. needles with such disease symptoms as chlorotic or necrotic patches, bands, or needle tips, as well as completely yellow or brown foliage taken from declining trees in a sparse stand. Needles from trees affected by needle-reddening disease were not included in this category, and
5. both green and symptomatic needles from trees affected by needle-reddening disease both in 1985 and 1986. From July to December 1986 samples were taken from those internodes that became affected by needle-reddening disease in the autumn of that year.

The needle samples were surface-sterilized in 0.15% $HgCl_2$, as described by Rack and Butin (1984). The needles were then cross-sectioned and the two resulting segments incubated on malt agar plates (3% malt extract, 1.5% agar). Only those fungi growing from cut edges were scored as endophytes. Their colonies were identified according to culture characters and the morphology of fruiting bodies and spores (Darker 1932, 1967; Rack and Butin 1984).

Alternatively, surface-sterilized needles were cross-sectioned into 0.4–0.7 mm long segments using a sterile razor blade. The resulting segments were incubated in serial order on malt agar. Each series of contiguous infected segments was scored as a single infection, termed a run of infection (Bernstein and Carroll 1977; see also Table 6, footnote).

2.2 Lophodermium piceae-Antiserum and Antibody Conjugates

Aereal mycelium of *L. piceae* was harvested from agar cultures, freeze-dried, and suspended in 0.9% NaCl. Portions of 2 mg mycelium in 1 ml saline were used for the immunization of a young rabbit according to the immunization scheme described by Huber and Hock (1985). Collected blood was stored overnight at 4°C and the coagulated part was separated from the serum.

Colloidal gold particles were prepared and conjugated with goat anti-rabbit IgG, as described by Slot and Geuze (1985).

2.3 Electron Microscopy

2.3.1 Scanning Electron Microscopy (SEM)

Needle samples were fixed in glutaraldehyde (3% in 0.05 M phosphate buffer, pH 6.9; 12h at 4°C) and dehydrated in a graded acetone series (25%, 50% plus 1% uranyl acetate, 70%, 96%, absolute acetone). They were then critical-point-dried and cut longitudinally, as described by Suske and Acker (1987a). The cut-edge surfaces of the needle samples were sputter-coated with gold and examined in a SEM, model Novascan 30 (Zeiss, Oberkochen).

2.3.2 Transmission Electron Microscopy (TEM)

Spruce needles were surface-sterilized and cross-sectioned into 0.4–0.7 mm long segments. Each second of the segments was fixed separately in a numbered vial containing glutaraldehyde (2.5% in 0.05 M phosphate buffer, pH 6.9; 12 h at 4°C). The remaining segments were incubated in serial order on agar medium. Fixed segments were dehydrated in a graded ethanol series (25%, 50%, 70%, 96%, absolute ethanol) and embedded in Lowikryl K4M resin (Chemische Werke Lowi, Waldkraiburg), as described by Altman et al. (1984).

For examination in the TEM, only those segments whose contiguous segments had revealed infection on malt agar were chosen. Thin sections were cut with an ultramicrotome, model Ultracut (Reichert, Vienna) and mounted on formvar-carbon-coated grids. For immunolabeling experiments, thin sections were indirectly gold-labeled as previously described in detail (Acker 1988), using the *L. piceae*-antiserum and antibody conjugates. Sections were stained with 2% aqueous uranyl acetate and lead citrate (REYNOLDS 1963), and finally examined in a TEM, model EM 109 (Zeiss, Oberkochen).

3 Needle Colonization by Fungi

3.1 The Isolation of Fungi from Green, Symptomless Needles

Different age-classes of green needles with no symptoms of damage, collected from both apparently healthy (Table 1) and declining (Table 2) trees, revealed a similar fungal spectrum and infection frequency. From 2- to 36-month old needles, *L. piceae* was isolated at similar frequencies. In foliage older than 36 months the infection frequency of the major endophytic colonizer of green, symptomless foliage, *L. piceae*, increased. Other endophytic species (Tables 1 and 2) were isolated only occasionally and were, therefore, not further considered in this study.

Table 1. Frequency of endophytic fungi in green, symptomless needles of apparently healthy trees

Needle age (months)	No. of samples	*Lophodermium piceae* (%)	Other species[a] (%)	Sterile (%)
2–12	244	0.4 ± 0.8 a	0.4 ± 0.8 a	99.2 ± 0.8 a
13–24	518	2.8 ± 2.5 a	0.5 ± 1.0 a	96.7 ± 2.5 b
25–36	541	5.8 ± 4.4 a	0.4 ± 0.8 a	93.8 ± 4.1 c
37–48	262	12.2 ± 6.8 b	3.6 ± 2.6 b	84.2 ± 5.4 d
49–60	281	33.4 ± 11.4 c	5.0 ± 3.4 b	61.6 ± 12.0 e

[a]Means followed by the same letter are not significantly different.

Table 2. Frequency of endophytic fungi in green, symptomless needles of declining trees[a]

Needle age (months)	No. of samples	*Lophodermium piceae* (%)	Other species[b] (%)	Sterile (%)
2–12	850	1.1 ± 1.7 a	0.1 ± 0.3 a	98.8 ± 1.6 a
13–24	1042	3.1 ± 1.9 b	0.7 ± 1.0 a	96.2 ± 2.0 b
25–36	430	5.3 ± 4.2 b	2.2 ± 2.3 a	92.5 ± 3.2.c
37–48	250	10.8 ± 3.3 c	2.6 ± 2.4 a	86.6 ± 3.5 d
49–60	263	29.5 ± 7.3 d	4.0 ± 2.2 a	66.5 ± 6.6 e

[a]Means followed by the same letter are not significantly different.
[b]No isolates of *R. kalkhoffii* are included in this column.

Permanently shaded needles (Table 3) had significantly higher infection frequencies by *L. piceae* than sun needles (significance level 0.001). While the infection frequency in the older needles increased, a large number of the needles in the shade were sterile.

3.2 The Isolation of Fungi from Needles Showing Disease Symptoms

Needle samples from trees with symptoms different from those described for needle-reddening disease revealed that most green needles with disease symptoms and completely yellow needles were sterile (Table 4). In addition, nearly 40% of the brown needles were also sterile. *L. piceae* was again the most frequently isolated species, but *R. kalkhoffii* was also found, particularly in brown needles.

The role of needle fungi with respect to needle-reddening disease (Table 5) shows that nearly 40% of the needles were colonized by endophytes in summer (July/August) before any visible symptoms could be observed. When needle-reddening disease became visibly apparent (September/October), *R. kalkhoffii* was isolated from ca. 10% of the samples. More than 50% of the needles were not infected. Only 55% infection was found in dead, brown needles in October. Further sampling in December showed that the *R. kalkhoffii* infection frequency had increased following needle death. All the needles examined in December had been colonized.

Table 3. Frequency of endophytic fungi in green, symptomless, but permanently shaded needles

Needle age (months)	No. of samples	*Lophodermium piceae* (%)	Other species[a] (%)	Sterile (%)
2–12	228	53.1 ± 4.5 a	0.8 ± 0.7 a	46.1 ± 5.6 a
13–24	258	58.0 ± 4.4 a	1.1 ± 0.7 a	40.9 ± 8.2 a
25–36	256	68.5 ± 7.9 a	1.6 ± 1.6 a	29.9 ± 8.2 a
37–48	266	87.5 ± 12.7 b	1.8 ± 2.1 a	10.7 ± 11.6 b

[a] Means followed by the same letter are not significantly different.

Table 4. Frequency of endophytic fungi in needles showing disease symptoms[a, b]

Needle condition	No. of samples	*Lophodermium piceae* (%)	*Rhizosphaera kalkhoffii* (%)	Other species (%)	Sterile (%)
Green with symptoms	568	8.6 ± 7.2 a	2.5 ± 1.9 a	1.0 ± 1.7 a	87.9 ± 7.9 a
Yellow	517	13.0 ± 5.6 a	6.0 ± 3.5 b	3.1 ± 3.6 a	77.9 ± 9.1 b
Brown	524	27.7 ± 12.6 b	21.2 ± 12.4 c	13.0 ± 4.0 b	38.1 ± 14.9 c

[a] Needle samples showing the symptoms of needle reddening disease were not included in Table 3.
[b] Means followed by the same letter are not significantly different.

Table 5. Fungal colonization of 4- to 5-year-old needles in connection with needle-reddening disease[a]

Needle condition/ Sampling date	No. of samples	*Lophodermium piceae* (%)	*Rhizosphaera kalkhoffii* (%)	Other species (%)	Sterile (%)
Symptomless/ July, Aug.	211	28.1 ± 4.6 a	0.0 ± 0.0 a	10.1 ± 3.8 a	61.8 ± 8.4 a
Yellow/ Sept., Oct.	200	34.0 ± 4.9 a	9.0 ± 2.2 b	5.0 ± 2.2 b	52.0 ± 6.5 a
Brown/ Oct.	200	27.0 ± 5.4 a	25.5 ± 8.3 c	2.5 ± 1.6 b	45.0 ± 11.2 a
Brown/ Dec.	412	30.5 ± 3.3 a	72.9 ± 9.8 d	0.0 ± 0.0 c	0.0 ± 0.0 b

[a] Means followed by the same letter are not significantly different.

The extent of individual needle colonization, i.e., the distribution of hyphae in the tissues of a given needle, affects the interpretation of the host-fungus interaction. Table 6 shows the number of separately isolated infections (runs of infection) of *L. piceae* in 4- to 5-year-old needles for different conditions. The lengths of these infections, expressed by the number of contiguous infected segments per run of infection, are listed in Table 7. Ca. 50% of the green and yellow needles only had a single isolated infection (Table 6) and nearly 90% of these were limited to a single segment (Table 7). From the remaining portion of these samples, two or three separate infections were isolated. Green needles from shaded internodes had a significantly higher percentage of multiple infections (Table 6). Runs of infection in these needles were longer on average than those

Table 6. Relation between the needle condition and the number of runs of infection[a] of *L. piceae*[a]

Needle condition	No. of infected needles	Number of needles with 1–5 runs of infection 1 run	2 runs	3 runs	4 runs	5 runs	Average number of runs/needle
Symptomless, shaded	50	4	10	11	19	6	3.2 ± 1.1 a
Symptomless	50	27	17	6	0	0	1.6 ± 0.7 b
Yellow	50	28	15	7	0	0	1.6 ± 0.7 b
Brown	50	46	4	0	0	0	1.1 ± 0.3 c

[a]The term "run of infection" was adapted from Bernstein and Carroll (1977) and designates a series of contiguous infected needle segments which was scored as a single infection. In the present report a run of infection can comprise 1 to all segments, and ends with the apex of the needle, the petiole end of the needle, or an uninfected segment.
[b]Means followed by the same letter are not significantly different.

Table 7. Relation between the needle condition and the length of runs of infection by *L. piceae*[a]

Needle condition	No. of runs of infection/ 50 needles	Segments per run of infection 1	2	3	4	5	6	9	all	Average number of infected segments/run
Symptomless, shaded	163	78	37	23	8	10	5	2	0	2.5 ± 1.6 a
Symptomless	79	67	12	0	0	0	0	0	0	1.1 ± 0.4 b
Yellow	79	69	10	0	0	0	0	0	0	1.1 ± 0.3 b
Brown	54	8	4	0	0	0	8	0	34	24.9 ± 15.7 c

[a]Means followed by the same letter are not significantly different.

in green needles collected from a sparse spruce stand (Table 7). In brown needles, runs of infection often extended over the whole needle length (Table 7) so that only a single run of infection was scored (Table 6).

3.3 Observation of Needle Fungi at the Cellular Level

Many questions concerning leaf fungi of Norway spruce needles cannot be answered from cultivation on agar only. For example, our culture studies provide little evidence for the exact location of the endophytic hyphae within the needle tissues and the host-fungus interaction. Therefore, microscopic observation can be a useful addition to culture studies.

On a cut-edge surface of an asymptomatic needle (Fig. 1) a single, unbranched hypha (*H*) can be observed in the mesophyll (*M*). Evidence based on combined SEM and culture observations has suggested that this hypha belongs to *L. piceae* (Suske and Acker 1987a). Only single hypha or very limited mycelia were observed in the SEM. This indicates that endophytic-infection loci in green

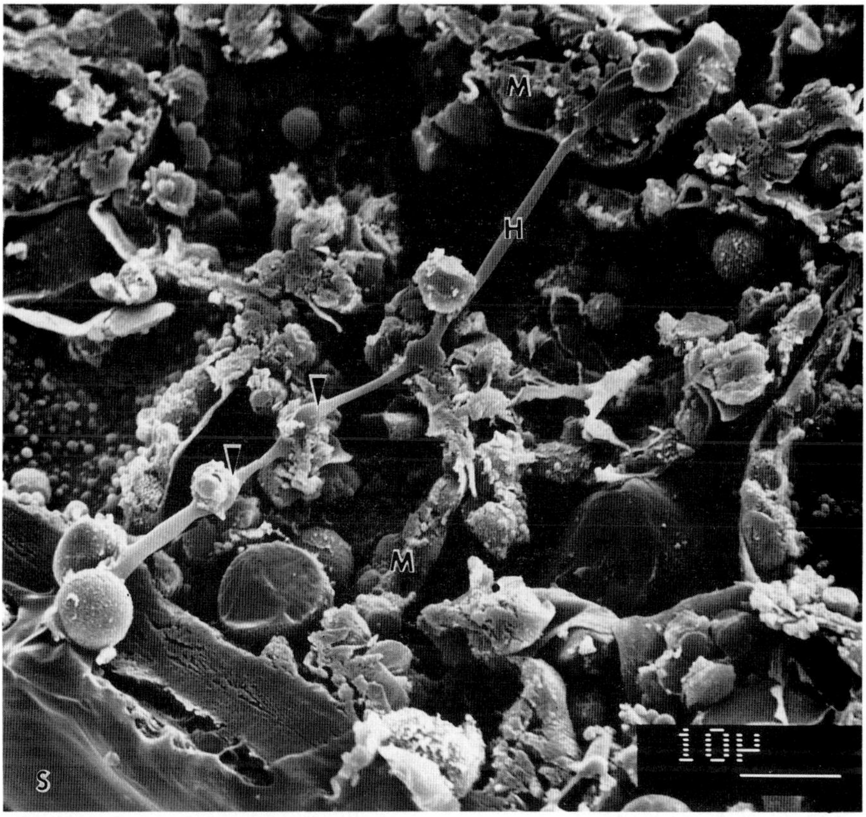

Fig. 1. Hyphae of *L. piceae* in near-surface tissue layers of a symptomless needle. Intracellular swellings (*arrowheads*) in close contact with mesophyll cells (*M*) can be observed along the hypha (*H*). Needle surface (*S*). Micrograph from Suske and Acker (1987a). SEM 5000x

needles were limited to small tissue areas. Irregularly shaped swellings along endophytic hyphae (Fig. 1, *arrow heads*) were frequently observed. In contrast to the green needles, dead, brown foliage (Fig. 2) was found to be heavily colonized by *L. piceae.*

The tissue portion actually examined by the thin-sectioning technique is so small that the transmission electron-microscopic detection of spruce needle cells colonized by endophytic fungi is very difficult. Therefore, we combined the thin-sectioning technique with culture studies to select those needle segments for subsequent TEM observation. Figures 3–6 were obtained from needle segments located in between two needle segments that had revealed infection by *L. piceae* after incubation on malt agar. No green needle was found with double or multiple different fungal specie infections. Immunolabeling experiments strongly indicated that the hyphae in the needle segments used to prepare thin sections were *L. piceae*, as the hyphae of this species showed a significantly higher labeling intensity (Fig. 4) than other endophytes labeled along with *L. piceae*-antiserum.

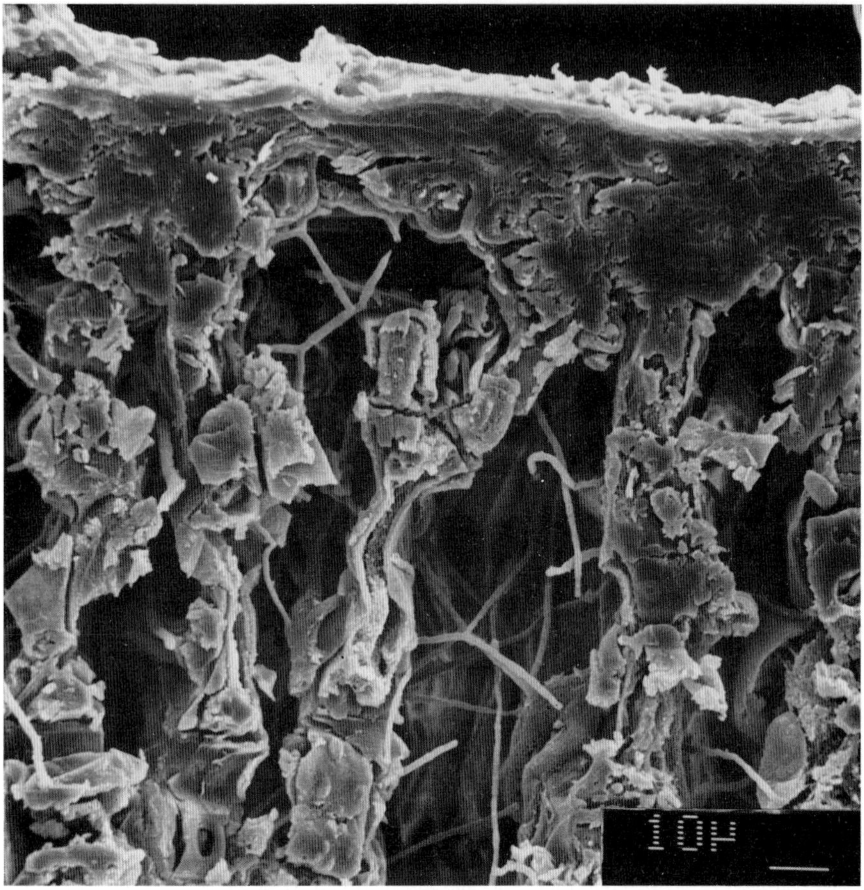

Fig. 2. Representative view of a cut-edge surface of a brown needle heavily colonized by *L. piceae*. SEM 5500x

In Figure 3, a few hyphae (*H*) can be seen within the cell lumen of an epidermal cell. Hyphal colonization of the hypodermis (*HY*) is demonstrated in Fig. 4. Near-surface mesophyll cells were invaded by both inter- and intracellular hyphae (Fig. 5, 6). While hyphae may contain endocells, i.e., cells present within another hyphal cell, the upper endocell seems to penetrate through a gap in the interrupted cell wall of the enclosing hypha (Fig. 5, EC, region between *arrowheads* tips). Outside the cell wall, this enclosing cell is surrounded by an electron-dense, capsule-like structure or sheath (*SH*) of unknown chemical composition and function. It is remarkable that the mesophyll cell walls of infected cells were often more electron-dense in the region close to hyphae (Fig. 5) when compared to the cell walls of uninfected mesophyll cells. The intracellular colonization of mesophyll cells was accompanied by the breakdown of host cell structures, as shown in Fig. 6. This micrograph also demonstrates that the shape

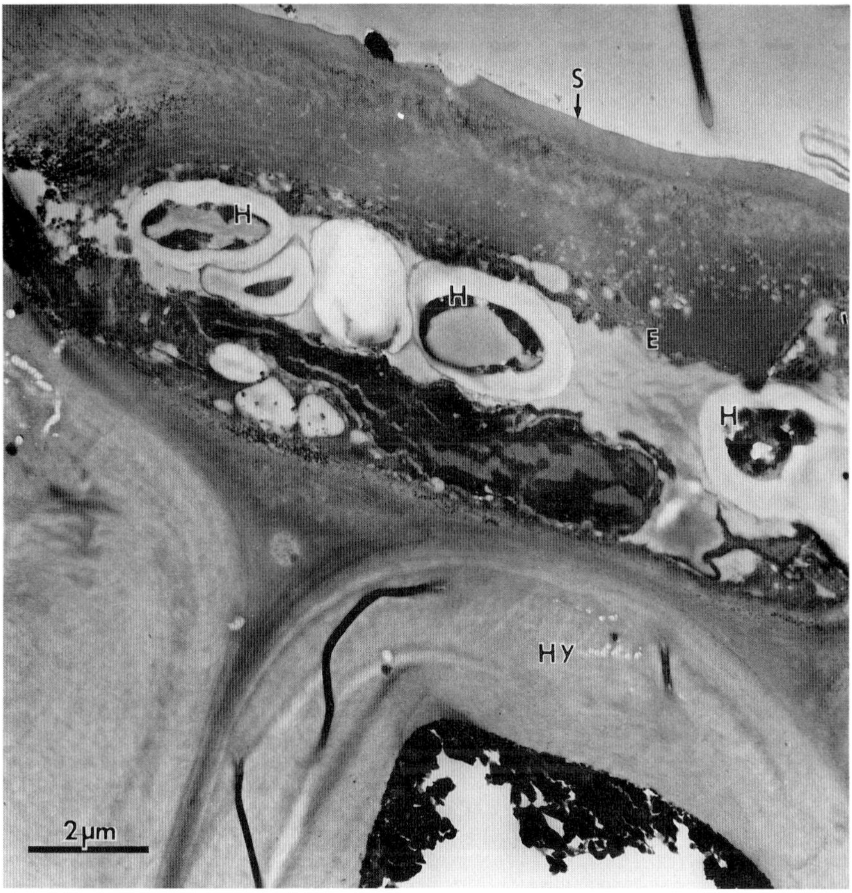

Fig. 3. Epidermal cell of an asymptomatic needle colonized by endophytic hyphae (*H*). Epidermis (*E*), hypodermis (*HY*), needle surface (*S*). TEM 8500 x

of endophytic hyphae can vary strongly and that hyphal cells can form an irregularly shaped cluster of cells. Again endocell formation can be observed. In general, the endophytic infection loci detected in asymptomatic needles were found to consist of only a few hyphae, localized in near-surface needle tissues. In contrast, brown needles taken from diseased trees and prepared in the same way were observed to be heavily colonized by *L. piceae* (Fig. 7).

4 Needle Fungi and Forest Decline

Green Norway spruce needles of all age classes examined were found to host endophytic fungi without any symptoms of disease. The major colonizer of these needles was *L. piceae*. Young needles revealed relatively low infection

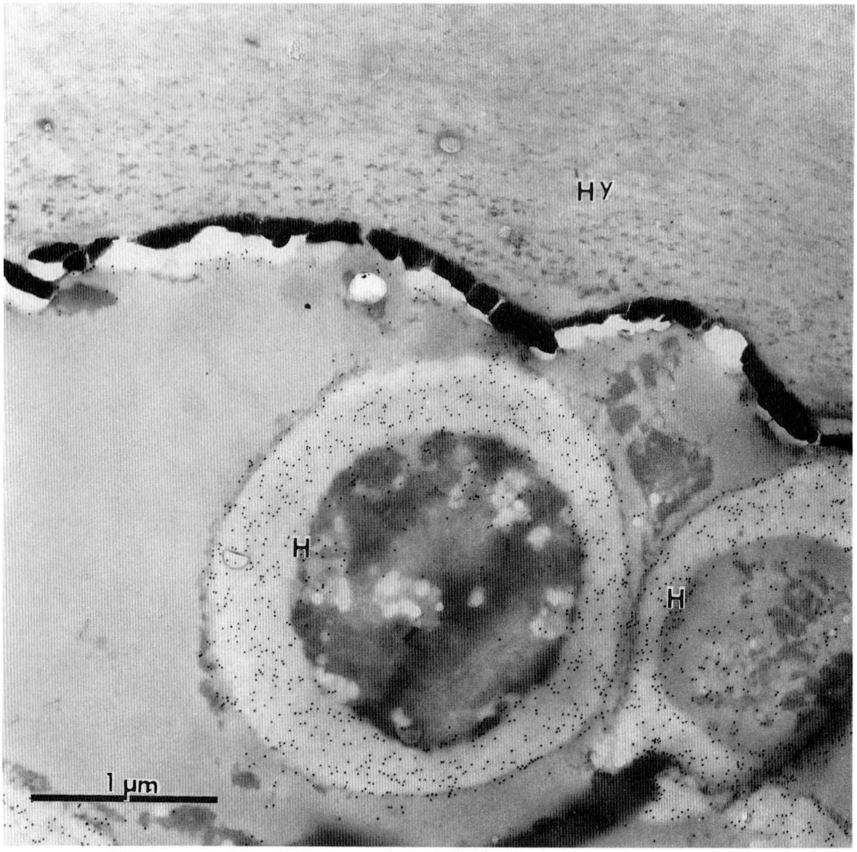

Fig. 4. Infected needle hypodermis (*HY*) after indirect immunogold labeling using *L. piceae*-antiserum. *Dark dots* represent gold particles. Hyphae (*H*). TEM 26500x

frequencies, whereas older foliage was more frequently infected. This has also been described for other endophytes (Bernstein and Carroll 1977; Petrini and Carroll 1981). In addition to needle age, the amount of exposure to light is important in the infection of green needles by *L. piceae*. Significantly higher infection rates were determined for symptomless needles harvested from permanently shaded internodes.

From both our culture studies and electron microscopic observations, we concluded that fungal infections by *L. piceae* in green needles were limited to small areas of near-surface tissues. Only dead, brown needles were heavily colonized by *L. piceae*. This suggests that the expansion of endophytic infection loci was limited in green needles in the course of the host-fungus interaction. Shaded needles seemed to provide more suitable conditions for fungal colonization than those exposed to sun. The longer runs of infection found in permanently shaded needles suggest that either the endophytic hyphae had

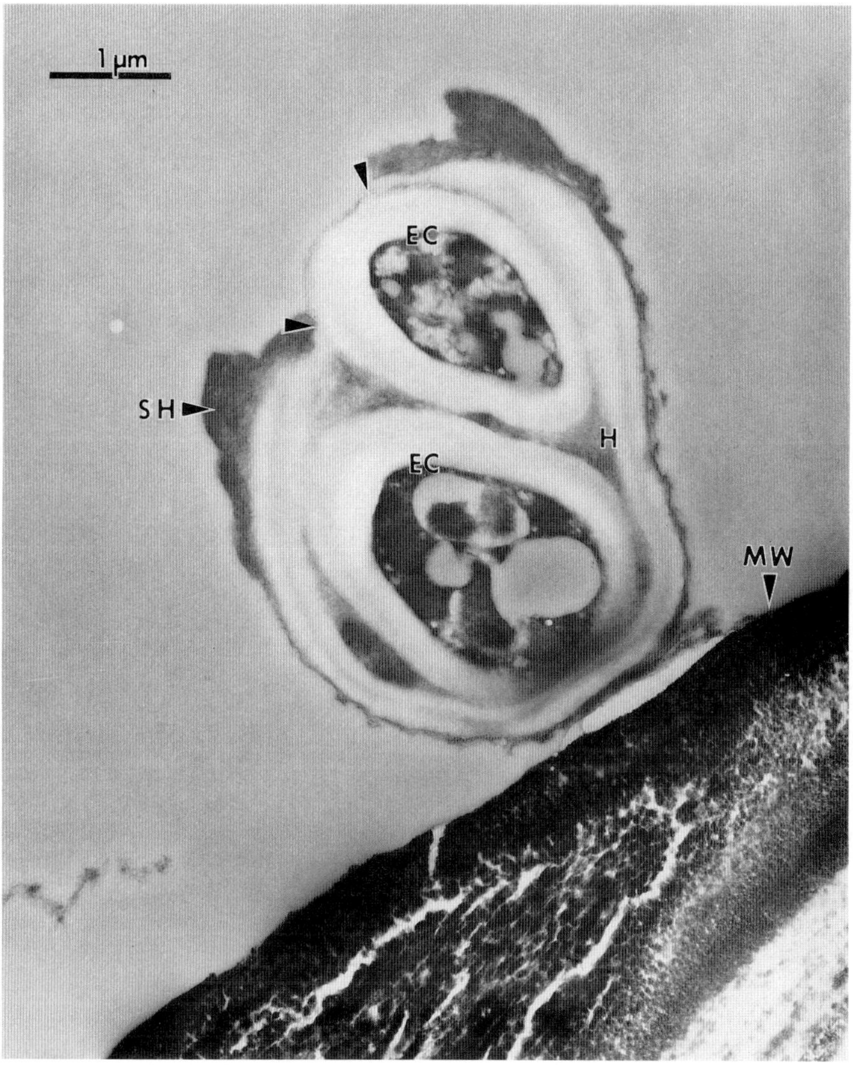

Fig. 5. Intercellular hypha (*H*) containing two endocells (*EC*). In the region between the arrowheads the hyphal cell wall is interrupted. Mesophyll cell wall (*MW*), electron-dense sheath (*SH*). TEM 17500x

begun to further spread in these needles, or the needles had been colonized by a large number of separate infections that could not be distinguished by the cultivation technique applied (Stone 1985).

The most interesting SEM observation concerns the shape and fine structure of endophytic hyphae with respect to irregularly shaped swellings along endophytic hyphae (Fig. 1, and Suske and Acker 1987a). In this sections these swellings probably correspond to endocell-containing hyphae (Fig. 5) or clusters of fungal

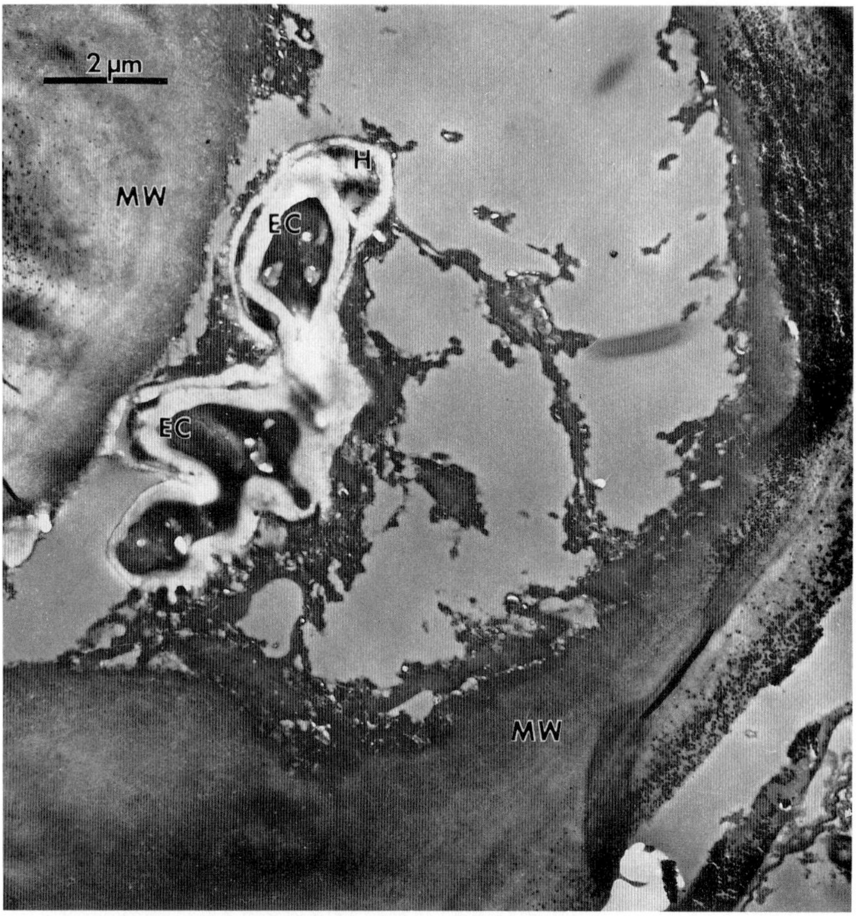

Fig. 6. Intracellular hyphae (*H*) forming an irregularly shaped cluster of cells surrounded by desintegrated mesophyll cell content. Endocells (*EC*), mesophyll cell wall (*MW*). TEM 8500x

cells (Fig. 6). Their shape and size were comparable when both techniques were applied. A similar relation between "dense packets of interwoven hyphae" and hyphal swellings with intrahyphal hyphae has been previously described for *Endogone pisiformis* (Link) Fr. (Jabaji-Hare and Charest 1987). Endocells or intrahyphal hyphae were observed for a number of fungi. Their initiation has been identified as a normal feature of growth (Hammill 1972) as well as to being related to increasing age of fungi (Chan and Stephen 1967) and microaerobic growth conditions (Miller and Anderson 1961; Calonge 1968; Kendrick and Molnar 1968). Most researchers agree that the occurrence of intrahyphal hyphae is a result of abnormal or stressful conditions, either from some deleterious action on fungal cells or very slow fungal growth (Pendland 1981; Benhamou and Ouellette 1986). Therefore, the observed endocells in spruce needles (Figs. 5, 6)

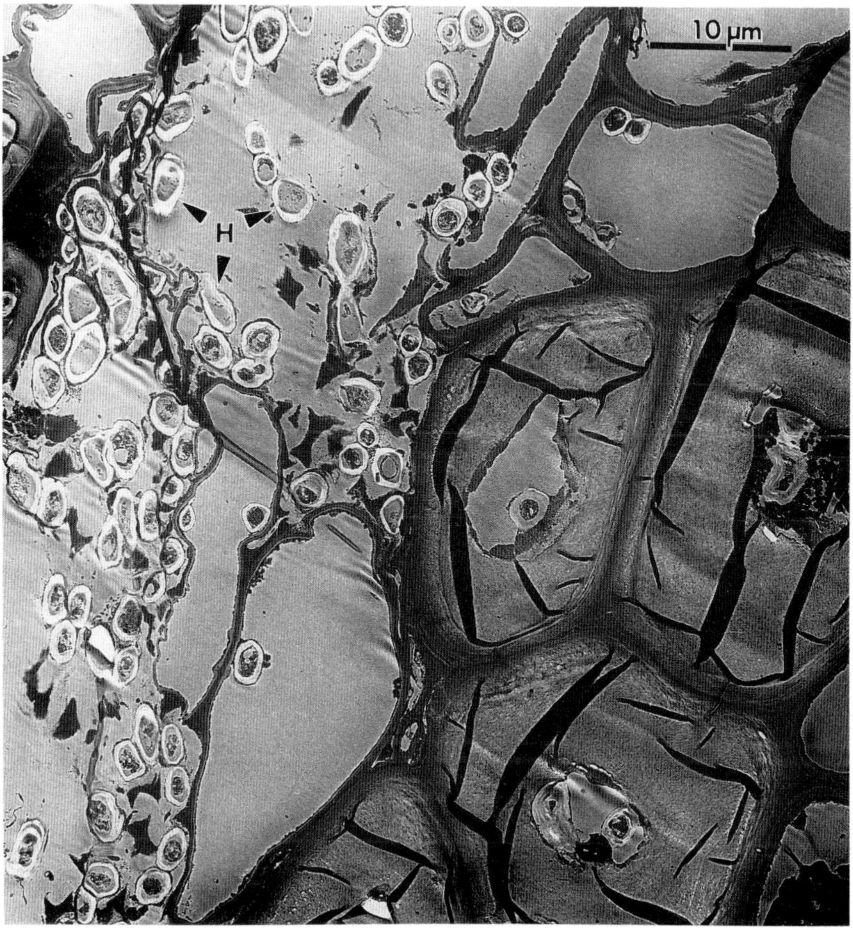

Fig. 7. *L. piceae* heavily colonizing central tissues of a brown needle. Hyphae (*H*). TEM 2000x

may represent resistance structures established as a response of the endophytic hyphae to the limitation of fungal growth in the plant tissues. Although cell structures of infected cells were disintegrated (Fig. 6), infected needles maintained their green, symptomless appearance. Damage caused by the sparcely distributed hyphae remained invisible even on needles multiply infected by *L. piceae*.

L. piceae was also isolated as the major colonizer of needles with different symptoms of disease (Table 3). However, the infection frequency of diseased foliage indicated that these symptoms appeared on needles irrespective of whether or not they had been infected by fungi. In addition to *L. piceae, R. kalkhoffii* was frequently observed. This species colonized only those needles with different disease symptoms or dead foliage. This supports Butin's (1986) observation that *R. kalkhoffii* is an early colonizer of dying or dead needles and does not infect green, healthy foliage.

Our studies of needle-reddening disease (Table 5) demonstrate that the majority of needles examined had already become affected by the disease before they were infected by leaf fungi. The major colonization of these needles occurred saprophytically by *R. kalkhoffii*, but only after the decline of foliage in late autumn when the infection ability of *R. kalkhoffii* is maximum (Diamandis 1978). Taken together, our isolation data suggest that the examined needle symptoms and needle-reddening disease were not caused by *L. piceae* nor *R. kalkhoffii*, and that these fungi alone were not responsible for the Norway spruce decline in the area under investigation. Similar results have already been reported for other localities (Butin and Wagner 1985, Schöpfer and Hradetzky 1984). It must be pointed out, however, that the impact of leaf fungi, especially of endophytes on Norway spruce needles, is still little understood, especially if the endophytic flora of Norway spruce needles plays a role in protecting needles from virulent pathogens (Matta 1971; Kuc and Hammerschmidt 1978) or herbivores (Carroll 1986). Whether needle endophytes are influenced by acid rain (Barklund and Rowe 1983), or whether endophytic fungi of Norway spruce may become virulent to trees once weakened by unfavorable environmental factors, influences stand vitality and ecological fitness.

5 Conclusions

By combining culture studies with ultrastructural and immunocytochemical observations, an additional level of information was obtained concerning the in vivo relationship between internal hyphae (endophytes) of *L. piceae* (Ascomycetes) and needle tissues of *Picea abies*.

Results from culture observations clearly demonstrate that *L. piceae* was the major colonizer of green, symptomless needles. The infection frequency of asymptomatic needles harvested from apparently healthy trees and from declining trees was similar, increasing with needle age from about 1% in the current year's needles to roughly 30% in 4- to 5-year old needles. The infection rate increased greatly in shaded needles from a dense stand ranging from 53% in the current year's needles to 87% in 3- to 4-year old needles. No correlation could be observed between fungal infections and the needle symptoms observed, including those for needle-reddening disease. *R. kalkhoffii* was found to be the most important colonizer of declining or dead foliage.

Intercellular and intracellular hyphae of *L. piceae* in green, symptomless spruce needles were observed for the first time in our laboratory by scanning electron microscopy (Suske and Acker 1987a). Irregularly shaped swellings along internal hyphae were identified in thin sections as containing endocells (intrahyphal hyphae). Endophytic infections consisted of only a few hyphae in near-surface needle tissues. Only brown needles were heavily colonized by *L. piceae*. The observation that in green needles the infection loci were limited to small tissue areas may explain why infected needles remained symptomless.

Immunogold staining, applied for specific identification of hyphae of *L. piceae* in needle tissues, contributed to the culture procedure used to identify this fungus. This technique provides a general procedure for detecting and localizing microbial constituents in infected host tissues, opening new perspectives in studying the poorly understood host-parasite interaction of this and other systems at the ultrastructural level.

References

Acker G (1988) Immunoelectron microscopy of surface antigens (polysaccharides) of gram-negative bacteria using pre- and post- embedding techniques. In: Mayer F (ed) Methods in microbiology: electron microscopy. Academic Press, Lond NY, 20:147–173

Altman LG, Schneider BG, Papermaster DS (1984) Rapid embedding of tissues in Lowikryl K4M for immunoelectron microscopy. J Histochem Cytochem 32:1217–1223

Barklund P, Rowe J (1983) Endophytic fungi in Norway spruce-possible use in bioindication of vitality. Aquilo Ser Bot 19:228–232

Benhamou N, Ouellette GB (1986) Ultrastructure study and cytochemical investigation, by means of enzyme-gold complexes, of the fungus *Ascocalyx abietina*. Can J Bot 65:168–181

Bernstein ME, Carroll GC (1977) Internal fungi in old-growth Douglas for foliage. Can J Bot 55:644–653

Butin H (1983) Krankheiten der Wald- und Parkbäume. Thieme, Stuttgart NY

Butin H (1986) Endophytische Pilze in grünen Nadeln der Fichte (*Picea abies* Karst.). Z Mykol 52:335–346

Butin H, Wagner C (1985) Mykologische Untersuchungen zur Nadelröte-Erkrankung der Fichte. Forstwiss Centralbl 3–4:178–186

Calonge FD (1968) Origin and development of intrahyphal hyphae in *Sclerotinia fructigena*. Mycologia 60:932–942

Carroll GC (1986) The biology of endophytism with particular reference to woody perennials. In: Fokkema NY, Van den Heuvel J (eds) Microbiology of the phyllosphere. Cambridge Univ Press, Lond, pp 205–222

Chan C, Stephen RC (1967) Intrahyphal hyphae in the genus *Linderina*. Can J Bot 45:1995–1998

Darker GD (1932) The hypodermataceae of conifers. Contrib Arnold Arbor Harvard Univ 1:1–131

Darker GD (1967) A revision of the genera of the hypodermataceae. Can J Bot 45:1399–1444

Diamandis S (1978) "Top-dying" of Norway spruce, *Picea abies* (L.) Karst. with special reference to *Rhizosphaera kalkhoffii* Bubak. Eur J For Pathol 8:345–356

Hammill TM (1972) Fine structure of annellophores. II. *Monotosporella shaerocephala*. Can J Bot 50:581–585

Huber SJ, Hock B (1985) A solid-phase enzyme immunoassay for quantitative determination of herbicide terbutryn. J Plant Dis Prot 92(2):147–156

Jabaji-Hare SH, Charest PM (1987) Ultrastructural and cytochemical observation on the somatic phase of *Endogone pisiformis* (*Endogonaceae*). Mycologia 79(3):433–444

Kandler O (1984) Woran der Wald krank ist. Die politische Meinung 29 (Nov/Dez):4–9

Kendrick WB, Molnar AC (1968) A new *Ceratocystis* and its *Verticicladiella* imperfect state associated with the bark beetle *Dryocoetes confusus* on *Abies lasiocarpo*. Can J Bot 43:339–43

Kuc J, Hammerschmidt R (1978) Acquired resistance to bacterial and fungal infection. Ann Appl Biol 89:313–317

Matta A (1971) Microbial penetration and immunization of uncongenial host plants. Annu Rev Phytopathol 9:387–410

Miller CV, Anderson NA (1961) Proliferations of conidiophores and intrahyphal hyphae in *Aspergillus niger*. Mycologia 53:433–436

Pendland JC (1981) Resistant structures in the entomogenous hyphomycete, *Nomuraea rileyi*: an ultrastructural study. Can J Bot 60:1569–1576

Petrini O (1986) Taxonomy of endophytic fungi of aerial plant tissues. In: Fokkema NY, Van den Heuvel J (eds) Microbiology of the phyllosphere. Cambridge Univ Press, Lond, pp 167–187

Petrini O, Carroll GC (1981) Endophytic fungi in foliage of some *Cupressaceae* in Oregon. Can J Bot 59:629–636

Rack K, Butin H (1984) Experimenteller Nachweis nadelbewohnender Pilze bei Koniferen. Eur J For Pathol 14:302–310

Rehfuess KE, Rodenkirchen H (1984) Über die Nadelröte-Erkrankung der Fichte in Süddeutschland. Forstwiss Centralbl 103:248–262

Rehfuess KE, Rodenkirchen H (1985) Über die Nadelröte-Erkrankung der Fichte in Süddeutschland. Forstwiss Centralbl 6:381–390

Reynolds DS (1963) The use of lead citrate at high pH as an electron opaque stain in electron microscopy J Cell Biol 17:208–212

Rostrup E (1891) Undersogelser over Snyltesvampes Angreb paa Skovpraeer. 1883–1888. Tidsskr Skovbrug 12:175–283

Schöpfer W, Hradezky J (1984) Der Indizienbeweis: Luftverschmutzung maßgebliche Ursache der Walderkrankung. Forstwiss Centralbl 103:231–248

Schütt P (1985) Das Waldsterben – eine Pilzkrankheit? Forstwiss Centralbl 104:169–177

Slot JW, Geuze HJ (1985) A new method of preparing gold probes for multiple-labeling cytochemistry. Eur J Cell Biol 38:87–93

Stone JK (1985) Foliar endophytes of *Pseudotsuga mensziesii* (Mirb.) Franco. Cytology and physiology of the host-endophyte relationship. PhD thesis, Eugene, Univ Oregon

Suske J, Acker G (1987a) Internal hyphae in young, symptomless needles of *Picea abies*: electron microscopic and cultural investigation. Can J Bot 65:2098–2103

Suske J, Acker G (1987b) Electron-microscopical investigation on the interaction of endophytic hyphae and tissues of Norway spruce needles. Eur J Cell Biol 44 (Suppl 19):56

2-C Soil Fauna Comparisons in Healthy and Declining Norway Spruce Stands

P. HARTMANN, M. SCHEITLER, and R. FISCHER

1 Introduction

The several hundred species of soil fauna which take part in different terrestrial decomposition processes either consume dead plant material (saprophagous), mutually interact with microbial organisms (microphytophagous), or prey upon other organisms (carnivorous). While soil fauna communities in the European spruce forests perform this general function, data on their specific structure and function are limited to some studies from Scandinavia (Abrahamsen 1972a,b; Huhta and Koskenniemi 1975; Nordström and Rundgren 1973; Petersen and Luxton 1982) and West Germany (Thiede 1977; Ellenberg et al. 1986; Funke 1986).

Researchers have therefore not yet understood whether and to what extent soil fauna communities or single populations in spruce forests are affected by forest decline. In 1985 and 1986 a comparative study was conducted in healthy and declining spruce forests in North Bavaria to examine the differences between the sites for indications of environmental changes possibly caused by acid deposition and soil acidification.

2 Sampling Procedure and Site Description

The sampling sites in the Fichtelgebirge were ca. 100-year-old spruce stands at either a healthy (Wülfersreuth) or a declining (Oberwarmensteinach) site on oligotrophic acid soils (geological subsoil: phyllite). On the forest floor, small unvegetated litter patches alternate with areas covered by *Deschampsia flexuosa, Vaccinium myrtillus* and *Sphagnum spp*.

Acid input is higher in the declining forest (for detailed data on a nearby 40-year-old spruce stand, see Chap. 4-I, this Vol.). Soil chemistry data on the litter layers at the two sites given in Table 1 indicate that pH and Ca were higher and K was lower at the healthy site.

For quantitative density estimations, different methods of extracting litter samples were used. Macrofauna, Mesofauna and Enchytraeidae were extracted according to Kempson et al. (1963), MacFadyen (1962), and O'Connor (1955), respectively. The number and size of random samples varied according to the groups of animals. Samples were taken monthly from May to October. The

E.-D. Schulze, O.L. Lange, R. Oren (Eds.)
Ecological Studies Vol. 77

Table 1. Soil chemistry data from the litter layer at the healthy and the declining sites

	Healthy site	Declining site
pH (in $CaCl_2/O_F$-layer)	3.2	2.8
S (mol l^{-1}) (total amount O_F-layer)	68.03	71.83
Ca (mol l^{-1}) (from percolation extracts)	24.2	12.0
Mg (mol l^{-1}) (from percolation extracts)	7.4	6.2
K (mol l^{-1}) (from percolation extracts)	16.4	34.0

Table 2. Number and size of samples for the estimation of meso- and macrofauna and annual emergence rates of pterygote insects

Organisms	No. of samples	Sample size
Macrofauna	6	Surface: 625 cm^2 (O_L- to O_H-layer)
Mesofauna	6	Depth: 4 cm Surface: 20 cm^2
Enchytraeidae	12	Depth: 9 cm (divided into 3 layers, each 3 cm) Surface: 20 cm^2
Pterygote insects (annual emergence rates)	6	1 m^2 (photoeclectors)

emergence rates of pterygote insects were determined by means of photoeclectors (Funke 1972; Table 2).

Comparative decomposition rates were determined for the healthy and declining sites by means of litterbags with rapidly decomposable leaves. In 1985 and 1986 these were *Urtica dioica* and *Alnus glutinosa*, respectively. The litterbags were embedded in the upper litter layers of unvegetated areas from July to October (Table 3). In doing this, we did not intend to investigate the actual decomposition rates of spruce litter, but to obtain data on the decomposition potential of the existing fauna, given the same easily digestible food supply. From these short-term experiments, information concerning the total decomposer activity and the efficiency of the different groups of organisms were derived.

The faunal taxa investigated were grouped according to trophic levels. An estimation of the faunal biomass was calculated from mean densities of individuals in the sites and average individual weights per taxon (mg dry wt.) based on published data (Table 4). These calculations are approximations, but the results are sufficient for the site comparisons.

Most soil fauna populations exhibited a very clumped distribution. Because errors of mean density estimates were usually rather high, all histograms were done without confidence intervals. The statistical comparisons are based on

Table 3. Different types of litterbags and decomposition effects

Mesh size (mm)	No. of samples	Decomposition
0.2	10	Abiotic + microbial
0.9	10	Abiotic + microbial + mesofaunal
5.0	10	Abiotic + microbial + total faunal

Table 4. Biomass (mg dry wt./Ind.) of investigated taxa

Trophic level	Taxa	Biomass (mg)	Reference[a]
Macrosaprophagous	Lumbricidae	17.0	1
	Diplopoda	1.4	7
Microphytophagous-saprophagous	Enchytraeidae	$32.0*10^{-3}$	1
	Cryptostigmata	$5.3*10^{-3}$	1
	Collembola	$2.7*10^{-3}$	1
	Protura	$0.2*10^{-3}$	2
	Symphyla	$8.0*10^{-3}$	2
Omnivores	Elateridae (Larvae)	2.1	4,5
Predators	Araneae	0.3	2
	Gamasina	$7.7*10^{-3}$	1
	Diplura	$6.2*10^{-3}$	2
	Lithobiidae	1.7	3
	Staphylinidae	0.25	6
	Cantharidae (Larvae)	0.3	6

[a] Authors of biomass data: 1 = Petersen and Luxton (1982); 2 = Persson and Lohm (1977); 3 = Poser (1988); 4 = Ellenberg et al. (1986); 5 = Strey (1972); 6 = Thiede (1977); 7 = Boller (1986).

robust nonparametric tests (Mann-Whitney U-Test; Sachs 1974). The aim of this study was not to determine absolute densities within the sites, but to look for consistent differences between them.

3 Species Composition and Density of Healthy and Declining Norway Spruce Stands

3.1 Macrosaprophagous Invertebrates

At both sites, macrophagous animals were rare. Diplopoda and Lumbricidae occurred with significantly lower densities ($p < 0.01$) at the declining site (Fig. 1).

The species spectra for Lumbricidae (*Lumbricus rubellus, Dendrobaena octaedra, Dendrobaena rubida*) and Diplopoda (*Orthochordeuma germanicum, Orobainosoma germanicum*) represent biological exceptions within their taxa. In contrast to the preponderance of basophilous species in the two taxa, these species are very acid-resistant. Nevertheless, they obviously cannot tolerate conditions at the declining site.

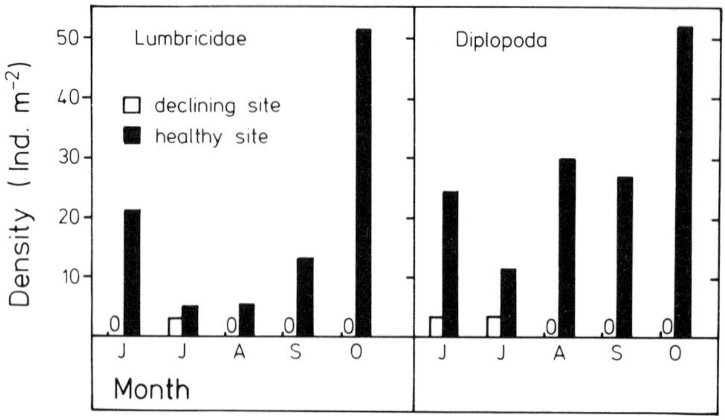

Fig. 1. Density fluctuations of macrosaprophagous soil invertebrates in healthy and declining spruce forests (Fichtelgebirge 1985)

3.2 Microphytophagous – Saprophagous Invertebrates

3.2.1 Enchytraeidae

Three species of Enchytraeidae (*Cognettia sphagnetorum, Mesenchytraeus pelicensis, Achaeta cf. brevivasa*) were found; *C. sphagnetorum* had the greatest biomass of all mesofaunal or macrofaunal species in the litter fauna of the Fichtelgebirge.

The enchytraeid fauna of oligotrophic acid spruce stands in Northern and Middle Europe are characterized by a poor species spectrum with a very high dominance of *C. sphagnetorum* (80–95%) (Petersen and Luxton 1982; Lundkvist 1982; Abrahamsen 1972a). The latter is an acidophilous species which reproduces asexually by fragmentation, thus enabling it to recover rapidly from population declines caused by unfavorable conditions (Lundkvist 1982).

Between August 1985 and October 1986, the density of Enchytraeidae in the declining stand was significantly ($p < 0.01$) higher than at the healthy site (Fig. 2) (Wagner 1987) as a result of a greater abundance of *C. sphagnetorum* and the subdominant *Achaeta spec.*

3.2.2 Acari – Cryptostigmata

As in most European forest soils, the Cryptostigmata constitute the main component of the acarine fauna in the Fichtelgebirge spruce forests with the highest population densities of all the microarthropods. Thus all other microphytophagous – saprophagous acarine taxa (5–15% of all Acari excl. Gamasina) were neglected in order to emphasize the importance of Cryptostigmata (= Oribatei).

Both sites in the Fichtelgebirge were very similar in their cryptostigmatic species spectra; of the 29 total species found, 25 species were in the declining stand, 24 species in the healthy stand, and 20 species common to both stands. The dominant species were *Tectocepheus velatus, Steganacarus spinosus, Suctobelba subtrigona, Nothrus silvestris, Chamobates cuspidatus, Camisia spinifer* and *Fuscozetes setosus.* All dominant species are known to occur in a great variety of

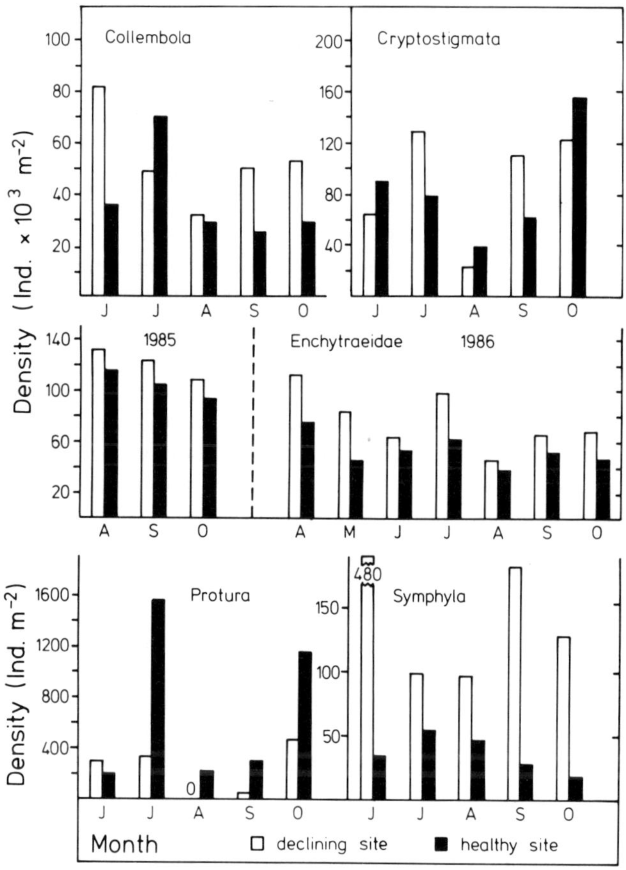

Fig. 2. Density fluctuations of microphytophagous – saprophagous soil invertebrates in healthy and declining spruce forests (Fichtelgebirge 1985/1986)

different habitats and are usually common in coniferous forests. According to Hagvar and Amundsen (1981) *T. velatus,* and *N. silvestris* are regarded as acidophilous. Most of the Cryptostigmata found (56–65% of the species) were microphytophagous species which feed on fungi and algae. Only 20–25% were macrophytophagous species which consume dead organic material. The remaining species (10–15%) were nonspecialized feeders (Schuster 1956). The present data of the standing crop of Cryptostigmata do not suggest differences ($p > 0.1$) between the two sites (Fig. 2).

3.2.3 Collembola

Based on data from May-September 1985, 36 total Collembola species were identified, with 26 species at the declining site, 29 species at the healthy site, and 18 species common to both sites.

The dominant species, *Folsomia quadrioculata* and *Tullbergia krausbaueri*, with ca. 50 and 18% respectively of total collembolan density, showed no significant differences between stands. Subdominant species with ca. 1–6% of total collembolan density in both stands were *Onychiurus armatus, Isotomiella minor* and *Tomocerus vulgaris*. All these species are very euryoecous and dominate in a wide range of terrestrial ecosystems. *F. mirabilis* and *M. pygmaea,* which are regarded as acidophilous (Hagvar and Abrahamsen 1984), were several times higher in density at the declining site. The density of the collembolan community in the declining stand shows a tendency ($p = 0.08$) towards greater abundance than at the healthy site (Fig. 2), but the total collembolan density is affected by fluctuations of *Folsomia quadrioculata.*

3.2.4 Symphyla – Protura

Little is known about Symphyla and Protura; in different spruce forests in North and South Germany, 12 Protura species have been recorded, 6 of them undescribed (Funke 1986). In our study, only individuals of the Genus *Eosentomon (E. transitorium* – group; see Nosek 1973) were identified.

Protura seem to be more or less specialized feeders on ectotrophic mycorrhiza (Sturm 1959) and they are supposed to be sensitive indicators of pollution deposition (Nosek 1982; Funke 1986). The mean density of Protura was significantly ($p < 0.02$) lower at the declining (254.5 Ind. m^{-2}) than at the healthy (855.1 Ind. m^{-2}) site (Fig. 2). Symphyla seem to be unspecialized feeders on dead organic material and microflora (Dunger 1983). In contrast to Protura, the Symphyla (species not yet identified) at the declining site had significantly higher densities ($p < 0.01$) than at the healthy site (Fig. 2).

3.2.5 Diptera

Litter-dwelling larvae of Diptera are important members of different soil-fauna food chains (Altmüller 1976; Hövemeyer 1985; Thiede 1977). Comparative density estimations in the study area were confined to annual emergence rates of Diptera-imagos from the soil.

As is typical for oligotrophic acid forest soils, Nematocera made up more than 90% of the total annual dipteran emergence rate (Brachycera < 10%) in both stands. A total of 27 Diptera families were identified: 25 families at the healthy site, 16 at the declining site, and 14 common to both sites. The populations of the generally microphytophagous – saprophagous *Sciaridae* and *Cecidomyidae* made up 75% of the total annual emergence density at the healthy site, which was 3–4 times higher than in the declining site. As shown in Fig. 3, this difference was caused by a lower hatching density of all dominant families (except *Mycetophilidae)* at the declining site. In comparison with available data, the emergence density at the declining site was very low (Hövemeyer 1985).

3.3 Omnivorous Invertebrates

Those animals classified as omnivores feed on living as well as dead organic materials. The only major representatives of this group were Elateridae – larvae

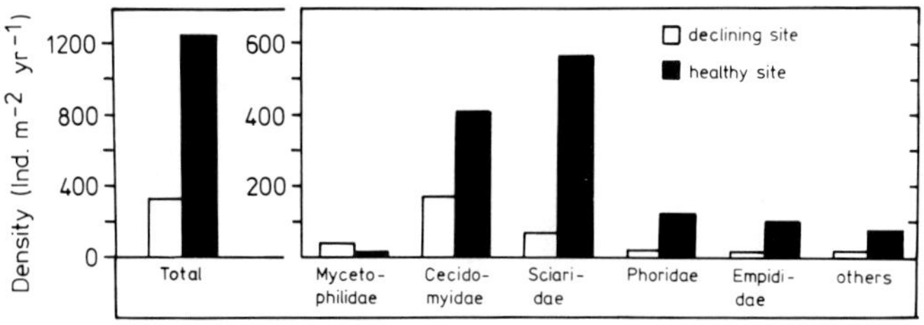

Fig. 3. Annual emergence rates of Diptera (imagos) in healthy and declining spruce forests (Fichtelgebirge 1985)

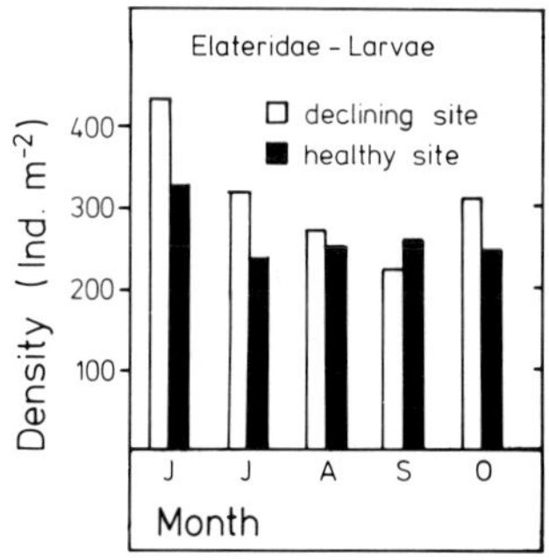

Fig. 4. Density fluctuations of the omnivorous Elateridae-larvae in healthy and declining spruce forests (Fichtelgebirge 1985)

(Ellenberg et al. 1986) which were dominated by the larvae of *Athous subfuscus* (in addition to single larvae of *Athous zebei* and *Dalopius marginatus*). *A. subfuscus* needs 5–7 years for its larval development (Strey 1972). Therefore, at any time, all the larval stages of several years coexist in the litter. Mean densities ranged from 200–300 Ind. m^{-2}), with emergence rates of adults less than 10 Ind. m^{-2}) yr^{-1}. Because of their relatively greater biomass and their ability to use different food resources, the Elateridae larvae occupy an important position within the soil macroinvertebrates of the spruce forests studied. They were found in significantly ($p = 0.02$) higher densities in the declining stand (Fig. 4).

3.4 Predators

At present the taxonomic analysis of predatory arthropods is still incomplete. For the site comparisons only higher taxa were used. The monthly density estimates (Fig. 5) showed that at the declining site, most of the carnivorous taxa were lower in density than at the healthy site. Significant differences were determined for Araneae ($p = 0.03$) and Lithobiidae ($p = 0.01$). Gamasina and Diplura appeared to be decreasing ($p = 0.06$ and $p = 0.08$) in the declining stand. Only Cantharidae – larvae (*Malthodes spp., Malthinus sp.)* occurred in greater numbers ($p = 0.02$) in the declining stand. There were no significant differences for the Geophilidae ($p > 0.1$) or Staphylinidae ($p > 0.1$).

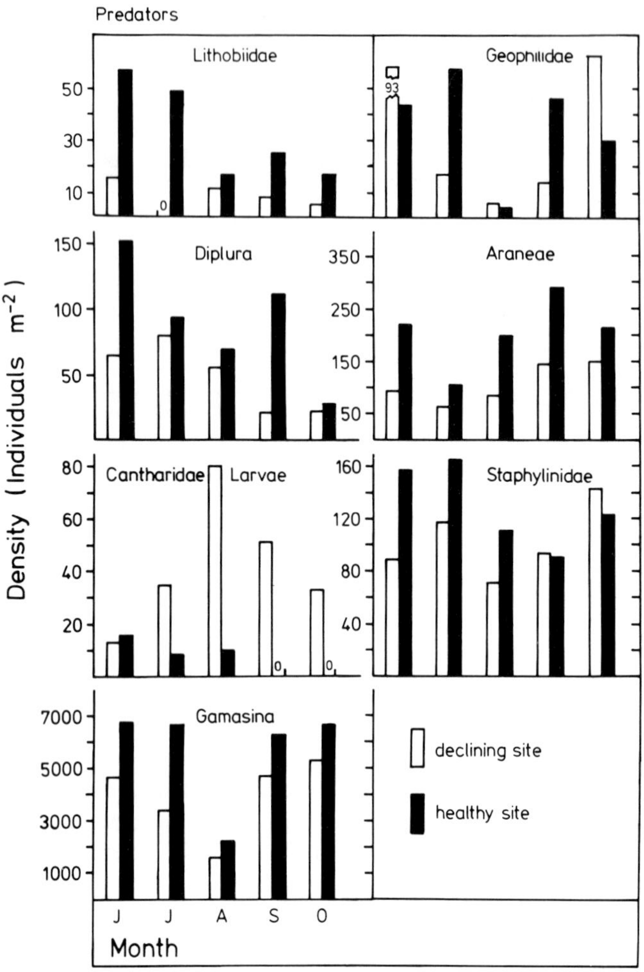

Fig. 5. Density fluctuations of carnivorous soil invertebrates in healthy and declining spruce forests (Fichtelgebirge 1985)

3.5 Comparison of Soil Fauna Biomass

The mean total biomass data (Table 5, Fig. 6) indicated a great dominance of microphytophagous – saprophagous invertebrates in the soil fauna of both spruce stands, with 63 and 77% of total biomass for the healthy and declining stands, respectively. Within this group, the mean biomass of Enchytraeidae was highest, followed by that of Cryptostigmata.

Macrophagous- and most of the carnivorous-invertebrate taxa were less well represented in the declining stand. This difference was more than compensated by the greater biomass of microphytophagous-saprophagous animals.

Table 5. Mean density (Ind m^{-2}, May-Sept.) and mean biomass (mg dry wt. m^{-2}) of soil invertebrates at the healthy and declining sites

Taxon	Density (Ind. m^{-2}) Declining stand	Healthy stand	Biomass (mg dry wt. m^{-2}) Declining stand	Healthy stand
Lumbricidae	1	24	17	408
Enchytraeidae	69427	47947	2222	1534
Collembola	55349	39299	149	106
Cryptostigmata	95096	91274	504	484
Gamasina	3823	5634	29	43
Symphyla	172	33	1	0.3
Lithobiidae	7	30	12	51
Geophilidae	41	34	74	61
Diplopoda	2	32	3	45
Protura	234	692	0.05	0.14
Diplura	44	80	0.3	0.5
Elateridae (Larvae)	308	260	647	546
Cantharidae (Larvae)	41	8	12	2
Staphylinidae	111	129	28	32
Araneae	117	210	35	63
Sum (mg dry wt. m^{-2})			3733	3377

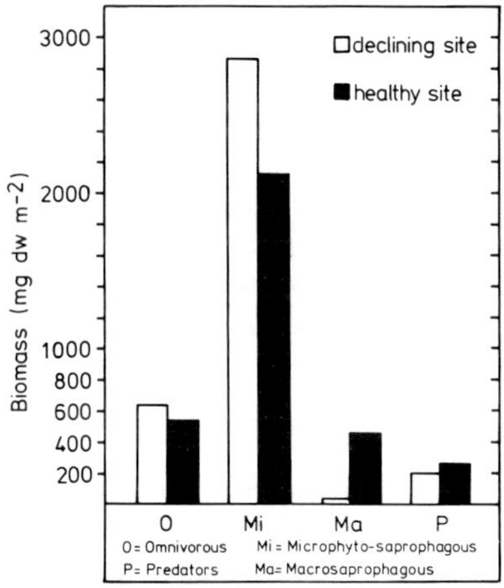

Fig. 6. Comparisons of mean soil fauna biomass (May – September) of healthy and declining spruce forests (Fichtelgebirge 1985)

Estimations of the greater mean total biomass for the declining site may not be significant, considering the calculation assumptions and the omission of several taxa (Protozoa, Nematodes, Diptera). From annual mean emergence rates of Diptera imagos for instance we would expect the Diptera-larvae density and biomass to be almost three times higher at the healthy site. Despite the uncertainties, the comparisons described here show that damage in the declining stand is not simply correlated with a decrease in total soil fauna biomass.

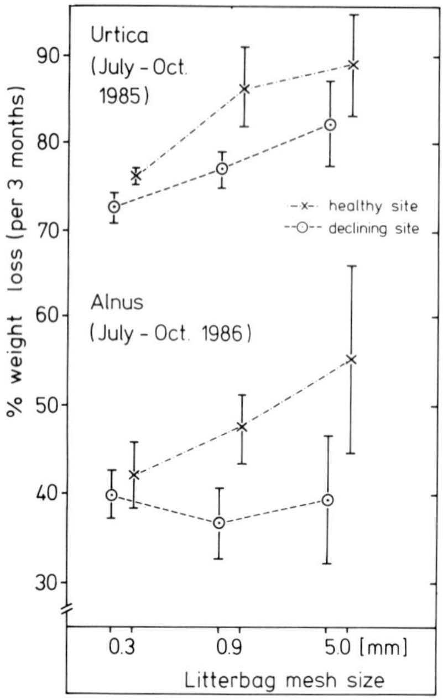

Fig. 7. Decomposition rates of *Urtica* and *Alnus* leaves (litterbags) in healthy and declining spruce forests (Fichtelgebirge 1985/1986)

3.6 Comparison of Decomposition Rates

Decomposer activities studied in 1985 with *Urtica* leaves and in 1986 with *Alnus* leaves indicated higher total decomposition rates at the healthy site (Fig. 7). Significant differences were apparent in litterbags with medium mesh-size as a result of the combined activity of microflora and mesofaunal invertebrates. The contribution of macrofaunal animals remained variable over three months because of highly aggregated distributions.

4 Acid Rain and Soil Fauna

If only the species spectra are considered, the soil fauna of both spruce stands are not very different. Both sites exhibit a composition which is typical for oligotrophic acid coniferous forests of Middle and Northern Europe (Petersen and Luxton 1982; Ellenberg et al. 1986; Huhta and Koskenniemi 1975). Such fauna are characterized by high mean densities of individuals but low total biomass as a result of the dominance of relatively small microphytophagous-saprophagous invertebrates, particularly Enchytraeidae, Cryptostigmata and Collembola. Within most taxa, acidophilous species particularly adapted to acid conditions in raw humus and mor soils dominate along with very euryoecous species, which usually occur in a wide range of terrestrial ecosystems, reaching their highest population densities in spruce forests. The main difference we found in our study was the lower existence of macrosaprophagous- and most carnivorous-invertebrate populations and distinctly higher density of microphytophagous-saprophagous animals in the declining site. In addition organic test material decomposed at the declining site significantly more slowly than at the healthy site.

With this descriptive comparison of healthy and declining spruce stands, the study examined differences in soil fauna composition which might indicate environmental changes from forest damage factors. The spruce litter at the declining site was affected by higher acid deposition (see Chap. 4-I) with less Ca but more K than at the healthy site.

This greater degree of soil acidification may be responsible for the faunal differences between the two stands. This hypothetical assumption is supported by data from experimental perturbations, which measured the response of soil animals to changes in soil chemistry through an increase in acidity, by artificial acid rain, or an increase in base saturation, by liming (Schaefer 1985).

Several studies show that Lumbricidae usually reach the limit of their ecological tolerance in oligotrophic acid spruce forests especially with respect to soil pH and Ca concentration (Abrahamsen 1972b; Nordström and Rundgren 1973). Lumbricidae seem to be dependent on a certain amount of Ca, which is probably necessary for buffering coelomatic fluids or intestinal food contents (Piearce 1972). Progressive acidification with increased loss of Ca and Mg Ions successively deteriorate environmental conditions for these animals. On the opposite, after liming, Lumbricidae increase their population density in spruce stands within 1 year (Hartmann et al. 1987). According to Boesener (1962) and Schauermann (1985), Lumbricidae are capable of changing raw humus layers or spruce litter to mull-like litter within 5–7 years after liming.

Many microphytophagous-saprophagous invertebrate populations respond positively to decreases in Ca. Enchytraeidae, particularly the dominant *Cognettia sphagnetorum*, and some Cryptostigmata and Collembola species increase their densities with artificial acidification even at pH 3.5 or lower (Abrahamsen et al. 1980; Hagvar and Amundsen 1981; Schauermann 1985). With still further acidification ($pH < 2.5$), most populations decline rapidly. In liming experiments, species with positive response to acidification have decreased their densities.

Diptera larvae seem to be sensitive to any perturbations, as emergence rates of Diptera imagos have declined following liming, as well as artificial acidification (Schauermann 1987; Funke 1986; Hartmann et al. 1987).

Carnivorous invertebrates do not seem to react consistently to acidification and available data do not allow generalizations. Decreasing rates of plant material decomposition have occurred in response to artificial acidification and decreased alkalinity (Abrahamsen et al. 1980; Baath et al. 1980; Hagvar and Kjondal 1981b). Only Lee and Weber (1983), using litter of hardwood, and Roberts et al. (1980), using pine needles, have reported higher decomposition rates after acidification. In our own litterbag-experiments with easily decayable plant material, liming always caused an increase in decomposition rates (Hartmann et al. 1987). Effects were most pronounced in litterbags of coarse and medium mesh size, probably as a result of enhanced mutual interactions of microflora (bacteria) and soil fauna.

5 Conclusions

Given the lack of information on the original state of soil fauna composition before forest decline, the differences between our sites appear to be mainly caused by a higher degree at soil acidification at the declining site. As the healthy site undergoes increasing acidification, we would predict that the density of Lumbricidae and the decomposition rate will decrease, while the numbers of some microphytophagous-saprophagous species will increase.

Specific predictions about the further development of the fauna at the declining site cannot be made at present. With the exception of the comparatively low density of Diptera, the composition of the soil fauna is quite similar to that typical for relatively stable mor soils in the cold part of the temperate zone (Petersen and Luxton 1982). Preliminary results from studies in other localities in North Bavaria have indicated that relatively poor fauna are also found in spruce stands on oligotrophic acid soils with only low acid deposition, as a result of natural acid conditions of coniferous raw humus on poor soils. In contrast, richer fauna are found in very damaged and polluted stands on more fertile soils. The chemical conditions of the soil seem to be important in the speed and extent of the effects of acidification. Under the influence of continuing acid deposition, any development of the soil fauna at the declining sites should surely be superimposed by effects of the total forest destruction through subsequent changes in the whole soil community.

Acknowledgements. This project was supported by the Bayrisches Staatsministerium für Landesentwicklung und Umweltplanung (Proj. Nr. 6480-653-49810).

References

Abrahamsen G (1972a) Ecological study of Enchytraeidae (Oligochaeta) in Norwegian coniferous forest soils. Pedobiologia 12:26–82

Abrahamsen G (1972b) Ecological study of Lumbricidae (Oligochaeta) in Norwegian coniferous forest soil. Pedobiologia 12:267–281

Abrahamsen G, Hovland J, Hagvar S (1980) Effects of artificial acid rain and liming on soil organisms and the decomposition of organic matters. In: Hutchinson TC, Howas M (eds) Effects of acid precipitation on terrestrial ecosystems. Plenum Press, NY, pp 341–362

Altmüller R (1976) Zum Energieumsatz von Dipterenpopulationen im Buchenwald (Luzulo-Fagetum). Diss Göttingen, 83 s

Baath E, Berg B, Lohm U, Lundgren B, Lundkvist H, Rosswall T, Söderström B, Wiren A (1980) Effects of experimental acidification and liming on soil organisms and decomposition in a Scots pine forest. Pedobiologia 20:85–100

Bösener R (1962) Die Lumbriciden gekalkter Fichtenbestände des Tharandter Waldes und ihre Bedeutung bei der Umwandlung der Humusauflage. Dtsch Akad Landwirtsch-Wissensch Berlin (Ost), Tagungsber 50:193–203

Boller F (1986) Diplopoden als Streuzersetzer in einem Lärchenwald. Nationalpark Berchtesgaden, Forschungsber 9:87 s

Dunger W (1983) Tiere im Boden. Neue Brehm Bücherei, Wittenberg – Lutherstadt, 3. Aufl, 280 s

Ellenberg H, Mayer R, Schauermann J (eds) (1986) Ökosystemforschung – Ergebnisse des Sollingprojekts 1966–1986. Ulmer, Stuttgart, 507 s

Funke W (1972) Energieumsatz von Tierpopulationen in Landökosystemen. Verh Dtsch Zool Ges Helgoland, s 95–106

Funke W (1986) Tiergesellschaften im Ökosystem "Fichtenforst"- Indikatoren von Veränderungen in Waldökosystemen. Projekt Eur Forschungszentrum (PEF) für Massnahmen zur Reinhaltung der Luft. Kernforschungszentrum Karlsruhe, KfK-PEF 9, 150 s

Hagvar S, Amundsen T (1981) Effects of liming and artificial acid rain on the mite (Acari) fauna in coniferous forests. Oikos 37:7–20

Hagvar S, Kjondal BR (1981) Decomposition of birch leaves: dry weight loss, chemical changes and effects of artificial acid rain. Pedobiologia 22:232–245

Hagvar S, Abrahamsen G (1984) Collembola in Norwegian coniferous forest soils. III. Relations to soil chemistry. Pedobiologia 27:331–339

Hartmann P, Scheidler M, Fischer R (1987) Auswirkungen der Kalkdüngung auf die Bodenfauna von Fichtenforsten. Verh Ges Ökol Göttingen (in press)

Hövemeyer K (1985) Die Zweigflügler (Diptera) eines Kalkbuchenwaldes: Lebenszyklen, Raum-Zeit-Muster und Nahrungsbiologie. Diss Göttingen, 280 s

Huhta V, Koskenniemi A (1975) Numbers, biomass and community respiration of soil invertebrates in spruce forests at two latitudes in Finland. Ann Zool Fenn 12:164–182

Kempson D, Lloyd M, Ghelardi R (1963) A new extractor for woodland litter. Pedobiologia 3:1–21

Lee JJ, Weber DE (1983) Effects of sulfuric acid rain on decomposition rate and chemical element content of hardwood leaf litter. Can J Bot 61(3):872–879

Lundkvist H (1982) Population dynamics of *Cognettia sphagnetorum* (Enchytraeidae) in a Scots pine forest in central Sweden. Pedobiologia 23:21–41

MacFadyen A (1962) Control of humidity in three funnel-type extractors for soil arthropods. In: PW Murphy (ed) Progress in Soil Zoology. Butterworth, Lond, pp 158–168

Nordstroem S, Rundgren S (1973) Associations of lumbricids in southern Sweden. Pedobiologia 13:301-326

Nosek J (1973) The european Protura (their taxonomy, ecology and distribution with keys for determination). Mus Hist Naturelle, Geneve, pp 1–345

Nosek J (1982) Indikationsbedeutung der Proturen. Pedobiologia 24:249–253

O'Connor FB (1955) Extraction of Enchytraeidae worms from a coniferous forest soil. Nature (Lond) 175:815–816

Persson T, Lohm U (1977) Energetical significance of the annelids and arthropods in a Swedish grassland soil. Ecol Bull/NFR 23, 211 pp

Petersen H, Luxton M (1982) A comparative analysis of soil fauna populations and their role in decomposition processes. Oikos 39 3:286–388

Piearce TS (1972) The calcium relations of selected Lumbricidae. J Anim Ecol 41:167–188

Poser T (1988) Chilopoden als Pradatore in linem Lanbwald. Pedobiologia 31:261–281

Roberts TM, Clarke TA, Ineson P, Gray TRJ (1980) Effects of sulfur deposition on litter decomposition and nutrient leaching in coniferous forest soils. In: Hutchinson TC, Havas M (eds) Effects of acid precipitation on terrestrial ecosystems. Plenum Publ NY, pp 381–393

Sachs L (1974) Angewandte Statistik. Springer, Berlin Heidelberg New York

Schaefer M (1985) Waldschäden und die Tierwelt des Bodens. Allg Forstz 27:676–679

Schauermann J (1985) Zur Reaktion von Bodentieren nach Düngung von Hainsimsen – Buchenwäldern und Siebenstern – Fichtenforsten im Solling. Allg Forstz 43:1159–1161

Schauermann J (1987) Tiergesellschaften der Wälder im Solling unter dem Einfluss von Luftschadstoffen und künstlichem Säure – und Düngereintrag. Verh Ges Ökol Giessen 1986, 16:53–62

Schuster R (1956) Der Anteil der Oribatiden an den Zersetzungsvorgängen im Boden. Z Morphol Ökol Tiere 45:1–33

Strey G (1972) Ökoenergetische Untersuchugen an *Athous subfuscus* Müll. und *Athous vittatus* (F.) (Elateridae, Col.) in Buchenwäldern. Diss Univ. Göttingen

Sturm H (1959) Die Nahrung der Proturen, Beobachtungen an *Acerentomon doderoi* Silv. und *Eosentomon transitorium*. Berl Naturwiss 46:90–91

Thiede U (1977) Untersuchungen über die Arthropodenfauna in Fichtenforsten (Populationsökologie, Energieumsatz). Zool Jahrb Syst 104:137–202

Wagner J (1987) Zur Enchytraeidenfauna in unterschiedlich stark belasteten Fichtenforsten. Staatsexamensarbeit (Thesis) Univ Bayreuth

Part 3. Direct Effects of Pollutants on Above-Ground Plant Parts

Preface

Plant damage from the direct impact of the smoke from such point sources as ore smelters and other coal-fired local industrial plants has been recognized for many centuries. As early as 1666, Evelyn first reported pollution damage to vegetation in his treatise "Fumigation or the Inconvenience of the Acer, and Smoake of London Dissipated" (see Bell 1984). At the end of the nineteenth century, Stöckhardt (1871) published his early scientific research, "Untersuchungen über die schädlichen Einwirkungen des Hütten- und Steinkohlenrauches auf das Wachstum der Pflanzen, insbesondere der Fichten und Tannen" ("Studies on the damaging influence of smelter and coal smoke on the growth of plants, especially spruce and fir"). At this time, botanists were also beginning to realize that lichens must be a very pollution-sensitive plant group, because they were disappearing from the central parts of European cities with polluted air. In 1866, Nylander described this disappearing lichen phenomenon in Paris and Arnold (1891, and following publications) later studied the phenomenon in Munich and its environments. During this period, scientists also recognized that evergreen coniferous trees were especially vulnerable to pollution in the cities. In 1901, Hesdörffer traced the cause of urban conifer damage to sulfur dioxide. According to Hesdörffer, sulfur dioxide was released from the numerous chimneys in the city center, generating sulfuric acid in the snow which covered conifer branches during winter. He concluded that the indirect generation of sulfuric acid from sulfur dioxide was thus responsible for killing the foliage of the trees. With the development of such assumptions, extensive analytical research began in the first decades of this century. This research combined the efforts of plant physiologists, who measured and analyzed the response of plants exposed to pollutants, and biochemists, who studied the metabolic processes involved. Today, extensive information has accumulated regarding the mechanisms which cause acute injury symptoms in plants subjected to high atmospheric pollution stress (see summaries such as Koziol and Whatley 1984; Winner et al. 1985; Hock and Elstner 1988).

In contrast to these observations, the forest decline which was observed in the 1970's in southern Germany was concentrated in areas previously regarded as relatively clean air regions, far removed from urban and industrial aggregations. The classical acute smoke damage studied around point sources and in cities seemed unlikely in these areas. Rather, long-term atmospheric pollutant loads with relatively low-level concentrations were more likely causes of the adverse

E.-D. Schulze, O.L. Lange, R. Oren (Eds.)
Ecological Studies Vol. 77

impacts on plant growth and vitality. As a result, the hypothesis that air pollutants directly affected aboveground plant parts was suggested, explaining forest decline (Forschungsbeirat Waldschäden 1984). Sulfur dioxide, nitrogen oxides, and ozone were considered to be the primary damaging agents.

Soon it became clear that the chronic impact of atmospheric pollutants might be a different mechanism of injury than the known acute impact of the same pollutants. However, structural, physiological, and biochemical responses of trees experiencing long-lasting exposure to pollutants are still poorly understood, because experimental research under controlled conditions is difficult to conduct, highly complex (Rehfuess and Bosch 1986), and almost impossible with mature trees. Thus, our current knowledge is based mainly on field experience. Part 3 of this Volume discusses the possible direct effects of pollutants on above-ground parts of the spruce trees at the experimental sites in the Fichtelgebirge.

The largest portion of the surface area of a tree directly exposed to the atmosphere, i.e., the foliage and the primary axes, is covered by a cuticle. The cuticle is a continuous extracellular membrane composed of biopolymer cutin and wax-like lipids, embedded within the cutin matrix or covering its outer surface. In addition to preventing excessive transpiration and loss of nutrients, the cuticle protects the inner tissue from penetration of substances and pathogens. Gaseous, wet, and particulate pollutants are primarily intercepted at this plant-atmosphere interface. Because of the enormous importance of the cuticle in protecting the plant, Chapter 3-A describes the conifer cuticle: its fine structure and chemical composition, as well as its transport properties.

Even if some gaseous pollutants, such as nitrogen dioxide, have a higher cuticular permeance than water vapor or carbon dioxide, the plant cuticle is still a very effective barrier for limiting the penetration of pollutants into the plant body. Plant cuticles have even proven totally impermeable to ozone. Moreover, the fine structure and the composition of cutin and the soluble cuticular lipids have not been found to be affected by air pollutants or acid precipitation in concentrations measured in the Fichtelgebirge. Despite the possible impact of air pollutants in environmental concentrations, the cuticle's transport properties do not seem to change. In contrast, some changes do seem to occur in conifer epicuticular lipids which form tufts of wax tubules, especially within the epistomatal chambers. It appears that air pollution and acid deposition might accelerate the age-dependent alteration of these structures (Chap. 3-A). However, no clear evidence for a direct pollution impact could have been established from such findings for the experimental sites in the Fichtelgebirge.

The most important organ in the uptake and the exchange of plant minerals and nutrients is the root system. However, some uptake and loss of ions also takes place through the above-ground plant parts. The cuticle, the bark, and especially the fissures in the superficial tissues are the sites of ion loss when the plant surface is moistened by rain, fog, or dew. Earlier studies have clearly shown that leaching of cations increases with the increasing acidity of the precipitation and is intensified by the impact of gaseous pollutants such as ozone. Since there is no doubt that some leaching always occurs under natural conditions, the crucial question is whether this process is intensified by the direct impact of pollutants to such an extent that forest decline due to nutrient loss is accelerated in the Fichtelgebirge. This problem is addressed in Chapter 3-C. The leaching and uptake of ions is experimentally described on different scales from single needles and branches to small trees. Canopy leaching is estimated by a canopy-balance model. In general, loss of such essential nutrients as magnesium and calcium through leaching seems to be relatively small under the conditions in the Fichtelgebirge. If the cation supply by the soil is adequate, the trees should easily compensate for a mineral leaching loss under natural conditions. However, if the supply of certain cations from the rhizosphere is low, any loss of these ions through leaching will aggrevate conditions for the trees. Thus, leaching to a certain extent might contribute to mineral deficiencies in the damaged trees, as discussed at the ecosystem level in Chapter 4-I.

Because they enter the interior of plants directly, some atmospheric pollutants may affect plant metabolism and vitality. A great number of organic micropollutants can be found in the air; many of these were identified in air samples at the experimental site at Oberwarmensteinach. In Chapter 3-B, a selection of potentially toxic, anthropogenic pollutants of low volatility including chlorinated pesticides, polycyclic aromatic hydrocarbons, polychlorinated biphenyls, and substituted phenols are discussed. These lipophilic compounds tend to accumulate over long exposure in the lipid layers, the cuticle and the epicuticular waxes of the needle epidermis. From this storage pool, they diffuse into the needle mesophyll. The amounts, distributions, and phytotoxicities of the persistent contaminants found in the spruce needles of the Fichtelgebirge are such that toxic effects could not be excluded. However, no direct relationship between organic-compound concentrations detected in the needles and the degree of damage could be established.

The gaseous anorganic atmospheric pollutants, namely, sulfur dioxide, nitrogen dioxides, and ozone, enter the spruce needles through the stomata. The amount of incorporated substances depends on the concentration gradients between the atmosphere and the needle interior, as well as on the plant conductance for diffusive transport. This conductance is controlled by the momentary stomatal aperture and the dynamic boundary layer around the spruce needles. In Chapter 3-D, pollutant uptake inside the needles and the possible impacts on plant metabolism are described. It is clear that the incorporation of sulfur dioxide and ozone by the spruce trees in the Fichtelgebirge must cause a certain burden on the detoxification mechanisms of the mesophyll cells. Latent injury may weaken the trees and make them more susceptible to other climatic or biotic stress factors such as drought or parasite attacks. However, such physiological properties of chronic injury are difficult to define and even more difficult to quantify. Physiological plant parameters such as needle photosynthesis, stomatal functioning, chloroplast pigment, and sulfur contents were measured in the field. Green trees at the healthy and declining sites and trees of different degrees of damage at the declining site were compared. There was no indication that metabolic damage such as depression of photosynthetic capacity or needle bleaching resulted from the direct impact of atmospheric pollution, which consequently could be excluded as the principal cause of injury. Rather, a correlation between damage and foliage mineral nutrition became obvious. This conclusion relates to the discussion in Part 4 of this Volume.

References

Arnold F (1891) Zur Lichenenflora von München. Ber Bayer Bot Ges 1:1–147

Bell JNB (1984) Air pollution problems in western Europe. In: Koziol MJ, Whatley FR (eds) Gaseous air pollutants and plant metabolism. Butterworth, Lond, pp 3–24

Evelyn J (1666) Fumifugium. Lond (cited after Bell 1984)

Forschungsbeirat Waldschäden (1984) Zwischenbericht. 1984 des Forschungsbeirat Literaturabteilung Kernforschungszentrum Karlsruhe, 368 pp

Hesdörffer M (1901) Fremdländische Nadelgehölze als deutsche Waldbäume. Daheim 6:15–18

Hock B, Elstner EF (1988) Schadwirkungen auf Pflanzen. Lehrbuch der Pflanzentoxikologie. Wissenschaftsverlag, Mannheim Wien Zürich

Koziol MJ, Whatley FR (1984) Gaseous air pollutants and plant metabolism. Butterworth, Lond

Nylander W (1866) Les lichenes du jardin du Luxembourg. Bull Soc Bot Fr 13:364–372

Rehfuess KE, Bosch C (1986) Experimentelle Untersuchungen zur Erkrankung der Fichte (*Picea abies* (L.) Karst.) auf sauren Böden der Hochlagen: Arbeitshypothese und Versuchsplan. Forstwiss Centralbl 105:201–206

Stöckhardt JH (1871) Untersuchungen über die schädliche Einwirkung des Hütten- und Steinkohlenrauches auf das Wachstum der Pflanzen, insbesondere der Fichten und Tannen. Tharandter Forstl Jahrb 21:218–254

Winner WE, Mooney HA, Goldstein RA (1985) Sulfur dioxide and vegetation. Stanford Univ Press, Stanford, California

3-A The Cuticles of Conifers: Structure, Composition and Transport Properties

M. Riederer

1 Introduction

The plant cuticle is frequently referred to when the direct effects of air pollutants on the above-ground parts of higher plants are discussed. It covers all primary parts of higher plants and makes up the largest portion of the plant surface area directly exposed to the atmosphere. Thus, gaseous, wet and particulate air pollutants are primarily intercepted at this plant-atmosphere interphase. The cuticle determines the rates of uptake of all substances reaching the leaf surface which are not able to enter the stomata by diffusion in the gas phase. It also limits the transport of material (e.g., inorganic ions, metabolites) from the interior of the plant to its surface.

The physical properties of this continuous, extracellular membrane are determined by its chemical composition and fine structure, which must be understood before the effects of air pollutants can be analyzed. It has been repeatedly speculated that air pollutants or acid precipitation may alter the structure and/or composition of plant cuticles, thereby increasing both the uptake of harmful substances and the loss of nutrient ions (Tamm and Cowling 1977; Kenk et al. 1983; Arndt et al. 1982; Zech et al. 1985).

In this chapter, current information on the fine structure and chemical composition of conifer cuticles will be summarized and evaluated, and an overview concerning the transport of water, gases, organic nonelectrolytes, and electrolytes across cuticles will be given. Special reference to the effect of environmental factors and pollutants will be made. The presentation of fine structure and chemical composition of conifer cuticles is based on the results of a number of studies, which will be supplemented by unpublished data of the Fichtelgebirge study from the author's laboratory. Due to experimental difficulties, little is known about the transport properties of conifer cuticles. Thus, the discussion of cuticular transport will have to rely heavily on results obtained with angiosperm cuticles. As sorption and transport in terms of properties of both cuticles and solutes have been generally described for angiosperm cuticles (Schönherr and Riederer 1989), these findings will be cautiously extrapolated to conifer cuticular membranes.

E.-D. Schulze, O.L. Lange, R. Oren (Eds.)
Ecological Studies Vol. 77

2 Fine Structure and Chemical Constitution of Cuticles

2.1 Fine Structure

All regions of the outer epidermal walls containing the biopolymer cutin are referred to as the "cuticular membrane" or simply the "cuticle" (Sitte and Rennier 1963; Holloway 1982a). This term includes both the peripheral regions made up exclusively of cutin ("cuticle proper") and those regions where the epidermal cell wall is encrusted to varying degrees with the cutin polymer ("cuticular layers"). Within the cuticle and often on its outer surface, solid deposits of waxlike substances are found (for a review on fine structure and histochemistry of the plant cuticle see Holloway 1982a).

In contrast to the wealth of scanning electron microscope (SEM) studies on the micromorphology of the surfaces of conifer needles, no transmission electron microscope (TEM) studies focusing specifically on the cuticular fine structure of the Pinaceae have apparently been conducted. What limited information is available is a by-product of studies on different topics. Electron (TEM) micrographs have been published of the needles of *Abies balsamea* (Chabot and Chabot 1977), *Picea sitchensis* (Holloway 1982a), *Pinus nigra* var. *maritima* (Campbell 1972) and *Pinus sylvestris* (Walles et al. 1973).

In all four cases, no distinction between a cuticle proper and a cuticular layer can be derived from the staining behaviour of the thin sections. As all regions of these cuticles are characterized by a reticulate pattern of electron dense fibrillae, the cuticles can be classified as Type 4, according to Holloway (1982a). However, Shovel and Alvin (1987) observed with the light microscope a two-layered organization of the cuticles of two species and one hybrid of *Tsuga*, although no clear identification of these two layers was made. The fibrillae observed in TEM micrographs form a three-dimensional network within type-4 cuticular membranes leading to a reticulate pattern in cross-sections. They presumably are of polysaccharide nature (Wattendorff and Holloway 1980) and appear to be embedded in a matrix of cutin, which is the major structural component of the cuticular membrane. Even after staining with various reagents, this cutinous matrix material has the appearance of an almost completely electron-translucent background.

TEM methods may reveal certain cuticular fine structures, but proved to be completely unsuited for the detection of the soluble cuticular lipids which, as wax-like substances, may be partially lost and melted during embedding and heat polymerization of epoxy embedding resins (Reed 1982). An alternative approach for detecting these deposits is polarized light microscopy which can localize the highly ordered wax crystallites by their birefringence. This was demonstrated in a study of cross-sections of cuticles of *Picea excelsa* and *Pinus sylvestris* in which a pronounced negative birefringence was observed in the outer regions of the cuticular membrane. Birefringence was removed by solvent extraction, which relates this special optical behavior to the presence of soluble cuticular lipids (Sitte and Rennier 1963).

In a study of the development of the cuticular fine structure of *Abies balsamea* needles, Chabot and Chabot (1977) found very thin epidermal cell walls (0.4 μm or less) and little cuticular development in needles from the earliest stages of epidermal differentiation, when the 2–3 mm long needles were still enclosed in the bud. During epidermal maturation, the cell wall and cuticle thickened, with the cuticle developing its reticulate structure. After 1 year, the differentiation had ceased. In a light-microscopy study, Lange and Schulze (1966) described the development of the cuticular membrane thickness of *Picea abies* needles. Just after flushing, the needles were covered by a very thin cuticle which increased in thickness when the needles had reached their final lengths, ca. 2 weeks after bud break. The final thickness of the cuticle was reached in October and amounted to ca. 3.4 μm at the flat sides and to ca. 4.3 μm at the four edges of the needle. The attachment of the cuticle to the epidermal cell wall changed even after the final thickness had been reached. Lendzian et al. (1986) reported that cuticular membranes could not be isolated by incubation in cellulase/pectinase solution from *Picea abies, Picea pungens, Picea mugo, Abies alba* and *Taxus baccata* when needles were older than 2 years. This phenomenon has tentatively been attributed to lignification of the cell wall/cuticle interface.

The lack of detailed information on the development of fine structure and chemical composition of cuticles of conifer needles is regrettable, as differences in cuticle thickness between needles from trees growing at low and high elevations play an important role in the hypotheses put forward to explain the alpine timberline (Baig and Tranquillini 1976; Tranquillini 1979). In most studies, a significant reduction in cuticle thickness with increasing elevation has been observed. An exception is a study by DeLucia and Berlyn (1984), who found a decrease in cuticle thickness of *Abies balsamea* from low altitudes to the Kampfzone, but a significant increase at the timberline.

2.2 The Composition of Cutin

The biopolymer cutin is the major structural component of the plant cuticle. At least partially encrusted with soluble cuticular lipids, this matrix is a three-dimensional polymer of mostly mono- and polyhydroxy n-alkanoic acids with predominant chain lengths of C_{16} and C_{18}. The cross-links between the monomers are ester and, in many species, presumably ether bonds. Presently, only reagents which can break ester bonds are available for the depolymerisation of cutin (Holloway 1984a). When employing these methods, however, only the ester cutin fraction can be analyzed, while the monomeric composition of cutin cross-linked by non-ester bonds is not yet accessible to analysis (Schmidt and Schönherr 1982; Riederer and Schönherr 1988b).

The composition of the ester cutin of a large number of plant species has been investigated (Holloway 1982b; Kolattukudy 1981) but cuticles of only four Pinaceae species have been analyzed. Detailed results were published for the leaf cutin of *Pinus sylvestris* (Hunneman and Eglington 1972) and for the stem cutin from green shoots of *Pinus radiata* (Franich and Volkman 1982). The major monomer of *Pinus sylvestris* cutin was 9,16-dihydroxyhexadecanoic acid followed by much smaller relative amounts of 9,10,18-trihydroxyoctadecanoic, 16-hydroxyhexadecanoic, hexadecanoic, 12-hydroxydodecanoic, and 14-hydroxytetradecanoic acid (in the order of decreasing frequency). In the stem cutin of *Pinus radiata*, a larger number of compounds was detected by improved analytical equipment. In the cutin of shoots grown under summer conditions, ω-hydroxyalkanoic acids dominated followed by α,ω-alkanedioic, dihydroxyalkanoic and alkanoic acids, while in winter cutin dihydroxyalkanoic acids were predominant. Both summer and winter cutin also contained very small amounts of the methoxy derivative of 9-epoxy-10,18-dihydroxyoctadecanoic acid.

The results of these studies must be viewed with caution for three reasons: (1) In both cases aqueous $ZnCl_2$-HCl was used to remove cellulose from the isolated cuticular membranes. Riederer and Schönherr (1988a) showed that in cuticles containing epoxy alkanoic acids, the use of aqueous HCl leads to the formation of chlorohydrin and *vic*-diol derivatives of this cutin monomer. Thus, the presence of epoxy monomers probably was totally overlooked in the study of *Pinus sylvestris* (Hunneman and Eglington 1972) and underestimated in the case of *Pinus radiata* stem cutin (Franich and Volkman 1982). In both cases, presumably excessive amounts of 9,10,18-trihydroxyoctadecanoic acid were found as this is the *vic*-diol derivative of 9-epoxy-10,18-dihydroxyoctadecanoic acid. (2) The $ZnCl_2$-HCl treatment may also lead to a substantial loss of unsaturated cutin monomers and to a limited dehydroxylation (Jones 1978). (3) The monomer composition of the cutins of both species was derived from area percentages of the gaschromatograms. Corrections for work-up losses and differential detector responses were not made. These corrections, however, are necessary as otherwise serious errors in estimating cutin monomer composition are unavoidable (Riederer and Schönherr 1986).

It remains unclear whether the cutins of conifers consist of both an ester and a non-ester fraction. However, the detection of epoxy alkanoic acid monomers in *Pinus radiata* stem cutin (Franich and Volkman 1982) and in *Picea abies* needle cutin (Riederer, unpublished results) suggests the presence of an additional non-ester fraction as shown for *Clivia miniata* cutin (Schmidt and Schönherr 1982; Riederer and Schönherr 1988b).

So far, no effects of air pollutants or acidic precipitation on cuticular fine structure or the chemical constitution of cutin have been reported. This is not surprising as the hydrolytic cleavage of aliphatic esters like the cutin polymer normally will take place only in the presence of bases. Even though the acid-catalysed hydrolysis of ester bonds is theoretically possible, very strong acids which would never occur under natural conditions are required for a reaction. Additionally, this reaction would proceed at a very slow rate (Isaacs 1987). The only direct impact of air pollutants on cuticular membranes described so far is the irreversible uptake of ^{15}NO and $^{15}NO_2$ by plant cuticles (Kisser-Briesack et al. 1987). However, mechanism, site, and effect of this reaction do not yet seem clear.

2.3 Soluble Cuticular Lipids

Numerous studies concerning the composition and the fine structure of conifer needle wax or soluble cuticular lipids (SCL) exist. Both terms are found in the literature and will be used here interchangeably. Cuticular waxes are of special importance as they constitute the main transport barrier (see Sect. 3.3). Two fractions of SCL can be distinguished: there are the intracuticular waxes, embedded within the cutin matrix, and the epicuticular waxes, covering the outer cuticular surface of a large number of plant species. The latter may form extensive wax blooms on the surface of leaves and primary stems.

2.3.1 Chemical Composition

Extraction. The cuticular waxes easily dissolve in a wide range of organic solvents, but are insoluble in any naturally occurring aqueous solution. The solubility in organic solvents has been frequently used as a simple and comprehensive operational definition of the lipid compounds embedded within the cutin and deposited on the outer surface of the cuticle. This property easily distinguishes the soluble cuticular lipids (SCL) from the second lipid fraction of the cuticle, the insoluble polymer cutin. In most studies, the SCL were extracted by dipping leaves in organic solvents. Only recently has the use of pyrolysis field-ionization mass spectrometry been proposed for the characterisation of intact needle surfaces (Schulten et al. 1986a). Nevertheless, the preparation of wax extracts remains necessary for a quantitative analysis, especially when minor compounds must be detected.

The extraction of SCL from needle surfaces should be achieved quantitatively without discriminating different substance classes, chain lengths, or sites of deposition within the cuticle. On the other hand, contaminating the wax extracts by lipids extracted from epidermal and mesophyll cells must be avoided. At the start of any analysis of amounts, location, and composition of cuticular waxes, the optimum solvent, temperature, and extraction time must be found. In the case of conifers, fresh needles have most frequently been immersed for a fairly short period of time (10–20 s, 1 min or not specified) in differing volumes of $CHCl_3$ (Schuck 1972; Schuck 1976; Corrigan et al. 1978; Niemann and Baas 1978; Günthardt and Wanner 1982a; Cape 1986; Schulten et al. 1986b). Other authors have used *n*-hexane (Tulloch 1987), petroleum ether (Beri and Lemon 1970) or aqueous ethanol (40–60%) followed by a hot 2:1 (v/v) benzene-ethanol mixture (von Rudloff 1959) as solvents. In some cases, the samples have been air-dried (Beri and Lemon 1970; Corrigan et al. 1978; Schulten et al. 1986b) or macerated (von Rudloff 1959) prior to extraction.

A critical examination of the time necessary for quantitative extraction of SCL from *Picea abies* needles led to the conclusion that the needles have to be immersed in $CHCl_3$ at 20–25°C for at least 6 h (Fig. 1). Shorter extraction times resulted in significantly lower yields of SCL. With immersions lasting only 10 s, not more than 50% of the total SCL could be extracted. After testing a range of

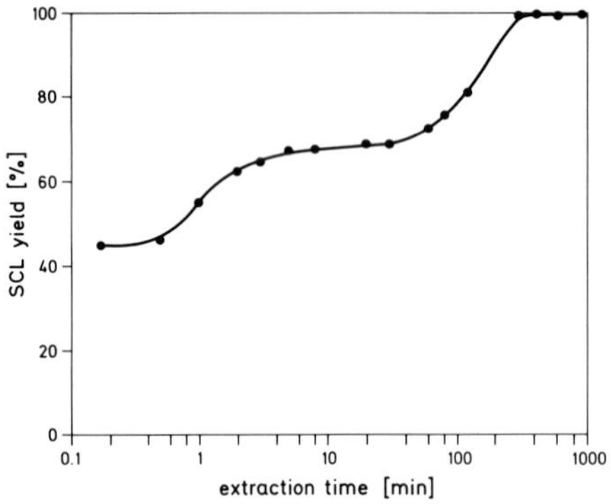

Fig. 1. Extraction curve of the soluble cuticular lipids (*SCL*) from needles of *Picea abies*. *Data points* are the means from five determinations each, and coefficients of variation are in the range of 20%. For each sample, 30 needles were carefully separated from the twigs by cutting their petioles and then extracted for varying periods of time with distilled $CHCl_3$ at 20°–25°C. The extracts were filtered through extracted glass wool, and the solvent removed at room temperature. SCL yields were determined gravimetrically

solvents differing in polarity, $CHCl_3$ was chosen on the basis of its maximum yield. A subsequent 1-hour wash with *n*-hexane was necessary for the complete extraction of the *n*-alkane fraction of SCL (Kurbasik 1989). The solvent extracts from *Picea abies* needles usually contain dark soot-like precipitates which have been identified as aerosol particles strongly attached to the needle surface. Simmleit et al. (1986) found that the elemental composition of these precipitates is comparable to that of ambient aerosols. In order to remove these and other particulate impurities, filtrating of the extract with thoroughly extracted glass wool is advisable.

The rather long period of time necessary for a quantitative extraction of SCL raises the question of whether the extract may become contaminated by lipids of intracellular origin. Using both chlorophyll and phospholipids as marker compounds, it has been demonstrated that no such contamination occurs even after 20 h (Kurbasik 1989). Apparently, the aqueous phase of the very thick outer epidermal cell walls represents a barrier sufficient for solvents of low miscibility with water. However, both chlorophyll and phospholipids have been found in CH_3OH extracts. Extensive tests have further demonstrated the necessity of very carefully preparing the needle samples in order to obtain uncontaminated extracts. The only satisfactory method developed so far is the separation of the needles from the twigs by cutting their short petioles with a razor blade, without injuring the leaves themselves. Any other method (e.g., picking the needles by forceps, or breaking them off after immersion of the twig in liquid nitrogen) results in an impressive coloration, and thus, contamination of the extract. Günthardt and Wanner (1982a) reported that their raw extracts from *Picea abies* needles contained up to 30 wt% of plant pigments.

In many cases it would be useful to distinguish between epi- and intracuticular waxes and to study their composition and amounts separately. In the literature, the term epicuticular waxes or lipids is used very loosely, as all lipids removed by short solvent extraction are thought to be epicuticular, even though there is no evidence showing conclusively that only the waxes from the cuticular surface, and not the ones of intracuticular origin, are removed by short-time dippings into organic solvents. Any method relying on the use of organic solvents for the extraction of waxes cannot discriminate between the epi- and intracuticular fractions, as the solvents penetrate the lipophilic cutin and dissolve some or most of the SCL sorbed to the cutin. A stripping method introduced by Haas and Rentschler (1984) seems much more promising. However, it remains to be shown conclusively that the relatively polar solvents used in this method do not dissolve and remove part of the intracuticular lipids, and that contamination by microorganisms growing on the leaf surface is irrelevant.

Total Amounts of SCL. Usually, the amounts of SCL recovered by solvent extraction have been related to the needle lengths (Günthardt and Wanner 1982a), fresh weights (Schuck 1972) or dry weights (Schuck 1976; Günthardt and Wanner 1982a; DeLucia and Berlyn 1984; Günthardt-Goerg 1986) of the needle samples extracted. The data produced, using needle weights as a reference, may be subject to variation due to changes in water and assimilate contents of the needles obviously not affecting the actual SCL coverage. In any case, weight- or length-related data on SCL amounts are of limited physiological relevance and should be replaced by area related determinations (e.g., Cape 1986).

This necessitates the determination of the surface area of conifer needles, which is somewhat difficult due to their irregular geometry. During the Fichtelgebirge study, a method for determining the surface areas of very large numbers of *Picea abies* needles was developed (Riederer et al. 1988). Using a computer-aided image-analysis system, the projected areas of about 60,000 needles were measured individually and subsequently converted into the actual areas by correction factors empirically determined from needle cross-sections.

The SCL surface coverage of needles of *Picea abies* from the Fichtelgebirge was studied at four locations using 48 trees ca. 130 years old (Table 1). Samples were taken in mid-June 1986 from the three youngest needle age-classes (1986, 1985, 1984) of twigs from the upper portions of the tree tops. From each age-class, five samples were randomly chosen, consisting of 30 needles each, and the SCL extracted as described above. After filtrating the extract, the solvent was evaporated at room temperature and the amount of SCL gravimetrically determined (Kurbasik 1989).

Because of the large sample sizes, the overall frequency distributions of the SCL coverage could be established by age-classes (Fig. 2). In all locations, the mean SCL coverage of each of the youngest age-class was about a factor of two lower than those of the two older age-classes (Table 2). Apparently, the SCL coverages had not yet been fully developed at the time of sampling. It could be shown independently, with ten trees ca. 30 years old, that the increase of wax coverages during the development of the needles reached a plateau in September of the first year.

At all four Fichtelgebirge locations, the means of the three-year-old needles (1984) were lower than those of the age-class 1985. The difference was quite

Table 1. Description of sampling sites for *Picea abies* needles in the Fichtelgebirge

Parameters	Sites			
Location	Oberwarmen-steinach	Selb	Schneeberg I	Schneeberg II
Elevation (m)	755	650	900	700
Growing season	May-Oct.	May-Oct.	June-Sept.	May-Oct.
Soil	Podsolic cambisol from phyllite	Eutrophic cambisol from basalt	Podsolic cambisol from granite	Podsolic cambisol from granite
Air pollution[a]				
SO_2 ($\mu g\ m^{-3}$)				
1985	14 (442)	21 (1071)	–	–
1986	23 (623)	32 (470)	–	–
NO (ppb)				
1985	1 (37)	9 (187)	–	–
1986	1 (13)	7 (105)	–	–
NO_2 (ppb)				
1985	5 (27)	10 (57)	–	–
1986	5 (27)	8 (44)	–	–
O_3 (ppb)				
1985	– (72)	19 (104)	–	–
1986	35 (85)	– (76)	–	–

[a] Atmospheric concentrations of selected pollutants were taken from Bayerisches Landesamt für Umweltschutz (1986, 1987). Mean summer concentrations are given together with maximum summer concentrations in parentheses.

drastic at Schneeberg (900 m) with a decrease of 23% (Table 2). This effect may be attributed to an increased erosion and/or a decreased biosynthesis of SCL in older needles. An age-related decrease in wax amounts has also been found with *Picea abies* trees grown at the timberline (Günthardt and Wanner 1982a), and with samples taken at different expositions (Schuck 1976), while constant amounts of SCL have been observed in *Pinus cembra* needles from 2 to 9 years old (Günthardt and Wanner 1982a).

Within age-classes there were highly significant differences between the SCL coverages at the four Fichtelgebirge locations. At the time of sampling, current-year needles from trees growing at Selb (lowest elevation, Table 1) had the lowest degree of SCL coverage while the needles from 1985 and 1984 did not differ from those grown at Oberwarmensteinach (Table 2). SCL coverage of older needles was higher at the two Schneeberg locations.

Analysis of SCL. Analyzing the SCL extracts from conifer needles can be problematic; the large number of compounds belong to different substance classes, and high molecular weight estolides are present. The classical work-up of the extract starts with a thin-layer or column chromatographic separation of substance classes (for a review of methods cf. Holloway 1984b). More recently, capillary gas chromatography/mass spectrometry has been used to identify individual compounds (Corrigan et al. 1978; Franich et al. 1978; Franich et al.

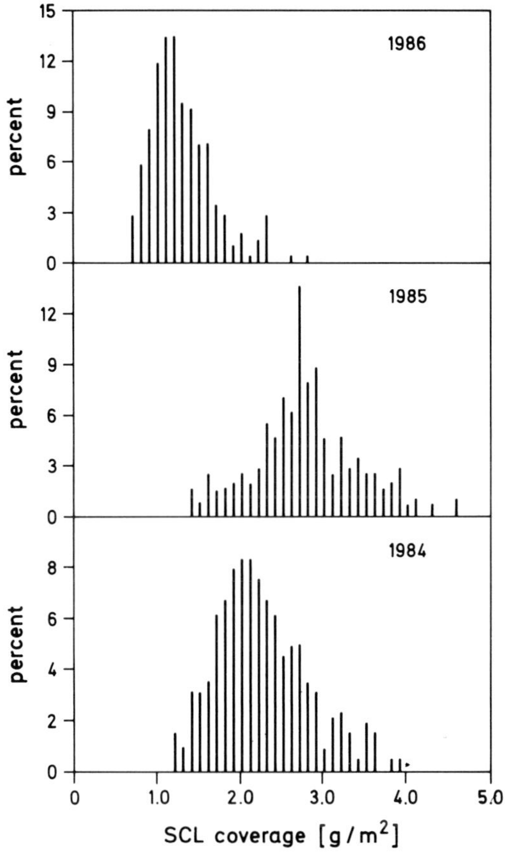

Fig. 2. Overall frequency distributions by age classes of the coverage by soluble cuticular lipids (SCL) of *Picea abies* needles sampled at four locations in the Fichtelgebirge. Samples consisting of five replications per age class (30 needles each) were taken during the second week of June 1986 from the upper crown sections of 48 ca. 130-year-old trees

Table 2. Soluble cuticular lipid coverages of *Picea abies* needles from four locations in the Fichtelgebirge

Site	Needle age class		
	1986[a] (g m^{-2})	1985[a] (g m^{-2})	1984[a] (g m^{-2})
Oberwarmensteinach	1.39 A	2.78 A	2.25 A
SD (N)	0.386 (121)	0.654 (116)	0.546 (123)
Selb	1.16 B	2.62 A	2.20 A
SD (N)	0.293 (50)	0.408 (47)	0.383 (47)
Schneeberg (900 m)	1.48 A	3.10 B	2.39 AB
SD (N)	0.413 (29)	0.758 (27)	0.807 (27)
Schneeberg (700 m)	1.33 A	3.04 B	2.52 B
SD (N)	0.341 (33)	0.413 (32)	0.454 (33)

[a] Means, standard deviations (SD) and number of samples (N) of SCL coverages of needles from 130-year-old trees. Each sample consisted of 30 needles. Samples were taken from 25, 10, 6, and 7 trees at Oberwarmensteinach, Selb, and two locations at Schneeberg, respectively (for site description, see Table 1). Within age classes, means which are not significantly different at the 95% level (Duncan's multiple range test) are followed by equal letters (A or B).

1979; Tulloch and Bergter 1981; Franich et al. 1985; Günthardt-Goerg 1986; Günthardt-Goerg 1987; Tulloch 1987) while in earlier work, analysis and identification were restricted mainly to *n*-alkanes and to other classes easily identified by reference compounds.

While perfectly appropriate for discovering and identifying minor and even exotic constituents of SCL, the methods relying on class preseparation by thin-layer or column chromatographic techniques are unsuitable for quantitatively analyzing the wax composition. Discriminations and losses during class separation are inevitable; they can be corrected only if the yield of the procedure is quantified for each compound separately. In view of the large number of compounds and the limited availability of commercially supplied pure standards, this is practically impossible.

In fact, no publication has yet appeared where correction factors for work-up losses have been used. The errors from sample work-ups can be avoided by analyzing the extracts without prior class separation (Kurbasik 1989). Conventionally, SCL composition is merely given as area percentages derived from the gas-chromatographic analysis of substance classes, and is not corrected for the differential response of the detector to different compounds.

Qualitative Composition of SCL. Qualitative data on the chemical composition of SCL are available for only a very limited number of Pinaceae species: *Abies balsamea* (Tulloch 1987), *Picea abies* (Günthardt-Goerg 1986), *P. glauca* (Tulloch 1987), *Pinus cembra* (Günthardt-Goerg 1986), *P. sylvestris* (Günthardt-Goerg 1986), and *P. radiata* (Franich et al. 1978; Franich et al. 1979; Franich et al. 1985). A comprehensive report on the composition of SCL was also published for *Juniperus scopulorum* (Tulloch and Bergter 1981). Some analyses have been restricted to the homologue distribution of *n*-alkanes (Herbin and Robins 1968; Schuck 1976; Trimble et al. 1982), while others have covered additional substance classes (von Rudloff 1959; Beri and Lemon 1970; Schuck 1972; Schütt and Schuck 1973; Corrigan et al. 1978; Niemann and Baas 1978).

So far, components belonging to 14 substance classes have been identified in the soluble cuticular lipids of Pinaceae leaves (Table 3). All compounds have in common a relatively large chain length (up to C_{64} in *n*-alkyl esters) and a long aliphatic portion. Soluble cuticular lipids of conifers are solids at physiological temperatures, with melting points at about 65–80°C (von Rudloff 1959, Däßler 1963, Sitte and Rennier 1963). The predominant single SCL component in all members of Pinaceae investigated is 10-nonacosanol (a C_{29} secondary *n*-alkanol). It makes up approximately one-half of the total mass of SCL extracts and is believed to constitute the tubules of the epicuticular wax fraction, especially within the epistomatal antechamber (see discussion below). This compound was identified first by Beri and Lemon (1970) using infrared, NMR and mass spectrometry.

Other characteristic and prominent components of conifer SCL are so-called estolides, which are polyesters of several *n*-alkanoic acid, hydroxyalkanoic acid, *n*-alkanol and alkanediol moeities. This rather heterogeneous class constitutes roughly 40% of *Abies balsamea* and *Picea glauca* SCL (Tulloch 1987) and 25–50%

Table 3. Substance classes identified in soluble cuticular lipids of members of Pinaceae

Substance class	Chain length
n-Alkanes	C_{21}–C_{33}
n-Alkanoic acids	C_{12}–C_{34} even
ω-Hydroxy *n*-alkanoic acids	C_{12}–C_{16} even
n-Alkanols	
Primary	C_{12}–C_{32} even
Secondary	C_{29}
n-Alkanediols	
Primary	C_{12}–C_{16} even
Secondary	C_{27}, C_{29}
n-Alkanals	C_{22}–C_{34} even
n-Alkyl esters	
Long-chain alkanol	C_{24}–C_{64} even
Methanol	C_{22}–C_{34} even
Estolides	
Diterpene acids	
Hydroxy diterpenes	
Triterpenoids	

of *Pinus cembra* cuticular wax (Günthardt-Goerg 1986), while much smaller amounts have been found in *Pinus radiata* (Franich et al. 1978), *Picea abies*, and *Pinus sylvestris* (Günthardt-Goerg 1986). In *Pinus radiata*, the molecular weights of five estolide fractions separated by thin-layer and column chromatography were about 800, 1100 and 1500 (Franich et al. 1978). This high molecular weight makes analyses very difficult and accounts for the very restricted attention most researchers have paid to this important class of substances. The composition of estolides has most thoroughly been studied by Franich et al. (1978), who found that *Pinus radiata* estolides are composed of a chain of interesterified ω-hydroxy alkanoic acids and α,ω-alkanediols terminated at each end with an *n*-alkanoic acid. There are also estolide fractions containing *n*-alkanols.

The remaining classes of substance found so far are not unique for conifer SCL, as they also occur in cuticular waxes of other plant species (Tulloch 1976; Baker 1982). Especially well documented are secondary *n*-alkanediols (C_{25}, C_{27}, C_{28}, C_{29}, C_{31}, substitution in positions 4,10, 5,10, 6,10, 7,10, 10,13 and 10,16; Franich et al. 1979), and *n*-alkyl esters (C_6 to C_{12} *n*-alkanols esterified with C_{24} to C_{36} *n*-alkanoic acids and C_{24} to C_{32} *n*-alkanols esterified with C_6 to C_{14} *n*-alkanoic acids; Franich et al. 1985; Tulloch 1987). Recently, the occurrence of methyl esters of C_{22} to C_{34} (mainly C_{30}) *n*-alkanoic acids and of long-chain *n*-alkanals in the SCL of *Abies balsamea* and *Picea glauca* has been demonstrated (Tulloch 1987). Though intensively studied (especially for taxonomic reasons, cf. Herbin and Robins 1968), *n*-alkanes seem to play only a minor role or are lacking entirely in conifer SCL. The same is true for free primary *n*-alkanols.

n-Alkanoic and ω-hydroxy *n*-alkanoic acids occur among the acidic fraction of the cuticular waxes of conifers. The diterpene acids (and hydroxylated diterpenes) which may be contained in minor amounts in SCL extracts from needles represent another analytical problem. In *Pinus radiata* SCL Franich et al. (1978) identified mainly pimaric, isopimaric, dehydroabietic and some hydroxy diterpene acids. These are the characteristic resin acids of conifer needles (Zinkel and Clarke 1985; Zinkel et al. 1985).

The recent discussion about potential effects of air pollutants, especially of photo-oxidants, on plant surfaces raises the question as to whether unsaturated compounds occur in connifer needle SCL. With the exception of minor amounts of unsaturated diterpene acids, no unsaturated compounds have been reported in the leaf SCL of any conifer species. This finding agrees with the relatively high melting points (60°C and higher) of the SCL (Däßler 1963; Sitte and Rennier 1963; von Rudloff 1959).

Only Beri and Lemon (1970) have found considerable amounts of *cis*-9-octadecenoic (9 area percent of total mixed acids), *cis*-9,12-octadecadienoic (14 area percent) and *cis*-9,12,15-octadecatrienoic acid (6 area percent) in the cuticular wax of *Picea mariana* and *Abies balsamea*. They extracted air-dried (for 5–6 days) needles with petroleum ether (boiling point between 30° and 60°C). This procedure inevitably leads to the drying of the outer epidermal walls, which consequently can no longer act as a solubility barrier to organic solvents immiscible with water (see above). Thus, the solvent can enter the epidermal and mesophyll cells and extract intracellular lipids for which these unsaturated alkanoic acids are characteristic. The authors' statement that after extraction and solvent removal they obtained "a green product" clearly shows that not only had the cuticle been extracted, but most likely, the unsaturated alkanoic acids originating from cells and organelles inside the needles.

Variation of SCL Composition. The first to describe variations in wax composition was von Rudloff (1959) who reported that only during the spring and summer months was an acidic fraction present in the SCL of *Picea pungens*. Some studies have been devoted to the intraspecific variation of the relative amounts of *n*-alkanes and ω-hydroxy alkanoic acids in *Pinus* (Herbin and Robins 1968), of *n*-alkanes, *n*-alkyl esters and *n*-alkanols in *Pinus sylvestris* (Schuck 1972) and *P. strobus* (Trimble et al. 1982) and of methanol and chloroform soluble extracts of *Larix leptolepis* (Niemann and Baas 1978). Herbin and Robins (1968) studying twelve trees of *Pinus engelmannii* found that the molar percentages (as calculated from area percentages) of three ω-hydroxy alkanoic acids varied by up to 49%. In *Pinus sylvestris* of five different provenances, the mean area percentages of various homologues within the series of *n*-alkanes, *n*-alkanols, and *n*-alkyl esters varied significantly (Schuck 1972).

The developmental stage of the foliage sampled has been identified as another factor influencing SCL composition. The primary and secondary needles, as well as needles from juvenile and mature trees of *Pinus radiata*, had SCL of different compositions. During growth and maturation, the 10-nonacosanol content of the SCL decreased, while the proportion of estolides and diterpenes increased (Franich et al. 1978; Franich et al. 1985). Needle age also affected the composition of cuticular waxes. Schütt and Schuck (1973) and Schuck (1976) reported differences due to needle age in *n*-alkane, *n*-alkanol and *n*-alkyl ester contents of *Pinus sylvestris*, and differences in *n*-alkane content of *Picea abies*. With increasing needle age, decreases of 10-nonacosanol and of 5,10-nonacosandiol were observed in *Pinus cembra*, while in *Picea abies* the amounts of 10-nonacosanol increased with age. No changes occurred in *Pinus sylvestris* SCL (Günthardt-Goerg 1986). In *Pinus cembra* and in *Picea abies*, increased amounts of estolides were found in older needles (Günthardt-Goerg 1986; Schulten et al. 1986b).

Whether air pollution and acid precipitation may affect the chemical composition of cuticular waxes is a matter of considerable interest and speculation. Compositional changes might be brought about by reactions of pollutants with constituents of SCL and/or indirectly by changing wax biosynthesis in a qualitative or quantitative manner. At this time, there is no conclusive evidence that the amounts of major SCL components of conifers are a function of immission levels.

2.3.2 Fine Structure of Epicuticular Waxes

A great deal of work has been devoted to documenting the epicuticular deposits of SCL on the needles of various coniferous species by scanning electron microscopy (SEM). This method is well suited to study the morphology of waxes on the outer surface of cuticles, while the fine structure of the intracuticular SCL deposits, which probably form the barrier for transport across the cuticle, unfortunately cannot be investigated by this method.

The SEM studies of wax morphology of conifer needles primarily focus on the presence or absence of so-called structural and amorphous epicuticular wax on different regions of the leaf. SCL deposits have been termed "structural" if discrete structures like tubules or rods are visible by SEM in contrast to "amorphous" deposits which lack clearly defined outlines, forming mostly irregular and more-or-less continuous plates covering the cuticle surface (Hanover and Reicosky 1971). This terminology is unsatisfactory; so-called amorphous waxes are probably not amorphous in the sense of lacking order on the microscopic level. The degree of microscopic order cannot be evaluated by SEM, as this method is unsuited for discriminating between crystalline and other solid states. Diffraction of light, X-rays, electrons, or neutrons would be the appropriate method for this purpose (Kreger 1958; Shipley 1986). With this provision, the terms "structural" and "amorphous" will be employed here in an exclusively morphological sense, without implying any specific molecular arrangement of the constituents of the observed structures.

Structural Wax. So-called structural wax has been found in all the species of Pinaceae examined so far, at least in the direct vicinity of the stomata. The stomatal complex of Pinaceae is characterized by subsidiary cells which overarch deeply sunken guard cells, thus forming a funnellike cavity called the epistomatal chamber or stomatal antechamber. The cuticle over the subsidiary and guard cells is covered by a dense tuft of tubular wax structures. This tuft fills the antechamber completely and forms a porous "plug" over the stomata, consisting of occasionally anastomosing wax tubules. This structure so far has been found in all members of the family Pinaceae with the exception of *Pinus cubensis*, where stomatal wax deposits are absent (Rhine 1924).

The special structure of the stomatal complex of conifers has attracted much attention. More-or-less detailed descriptions and SEM micrographs of the "plugs" in the stomatal antechambers of the following species have been published: *Abies alba* (Grill 1973a; Schmitt et al. 1987; Rinallo et al. 1986; Bermadinger et al. 1988), *A. balsamea* (Chabot and Chabot 1977; DeLucia and Berlyn 1984), *A. concolor* (Hanover and Reicosky 1971), *Picea abies* (Grill 1973a,b; Günthardt and Wanner 1982b; Günthardt 1985; Kim 1985; Magel and Ziegler 1986; Rinallo et al. 1986; Sauter and Voß 1986; Grill et al. 1987; Schmitt et al. 1987; Sauter et al. 1987; Bermadinger et al. 1988), *P. pungens* (Hanover and Reicosky 1971; Reicosky and Hanover 1976; Brunegger et al. 1982), *P. glauca* (Hanover and Reicosky 1971; Jeffree et al. 1971), *P. sitchensis* (Johnson and Jeffree 1970), *Pinus banksiana* (Johnson and Riding 1981; Riding and Percy 1985), *P. cembra* (Grill 1973a; Günthardt and Wanner 1982b; Günthardt 1985), *P. mugo* (Grill 1973a), *P. nigra* (Rhine 1924; Grill 1973a), *P. radiata* (Rook et al. 1971; Wells and Franich 1977; Franich et al. 1977) *P. strobus* (Johnson and Riding 1981; Percy and Riding 1978; Trimble et al. 1982; Krause and Houston 1983; Krause and Dochinger 1984; Riding and Percy 1985), *P. sylvestris* (Hanover and Reicosky 1971; Grill 1973a; Huttunen and Laine 1981; Huttunen and Laine 1983; Cape 1983; Crossley and Fowler 1986) and *Pseudotsuga menziesii* (Thair and Lister 1975; Lister and Thair 1981). The distribution pattern of epicuticular wax and the shape of the sometimes protruding mouths of the stomatal antechambers (so-called Florin ring) of 51 species of the genus *Pinus* have been described, classified, and related to taxonomic entities by Yoshie and Sakai (1985).

With the exception of glaucous species, the occurrence of structural wax is generally restricted to the stomatal complex and its vicinity. On the needles of *Picea abies*, as well as of *Pinus nigra* and *P. mugo*, tufts formed by wax tubules were found only in the stomatal antechamber and the circumference of its mouth. Tubules were absent from the remaining needle surface, which has a dark, specular appearance due to "amorphous" wax deposits. In contrast to this, on *Pinus sylvestris* needles, structural wax has also been found in the space between the stomata, forming thin, macroscopically visible, longitudinal strips. The most extensive structural wax covering can be observed on the hypostomatic needles of *Abies alba*. There, wax tufts spread all over the abaxial needle surface with the exception of the midrib and the needle margins.

Most authors have referred to the wax structures found on the surfaces of conifer cuticles as tubes, tubules or tubular in form, while their exact geometric shape has been subject to dispute for some time. The use of these terms has not only implied that these structures have a cylindrical shape but also that they are hollow. This feature was discounted by Martin and Juniper (1970) as "almost certainly" an electron-optical artifact. However, conclusive evidence has been presented that the epicuticular wax structures of *Picea sitchensis* (Johnson and Jeffree 1970), *Pseudotsuga menziesii* (Thair and Lister 1975; Lister and Thair 1981), and *Picea pungens* (Brunegger et al. 1982) are indeed hollow. From TEM examination of replica preparations of *Pseudotsuga menziesii* structural wax, the exact dimensions of the tubules could be determined: 2.3 μm length, 136 nm outer diameter and 35–40 nm wall thickness (Thair and Lister 1975). The surface of the tubules had a striated appearance, with the striae oriented helically. Structures with similar habit could be produced in vitro when chloroform extracts of needles were subjected to slow evaporation (Lister and Thair 1981).

The morphology of surface wax deposits may be altered during sample preparation for SEM. Various authors have stressed the necessity of following a strictly standardized method for the treatment of needle samples prior to their examination by SEM. Others have not even given details of the procedures used in sample preparation. Kim (1985) has observed alteration of the morphology of structural wax when the needles of *Picea abies* dried at 55°C, but no morphological differences between the wax structures of needles dried at room temperature and at 45°C. Drying at 45°C or at room temperature over P_2O_5, however, led to the formation of fissures in the wax plug (Kim 1985, Schmitt et al. 1987). Another critical problem has been the formation of ice crystals on the cuticle surface when frozen plant material has been used and the object chamber of the SEM has not been refrigerated sufficiently (Brunegger et al. 1982). Ice crystals may cover structural wax and thus mimic the presence of amorphous wax deposits. Another artifact may be the closing of the hollow centre of the tubulus of structural wax during sputter-coating (Crossley and Fowler 1986).

Variation of Morphology. Changes in the habit of structural surface waxes have been related to aging, environmental factors, and air pollution. The ontogenetic development of the wax deposits in the stomatal antechamber has been studied with *Abies balsamea* (Chabot and Chabot 1977), *Picea abies* (Günthardt 1985), *P. pungens* (Reicosky and Hanover 1976), *Pinus cembra* (Günthardt 1985), *P. strobus* and *P. banksiana* (Johnson and Riding 1981; Riding and Percy 1985). In all cases, structural wax could be detected in the epistomatal chambers of needles which were still covered by the bud scales or the needle fascicle. At the time of flushing, the stomatal antechambers were filled with wax tubules. In *Picea pungens*, the structural wax in the nonstomatal areas of the needle also started to develop while still in the bud (Reicosky and Hanover 1976). These studies have also demonstrated that the structural wax in the stomatal antechamber protrudes from the surface of both the guard cells and the subsidiary cells. The bases of the tubules have been observed to merge with the cuticle over these areas.

During later stages of the life of a conifer needle, a gradual change in epicuticular wax habit from structural to amorphous has been observed in all species investigated. The tubules on the outer surface of the stomatal antechambers of *Picea pungens* (Reicosky and Hanover 1976), *P. abies* (Huttunen and Laine 1981; Huttunen and Laine 1983; Günthardt and Wanner 1982b), *Pinus cembra* (Günthardt and Wanner 1982b), and *P. sylvestris* (Crossley and Fowler 1986) started to fuse during the second vegetation period. The fusion process proceeded from the borders of the antechamber mouth inward. The transition from structural to amorphous deposits resulted in continuous scale-like coverings of considerable parts of the visible surface of the stomatal plugs. No information is available as to whether or not this morphological alteration also proceeds into deeper areas of the plugs. In very old needles, the fusion process seems to reach a final stage. This has been shown with samples of *Pinus cembra* and *Picea abies* needles from the timberline, where wax layers of compact appearance covered the stomata of 6- to 8-year-old needles (Günthardt and Wanner 1982b).

In the glaucous *Picea pungens* a change of structural wax habit has also been observed in nonstomatal regions (Reicosky and Hanover 1976). Depending on the variety of *Picea pungens* studied, "degradation" of structural SCL started either by mid-November of the first year or during the second vegetation period.

The changes of structural wax morphology are usually accompanied by the erosion of the deposits both in stomatal and nonstomatal regions of the needle (Reicosky and Hanover 1976; Huttunen and Laine 1981; Huttunen and Laine 1983). The degree of erosion is a function of age and of exposition and is probably due to the mechanical impact of rain. It has been shown that appreciable amounts of epicuticular wax can be removed from the surfaces of leaves by natural and simulated rain (Baker and Hunt 1986; Mayeux and Jordan 1987).

The so-called degradation of structural SCL must be viewed as a natural phenomenon following from the physical and chemical properties of the structures in question. As to chemical composition, there is good reason to assume that the wax tubules on the surface of *Picea abies* and other conifer needles consist exclusively of the saturated secondary *n*-alkanol 10-nonacosanol. Günthardt (1985) has been able to obtain tubular structures by recrystallizing 10-nonacosanol, while no other fraction of extracted needle SCL yielded tubules.

The tubular morphology of these structures is in contrast to the orthorhombic crystal lattice, which is the most stable and therefore the most widespread arrangement of long-chain, odd-numbered alkanols at physiological temperatures (Shipley 1986, Small 1986). An orthorhombic crystal consists of unit cells whose three axes are of different lengths and meet each other at right angles. Under certain conditions of crystal growth, however, solids may form flat or curled ribbons, twisted fibers, helical strands or complete tubules instead of regular crystals. This arrangement is favored by the presence of a polar head-group, a center of asymmetry or tightly bound solvent since adjacent molecules thus easily fail to align themselves in exact juxtaposition. As a consequence screw dislocations, i.e., curved lattice planes, result. Screw dislocations, leading to spontaneous spiral or helical crystal growth, are very common in a wide range of inorganic and organic materials (Chichakli and Jessen 1967; Mutaftschiev 1977; Adamson 1982).

In comparison to the corresponding regular crystal, structures like twisted fibers or tubules have very large specific surface areas. As for energetic reasons any structure tends to minimize its area in contact with the surrounding medium, the gradual transformation of complex structures into simple regular crystals will occur spontaneously. From these considerations it becomes clear that the wax tubules on the surfaces of conifer needles are by themselves rather labile structures, which naturally are prone to degradation. The rearrangement of lipid structures easily proceeds at room temperature, as has been observed with long-chain *n*-alkane waxes (Fox 1958).

The alterations of structural wax habit over time are therefore not at all surprising. In contrast, the really intriguing question is how the plant can preserve the tubular SCL structures on the surface of its leaves over many months or even years. This is especially necessary within the epistomatal chamber, otherwise the concomitant loss of wax deposit porosity would impede gas exchange. Presumably the tubular habit is maintained by continually producing new tubes, replacing degraded or eroded ones. This hypothesis of continuous wax production even in fully grown needles is supported by the observation that the wax deposits in the epistomatal chambers of *Picea* needles were completely reconstituted within 6 weeks after they had been removed by solvent extraction (Kurbasik 1989). Natural degradation of wax tubes may become visible only when an aging needle can no longer allocate the input of metabolic energy necessary to continue wax production and replacement.

Pollutants and Wax Morphology. Studies increasingly show that under the influence of air pollution the natural aging process of the structural epicuticular wax of most conifer species is accelerated. Comparing *Picea abies* needles from areas with visible SO_2 vegetation damage with needles from areas without damage, Grill (1973a) demonstrated for the first time a premature agglomeration of structural wax in the epistomatal chambers under polluted conditions. Subsequent studies using more or less identical experimental designs (no emission levels specified) were able to qualitatively reproduce that finding (Østensen and Gullvåg 1985; Gullvåg and Østensen 1986). Sauter and Voß (1986) found a greatly accelerated formation of scalelike structures covering the mouth of the epistomatal chamber of *Picea abies* needles in declining trees. Needles were shed when 80–90% of the stomata were covered by such a scale. The degree of epicuticular wax tubule fusion decreased from the top to the bottom of a tree and was higher in sun-exposed than in shadowed needles (Kim 1985). Under equal pollution levels, trees with altered structural wax and trees without any morphological changes were both found (Grill et al. 1987). Both did not differ in mean needle length or dry weight per 100 needles, but trees with degraded structural wax had thinner crowns.

Sauter et al. (1987) recently showed that the degradation of epistomatal wax deposits of *Picea abies* needles was also accelerated by a 20-week exposure at the edge of a motorway, as well as by a very short but intensive fumigation with the exhaust of an Otto motor (560 ppm NO_x, 0.3% CO, Ca. 250 ppm hydrocarbons). Fine dust from a magnesite plant containing MgO was also found to cause premature degradation of the structural wax of *Picea abies* needles (Bermadinger et al. 1988).

The structural changes in *Picea abies* wax could be reproduced under controlled conditions in growth chambers using fairly young trees. Weekly sprays with pH 3.5 solutions of H_2SO_4, HNO_3, and a combination of both, resulted in extensive erosion and fusion of structural wax after 30 weeks (Rinallo et al. 1986). Partial regeneration of wax morphology was observed after replacement of the acid by a bidistilled water treatment. Using 5-year-old clones of *Picea abies*, Magel and Ziegler (1986) observed a high frequency of aggregated SCL deposits in the stomatal antechambers of young needles exposed in growth chambers for about 5 months to O_3 at environmental concentrations, with peaks up to 130 ppb, and to acidic fog. For about 30 months Schmitt et al. (1987) applied various combinations of O_3 (approx. 100 $\mu g/m^3$), SO_2 (winter: 80–100 $\mu g/m^3$, summer: 50–80 $\mu g/m^3$), and acidic rain (pH 3.8–4.0) to 4- to 5-year-old seedlings of *Picea abies* growing in chambers. While acidic precipitation alone or in combination with O_3 caused agglomerations only in some stomatal chambers, widespread fusion was observed under the influence of a combination of O_3, SO_2, and acidic rain. The most effective combination was SO_2 and acidic rain, which led to encrustations covering the whole mouth of the stomatal antechamber.

Ample evidence for the influence of air pollution on the morphology of structural wax is available for *Pinus sylvestris*. Again, needle samples from areas differing in air pollution levels have been compared. In all cases, increased erosion and fusion of epicuticular wax tubules have been observed (Huttunen and Laine 1981; Huttunen and Laine 1983; Crossley and Fowler 1986). In polluted areas (SO_2 concentration: winter, 40–55 $\mu g/m^3$; summer, 10–25 $\mu g/m^3$), erosion of structural wax started immediately after both flushing and needle elongation had terminated, and proceeded at a rate seven to ten times higher than under unpolluted conditions (Huttunen and Laine 1983). Crossley and Fowler (1986) have established scoring criteria in order to quantify degradation, and have found increased "weathering rates" under polluted conditions (SO_2 concentration ca. 40 $\mu g/m^3$). The degree of erosion could be related to the sulfur content of the needles (Huttunen and Laine 1981) and was aggravated by the presence of particles on the leaf surface (Huttunen and Laine 1983). The most rapid degradation of the tubules of 2- to 5-year-old needles occurred during winter (Huttunen and Laine 1983). The surfaces of *Pinus sylvestris* needles from polluted sites (40 μg SO_2/m^3) had lower contact angles for water than from unpolluted sites [(10 μg SO_2/m^3) (Cape and Fowler 1981; Cape 1983)].

Similar results have been reported for *Pinus strobus*; site comparisons have again shown accelerated degradation of wax morphology under polluted conditions (Percy and Riding 1978; Krause and Dochinger 1984). These phenomena have also been reproduced under controlled conditions (Riding and Percy 1985). Riding and Percy (1985) found a reduced rate of wax deposition during needle development when the plants were fugimated for 11 weeks from bud-break, with the relatively high concentration of 132 μg SO_2/m^3. This fugimation led to a 5-week delay in the complete filling of the epistomatal chambers with tubular wax. Ambient air with SO_2 and O_3 at background levels had no effect on the structure of the epistomatal plugs of 10 SO_2-tolerant clones of *Pinus strobus*, whereas fissures occurred in 12 SO_2-sensitive clones (Krause and Houston 1983). Charcoal-filtered air was used as a control and had no effect on either type of clone. No impact of O_3 fumigation (0.30 ppm) on the structure of the epistomatal plug, and no correlation between wax morphology and O_3 sensitivity of two clones of *Pinus strobus* were found (Trimble et al. 1982). In the case of *Pinus banksiana*, even fumigation with 132 μg SO_2/m^3 for 11 weeks beginning at bud-break had no effect on the structure of epicuticular wax and on its deposition during ontogeny (Riding and Percy 1985).

Appreciably less information can be found on the action of air pollution on the structural wax of *Abies alba*. Rinallo et al. (1986) subjected 5-year-old seedlings to the same treatments of acidic sprays as described above for *Picea abies*. Again, agglomeration of tubular wax and encrustations were observed, although at a somewhat reduced frequency. Acidic precipitation alone and a combination of O_3 and acidic rain had no effects on the wax morphology of needles of *Abies alba* seedlings grown under controlled conditions (treatments specified above; Schmitt et al. 1987). Fusion of tubules occurred after treatments with a combination of SO_2 and acidic rain, as well as when a combination of SO_2, O_3 and acidic rain had been applied. However, this was the case only in very flat stomatal antechambers. The tubes filling deeply sunken epistomatal chambers were not affected at all.

Thus, it appears that air pollution does accelerate the alteration of the wax tubules and the erosion of needle surface waxes under both natural and controlled conditions. However, a critical re-evaluation of the reports about the occurrence of fissures in the wax plug over the stomata seems necessary to exclude the possibility of artifacts. Several authors have been able to show that inappropriate sample-preparation methods result in this type of structure.

While the fact of structural changes is unquestionable, the hypotheses put forward so far to explain the phenomenon are lacking experimental and analytical evidence. A recurring theme is the rather general idea that chemical factors may be the immediate cause, or that atmospheric gases may react with the wax altering its chemical nature. Cape (1986) has attributed morphological changes of the wax of *Pinus sylvestris* to age-dependent shifts in chemical composition, and Franich et al. (1978) have suggested that increased synthesis of diterpene acids in older needles may be responsible.

Recently, Sauter and Voß (1986) have suggested that a NO_x derived formation of O_3 would cause the structural breakdown of unsaturated components of conifer cuticular waxes. These authors have based their argument on a report by Beri and Lemon (1970), which states that unsaturated alkanoic acids are common constituents of the SCL of conifer needles. However, the unsaturated lipids they found were most likely extracted from the interior of cells, rather than from the cuticle (see discussion of SCL extraction). No other studies on the composition of conifer SCL have confirmed the presence of unsaturated alkanoic acids. Further, there is no evidence supporting the assumption that 10-nonacosanol, which probably forms the structural wax of conifers, may be degraded or substituted by a different compound in polluted environments. These considerations lead to the conclusion that it is very unlikely that the impact of SO_2, O_3, NO_x or acidic precipitation on the surface deposits of SCL should directly result in the widely observed structural changes.

However, an alternative hypothesis based on the physico-chemical properties of the tubular wax structures can be proposed. It is possible that the same combination of chemical, physical and physiological factors, presumably resulting in the natural degradation of structural wax morphology during the aging of the needles, also causes accelerated deterioration under polluted conditions. Air pollutants may affect structural wax morphology of conifer needle surfaces in one or both of the following two ways:

1. Degradation results from the principal instability of the wax tubules which implies that continuing maintenance of these structures is necessary. On the leaf surface, this can only be achieved by replacing degraded wax tubules by newly synthesized ones. Consequently, any process or substance that significantly retards the metabolism, and thus the wax biosynthesis in the needles, is likely to result in an accelerated appearance of degraded wax structures. This hypothesis is supported by SEM findings: the alterations resulting from air pollutants and acidic precipitation do not differ from those attributed to natural aging. Further, circumstantial evidence is provided by the observation that different conifer species (and even individuals from the same species) vary in their susceptibility to structural degradations when exposed to identical air pollutants, even though their wax tubules are very likely of identical composition. Huttunen and Laine observed (1983) that the most rapid degradation of SCL morphology occurred during winter, further supporting the view that metabolism is involved.

The transformation of the tubules into larger plates is favored by elevated temperatures as the mobility of the constituent molecules increases. Consequently, a larger degree of morphological degradation should be expected in needles subjected to higher temperatures. The gradients of structural degradation observed with needles of *Picea abies* which have been directly related to the exposition to the sun (Kim 1985) are in accordance with this suggestion.

2. Another way by which air pollutants may alter epicuticular wax morphology could be the sorption of foreign volatile molecules on the surface and in the interior of the tubular structures. These volatiles could be either of atmospheric origin or be produced in the mesophyll of conifer needles, possibly in rates proportional to their susceptibility to air pollutants. The stomata not only are the sites of gas exchange, but also serve as the ports through which considerable amounts of organic vapors (e.g., monoterpenes) originating from the metabolism of needles are emitted. Due to the unstirred environment within the voids of the epistomatal wax plug, prolonged and elevated concentrations of these endogeneous volatiles may build up in the immediate vicinity of the wax tubules. Sorption in the wax tubules can easily occur and result in a swelling of the molecular arrangement. Swelling loosens molecular interactions and thus strongly favors the rearrangement of molecules. This eventually can lead to the transition into the most stable form of solid *n*-alkanols, namely, crystalline plates.

3 Transport Properties of Plant Cuticles

3.1 The Description of Cuticular Penetration

3.1.1 Nonelectrolytes

The rules governing the cuticular penetration of uncharged species have been extensively reviewed for water by Schönherr (1982) and for organic solutes by Schönherr and Riederer (1989). The sequence of events leading to the continuous transport of a nonelectrolyte across the cuticle may be viewed as the combination of three separate processes: (1) the sorption (dissolution) of the compound in the cuticle phase; (2) its diffusion across the cuticle; and (3) its desorption from the opposite surface of the cuticle. This basic concept was introduced by Crafts and Foy (1962) for the transport of nonelectrolyte solutes, but can be modified to apply to any type of molecule. Steps (1) and (3) can be characterized by the concentration of the permeant in the cuticle in relation to the adjacent medium, while step (2) depends on the mobility of the sorbed molecules within the cuticle phase.

The degree of sorption of a compound in the cuticle phase can be determined by equilibrating a given amount of isolated cuticular membrane with a precisely known volume of solution, gas or vapor. After equilibrium has been reached, the concentrations of the compound in question are determined in both the cuticle and the surrounding medium (Riederer and Schönherr 1984; Kerler and Schönherr 1988a; Schönherr and Riederer 1989). Then, the partition coefficient (K) can be calculated as the ratio of the molal concentration in the cuticle phase (C_c) and the molal concentration in the supernatant (C_o) according to

$$K = C_c/C_o. \tag{1}$$

The partition coefficient, a dimensionless figure, expresses the cuticular concentration of a sorbed permeant relative to its concentration in the outside medium. Sorption is strictly reversible, even though the period of time needed for attaining sorption equilibrium is a matter of the mobility of the solute in both phases. Partition coefficients may not only be used for analyzing the transport properties of cuticles, but also for describing quantitatively the accumulation potential of the lipophilic cuticle for nonpolar molecules from the environment and the interior of the plant (Riederer and Schönherr 1984; Schönherr and Riederer 1989).

The time needed to establish equilibrium depends only on the mobility of the (solute) molecules in the cuticle, provided the outside solution is well stirred. The mobility of a species within the cuticle can be described quantitatively by the diffusion coefficient (D in $m^2\ s^{-1}$). At constant temperature and pressure, the diffusion coefficient is only a function of the properties of the permeant and the membrane and ideally does not depend on the concentration of sorbed molecules (Barrer 1968; Cussler 1984).

The flow of material across a membrane increases with the amount of permeant present within the membrane relative to the outside concentration (high partition coefficient), and the faster sorbed molecules move within the membrane (high diffusion coefficient). This qualitative statement can be expressed quantitatively by

$$p = K\,D \tag{2}$$

where p (in $m^2\ s^{-1}$) is the permeability coefficient of a membrane of unit thickness for a given chemical (Hartley and Graham-Bryce 1980; Crank 1975). This simple relationship applies only to homogeneous membranes where neither K nor D vary in the direction of transport (Barrer 1968). Even though this assumption may not be valid for cuticular membranes (Schönherr and Riederer 1989), Eq. (2)

nevertheless expresses the fundamental relationship determining the transport across cuticles in qualitative terms.

The permeability coefficient p characterizes the transport properties of a membrane of unit thickness, thus serving well for comparisons between membranes consisting of different types of material. Usually, instead of the permeability coefficient, a mass transfer coefficient (permeance, P, in m s^{-1}) is used to characterize the transport across a membrane of given composition and thickness which may not be known precisely. The permeance is obtained from the permeability coefficient according to

$$P = p/l \tag{3}$$

with l being the membrane thickness (in m). Permeances are related to resistances (R in s m^{-1}) by

$$R = 1/P. \tag{4}$$

Resistances are additive and thus can be used to calculate the overall resistance of several barriers in series. The permeance of a given membrane can be measured from steady-state diffusion experiments without making any assumptions concerning the variation of partition or diffusion coefficients. In a typical experiment, the amount of substance (M) that diffuses across a membrane of area (A) in a certain period of time (t) is determined and is related to the driving force acting. Generally speaking, the driving force is the difference between the chemical potentials of the phases at each side of the membrane ($\triangle\mu$):

$$\triangle\mu = R\,\mathrm{T}\ln(a^o/a^1) \tag{5}$$

R is the ideal gas constant, T the absolute temperature (in K) and a the activity of the permeant in the two bulk phases (i and o) adjacent to the membrane. Under isothermal conditions, the driving force is actually determined by the gradient of activities. Since the ratio of activity coefficients for nonelectrolyte solutes at low concentrations in the cuticle and the aqueous solution can be assumed to be close to unity, the activities may be replaced by the concentrations in the phases bordering the inner and outer surface of the cuticle. In the cases of vapors or gases, the difference of vapor pressure or mol fractions may be used as driving force.

After a certain time lag, the flow rate (F) of a permeating species becomes independent of time: the system has reached the steady state (Crank 1975; Kerler et al. 1984; Schönherr and Riederer 1989). The lag time may serve to estimate the diffusion coefficient if homogeneity of the membrane can be assumed (Barrer 1968). The steady-state flow rate is related to the permeance according to

$$P = \frac{F}{A \quad \triangle C} \tag{6}$$

with $\triangle C$ (in mol m^{-3}) being the concentration difference between both adjacent phases. Equation (6) is of great practical importance, as it allows the calculation of the flow rate of a compound across a given cuticle if permeance and driving force are known. The flow of nonelectrolytes is independent of the presence of additional compounds in the system. For situations where the flow rate (F) is not invariable with time, different approaches have to be used (see Hartley and Graham-Bryce 1980).

In a limited number of cases (e.g., transport of gases or vapors over an interphase separating two gas-phase compartments) the use of differences of mole fractions as driving forces has been preferred for practical reasons. In such cases, for an identical flow rate, permeances with a different numerical value and dimension (mol m^{-2} s^{-1}) are obtained. These permeances can be easily converted into concentration difference based values by multiplication with the molar volume of air at the given temperature and pressure.

3.1.2 Electrolytes

The general principles of nonelectrolyte flow across cuticular membranes can also be applied to the transport of electrolytes. Again, the flow rate of a chemical species crossing the cuticle is a function of its concentration and mobility in the cuticle and the driving force acting on it. In electrolyte transport, however, the coupling of flows of all mobile species present in the system has to be taken into account and the correct driving forces must be used. Simple analytical solutions of these problems are available only for systems containing a small number of ionic species, while multi-ion systems are rather complicated to treat (Helfferich 1962).

The presence of dissociable groups affects the transport of ions across cuticles. The cuticle is a polyelectrolyte with an isoelectric point around 3.0 (Schönherr and Huber 1977). Fixed negative charges are due to carboxyl groups contributed by cutin and polyuronic acids and proteins (Schönherr and Bukovac 1973). Fixed positive charges are thought to originate from residues of basic amino acids contained in a nonextractable form within the cuticle. The total cation exchange capacity of the cuticle obtained from various plant species has been found to range from 250–500 meq kg^{-1} while the fixed positive charges have amounted to 12–25 meq kg^{-1} (Schönherr 1982).

The ionization of fixed charges depends on the pH. Under physiological conditions ($pH > 3$), the cuticle is a weakly acidic cation exchanger whose negative charges probably are compensated by Ca^{2+} counterions. When it is brought into contact with a dilute electrolyte solution (cell wall water at the inner, or rain and fog at the outer surface), the counterions can be exchanged for different cations. The anions contained in the outside solutions will be excluded practically completely from the cuticle due to the repelling force of the Donnan potential. Thus, the polyelectrolyte nature of plant cuticles has two far-reaching consequences: with dilute solutions bordering the surfaces of the cuticle only cations can be taken up by ion exchange and transported, while anions remain excluded. Cuticles are permselective membranes. Secondly, the concentration of a permeating cation is directly determined by the concentration of fixed negative charges within the cuticle and not by the concentration in the outside medium. The exchange capacity is only a function of the pH of the bathing solution. At solution concentrations equal to and higher than the concentration of the fixed charges in the cuticle, increasing amounts of sorbed electrolytes (cations and anions) within the membrane have to be expected as Donnan exclusion becomes less effective (Helfferich 1962). In any case, the intracuticular concentration of permeant cannot be simply described by a partition coefficient as for nonelectrolytes.

A second complication encountered with electrolytes concerns the formulation of the driving force acting on the molecules on both sides of and within the membrane. For the diffusion of nonelectrolytes it was sufficiently precise to consider the difference of activities and even of concentrations between the two sides of the cuticle. When analyzing the transport of a certain electrolyte species (say K^+) across the cuticle, however, measuring the potassium concentrations or activities on both sides of the membrane cannot determine whether a net K^+ transport will occur and in which direction ions will flow. With ionic species, fluxes are always coupled, because no net charge transfer can occur and electroneutrality must be maintained in any part of the system. Whenever an ion crosses the cuticle, either an ion of identical charge must move in the opposite direction, or an ion of opposite charge must follow. This is almost impossible as long as Donnan exclusion is effective.

The factor coupling the fluxes of different ionic species is the electrical potential across the membrane. Thus, an additional term must be added to the chemical potential according to Eq. (5) in order to express the difference of chemical potentials as the driving force for the transmembrane transport of an ionic species j. The difference of the electrochemical potential ($\Delta\tilde{\mu}_j$) is given by

$$\Delta\tilde{\mu}_j = R\,\mathrm{T} \ln (a_j^o/a_j^i) + z_j\,\mathrm{F}\,\Delta\mathrm{E} = \Delta\mu_j + z_j\,\mathrm{F}\,\Delta\mathrm{E} \qquad (7)$$

with z_j being the charge number (e.g., for K^+: +1, for Ca^{2+}: +2), F the Faraday constant, and ΔE the difference of the electrical potential between the two sides of the cuticle.

The influence of the electrical term in Eq. (7) on the driving force is considerable: a –18 mV difference of the electrical potential exactly compensates an activity difference of the species in question of a factor of two between the solutions adjacent to the cuticle. Thus, a single electrolyte species may even flow against the direction of the concentration gradient if the driving force is dominated by the electrical term. Transport of an electrolyte species j proceeds as long as $\Delta\tilde{\mu}_j \neq 0$. At equilibrium ($\Delta\tilde{\mu}_j = 0$), flow ceases because the two terms on the righthand side of Eq. (7) cancel each other, i.e., the driving force exerted by the activity gradient is exactly balanced by the force derived from the electric potential difference.

Electrolyte transport across cuticles, and ion exchange on their surfaces, play important roles in the leaching of nutrients from the aerial parts of plants and in the uptake of foliar fertilizers. The leaching phenomena observed in a large number of plant species (Tukey 1970) need to be analyzed according to the principles outlined above. It is obvious, however, that the quantitative analysis of the cuticular transport of mixtures of nutrient ions of different valencies, occurring at different activities, is a rather formidable task.

3.2 Transport Parameters

Few quantitative data on the transport properties of plant cuticular membranes are available. In the past, most studies on this subject were only qualitative. For a critical evaluation of methods see Schönherr (1982), Kerler et al. (1984) and Schönherr and Riederer (1989). A short outline of the transport parameters of plant cuticles will be presented. It may be useful in assessing the direct impact of environmental factors and pollutants on the above-ground parts of plants.

Water permeability of leaf and fruit cuticles has been measured for a number of species (Schönherr 1982). Permeances based on the gas-side concentration difference as driving force ranged from 8.19×10^{-7} (*Schefflera*) to 1.22×10^{-5} (*Pyrus*) for leaf cuticles and from 2.23×10^{-5} (*Solanum*) to 1.42×10^{-4} (*Lycopersicon*) m s^{-1} for fruit cuticles. Apparent diffusion coefficients, which assumed that the cuticles were homogeneous in direction of transport, were in the range from 1.57×10^{-15} (*Schefflera*) to 4.96×10^{-14} (*Lycopersicon*) m^2 s^{-1}. Permeances tended to increase with increasing membrane thickness. When comparing different species, thin cuticles were better water barriers than thick ones, while within a species, no correlation between thickness and permeability could be observed (Becker et al. 1986).

So far, transport parameters for gases have been determined only for cuticles isolated from the leaves of *Citrus* and the fruits of *Lycopersicon* and *Capsicum*. Permeances for O_2 with gas-side concentration differences used as the driving forces amounted to 8.98×10^{-9} (*Citrus*), 3.33×10^{-8} (*Lycopersicon*) and 4.01×10^{-8} (*Capsicum*) m s^{-1}. Mean overall diffusion coefficients were in the range of 2.5 $\times$ 10^{-14} (*Citrus*) m^2 s^{-1} (Lendzian 1982). Gas-side permeances for *Citrus* leaf cuticles were 5.40×10^{-8} m s^{-1} (CO_2 and 2.01×10^{-7} m s^{-1} (SO_2). The permeance for NO_2 has been found to be much higher than that of O_2 or SO_2 (Lendzian 1987). Plant cuticles have proven to be impermeable to O_3, as this gas is apparently destroyed during its passage across the cuticle (Kerstiens 1988).

More data exist on the transport of organic nonelectrolytes across cuticular membranes. This special interest is due to the importance of the cuticle during the foliar penetration of agrochemicals (Chamel 1986) and environmental pollutants. Permeances based on concentration differences in the liquid phase as the driving forces for organic nonelectrolyte solutes range from 1×10^{-10} m s^{-1} (*Ficus*, nondissociated species of 2,4-dichlorophenoxyacetic acid) to 8.6×10^{-7} m s^{-1} (*Citrus*, hexachlorobenzene) (Riederer and Schönherr 1985; Kerler and Schönherr 1988b). The permeance of *Citrus* leaf cuticles have been estimated with an accuracy sufficient for practical purposes from the 1-octanol/water partition coefficient ($K_{o/w}$) according to the following equation: $\log P = 0.704 \log K_{o/w} - 11.2$ ($r = 0.91$) (Kerler and Schönherr 1988b). Due to membrane heterogeneity, only rough estimates of diffusion coefficients can be given. They are in the range of 10^{-14} to 10^{-15} m^2 s^{-1} for the polymer matrix, the cuticle where the soluble cuticular lipids has been removed by solvent extraction, and of the order of 10^{-16} to 10^{-17} m^2 s^{-1} (or even less) in the outer layer of the cuticles (Schönherr and Riederer 1988).

The partition coefficients determined for organic nonelectrolytes in the system cuticle/water ranged over more than eight orders of magnitude. The extreme values were 0.048 for methanol and 2×10^7 for diethylhexylphthalate (Kerler and Schönherr 1988a; Schönherr and Riederer 1989). Cuticle/water partition coefficients ($K_{c/w}$) can also be predicted from both the $K_{o/w}$ values and the aqueous solubilities of the solutes (S_w in mol l^{-1}) according to the linear equation: $\log K_{c/w} = 0.057 + 0.970 \log K_{o/w}$ ($r = 0.987$) and $\log K_{c/w} = 1.118 - 0.596 \log S_w$ ($r = 0.978$), respectively (Schönherr and Riederer 1989).

Quantitative information concerning the transport of electrolytes across plant cuticles is lacking completely, as no studies appear to have been conducted where the electrical gradient across the membrane has been measured. Permeances cannot be calculated and flows cannot be interpreted if the electrical potential gradient across the membrane is unknown. Single-ion permeabilities for extracted *Citrus* leaf cuticles have been measured in self-diffusion experiments with no concentration and electrical gradients for $^{24}N^+$ and $^{32}Br^-$ (Schönherr and Huber 1977). It is not possible, however, to estimate leaching or ion uptake under natural conditions from these data.

3.3 Factors Affecting Cuticular Transport and Accumulation

Various factors influencing transport and accumulation in plant cuticles have been identified. The most striking effect is that caused by the soluble cuticular lipids (SCL). Solvent extraction has been found to increase the permeance for water of *Lycopersicon esculentum* fruit, *Citrus aurantium* leaf, and *Allium cepa* bulb scale cuticles by factors of 20,233, and 1542, respectively (Schönherr 1982). Findings for the extraction of SCL showed less effect on gas transport: permeances of *Citrus* leaf cuticles for O_2 (Lendzian 1982), SO_2, and CO_2 (Lendzian 1984) increased by factors of only 15, 17, and 26, respectively. The effect of the SCL on organic nonelectrolyte permeance of plant cuticles depended on the polarity of the permeating molecule. With decreasing polarity, the effect of SCL extraction decreased (Schönherr and Riederer 1989). After extraction, the permeance of 4-nitrophenol was 1330 times higher, while the value for the very nonpolar diethylhexylphthalate increased by a factor of only 9.3 (Kerler and Schönherr 1988a).

The soluble cuticular lipids can also affect sorption. Their effect, however, is much smaller than that on permeability. With 2,4-dichlorophenoxyacetic acid and the cuticles from twelve plant species, significant increases of the cuticle/water partition coefficients ranging from 1.2–2.6 have been found in all cases (Riederer and Schönherr 1984). The cuticle/water partition coefficients of eleven compounds strongly differing in lipophilicity were influenced to varying degrees by SCL extraction. The factors of change ranged from about 0.9–2.4 (Kerler and Schönherr 1988b).

The large effect of SCL extraction on permeability evidently demonstrates the decisive function of soluble cuticular lipids as the main transport limiting barrier in plant cuticles. The low permeances of the cuticular membranes are due to crystalline deposits of the long chain aliphatic compounds within the cutin matrix and/or on the surface of the cuticle, which appear to be inaccessible even for slightly to intermediately nonpolar organic nonelectrolytes. Diffusion is assumed to take place exclusively in submicroscopic defect structures between more or less extended crystallites (Schönherr and Riederer 1989). The limited effect on sorption shows that only a small portion of the sorption sites of cutin is occupied by the SCL.

Structural features of the cuticle also affect the rate of desorption observed when a plant cuticle preloaded with solute is washed simultaneously on both surfaces. This is a situation naturally occurring after a dose of chemical has been taken up by the cuticle and desorption starts from both sides. Under bilateral desorption conditions, only 2–3% of the amount of 2,4-dichlorophenoxyacetic acid contained in cuticular membranes has been reported as desorbed from the outer surface, while 86–92% desorbed from the inner surfaces within 6 h (Schönherr and Riederer 1988). This high assymetry has been found to be especially pronounced in leaf cuticles and has been attributed to the presence of apolar and crystalline SCL on the surface and in the outer layers of the cuticles. As a consequence, the wash-out of (lipophilic) solutes once sorbed within the cuticle is slow and quantitatively insignificant. Most of the sorbed chemical will diffuse to the interior of the organ.

These factors influencing transport across and accumulation within the cuticle depend only on the properties of the solutes and the cuticles. The most important exterior factor affecting transport and accumulation is the driving force acting between the two media adjacent to the cuticle. According to Eqs. (5) and (6) for nonelectrolytes, it is the ratio of the activities or concentrations on both sides of the membrane which causes a net flow across the membrane. Thus, the driving force is not only a function of the outside concentration, but a direct result of the concentration which a solute can build up in the aqueous phase of the apoplast. This internal concentration can be kept low if transport and metabolism of the permeant occur, but it will rise to an equilibrium value if both mechanisms of withdrawal are absent. As a consequence, high cuticular concentrations of pollutants are probably good indicators only for the lack of withdrawal reactions and are less suited for indicating the phytotoxic potential of the chemical (Schönherr and Riederer 1989).

Another factor affecting the uptake of solute into the leaf is the volume of the solution droplet reaching the cuticle surface. When the solute is taken up by the cells and metabolized and/or translocated rapidly, the time needed for the penetration of 50% of the solute content of a sessile droplet can be estimated as a function of the droplet size and the permeance of the cuticle (Schönherr and Riederer 1989). For a hemispherical droplet having a radius of 1 μm and for a permeance of 1×10^{-6} m s^{-1}, only 0.46 s are needed for 50% penetration. This time increases by a factor of 10 for every decrease of P by one order of magnitude. Seventy-seven min are required when P is only in the range of 1×10^{-10} m s^{-1}. Keeping the permeance constant, the time needed for 50% penetration increases proportionally to the droplet size. Schönherr and Riederer (1989) have estimated

77 min, 12.8 h, 5.3 days and 7.6 weeks for droplet radii of 1, 10, 100 and 1000 μm, respectively, and a permeance of 1×10^{-10} m s^{-1}.

When equilibrium between sessile droplet and cuticle is attained because no withdrawal by metabolism or translocation occurs, the relative distribution of the solute formerly contained in the droplet can be estimated. About 99.96, 99.6, 96.0, and 70.4%, respectively, of a solute with a cuticle/water partition coefficient of 10^3 would be contained at equilibrium in a 2 μm thick cuticle, when taken up from sessile droplets whose radii in the spherical state had been 1, 10, 100, and 1000 μm (Schönherr and Riederer 1989). Both estimates underline the importance of droplet size for solute uptake. They predict that uptake will be faster and accumulation will proceed to a larger degree if a given volume of precipitation arrives as small droplets, instead of large drops, at the cuticle surface.

It is doubtful whether air pollutants directly affect the permeability of plant cuticles. Kerstiens (1988) has studied the effect of a 1-hour treatment with nitrogen containing an unnaturally high dose of 2.5% O_3 on the water permeability of isolated leaf and fruit cuticular membranes from several plant species. The effects observed were somewhat contradicting as they ranged from a decrease of permeance by ca. 20% in *Ficus* leaf cuticles to an increase by 200% in *Capsicum* fruit cuticles. As a general rule, the O_3 effect increases with the initial water permeance. Treating seedlings of *Fagus sylvatica* for 4 weeks, starting at the time of flushing with environmental concentrations of O_3 (peak concentrations up to 130 ppb), had no effect on the water permeance of astomatous cuticular membranes measured in situ (Kerstiens 1988).

3.4 The Role of Cuticles in Transport

The question concerning the contribution of cuticular transport to the total uptake, loss, and exchange of liquids, gases and solutes has remained open so far. The combined surface area of the above ground organs of plants usually exceeds the ground area they are growing on. Leaf area indices of up to 20 have been observed (Schulze 1982) while in the Fichtelgebirge, leaf area indices for *Picea abies* stands of about 10 have been estimated (Oren et al. 1987). Due to the large surface area, most airborne pollutants are intercepted by leaf surfaces rather than by the soil surface. On the other hand, material originating from within the plant also encounters the cuticular barrier if it is to be removed via aboveground organs by leaching.

The cuticle controls the bidirectional flow across the interphase to varying degrees, depending on the physical state of the material reaching its inner or outer surface. For material in the liquid state and for nonvolatile solutes, cuticular control is complete as no alternative pathways exist. It has been shown by Schönherr and Bukovac (1972) that stomatal pores cannot be penetrated by any naturally occurring aqueous solution, due to the geometry and the highly hydrophobic surface of the stomata. In the absence of external pressure, the development of an aqueous continuum across the stomatal pore is impossible. Thus, any bulk flow of liquids or solutions from the surface of a leaf into the interior and vice versa is excluded. Liquid molecules and solutes have to take the detour across the cuticle.

The situation is less straightforward with gases and vapors. Here, two parallel pathways across the leaf/air interphase exist: the pores of open stomata and the cuticle. The resistance of the stomatal route to the transport of material in the gas phase is several orders of magnitude lower than the resistance of the cuticular pathway. Thus, if the stomata are open, most of the exchange of gases and vapors between the interior of a leaf and the surrounding atmosphere takes place via the stomatal pores, while cuticular transport plays only a negligible role. When the stomata are closed, however, the flow of gases and vapors across the leaf/air interphase is determined exclusively by the cuticle. With most plant species, the stomatal and cuticular pathways for the transport of gases and vapors are completely independent. This is not the case, however, with most members of the family of Pinaceae which develop large deposits of structural wax in the stomatal antechamber. Here, cuticular structures interfere with the gas transport across the stomata.

While it is clear that the cuticle is the first barrier in the foliar uptake of solutes and liquids with a low vapor pressure, the relative contribution of the cuticular membrane to the overall resistance of the leaf is frequently a matter of discussion. Before a molecule can reach or leave the cytoplasm of the epidermal cells, it has to cross the plasmalemma and the epidermal cell wall in addition to the cuticle. The question is whether the former two resistances acting in series with the cuticle will significantly contribute to the total resistance. Schönherr and Riederer (1989) have estimated the resistances of cuticles, cell walls, and the plasmalemma, using the appropriate permeances, diffusion coefficients, partition coefficients, and path lengths. From measured permeances, a range of resistances of cuticular membranes of 10^6–10^{10} s m^{-1} depending on the lipophilicity of the permeating molecule has been calculated. The resistance of a 5-μm-thick cell wall is estimated to be not larger than 5×10^4 s m^{-1}. Estimates for the resistance of the plasmalemma have been in the range of 50–5×10^5 s m^{-1} for solutes with 1-octanol/water partition coefficients between 10^6 and 10^2, respectively.

In both the cuticle and the plasmalemma, resistances decrease with increasing lipid solubility of a solute. Thus, when comparing the resistances of the two types of membrane, one has to look at the resistances for a solute of equal lipophilicity, i.e., one must compare the lowest with the lowest and the highest with the highest values of the resistances given above. These considerations clearly show that for all solutes with cuticular permeances smaller than 10^{-6} m s^{-1}, it is the cuticle which represents the only transport barrier. This conclusion is valid for both the uptake into and for the leaching of substances from leaf surfaces. The condition $P < 10^{-6}$ m s^{-1} is met by all cuticles and substances studied so far.

3.5 Gas Diffusion in the Stomatal Antechamber of Pinaceae

The presence of tubular deposits of SCL in the stomatal antechambers of the needles of most Pinaceae species has stimulated much speculation about the function of this structure. This interest has been further advanced by SEM observations that the morphology of the outer surface of these porous "plugs" is altered with increasing needle age and that these changes are accelerated in the presence of various air pollutants and acid precipitation. This observation has led many authors to the speculative conclusion that gas exchange through the stomata is impaired due to the observed degradation of the plug surface.

However, positive evidence linking changes of wax morphology to reduced stomatal conductances for gases or water vapor are lacking. Until direct experimental evidence is available, a theoretical consideration of the role of the epistomatal SCL deposits in gas transport may be useful in order to assess the potential effects of structural changes. Jeffree et al. (1971) have investigated the morphology of the epistomatal deposits of *Picea sitchensis* needles and derived some quantitative estimates concerning the possible effect on the exchange of water vapor and CO_2. Under the assumption that the wax tubules (1 μm long, 150 nm outer diameter) occupy about half of the cross-sectional area at the mouth of the antechamber and that the diffusion coefficient in the voids reaches only 70% of its value in the free gas phase, they concluded that the wax deposits contributed 37% to the total resistance of the stomatal complex. Considering total pathways, they estimated that the wax-filled antechambers contributed 66 and 32% of the entire resistance toward the diffusion of water vapor and CO_2, respectively.

Generally speaking, the diffusive resistance of a porous material like the epistomatal "plugs" is increased by the following features: (1) the reduction of cross-sectional area available for diffusion; (2) the increased diffusion paths due to tortuous voids; and (3) the reduced diffusion coefficients in the narrow interstitial voids (Cussler 1984). While (1) and (2) are almost universal for porous materials, feature (3) is restricted to material with pores whose diameters are much smaller than the mean free path of the diffusing molecules (Knudsen regime). At ambient temperatures and pressures, the mean free paths of CO_2, O_2, and H_2O are 62, 70 and 90 nm, respectively (Cussler 1984). Thus, pore diameters less than ca. 30–45 nm would be necessary to reduce diffusion coefficients considerably. The voids between the wax tubes have been estimated to have diameters of ca. 2 μm (Jeffree et al. 1971). Consequently, an effect of pore diameter on the diffusion coefficient of gases in the voids is quite unlikely.

The epistomatal deposits of SCL inevitably lead to a reduction of cross-sectional area and to an increase of the diffusion path length in the stomatal antechamber. This effect of tortuous pathways can be expressed phenomenologically by a tortuosity factor, by which the diffusion coefficient is divided in order to correct for the longer effective path length travelled by the diffusing molecule. Tortuosity factors for a wide variety of porous materials have been found to be in the range of 2–6 with a mean value around 3 (Cussler 1984).

Using the dimensions of the stomatal antechamber measured by Jeffree et al. (1971) and applying their formulas for the calculation of total resistance, the effects on total resistance of varying areas occluded by compact wax and of different tortuosities can be shown. Such an analysis leads to the conclusion that (1) increasing tortuosity is much more efficient in increasing resistance than filling increasing portions of the cross-sectional area with SCL, and (2) that the effect of a reduced diffusion cross-section is higher the more area is already occluded. A degradation of plug structure, while conventionally seen only as a reduction of cross-sectional area at the mouth of the antechamber, may in fact have both consequences. Relatively small structural alterations in the interior of the plug may lead to a greatly increased tortuosity and thus to an increase in resistance. These considerations may explain the 50% reduction in leaf conductance observed in 5-year-old *Picea abies* needles as compared to 1-year-old ones (Zimmermann et al. 1988).

4 Conclusions

With respect to the direct impact of air pollution on the aboveground parts of conifers, the following conclusions can be drawn regarding the structure, the composition and the transport properties of cuticles:

1. The cuticle covers all aboveground primary parts of plants. It is a continuous extracellular membrane primarily composed of the biopolymer cutin and waxlike lipids embedded within the cutin matrix and/or covering the outer surface of the cuticle.
2. The cutin polymer is insoluble in any solvent and can only be degraded by saponification or transesterification. The cuticular waxes, which are solids at ambient temperatures, dissolve only in organic solvents and therefore are called soluble cuticular lipids (SCL). They are saturated, long-chain compounds belonging to several substance classes. Estolides (polyesters of hydroxyalkanoic acids and alkanols) and secondary alkanols are characteristic of conifer SCL.
3. So far, no effects of air pollutants or acidic precipitation on the fine structure or the composition of cutin and SCL have been observed in any species of conifers.
4. Epicuticular SCL of conifers form tufts, consisting of wax tubules especially within and in the vicinity of the epistomatal chambers. The morphology of these deposits varies with needle age and in the presence of air pollutants or acidic precipitation. The morphological changes very likely are no consequence of the direct impact of air pollution. An alternative hypothesis explaining this phenomenon on the basis of the physicochemical properties of those structures is presented.
5. During the uptake into or the loss of substances from leaves, the cuticle is the transport limiting barrier for any substance which is not in the gaseous state. The cuticular transport properties for water, gases, organic solutes and ions are discussed. The uptake of substances across the cuticle is influenced by the presence of SCL, the polarity of the molecules, and the concentration gradient. The size of solution droplets in rain or fog also influences uptake.
6. Currently, there is no evidence for any alteration of the transport properties of the cuticles of any plant species by the impact of air pollutants in environmental concentrations.

References

Adamson AW (1982) Physical chemistry of surfaces. 4th edn, Wiley, NY

Arndt U, Seufert G, Nobel W (1982) Die Beteiligung von Ozon an der Komplexkrankheit der Tanne (*Abies alba* Mill.) – eine prüfenswerte Hypothese. Staub Reinh Luft 42:243–246

Baig MN, Tranquillini W (1976) Studies on upper timberline: morphology and anatomy of Norway spruce (*Picea abies*) and stone pine (*Pinus cembra*) needles from various habitat conditions. Can J Bot 54:1622–1632

Baker EA (1982) Chemistry and morphology of plant epiticular waxes. In: Cutler DF, Alvin KL, Price CE (eds) The plant cuticle. Academic Press, Lond NY, pp 139–165

Baker EA, Hunt GM (1986) Erosion of waxes from leaf surfaces by simulated rain. New Phytol 102:161–173

Barrer RM (1968) Diffusion and permeation in heterogeneous media. In: Crank J, Park GS (eds) Diffusion in polymers. Academic Press, Lond NY, pp 165–217

Bayerisches Landesamt für Umweltschutz (1986) Lufthygienischer Jahresberich 1985. Schriftenr Bayer Landesamt Umweltsch 70:1–223

Bayerisches Landesamt für Umweltschutz (1987) Lufthygienischer Jahresberich 1986. Schriftenr Bayer Landesamt Umweltsch 76:1–238

Becker M, Kerstiens G, Schönherr J (1986) Water permeability of plant cuticles: permeance, diffusion and partition coefficients. Trees 1:54–60

Beri RM, Lemon HW (1970) Chemical examination of the wax from needles of black spruce (*Picea mariana*) and balsam fir (*Abies balsamea*). Can J Chem 48:67–69

Bermadinger E, Grill D, Golob P (1988) The different influence of magnesite emissions on the surface waxes of Norway spruce and silver fir. Can J Bot 66:125–129

Brunegger A, Grasenick F, Jakopic E, Waltinger H (1982) Ein Beitrag zur Präparation für elektronenmikroskopische Untersuchugen von Wachsstrukturen auf Pflanzenblättern. Mikroskopie 39:154–166

Campbell R (1972) Electron microscopy of the epidermis and cuticle of the needles of *Pinus nigra* var. *maritima* in relation to infection by *Lophodermella sulcigena*. Ann Bot 36:307–314

Cape JN (1983) Contact angles on needles of scots pine (*Pinus sylvestris*) growing in polluted atmospheres. New Phytol 83:293–299

Cape JN (1986) Effect of air pollution on the chemistry of surface waxes of scots pine. Water Air Soil Pollut 31:393–399

Cape JN, Fowler D (1981) Changes in epicuticular wax of *Pinus sylvestris* exposed to polluted air. Silva Fenn 15:457–458

Chabot JF, Chabot BF (1977) Ultrastructure of the epidermis and stomatal complex of balsam fir (*Abies balsamea*). Can J Bot 55:1064–1075

Chamel A (1986) Foliar absorption of herbicides: study of the cuticular penetration using isolated cuticles. Physiol Vég 24:491–508

Chichakli M, Jessen FW (1967) Crystal morphology in hydrocarbon systems. Ind Eng Chem 59:86–98

Corrigan D, Timoney RF, Donnelly DMX (1978) *n*-Alkanes and ω-hydroxyalkanoic acids from the needles of twenty-eight *Picea* species. Phytochem 17:907–910

Crafts AS, Foy CL (1962) The chemical and physical nature of plant surfaces in relation to the use of pesticides and to their residues. Residue Rev 1:112–139

Crank J (1975) Mathematics of diffusion. Clarendon Press, Oxford

Crossley A, Fowler D (1986) The weathering of scots pine epicuticular wax in polluted and clean air. New Phytol 103:207–218

Cussler EL (1984) Diffusion. Cambridge Univ Press, Cambridge

Däßler HG (1963) Untersuchungen an Nadelwachs rauchgeschädigter Tannen. Z Chem 3:424–425

DeLucia EH, Berlyn GP (1984) The effect of increasing elevation on leaf cuticle thickness and cuticular transpiration in balsam fir. Can J Bot 62:2423–2431

Fox RC (1958) The relationship of wax crystal structure to the water vapor transmission rate of wax films. TAPPI 41:283–289

Franich RA, Volkman JK (1982) Constituent acids of *Pinus radiata* stem cutin. Phytochem 21:2687–2689

Franich RA, Wells LG, Barnett JR (1977) Variation with tree age of needle cuticle topography and stomatal structure in *Pinus radiata* D. Don. Ann Bot 41:621–626
Franich RA, Wells LG, Holland PT (1978) Epicuticular wax of *Pinus radiata* needles. Phytochem 17:1617–1623
Franich RA, Gowar AP, Volkman JK (1979) Secondary diols of *Pinus radiata* needle epicuticular wax. Phytochem 18:1563–1564
Franich RA, Goodin SJ, Volkman JK (1985) Alkyl esters from *Pinus radiata* foliage epicuticular wax. Phytochem 24:2949–2952
Grill D (1973a) Rasterelektronenmikroskopische Untersuchungen an SO_2-belasteten Fichtennadeln. Phytopathol Z 78:75–80
Grill D (1973b) Rasterelektronenmikroskopische Untersuchungen an Nadeln einiger Pinaceen, Cupressaceen und Taxaceen. Mikroskopie 29:348–35
Grill D, Pfeifhofer H, Halbwachs G, Waltinger H (1987) Investigations on epicuticular waxes of differently damaged spruce needles. Eur J For Pathol 17:246–255
Günthardt MS (1985) Entwicklung der Spaltöffnungen und der epicuticulären Wachsschicht bei *Pinus cembra* and *Picea abies*. Bot Helvet 95:5–12
Günthardt-Goerg MS (1986) Epicuticular wax of needles of *Pinus cembra, Pinus sylvestris* and *Picea abies*. Eur J For Pathol 16:400–408
Günthardt-Goerg MS (1987) Epicuticular wax formation on needles of *Picea abies* and *Pinus cembra*. In: Stumpf PK, Mudd JB, Nes WD (eds) The metabolism, structure, and function of plant lipids. Plenum, NY, pp 557–559
Günthardt MS, Wanner H (1982a) Die Menge des cuticulären Wachses auf Nadeln von *Pinus cembra* L. und *Picea abies* (L.) Karsten in Abhängigkeit von Nadelalter und Standort. Flora 172:125–137
Günthardt MS, Wanner H (1982b) Veränderungen der Spaltöffnungen und der Wachsstruktur mit zunehmendem Nadelalter bei *Pinus cembra* (L.) und *Picea abies* (L.) Karst. an der Waldgrenze. Bot Helvet 92:47–60
Gullvåg BM, Østensen H (1986) Wax layer erosion in spruce needles – an indicator of air-borne pollution. J Ultrastruct Res 94:280
Haas K, Rentschler I (1984) Discrimination between epicuticular and intracuticular wax in blackberry leaves: ultrastructural and chemical evidence. Plant Sci Lett 36:143–147
Hanover JW, Reicosky DA (1971) Surface wax deposits on foliage of *Picea pungens* and other conifers. Am J Bot 58:681–687
Hartley GS, Graham-Bryce IJ (1980) Physical principles of pesticide behaviour. Academic Press, Lond NY
Helfferich F (1962) Ion exchange. McGraw-Hill, NY
Herbin GA, Robins PA (1968) Studies on plant cuticular waxes. III. The leaf wax alkanes and ω-hydroxy acids of some members of the Cupressaceae and Pinaceae. Phytochem 7:1325–1337
Holloway PJ (1982a) Structure and histochemistry of plant cuticular membranes: an overview. In: Cutler DF, Alvin KL, Price CE (eds) The plant cuticle. Academic Press, Lond NY, pp 1–32
Holloway PJ (1982b) The chemical constitution of plant cutins. In: Cutler DF, Alvin KL, Price CE (eds) The plant cuticle. Academic Press, Lond NY, pp 45–85
Holloway PJ (1984a) Cutins and suberins, the polymeric plant lipids. In: Mangold HK, Zweig G, Sherma J (eds) CRC Handbook of chromatography, Lipids, vol 1. CRC Press, Boca Raton, pp 321–345
Holloway PJ (1984b) Surface lipids of plants and animals. In: Mangold HK, Zweig G, Sherma J (eds) CRC Handbook of chromatography, Lipids, vol 1. CRC Press, Boca Raton, pp 347–380
Hunneman DH, Eglington G (1972) The constituent acids of gymnosperm cutins. Phytochem 11:1989–2001
Huttunen S, Laine K (1981) The structure of pine needle surface (*Pinus sylvestris* L.) and the deposition of air-borne pollutants. Arch Ochr Srodowiska 2–4:29–38
Huttunen S, Laine K (1983) Effect of air-borne pollutants on the surface wax structure of *Pinus sylvestris* needles. Ann Bot Fenn 20:79–86
Isaacs NS (1987) Physical organic chemistry. Longman, Harlow
Jeffree CE, Johnson RPC, Jarvis PG (1971) Epicuticular wax in the stomatal antechamber of Sitka spruce and its effects on the diffusion of water vapour and carbon dioxide. Planta 98:1–10

Johnson RW, Jeffree CE (1970) Negative stain in wax tubes from the surface of Sitka spruce leaves. Planta 95:179–182

Johnson RW, Riding RT (1981) Structure and ontogeny of the stomatal complex in *Pinus strobus* L. and *Pinus banksiana* Lamb. Am J Bot 68:260–268

Jones JH (1978) Chemical changes in cutin obtained from cuticles isolated by the zinc chloride-hydrochloric acid method. Plant Physiol 62:831–832

Kenk G, Evers F, Unfried P, Schröter H (1983) Düngung als Therapie gegen Immissionswirkungen in Tannen-Fichten-Beständen? AFZ 154:153–170

Kerler F, Schönherr J (1988a) Accumulation of lipophilic chemicals in plant cuticles: prediction from octanol/water partition coefficients. Arch Environ Contam Toxicol 17:1–6

Kerler F, Schönherr J (1988b) Permeation of lipophilic chemicals across plant cuticles: prediction from partition coefficients and molecular volumes. Arch Environ Contam Toxicol 17:7–12

Kerler F, Riederer M, Schönherr J (1984) Non-electrolyte permeability of plant cuticles: a critical evaluation of experimental methods. Physiol Plant 62:599–606

Kerstiens G (1988) Funktionelle Veränderungen der pflanzlichen Kutikula durch Ozon. Doct Diss Tech Univ München

Kim YS (1985) REM-Beobachtungen immissionsbeschädigter Fichtennadeln. Centralbl Gesamte Forstwes 102:96–105

Kisser-Briesack GM, Scheunert J, Gnatz G, Ziegler H (1987) Uptake of $^{15}NO_2$ and ^{15}NO by plant cuticles. Naturwissenschaften 74:550–551

Kolattukudy PE (1981) Structure, biosynthesis, and biodegradation of cutin and suberin. Annu Rev Plant Physiol 32:539–567

Krause CR, Dochinger LS (1984) Surface changes to pine needles induced by ambient particles. Phytopathol 74:870

Krause CR, Houston DB (1983) Morphological variation in epicuticular wax of SO_2-sensitive and -tolerant Eastern white pine clones. Phytopathol 73:1266–1269

Kreger DR (1958) Wax. In: Ruhland W (ed) Handbuch der Pflanzenphysiologie. Bd 10, Springer, Berlin Göttingen Heidelberg, S 249–269

Kurbasik K (1989) Lösliche kutikuläre Lipide von Fichtennadeln (*Picea abies* L. [Karst.]). Doct Diss Tech Univ München

Lange OL, Schulze ED (1966) Untersuchungen über die Dickenentwicklung der kutikularen Zellwandschichten bei der Fichtennadel. Forstwiss Centralbl 85:27–38

Lendzian KJ (1982) Gas permeability of plant cuticles. Planta 155:310–315

Lendzian KJ (1984) Permeability of plant cuticles to gaseous air pollutants. In: Koziol MJ, Whatley FR (eds) Gaseous air pollutants and plant metabolism. Butterworths, Lond, pp 77–81

Lendzian KJ (1987) Aufnahme und zellphysiologische Wirkungen von Luftschadstoffen. Naturwissenschaften 74:282–288

Lendzian KJ, Nakajima A, Ziegler H (1986) Isolation of cuticular membranes from various conifer needles. Trees 1:47–53

Lister GR, Thair BW (1981) In vitro studies on the fine structure of epicuticular leaf wax from *Pseudotsuga menziesii*. Can J Bot 59:640–648

Magel E, Ziegler H (1986) Einfluß von Ozon und saurem Nebel auf die Struktur der stomatären Wachspfropfen in den Nadeln von *Picea abies* (L.) Karst. Forstwiss Centralbl 105:234–238

Martin JT, Juniper BE (1970) The cuticle of plants. Arnolds, Lond

Mayeux HS, Jordan WR (1987) Rainfall removes epicuticular waxes from *Isocoma* leaves. Bot Gaz 148:420–425

Muftaschiev B (1977) Adsorption and crystal growth. In: Vanselow R, Tong SY (eds) Chemistry and physics of solid surfaces. CRC Press, Cleveland, pp 73–86

Niemann GJ, Baas WJ (1978) Intra-species variation in the chemical composition of needles of *Larix leptolepis*. Acta Bot Neerl 27:229–238

Østensen H, Gullvåg B (1985) The wax layer of *Picea abies* needles studied from different German and Norwegian localities subjected to airborne pollution. J Ultrastruct Res 91:268

Oren R, Schulze ED, Werk KS, Meyer J, Schneider BU, Heilmeier H (1988) Performance of two *Picea abies* (L.) Karst. stands at different stages of decline. I. Carbon relations and stand growth. Oecologia (Berl) 75:25–37

Percy KE, Riding RT (1978) The epicuticular waxes of *Pinus strobus* subjected to air pollutants. Can J For Res 8:474–477

Reed DW (1982) Wax alteration and extraction during electron microscopy preparation of leaf cuticles. In: Cutler DF, Alvin KL, Price CE (eds) The plant cuticle. Academic Press, Lond NY, pp 181–195

Reicosky DA, Hanover JW (1976) Seasonal changes in leaf surface waxes of *Picea pungens*. Am J Bot 63:449–456

Rhine JB (1924) Clogging of stomata of conifers in relation to smoke injury and distribution. Bot Gaz 78:226–232

Riding RT, Percy KE (1985) Effects of SO_2 and other air pollutants on the morphology of epicuticular waxes on needles of *Pinus strobus* and *Pinus banksiana*. New Phytol 99:555–563

Riederer M, Schönherr J (1984) Accumulation and transport of (2,4-dichlorophenoxy)acetic acid in plant cuticles: I. Sorption in the cuticular membrane and its components. Ecotoxicol Environ Safety 8:236–247

Riederer M, Schönherr J (1985) Accumulation and transport of (2,4-dichlorophenoxy)acetic acid in plant cuticles: II. Permeability of the cuticular membrane. Ecotoxicol Environ Safety 9:196–208

Riederer M, Schönherr J (1986) Quantitative gas chromatographic analysis of methyl esters of hydroxy fatty acids derived from plant cutin. J Chromatogr 360:151–161

Riederer M, Schönherr J (1988a) Covalent binding of chemicals to plant cuticles: Quantitative determination of epoxide contents of cutins. Arch Environ Contam Toxicol 17:21–25

Riederer M, Schönherr J (1988b) Development of plant cuticles: fine structure and cutin composition of *Clivia miniata* Reg. leaves. Planta 174:127–138

Riederer M. Kurbasik K, Steinbrecher R, Voss A (1988) Surface areas, lengths and volumes of *Picea abies* (L.) Karst. needles: Determination, biological variability and effect of environmental factors. Trees 2:165–172

Rinallo C, Raddi P, Gellini R, Di Lonardo V (1986) Effects of simulated acid deposition on the surface structure of Norway spruce and silver fir needles. Eur J For Pathol 16:440–446

Rook DA, Hellmers H, Hesketh JD (1971) Stomata and cuticular surfaces of *Pinus radiata* needles as seen with a scanning electron microscope. J Arizona Acad Sci 6:222–224

Sauter JJ, Voß JU (1986) SEM-observations on the structural degradation of epistomatal waxes in *Picea abies* (L.) Karst. and its possible role in the "Fichtensterben". Eur J For Pathol 16:408–423

Sauter JJ, Kammerbauer H, Pambor L, Hock B (1987) Evidence for the accelerated micromorphological degradation of epistomatal waxes in Norway spruce by motor vehicle emissions. Eur J For Pathol 17:444–448

Schmidt HW, Schönherr J (1982) Development of plant cuticles: occurrence and role of non-ester bonds in the cutin of *Clivia miniata* Reg. leaves. Planta 156:380–384

Schmitt U, Ruetze M, Liese W (1987) Rasterlektronenmikroskopische Untersuchungen an Stomata von Fichten- und Tannennadeln nach Begasung und saurer Beregnung. Eur J For Pathol 17:118–124

Schönherr J (1982) Resistance of plant surfaces to water loss: Transport properties of cutin, suberin and associated lipids. In: Lange OL, Nobel PS, Osmond CB, Ziegler H (eds) Physiological plant ecology II, Encyclopedia of plant physiology, vol 12B. Springer, Berlin Heidelberg New York, pp 153–179

Schönherr J, Bukovac MJ (1972) Penetration of stomata by liquids. Dependence on surface tension, wettability, and stomatal morphology. Plant Physiol 49:813–819

Schönherr J, Bukovac MJ (1973) Ion exchange properties of isolated tomato fruit cuticular membrane: exchange capacity, nature of fixed charges and cation selectivity. Planta 109:73–93

Schönherr J, Huber R (1977) Plant cuticles are polyelectrolytes with isoelectric points around three. Plant Physiol 59:145–150

Schönherr J, Riederer M (1988) Desorption of chemicals from plant cuticles: evidence for asymmetry. Arch Environ Contam Toxicol 17:13–19

Schönherr J, Riederer M (1989) Foliar penetration and accumulation of organic chemicals in plant cuticles. Rev Environ Contam Toxicol 108:1–70

Schuck HJ (1972) Die Zusammensetzung der Nadelwachse von *Pinus sylvestris* in Abhängigkeit von

Herkunft und Nadelalter sowie ihre Bedeutung für die Anfälligkeit gegenüber *Lophodermium pinastri* (Schrad.) Chev. Flora 161:604–622

Schuck HJ (1976) Quantitative Variation der Wachsauflage und der cuticulären Kohlenwasserstoffe bei *Picea abies*-Nadeln. Flora 165:303–314

Schütt P, Schuck HJ (1973) Jahreszeitliche Schwankungen in der Zusammensetzung der Cuticularwachse von *Pinus sylvestris*. Flora 162:206–214

Schulten HR, Simmleit N, Rump HH (1986a) Forest damage: characterization of spruce needles by pyrolysis field ionization mass spectrometry. Int J Environ Anal Chem 27:241–264

Schulten HR, Simmleit N, Rump HH (1986b) Soft ionization mass spectrometry of epicuticular waxes isolated from coniferous needles. Chem Phys Lipids 41:209–224

Schulze ED (1982) Plant life forms and their carbon, water and nutrient relations. In: Lange OL, Nobel PS, Osmond CB, Ziegler H (eds) Physiological plant ecology II, Encyclopedia of plant physiology, vol 12B. Springer, Berlin Heidelberg New York, pp 615–676

Shipley GC (1986) X-ray crystallographic studies of aliphatic lipids. In: Small DM (ed) The physical chemistry of lipids. Handbook of lipid research, vol 4. Plenum Press, Lond NY, pp 97–147

Shovel KS, Alvin KL (1987) Patterns of cuticular organization in the hybrid *Tsuga* × *jeffryi* (Henry) Henry and its putative parents. Bot J Linn Soc 94:373–383

Simmleit N, Rump HH, Schulten HR (1986) Nichtabwaschbare Aerosolteilchen auf den Oberflächen von Koniferennadeln. Staub Reinhalt Luft 46:256–258

Sitte P, Rennier R (1963) Untersuchungen an cuticularen Zellwandschichten. Planta 60:19–40

Small DM (1986) The physical chemistry of lipids. Handbook of lipid research, vol 4. Plenum Press, Lond NY

Tamm CO, Cowling EB (1977) Acidic precipitation and forest vegetation. Water Air Soil Pollut 7:503–511

Thair BW, Lister GR (1975) The distribution and fine structure of the epicuticular leaf wax of *Pseudotsuga menziesii*. Can J Bot 53:1063–1071

Tranquillini W (1979) Physiological ecology of the alpine timberline. Springer, Berlin Heidelberg New York

Trimble JL, Skelly JM, Tolin SA, Orcutt DM (1982) Chemical and structural characterization of the needle epicuticular wax of two clones of *Pinus strobus* differing in sensitivity to ozone. Phytopathol 72:652–656

Tukey HB (1970) The leaching of substances from plants. Annu Rev Plant Physiol 21:305–324

Tulloch AP (1976) Chemistry of waxes of higher plants. In: Kolattukudy PE (ed) Chemistry and biochemistry of natural waxes. Elsevier, Amst, pp 235–287

Tulloch AP (1987) Epicuticular waxes of *Abies balsamea* and *Picea glauca*: occurrence of long-chain methyl esters. Phytochem 26:1041–1043

Tulloch AP, Bergter L (1981) Epicuticular wax of *Juniperus scopulorum*. Phytochem 20:2711–2716

Von Rudloff E (1959) The wax of the leaves of *Picea pungens* (Colorado spruce). Can J Chem 37:1038–1042

Walles B, Nyman B, Aldén T (1973) On the ultrastructure of needles of *Pinus sylvestris* L. Stud For Suec 106:1–26

Wattendorff J, Holloway PJ (1980) Studies on the ultrastructure and histochemistry of plant cuticles: the cuticular membrane of *Agave american* L. in situ. Ann Bot 46:13–28

Wells LG, Franich RA (1977) Morphology of epicuticular wax on primary needles of *Pinus radiata* seedlings. N Z J Bot 15:525–529

Yoshie F, Sakai A (1985) Types of Florin rings, distributional patterns of epicuticular wax, and their relationships in the genus *Pinus*. Can J Bot 63:2150–2158

Zech W, Suttner T, Popp E (1985) Elemental analysis and physiological responses of forest trees in SO_2-polluted areas of NE-Bavaria. Water Air Soil Pollut 21:175–183

Zimmermann R, Oren R, Schulze ED, Werk KS (1988) Performance of *Picea abies* (L.) Karst. stands at different stages of decline. II. Photosynthesis and leaf conductance. Oecologia (Berl) 76:513–518

Zinkel DF, Clarke WB (1985) Resin acids of *Pinus resinosa* needles. Phytochem 24:1267–1271

Zinkel DF, Magee TV, Walter J (1985) Major resin acids of *Pinus nigra* needles. Phytochem 24:1273–1277

3-B Organic Micropollutants and Plants

A. REISCHL, M. REISSINGER, and O. HUTZINGER

1 Introduction

Current studies on the causes of forest decline are focused mainly on the role of inorganic gases and the resulting acidification of the plant environment which may lead to direct or indirect injury. In the last few years, however, several studies have identified other environmental stress factors acting on plants, such as bleaching by the photochemical activation of volatile halogenated hydrocarbons (Frank and Frank 1985), and the influence of monoterpenes on the photosynthetic apparatus (Wagner et al. 1987).

Those organic chemicals produced and emitted by man and their subsequent atmospheric reaction products have not been intensively examined for effects on forest trees. Limited information is available on the phytotoxic effects of organic air constituents on plants, other than those of agricultural importance. In particular, little is known about the effects resulting from low levels, long exposure times, or compound combinations. In addition, there are surprisingly few studies on uptake routes and translocation of low-level organic contaminants in non-agricultural plants. Therefore, the stress potential of these substances cannot be assessed at the present time (Gaffney et al. 1987).

The large number of organic substances (ca. 4 million according to several authors) has often been used to show the impossibility of checking each group of compounds for possible phytotoxic effects. However, most organic compounds are of limited environmental importance because of low stability, low vapor pressure, or small or even no industrial production. In a literature survey Graedel et al. (1986) were able to find only 2000–3000 organic species that have been identified in the air. Table 1 lists those organic compounds which could be identified in air samples from the Fichtelgebirge, FRG. The number of substances identified by Geschke (1986) is limited because of analytical methodology and ambient air concentrations. An analysis of forest air similar to that shown in Table 1 was undertaken by Jüttner (1986) in damaged forests in the southern Black Forest, FRG. He identified significant concentrations of toluene and other aromatics, as well as various other chlorinated and unchlorinated solvents.

This chapter focuses on some anthropogenic organic pollutants of low volatility, namely, chlorinated pesticides, polycyclic aromatic hydrocarbons (PAH), polychlorinated biphenyls (PCB) and substituted phenols. These low-volatility substances have been called "semi-volatile organic pollutants (SVOP)" (Harkov 1986), and defined as substances with a vapor pressure of less than 1kPa

E.-D. Schulze, O.L. Lange, R. Oren (Eds.)
Ecological Studies Vol. 77

Table 1. Organic compounds occurring in selected air samples taken at Oberwarmensteinach, Fichtelgebirge[a]

Substance	Possible source
Alkanes	
$C_{10}-C_{32}$	Anthropogenic,natural
Aliphatic Alcohols	Natural
Aliphatic Aldehydes	
n-Nonanal	Natural
n-Dodecanal	Natural
n-Hexadecanal	?
Aliphatic Ketones	
Trimethyl-2- pentadecanone	?
Aliphatic Acids	
Palmitic acid	Natural,anthropogenic
Stearic acid	Natural,anthropogenic
Terpenoids	Natural
Cineol	
Camphene	
5-(1-Methylethyl) bicyclo[3,1,0]-hexane-2-one	
Borneol	
Endobornyl acetate	
Junipen	
6,6-Dimethyl-bicyclo[3,1,1]-2-heptanone	
Alkylbenzenes	
Ethyldimethylbenzene	Anthropogenic
Cymol	Natural
Tetramethylbenzene	Anthropogenic
Methylpropylbenzene	Anthropogenic
Oxygen-containing benzene derivatives	
Phenol	Natural,anthropogenic
Ethylphenols	Anthropogenic
Acetophenone	Natural,anthropogenic
Ethylacetophenone	Natural,anthropogenic
Dimethylbenzaldehydes	Anthropogenic
Ethylbenzaldehyde	Anthropogenic
Aromatic acids	
Benzoic acid	Anthropogenic,natural
Ethylbenzoic acid	Anthropogenic
Aromatic esters	
Benzoic acid benzylester	Natural
Methyl salicylate	?
Benzyl salicylate	?
Phthalic acid esters	Anthropogenic
Phthalic acid - di(2-ethylhexyl) ester	
Phthalic acid - dibuthyl ester	
Polycyclic aromatic hydrocarbons	
	Mainly anthropogenic

Table 1. *Cont'd.*

Substance	Possible source
Naphthalene	
Methylnaphthalenes	
Biphenylene	
Fluorene	
9-Fluorenon	
Phenanthrene,Anthracene	
Fluoranthene,Pyrene	
Heterocyclics	
Dibenzofuran	

[a]The substances listed were identified in two successive samples (29.08.1985 to 02.09.85). The sampling system does not differentiate between the particle phase and the gaseous phase (Geschke 1986).

at 20°C and 100 kPa. As these atmospheric organic trace constituents include chemicals of widely ranging physicochemical properties, the distribution of different compounds between the atmospheric phases will be quite different and, as a consequence, the path of entry into the forest ecosystem. These factors in turn affect how air pollutants exert phytotoxic effects, as will be shown below.

The distribution of organic air constituents depends not only on the chemical properties of the compounds, but also on the type and amount of atmospheric particulate matter. Organic compounds which are found mainly in the particle phase in regions with a high particle load are also found mainly in the gaseous state in areas of low particle concentration (Wittlinger and Ballschmitter 1987; Eisenreich et al. 1981). Those compound classes which typically represent different direct pathways from the atmosphere to aboveground plant parts and are also of phytotoxic interest will be briefly discussed.

Polycyclic aromatic hydrocarbons (PAH) are not only produced and emitted by man but are also part of the natural biochemical and geochemical cycles. However, anthropogenic processes may now serve as the main contributors to atmospheric PAH-concentrations (Sims and Overcash 1983). PAH are mainly emitted by combustion processes in which adsorption on soot or fly ash particles results in PAHs emerging bound to particulate matter (Lee and Schuetzle 1983; Grimmer 1983). Particle-bound emission is especially true for the PAH compounds of higher molecular weight (> 250), e.g., benzopyrenes or benzofluoranthenes (Cautreels and Van Cauwenberghe 1978; Broddin et al. 1980). Binding of these higher weight PAH to an "accumulation mode particle" (with a diameter of 0.1–2 μm) provides high residence times, and thus the possibility of long range transport in the atmosphere (Müller 1984). Many studies have examined the physical site of particle deposition onto plants and wet scavenging of particles from the atmosphere onto vegetation (Pruppacher et al. 1983; Georgii and Pankrath 1982; Hosker and Lindberg 1982). Possible effects on forest decline have been discussed (Jonas and Heinemann 1985; Simmleit et al.

1986). However, only a few studies have investigated the deposition fluxes of organic trace substances via particulates onto forests (Matzner 1984; Thomas et al. 1983). PAH accumulation has been intensively studied for different plant species (Thomas et al. 1984; Herrmann and Baumgartner 1987). Sims and Overcash (1983) suggested that above-ground plant parts, especially the leaves of trees, exhibit the highest PAH concentrations (up to 88 ppb for some PAH compounds). However, the phytotoxic effects of PAH range from growth stimulation to growth depression, varying according to the specific compound, the target species and the PAH concentration (Sims and Overcash 1983).

Besides direct toxic action, PAH may serve as precursors for peroxides. They have been detected on needles of *Picea abies* and there has appeared to be a relationship between peroxide concentration and the degree of needle injury (Stärk and Stauff 1986). Since scientific effort has concentrated mainly on a few PAH demonstrating carcinogenic potential (e.g., Benzo[a]pyrene) and limited attention has been given to toxic effects on coniferous trees, the role PAH play in forest decline cannot be estimated at the present time.

In addition to dry particle deposition and washout, the wet scavenging of gaseous substances may also be important. The vapor-particle distribution and Henry's-law constant of a compound strongly influence the way in which it may be removed from the atmosphere (Bidleman et al. 1986; Eisenreich et al. 1981). According to Pankow et al. (1984), gas-scavenging will dominate if

$$R \cdot T/H(T) \gg W\rho\phi/(1-\phi)$$

T — Temperature[K]
H(T) — Henry's law constant, temperature-dependent[$atm \cdot mol/m^3$]
$W\rho$ — Washout ratio for the aerosol particles
ϕ — fraction of compound associated with aerosol phase after Junge (1977)

This is the case for phenols as Leuenberger et al. found in 1985 (Table 2).

Because of their phytotoxic potential, nitrated phenols and related compounds which occur in rather high amounts (see Table 2) in samples of wet deposition are of particular interest (Rippen et al. 1987). Nitrated phenols are

Table 2. Mean concentrations of phenol and some of its methyl-, methoxy and nitro-derivatives in air and rain from Portland, Oregon (Leuenberger et al. 1985)

Compound	Concentration range In rain (ng l^{-1})	In air (ng m^{-3})
Phenol	> 280	320
2-Methylphenol	>1000	71
3- + 4- Methylphenol	>1100	130
2-Methoxyphenol	>2100	150
2,6 Dimethylphenol	140	11
2-Nitrophenol	59	24
2,4+2,5-Dimethylphenol	820	33
3,5-Dimethylphenol	490	20
3,4-Dimethylphenol	130	4

probably formed during atmospheric transport, as well as from direct emissions (Nojima et al. 1983; Grosjean 1984). Rippen and coworkers (1987) pointed out that possible precursors of nitrophenols, mainly benzene, toluene and NO_x, are emitted in amounts of many thousands of tons a year in the FRG.

The permeability of isolated cuticles for some mono-nitrophenols has been investigated (Shafer and Schönherr 1985; Kerler and Schönherr 1988b) and a relationship between the octanol- water-partition coefficient and the cuticle-water-partition coefficient has been proposed. However, there is a lack of data on the accumulation of nitrophenols on tree leaf surfaces under environmental conditions. This information is all the more necessary as most of the nitro- and/or methyl-substituted phenols do not have an octanol-water distribution ratio greater than 100 (Schwarzenbach et al. 1988), so that they exhibit only moderate bioaccumulation in addition to biotic and abiotic degradability (Rippen 1984).

The exposure of aboveground plant surfaces to organic chemicals in the vapor state may be most important in areas with low particle burden and semivolatile organics with high Henry constants. Even p,p′-DDT or PCB mixtures with low vapor pressures should occur mainly in the vapor phase, in a clean atmosphere (Eisenreich et al. 1981; Wittlinger and Ballschmitter 1987). These lipophilic compounds should tend to accumulate on the lipid layer of leaf surfaces. The high affinity of plant surfaces to adsorb airborne organics has been confirmed in many studies using different plant species (Buckley 1982; Gaggi and Bacci 1985; Thomas et al. 1984; Reischl et al. 1987a). Concentration factors obtained in several of these studies are summarized in Table 3. To our knowledge, the only studies which examine the accumulation kinetics of a plant/vapor-phase system for brief periods are the experiments from Bacci and Gaggi (1987).

The factors proposed by Thomas et al. (1983) and Reischl (unpublished) which are listed in Table 3 are not necessarily exact bioconcentration factors, especially in the case of the highly lipophilic substances; equilibrium was uncertain. Bacci and Gaggi (1987) also failed to reach equilibrium conditions after a period of 800 h but calculated the bioconcentration factors from a two-compartment model. However, the listed values show the tendency of organic chemicals to concentrate on leaves in amounts which magnify the dose to which plants are exposed.

Calamari et al. (1987) proposed the following calculation from the data of Bacci and Gaggi (1987) for the accumulation of some chlorinated hydrocarbons in Azalea leaves (see Table 3):

$$logFCF = 1.25 \cdot logH + 4.06$$

FCF = concentration in foliage/concentration in air [ng/g w.w. / ng/g]
H = Henry's-law constant $[Pa \cdot m^3 mol^{-1}]$

However, this calculation is limited to substances possessing physicochemical properties similar to those used for the correlation, and may lead to incorrect results for other plant species (Riederer and Schönherr 1984). It should be noted that organic compounds invade living plant tissue not only by accumulation and cuticular penetration or stomatal uptake of the aboveground plant parts, but by the deposition of organics on soil, their subsequent downward transport, and their

Table 3. Concentration factors (atmosphere-plant) of organic pollutants for different plant species[a]

Compound	Type of impact	Factor	Plant species	Reference
	Bulk precipitation		*Hypnum cupressiforme*	Thomas et al. (1983)
Benzo-a-pyrene		2573		
Fluoranthene		1325		
Benzo(ghi)-perylene		1845		
HCB		3156		
α -HCH		709		
γ -HCH		897		
PCB-mixture		500		
	Particulates			
Benzo-a-pyrene		180		
Fluoranthene		46		
Benzo(ghi)-perylene		141		
γ -HCH		58		
	Gas phase		*Azalea indica*	Bacci and Gaggi (1987)
α -HCH		4600		
γ -HCH		3400		
p,p′-DDE		134 000		
p,p′-DDT		192 500		
PCB-mixture		85 400		
	Gas/particle phase		*Picea abies*	Reischl (Unpublished)
PCDF		60 000-158 000		
PCDD		53 300-326 000		
γ -HCH		6600		
HCB		20 090		
p,p′-DDT		149 000		

[a] The values taken from Thomas et al. 1983 were calculated from environmental measurements; Bacci and Gaggi (1987) did some glasshouse experiments. The values given for *Picea abies* and PCDD/PCDF were calculated from air measurements done by Buck and Kirschmer (1986) and from our own measurements of spruce needle concentrations from Northrhine-Westfalia. The factors given for γ-HCH, HCB and p,p′-DDT originate from our own air/needle- measurements at Oberwarmensteinach, Fichtelgebirge.

interaction with the plants' root systems. However, few studies have examined the entry into perennial plants via the root system.

The uptake of a PCB-mixture, containing low chlorinated PCBs (2–3 chlorine atoms per molecule) into three-year-old *Picea abies* seedlings was studied by Moza et al. (1979). The PCBs were applied in sewage sludge. After an exposure time of 4 years, 0.8% of the total applied PCBs, including metabolites, could be detected in stems and needles. The mode of PCBs application in this study may have strongly influenced the distribution behavior of these substances, and it is not clear if absorption occurred through the below- or above-ground plant parts.

Translocation studies of lindane in coffee plants showed rapid movement of lindane through roots and stems into leaves (Rüegg 1977). Nash (1970) studied the uptake routes of p,p′-DDT, Dieldrin, Endrin and Heptachlor in soybean plants and found quite different uptake behavior of these compounds. While p,p′-DDT was mainly taken up via air, the greatest portions of Dieldrin, Endrin and Heptachlor were translocated through the stem.

A study of the uptake of PCBs from the surrounding atmosphere via bark demonstrated PCB accumulation in bark and wood (Meredith and Hites 1987). Uptake of pesticides via bark has also been demonstrated from stem-spraying applications (Müller 1986). Studies of the uptake routes of a PCB-mixture into different vegetables, on the other hand, found that air was the main uptake route, while soil was a secondary route (Bacci and Gaggi 1985). The portion of a chemical taken up via air on the one hand, and soil on the other, strongly depends on the physicochemical properties of the chemicals and the environment of the trees. Simplified laboratory investigations may lead to errors as well as neglect of the ultimate steady-state concentrations.

2 Occurrence and Distribution of Some Chlorinated Micropollutants in Norway Spruce Needles

Our studies in the Fichtelgebirge project concentrated on the occurrence, accumulation behavior and distribution of some semivolatile chlorinated organic compounds in the needles of *Picea abies*. Since we had done some basic studies on airborne pollutants and their behavior in spruce needles, no systematic sampling was necessary. The sampling location, near Oberwarmensteinach in the Fichtelgebirge, is described elsewhere in this volume.

Spruce trees (*Picea abies*) which were about 15 years old, showing no obvious damage, were used. The samples were taken at a height of 150–200 cm. The analytical procedure used is described elsewhere (Reischl et al. 1987b).

Plant leaves are known to adsorb lipophilic substances, since they are surrounded by a lipid layer which prevents uncontrolled loss of plant water (Martin and Juniper 1979). Lipids, which are soluble in organic solvents, are either embedded in the cutin or occur on the leaf surface as epicuticular wax (Baker 1982). A closer description of the cuticle of conifers and its properties is presented by Riederer in this volume. The compounds chosen for systematic studies were selected on the basis of their long half-lives in the environment, because their persistence allows accumulation studies for longer periods of time, without having to consider rapid degradation. Thus, for example, rapid degradation makes PAHs unsuitable for such studies (Butler and Crossley 1981; Sims and Overcash 1983; Stärk and Stauff 1986). Some physicochemical properties of the substances used in our studies are shown in Table 4.

As shown in Fig. 1, the values measured for all substances were in the lower ppb-range. The concentrations increased with the age of the needles. The detected concentrations were in the same range as that reported by Gaggi and Bacci (1985) for pine needles. p,p′-DDT did not seem to have reached steady state for the 5 years of needle age studied. The tendency of plant leaves to continue accumulating organic micropollutants over an excess of 2 years has been de-

Table 4. Some physicochemical properties of the substances used in our studies[a]

Compound	lg p	Solubility (H_2O) mg l^{-1} at 25°C	Vapor pressure at 20°C[Pa]	Le Bas molar volume m^3 mol^{-1}
α-HCH	3.81_a	$1.2–2_a$	0.0033_a	258.6
γ-HCH	3.72_a	7.52_a	0.00125_b	258.6
HCB	6.18_a	0.006_a	0.00145_a	236.4
p,p′-DDE	5.69_a	0.014_a	0.00087_a	335.2
p,p′-DDT	6.01_a	0.005_a	0.00002_a	363.5

[a]lg p is the log $_{10}$ of the octanol/water-partition coefficient. The molar volume was calculated according to Le Bas (Lyman et al. 1982). The letters behind data stand for references used: *a* - U.S.EPA 1979; *b* - Lyman et al. 1982

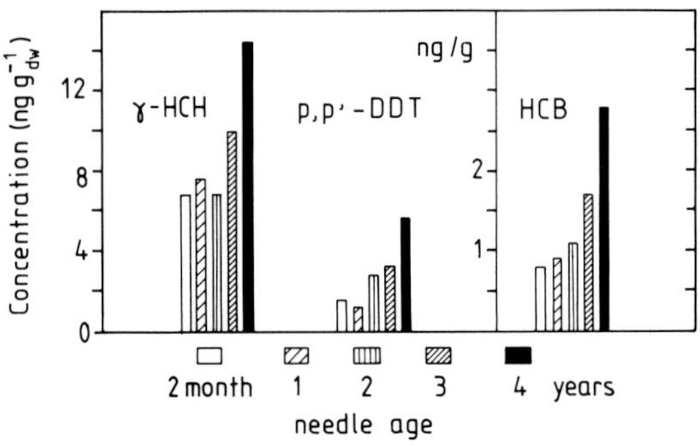

Fig. 1. Concentrations of some chlorinated hydrocarbons in needles of 15-year-old *Picea abies* trees (ng g^{-1} needle dry wt) from Oberwarmensteinach, Fichtelgebirge

scribed by different authors and for different plant species (Gaggi and Bacci 1985; Thomas et al. 1984; Herrmann and Baumgartner 1987). However, studies by Riederer, Kerler and Schönherr on the accumulation behavior of some semivolatile substances from aqueous solutions have shown that isolated cuticles of different plant species reach equilibrium within a few hours (Riederer and Schönherr 1984; Kerler 1986). It may be expected that isolated cuticles of spruce needles will show similar uptake kinetics.

Assuming that the mean ambient air concentrations of the substances investigated were constant within a time range of a few years, the accumulation behavior of spruce needles may signify that the cuticle is the first part of the needle to reach equilibrium with ambient air concentrations. Then, due to reduced transport into intracuticular needle parts, the needle interior reaches equilibrium concentration much more slowly. This would explain the observed long-term accumulation. Thus, it would appear that the cuticle does not protect the inner leaf compartments from lipophilic pollutants, but instead, may function as a storage pool which forces the flux of accumulated substances into inner leaf parts.

In addition to penetration via the cuticle, micropollutants in the vapor-state may also enter the needle through the stomata. This is not discussed in this chapter, but Riederer (this Vol.) gives a brief discussion of the importance of stomatal entry of foreign compounds. Therefore, the next question of phytotoxicological interest is how much organic pollutant can penetrate the cuticle and enter the interior of the plant. Additional studies concentrated on how the substances listed in Table 4 are partitioned between different spruce needle compartments.

3 Distribution Between the Cuticular Wax and the Needle

Some cuticular components can be easily washed off with organic solvents and are therefore labeled "Soluble Cuticular Lipids" (SCL) (see also Riederer, this Vol.). This solubility was used to try to understand the distribution of the compounds within the needle better. A two-step extraction procedure similar to that used in studies of the application of flamprop-methyl on wheat (Hamilton et al. 1982) was applied to the needles, described as follows:

1. Removal of the soluble cuticular lipids: Soluble lipids were removed by washing with dichloromethane (Kollatakudy 1976). To minimize solvent penetration into inner parts of the needle, two short consecutive washing steps, each consisting of 0.5 min shaking, were used. The soluble lipid extract was cleaned and analyzed as described elsewhere (Reischl et al. 1987b).
2. Extraction of the rest-needle (RN)[1] which remained after removing the soluble lipids. The needle was macerated and the extract cleaned and analyzed. The concentrations in the two needle compartments are shown in Table 5.

The listed quotients (Table 5) show great differences for p,p′-DDT and p,p′-DDE on the one hand, and HCB and the HCHs on the other. The concentration behaviour in the two needle compartments supports the hypothesis that the first group (p,p′-DDE, p,p′-DDT) of chemicals enters plants via atmosphere, but

Table 5. Concentrations and concentration ratios of some chlorinated pesticides in the SCL/RN of 1- to 4-year-old needles[a]

Needle age (yr)	1	2	3	4
α-HCH	0.82/4.1	0.69/4.4	1.08/3.75	1.23/6.98
Quotient	0.20	0.16	0.29	0.18
γ-HCH	1.8/11.1	0.32/6.4	2.18/5.2	3.23/6.89
Quotient	0.16	0.05	0.42	0.47
HCB	0.31/0.78	0.31/0.78	0.37/0.98	0.37/1.61
Quotient	0.39	0.40	0.38	0.23
p,p′-DDT	2.2/0.53	1.5/0.95	2.37/1.12	2.6/3.77
Quotient	4.15	1.58	2.12	0.69
p,p′-DDE				
Quotient	5.68	1.89	0.97	0.63

[a] Concentrations in ng/g needle dry weight. For p,p′-DDE no absolute quantification was done.

[1] Body of needle after removal of soluble cuticular lipids is referred to as the rest-needle (RN).

without rapid diffusion into inner plant parts, while the second substance group (HCHs,HCB) is more mobile and possibly enters inner needle parts.

Refined distribution studies using glass columns packed with spruce needles and eluted with dichloromethane for successively prolonged times were undertaken for different needle ages. The resulting pollutant concentrations in the single fractions are shown in Figs. 2–4.

Figure 2 confirms that p,p′-DDT remains in the outer layer of the needle, while the HCHs and HCB can still be detected in the later fractions. Similar results were obtained from elution of 1- to 4-year-old needles. In these experiments, the needle which remained at the end of the whole elution process was treated as previously described for the RN. Only γ-HCH, p,p′-DDT and HCB are discussed further (Figs. 3, 4, 5).

HCB, as well as the HCHs, were detected in the later elution-fractions, but p,p′-DDT was detected only up to the third fraction. This would indicate that p,p′-DDT does not diffuse as deeply into the cuticle as the other compounds. The behavior of HCB could be explained by its very good sorption and diffusion into the cuticle. This has been observed for isolated cuticles of other plants (Kerler 1986). Rapid diffusion through the cuticle could also be an explanation for the elution behavior of γ-HCH. The anomalous behavior of p,p′-DDT could be explained by its high molar volume, which hinders diffusion, or its low vapor pressure and resulting particle-bound accumulation on the needle surface. The differentiation between particle-bound deposition and gaseous adsorption cannot be assessed with the methods used. Organic compounds bound on deposited particles may stay mainly on these particles and may not significantly diffuse into the cuticle. Therefore, as needle surface waxes are removed, particles containing adsorbed compounds will also be removed (Simmleit et al. 1986). Nevertheless, the results presented above clearly show that the main portion of p,p′-DDT is unable to enter inner needle parts. This may be due to the manner of deposition onto the plant surfaces, as well as to slow diffusion.

Kerler and Schönherr (1988b) found that cuticular permeance coefficients (for cuticles of *Citrus aurantium*) could best be predicted from cuticle-water-

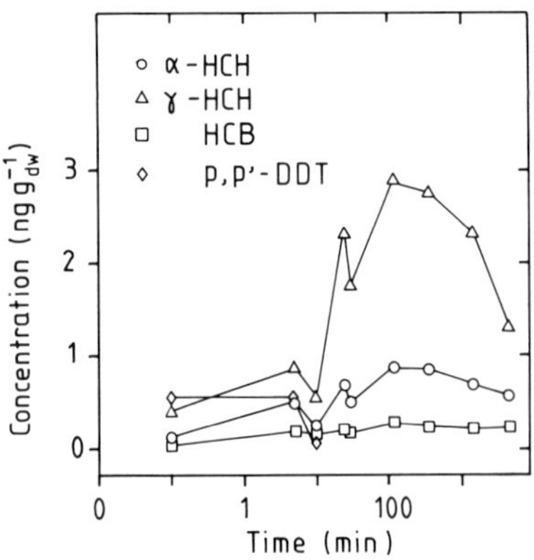

Fig. 2. Elution diagram for ca. 4-month-old needles. The concentration units are ng g^{-1} needle dry wt

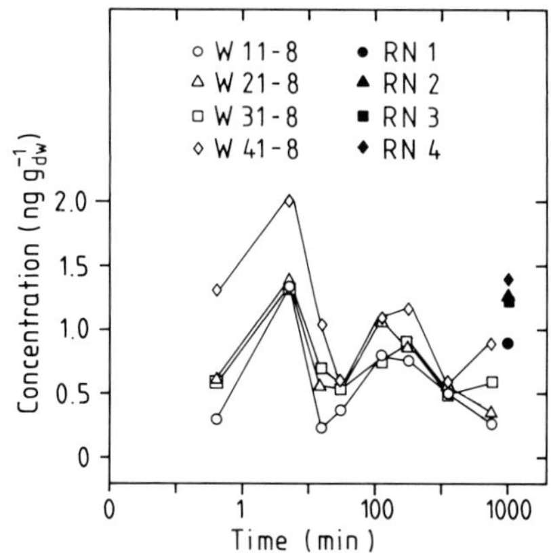

Fig. 3. Elution diagram for 1- to 4-year-old needles for $\gamma-$HCH . Wnl-8 (n = 1–4 years needle age) are the single elution fractions. The last value represents the concentration of the remaining needle material (*RN*) for the different needle ages (n = 1–4 years needle age). Concentrations are given in ng g^{-1} needle dry wt

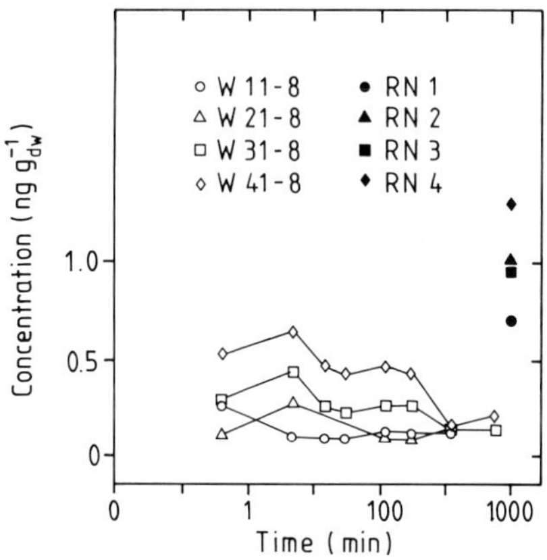

Fig. 4. Elution diagram for 1- to 4-year-old needles for HCB. Wnl-8 (n = 1–4 years needle age) are the single elution fractions. The last value represents the concentration of the remaining needle material (*RN*) for the different needle ages (n = 1–4 years needle age). Concentrations are given in ng g^{-1} needle dry wt

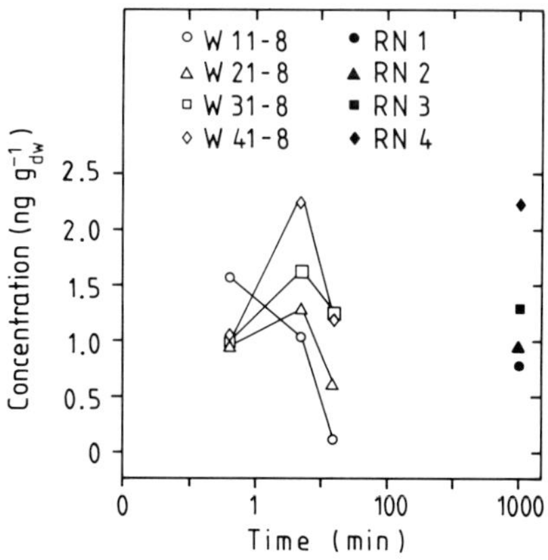

Fig. 5. Elution diagram for 1- to 4-year-old needles for p,p'-DDT. Wn1-8 (n = 1–4 years needle age) are the single elution fractions. The last value represents the concentration of the remaining needle material (*RN*) for the different needle ages (n = 1–4 years needle age). Concentrations are given in ng g^{-1} needle dry wt

partition coefficients and Le Bas molar volume (see also Riederer, this Vol.). Although similar molar volumes may be calculated for HCB and HCHs which may accumulate on plants through means other than particulate matter, it should be noted that these compounds have quite different octanol-water-distribution ratios (see Table 4). This should be further investigated.

Certain amounts of accumulated pollutants remain in the unmacerated needle after an exposure time of about 100 h (Figs. 3–5) and increase with needle age.

According to studies on the desorption kinetics of 2,4-D from isolated cuticles (Schönherr et al. 1988) these amounts at least might represent that portion of organic pollutants which could enter the needle interior and develop phytotoxicity.

Kerler and Schönherr (1988a) calculated the equilibrium mass ratios between the cuticle and plant cells and tissues for some organic pollutants, but the calculation model treats cells and tissues as pure water. According to these calculations, compounds with octanol/water-partition coefficients greater than 10^4 should not occur in living plant tissue. This results in the case of *Picea abies* in a strong underestimation of the concentrations of highly lipophilic substances in the interior needle parts, as indicated by our studies. However, according to these calculations compounds with lower lg p, such as nitrated phenols (lg p ca. 2) can be expected to be mainly present in the plant interior. In combination with their strong phytotoxic properties (Shea et al. 1983), these and other substituted phenolics could act as important stress factor.

4 Phytotoxicity of Organic Atmospheric Micropollutants: A Summary

Organic pesticides which are applied to plant surfaces can enter the plant interior. This fact is widely known from agricultural applications (Müller 1986; Hartley and Graham-Bryce 1980). However, the behavior of organic pollutants which accumulate on needle surfaces may be quite different. The much lower concentrations and the long needle life are two factors that may effect penetration behavior.

Mutagenicity and carcinogenic potential of airborne particulate samples have frequently been investigated (Takeda et al. 1984; Harkov 1982), but little attention has been given to the phytotoxic potential of atmospheric organics. Most work dealing with phytotoxicity of organic contaminants in trace amounts has been done using algal cell cultures as test systems, which may offer a relative measure of the toxic potential of substances on photosynthetic organisms in general (Table 6). However, algal-test toxicity data may not be directly used to predict the toxic effects of measured substance-concentrations in the intact leaves of higher plants, since transport and distribution phenomena inside the leaf strongly influence the effective dose (Krell and Sandermann 1985; Kerler and Schönherr 1988a; Huber and Huber 1984). A barrier like the cuticle does not exist in algae, but serves in the uptake and storage behavior of higher plants. This

Table 6. Toxic concentration ranges of some organic compounds for different algal organisms and the different toxicity indications used[a]

Compound	Organism	Toxic concentration (mg l^{-1})	reference
4,4′-DDT	*Chlorella fusca*	0.1	Goulding and Ellis (1981)
2,4,6-Trichlorophenol	*Scenedesmus subspicatus*	1.1 (EC_{10})	Geyer et al. (1985)
4-Chloraniline	"	0.4 "	"
PCP	"	0.03 "	"
Atrazine	"	0.04 "	"
Lindane	"	0.5 "	"
4-Nitrophenol	"	8 "	"
4,5-Dichloroguaiacol	*Selenastrum capricornutum*	3.1 (LC_{50})	Kuivasniemi et al. (1985)
3,4,5-Trichloroguaiacol	"	0.8 "	"
3,4-Dichlorocatechol	"	0.3 "	"
3,4,6-Trichlorocatechol	"	0.2 "	"
Tetrachlorocatechol	"	0.08 "	"
9,10-Phenanthraquinone	*Dunaliella bioculata*	0.38 (EC_{50})	Heldal et al. (1984)
1,2-Naphthoquinone	"	1.0 "	"
1,4-Naphthoquinone	"	0.011 "	"
Fluorence	"	15.5 "	"
9-Fluorenone	"	5.7 "	"
9-Fluorenhydroperoxide	"	0.5 "	"

[a]The selected chemicals occur in the environment.

means that the physiologically active dose may be much lower than the concentration measured. However, in addition to the two-compartment differentiation described above, pollutant concentrations in different leaf subcompartments cannot be analytically resolved at the present time. The modeling of pollutant distribution within the tree, as attempted by Schramm et al. (1987), may be useful in estimating the physiologically active substance concentrations.

The values presented in Table 6 indicate that organic substances show effects on algal organisms in concentration ranges which, in some cases, are far higher than that found in samples from such environmental compartments as plant leaves (see also above). However, leaves accumulate not only one but many hundreds of potential phytotoxic compounds, which combine toxic effects in a way which cannot be estimated. Instead of single substance-testing, complex environmental sample testing for phytotoxic effects may make rapid estimation of phytotoxic potentials possible. Figure 6 shows the effects of an organic solvent extract from atmospheric particles on a tomato cell-culture.

As shown in Fig. 6, phytotoxicity testing of a complex mixture of atmospheric pollutants gives an impression of the phytotoxic potential of atmospheric organics. However, the impact on higher plants can not be estimated from these data, since the single compounds may possess quite different transport properties once deposited on the leaf surface. Identification of substance groups primarily

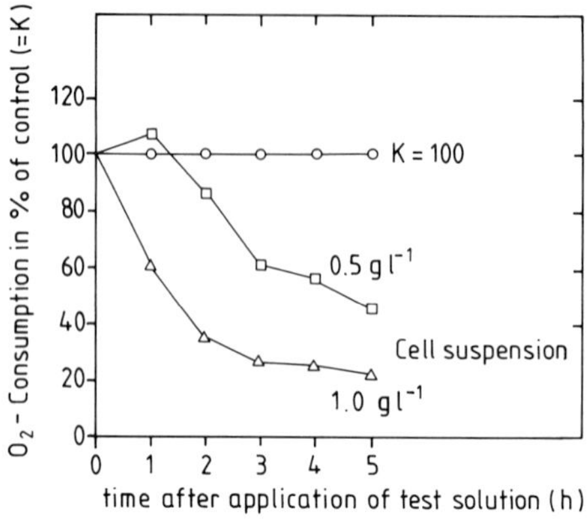

Fig. 6. Oxygen consumption of a tomato-callus cell culture after application of an acetone extract from air particles. The particle sample was taken on the Ochsenkopf, Fichtelgebirge (Geschke 1986)

responsible for the phytotoxic effects of ambient air, and deposition samples on algal cell or higher plant-cell cultures, may allow further focus on the distribution and uptake behavior of these specific compound groups.

5 Conclusions

Plant surfaces, especially leaves, accumulate lipophilic compounds to a considerable extent. The accumulation behavior and the distribution of selected persistent chlorinated compounds in *Picea abies* needles indicates that diffusion occurs into interior needle parts. This raises the question as to what extent the amounts diffused may cause toxic effects.

A direct relationship between organic compound concentrations detected in *Picea abies* needles and the degree of damage has not yet been established. However, the amounts, accumulation, distribution and phytotoxicity of the persistent contaminants found in *Picea abies* needles in the Fichtelgebirge give good reason to further investigate the hypothesis that organic pollutants contribute to forest decline. Organic substances such as substituted phenols, polycyclic aromatic hydrocarbons and related compounds, should be included in further studies.

Acknowledgements. We thank Mr. Wittlinger and Mr. Ballschmiter for allowing us to use as yet unpublished data and information in this chapter and Mr. McLachlan for helpful discussions.

References

Bacci E, Gaggi C (1985) Polychlorinated biphenyls in plant foliage: translocation or volatilization from contaminated soil. Bull Environ Contam Toxicol 35:673

Bacci E, Gaggi C (1987) Chlorinated hydrocarbon vapors and plant foliage: kinetics and application, Chemosphere 16:2515

Baker EA (1982) Chemistry and morphology of plant epicuticular waxes. In: Cutler DF, Alvin KL, Price CE (eds) The plant cuticle. Academic Press, Lond NY

Bidleman TF, Billings WN, Foreman WT (1986) Vapor-paricle partitioning of semivolatile organic compounds:estimates from field collections. Environ Sci Technol 20:1038

Broddin G, Cautreels W, Van Cauwenberghe K (1980) On the aliphatic and polyaromatic hydrocarbon levels in urban and background aerosols from Belgium and the Netherlands. Atmos Environ 14:895

Buck M, Kirschmer P (1986) Immissionsmessungen polychlorierter Dibenzo-p-Dioxine und Dibenzofurane in Nordrhein-Westfalen. LIS-Ber 62:1–47

Buckley EH (1982) Accumulation of airborne polychlorinated biphenyls in foliage. Science 216:520

Butler JD, Crossley P (1981) Reactivity of polycylic aromatic hydrocarbons adsorbed on soot particles Atmos Environ 15:91

Calamari D, Vighi M, Bacci E (1987) The use of terrestrial plant biomass as a parameter in the fugacity model. Chemosphere 16:2359

Cautreels W, Van Cauwenberghe K (1978) Experiments on the distribution of organic pollutants between airborne particulate matter and the corresponding gas phase. Atmos Environ 12:1133

Eisenreich SJ, Looney BB, Thornton JD (1981) Airborne organic contaminants in the Great Lakes ecosystem. Environ Sci Technol 15:30

Frank H, Frank W (1985) Chlorophyll-bleaching by atmospheric pollutants and sunlight. Naturwissenschaften 72:139

Gaffney JS, Streit GE, Spall WD, Hall JH (1987) Beyond acid rain. Environ Sci Technol 21:519

Gaggi C, Bacci E (1985) Accumulation of chlorinated hydrocarbon vapors in pine needles. Chemosphere 14:451

Georgii HW, Pankrath J (eds) (1982) Deposition of atmospheric pollutants. Reidel, Dordrecht

Geschke A (1986) Organische luhaltsstoffe der Waldluft als inogliche Mitrerursacker der Waldschaden: Probennalure und GC/OIS-Analytics Diplomarbeit Univ Bayreuth, Lehrstuhl Ökologische Chemie Geochemie

Geyer H, Scheunert I, Korte F (1985) The effects of organic environmental chemicals on the growth of the alga *Scenedesmus subspicatus*: a contribution to environmental biology. Chemosphere 14:1355

Goulding KH, Ellis SW (1981) The interaction of DDT with two species of freshwater algae Environ Pollut Ser A 25:271

Graedel TE, Hawkins DF, Claxton LD (1986) Atmospheric chemical compounds. Academic Press, Lond NY

Grimmer G (1983) Profile analysis polycyclic aromatic hydrocarbons in air. In: Bjorseth A (ed) Handbook of polycyclic aromatic hydrocarbons. Dekker, NY

Grosjean D (1984) Atmospheric reactions of ortho cresol:gas phase and aerosol products Atmos Environ 18:1641

Hamilton RJ, McCann AW, Sewell PA, Merrall GT (1982) Foliar uptake of the wild oat herbicide flamprop-methyl by wheat. In: Cutler DF, Alvin KL, Price CE (eds) The plant cuticle. Academic Press, Lond NY

Harkov R (1982) Toxic air pollutants-assessing their importance. Sci Total Environ 26:67

Harkov R (1986) Semivolatile organic compounds in the atmosphere: a review. J Environ Sci Health A 21(5):409

Hartley GS, Graham-Bryce IJ (1980) Physical principles of pesticide behaviour. Academic Press, Lond NY

Heldal M, Norland S, Lien T, Knutsen G (1984) Toxic responses of the green alga *Dunaliella bioculata* (Chlorophycea, Volvocales) to selected oxidised hydrocarbons. Environ Pollut Ser A 34:119

Herrmann R, Baumgartner I (1987) Regional variation of selected polyaromatic and chlorinated hydrocarbons over the South Island of New Zealand, as indicated by their content in *Pinus radiata* needles. Environ Poll 46:63

Hosker RP, Jr.,Lindberg SE (1982) Review: atmospheric deposition and plant assimilation of gases and particles. Atmos Environ 16:889

Huber W, Huber A (1984) Schadstoffbelastungen für Wasserpflanzen. In: Hock B, Elstner E (eds) Pflanzentoxikologie. Bibliogr Inst, Mannheim, pp 118–131

Jonas R, Heinemann K (1985) Schädigung von Pflanzen durch abgelagerte Schadstoffe. Staub Reinh Luft 45:112

Jüttner F (1986) Analysis of organic compounds (VOC) in the forest air of the southern Black Forest. chemosphere 15:985

Junge CE (1977) Basic considerations about trace constituents in the atmosphere as related to the fate of global pollutants. In: Suffet IH (ed) Fate of pollutants in the air and water environments. Wiley, NY

Kerler F (1986) Doctoral Thesis, TU München, Lehrstuhl Bot

Kerler F, Schönherr J (1988a) Accumulation of lipophilic chemicals in plant cuticles: prediction from octanol water partition coefficients. Arch Environ Contam Toxicol 17:1

Kerler F, Schönherr J (1988b) Permeation of lipophilic chemicals across plant cuticles: prediction from partition coefficients and molar volumes. Arch Environ Contam Toxicol 17:7

Kollatukudy PE (1976) Chemistry and biochemistry of natural waxes. Elsevier, Amst

Krell HW, Sandermann H (1985) Compartimentation of the plant metabolic enzymes for the persistent plasticizer chemical, Bis-(2-Ethyl-Hexyl)Phthalate. Plant Sci 40:87

Kuivasniemi K, Eloranta V, Knuutinen J (1985) Acute toxicity of some chlorinated phenolic compounds to *Selenastrum capricornutum* and Phytoplankton. Arch Environ Contam Toxicol 14:43

Lee FSC, Schuetzle D (1983) Sampling, extraction and analysis of polycyclic aromatic hydrocarbons from internal combustion engines. In: Bjorseth A (ed) Handbook of polycyclic aromatic hydrocarbons. Dekker, NY

Leuenberger C, Ligocki MP, Pankow JF (1985) Trace organic compounds in rain: 4. Identities, concentrations, and scavenging mechanisms for phenols in urban air and rain. Environ Sci Technol 19:1053

Lyman WJ, Reehl WF, Rosenblatt DH (1982) Handbook of chemical property estimation methods. McGraw-Hill Book Co, NY

Martin JT, Juniper BE (1979) The cuticles of plants. Arnold, Lond

Matzner E (1984) Annual rates of deposition of polycyclic aromatic hydrocarbons in different forest ecosystems. Water Air Soil Pollut 21:425

Meredith ML, Hites RA (1987) Polychlorinated biphenyls: accumulation in tree bark and wood. Environ Sci Technol 21:709

Moza PN, Scheunert I, Klein W, Korte F (1979) Long term uptake of lower chlorinated biphenyls and their conversion products by spruce trees (*Picea abies*) from soil treated with sewage sludge. Chemosphere 8:373

Müller F (1986) Phytopharmakologie. Ulmer, Stuttgart

Müller J (1984) Atmospheric residence time of carbonaceous particles and particulate PAH-compounds. Sci Total Environ 36:339

Nash RG, Beall ML (1970) Chlorinated hydrocarbon insecticides: root uptake versus vapor contamination of soybean foliage. Science 168:1109

Nojima K, Kawaguchi A, Ohya T, Kanno S, Hirobe M (1983) Studies on photochemical reaction of air pollutants. X. Identification of nitrophenols in suspended particles. Chem Pharm Bull 31(3):1047

Pankow JF, Isabelle LM, Asher WE (1984) Trace organic compounds in rain. 1. Sampler and analysis by adsorption/thermal desorption (ATD). Environ Sci Technol 18:310

Pruppacher HR, Semonin RG, Slinn WGN (eds) (1983) Precipitation scavenging, dry deposition and resuspension. vol 1 + 2, Elsevier, Amst

Reischl A, Thoma H, Reissinger M, Hutzinger O (1987a) PCDD und PCDF in Koniferennadeln. Naturwissenschaften 74:88

Reischl A, Reissinger M, Hutzinger O (1987b) Occurrence and distribution of atmospheric organic micropollutants in conifer needles. Chemosphere 16:2647

Riederer M, Schönherr J (1984) Accumulation and transport of (2,4-dichlorophenoxy) acetic acid in plant cuticles: I. Sorption in the cuticular membrane and its components. Ecotoxicol Environ Safety 8:236

Rippen G (1984) Handbuch der Umweltchemikalien. Ecomed, Landsberg

Rippen G, Zietz E, Knacker T, Klöepffer W (1987) Do airborne nitrophenols contribute to forest decline? Environ Technol Lett 8:475

Rüegg EF, Lard KA, Mesquita TB (1977) Uptake and movement of $_{1}$4C-lindane in coffee plants. Arq Inst Biol Sao Paulo 44:235

Schönherr J, Kerler F, Riederer M (1984) Cuticular lipids as interface between plant and environment. Dev Plant Biol 9:491

Schönherr J, Kerler F, Riederer M (1988) Desorption of chemicals from plant cuticles: evidence for asymmetry. Arch Environ Contam Toxicol 17:13

Schramm KW, Reischl A, Hutzinger O (1987) UNITTree-A multimedia compartment model to estimate the fate of lipophilic organic chemicals in plants. Chemosphere 16:2653

Schwarzenbach RP, Stierli R, Folsom BR, Zeyer J (1988) Compound properties relevant for assessing the environmental partitioning of nitrophenols. Environ Sci Technol 22:83

Shafer WE, Schönherr J (1985) Accumulation and transport of phenol, 2-Nitrophenol and 4-Nitrophenol in plant cuticles. Ecotoxicol Environ Safety 10:239

Shea PJ, Weber JB, Overcash MR (1983) Biological activities of 2,4-dinitrophenol in plant-soil systems. Res Rev 87:1

Simmleit N, Rumpp HH, Schulten HR (1986) Nichtabwaschbare Areosolteilchen auf den Oberflächen von Koniferennadeln. Staub Reinh Luft 46:256

Sims RC, Overcash MR (1983) Fate of polynuclear aromatic compounds (PNAs) in soil-plant systems. Res Rev 88:1

Stärk G, Stauff J (1986) Anreicherung hochmolekularer Peroxide auf Fichtennadeln in Reinluftgebieten Staub Reinh Luft 46:396

Takeda N, Teramishi K, Hamada K (1984) Mutagenicity of air pollutants collected at industrial, urban-residential and rural areas. Bull Environ Contam Toxicol 32:688

Thomas W, Riess W, Hermann R (1983) Processes and rates of deposition of air pollutants in different ecosystems. In: Ulrich B, Pankrath J (ed) Effects of accumulation of air pollutants in forest ecosystems. Reidel Publ Co, Dordrecht

Thomas W, Rühling A, Simon H (1984) Accumulation of airborne pollutants (PAH, chlorinated hydrocarbons, heavy metals) in various plant species and humus. Environ Pollut Ser A 36:295

USEPA (1979) Water-related environmental fate of 129 priority pollutants. vol I, II, EPA-440/4-79-029a,b

Wagner E, Vollbrecht P, Janistyn B, Gross K, Woerth J (1987) Monoterpen-vermittelte Zerstörung des Photosyntheseapparates von Waldbäumen. AFZ 27 28 29:705

Wittlinger R, Ballschmiter K (1987) Global baseline pollution studies XI: Congener-specific determination of polychlorinated biphenyls(PCB) and occurrence of alpha and gamma-hexachlorocyclohexane(HCH),4,4′-DDE and 4,4′-DDT in continental air. Chemosphere 16:2497–2513

3-C Leaching and Uptake of Ions Through Above-Ground Norway Spruce Tree Parts

O. Klemm (Coordinator)
with contributions by O. Klemm (Sects. 1 and 2);
U. Kuhn, and E. Beck (Sect. 3);
C. Katz, R. Oren, E.-D. Schulze, and E. Steudle (Sect. 4);
E. Mitterhuber, H. Pfanz, and W.M. Kaiser (Sect. 5);
M. Kaupenjohann, B.-U. Schneider, R. Hantschel, R. Horn, and W. Zech (Sect. 6); R. Hantschel, R. Horn, M. Kaupenjohann, W. Zech, and B.-U. Schneider (Sect. 7)

1 Introduction

Increasing air pollution in Central Europe during the last century has resulted in a substantial input of inorganic ions into the forest ecosystem. Deposition mainly occurs through snow and rainfall, fog and cloudwater impaction, dry deposition of particles, and upon solution of acidic gases in the water films of wet surfaces, e.g., after dewfall (see Part 1).

In addition to acid rain interaction with plants through the soil root system (Chap. 4-E), the uptake and loss of ions through above-ground plant parts (cuticles and bark) have to be considered. Comprising about 50% of the surface of a conifer exposed to the atmosphere, cuticles are nonliving membranes which protect the plant from losses of water, dissolved ionic species, and gases. Diffusion of uncharged molecules through cuticles can be described by Fick's first law (Schönherr 1976a). The molecular diffusivities of ions, however, vary significantly, so that diffusion potentials across membranes have to be considered as additional driving forces (Reed and Tukey 1982; Riederer Chap. 3-A, this Vol.). Hence, experimental evaluation of ionic diffusion must be based on determining both the concentration gradients of all ions and the electric potentials through a cuticle in vivo. This cannot be measured with current techniques.

In experiments with isolated cuticles, the electric potential across the membrane must also be taken into consideration (cf. Riederer Chap. 3-A, this Vol.).

The diffusivity of a cation seems to be inversely reciprocal to the size of its hydrated species and decreases in the following sequence (Kannan 1986; McFarlane and Berry 1974):

$$Cs^+ = Rb^+ > K^+ > Na^+ = Ba^{2+} > Sr^{2+} > Ca^{2+}$$

From the size of the hydrated species of Mg^{2+}, it is followed that its diffusitity is even lower than that of Ca^{2+}. In leaves, the most abundant cations are K^+, Mg^{2+}, and Ca^{2+}. There is strong evidence that leaching of these ions (Tukey 1970; Kreutzer and Bittersohl 1986; Skiba et al. 1986) as well as of Mn^{2+} (Cole and Johnson 1977), and Zn^{2+} (Mengel et al. 1987) is accomplished via exchange for protons through wet leaf surfaces. The degree of leaching usually increases with increasing precipitation acidity (Wood and Bormann 1975; Hindawi et al. 1980;

E.-D. Schulze, O.L. Lange, R. Oren (Eds.)
Ecological Studies Vol. 77

Horndtvedt et al. 1980; Johnson et al. 1982; Abrahamsen 1983; Scherbatskoy and Klein 1983; Cronan 1984; Skeffington and Roberts 1985; Skiba et al. 1986; Mengel et al. 1987).

Dilute acid applied to leaf surfaces can be significantly buffered by several plant species (Lepp and Dickinson 1976; Oertli et al. 1977; Wood and Borman 1977; Adams and Hutchinson 1984; Larsen 1986; Adams and Hutchinson 1987; Hutchinson and Adams 1987). Though not all the buffering mechanisms (e.g., dissolution of deposited particles, ammonia in greenhouses, cation exchange through the cuticle) have been determined, the suggestion that cation exchange is a buffering reaction is supported by histological investigations. Precipitation acidity may lead to erosion of the epicuticular waxes (Evans 1984), which form the most effective layer to prevent diffusion (Schönherr 1976b; Dreyer et al. 1981; Evans 1984; Kannan 1986). Leaf injuries and necroses can be observed as a response to acid rainwater at a threshold pH between 3 and 4 in broad-leaved plants (Wood and Bormann 1975; Evans and Curry 1979; Hindawi et al. 1980; Jacobson 1980; Norby and Luxmoore 1983; Proctor 1983; Evans 1984; Paparozzi and Tukey 1984; Evans et al. 1985; Caporn and Hutchinson 1986; Johnston et al. 1986; Adams and Hutchinson 1987) or below 3 for conifers (Horntvedt et al. 1980). Although a pH considerably below 4 is a rare event in Fichtelgebirge precipitation (Chap. 1-B), such pH certainly can occur on leaf surfaces because evaporation of acid precipitation water leads to a substantial drop in the pH (Evans and Curry 1979; Teuchert and Teuchert 1983; Frevert and Klemm 1984; Unsworth 1984; Klemm et al. 1987). Lesions have been found (Evans et al. 1977a) mainly near the bases of trichomes and around stomata (Evans et al. 1977b; Evans 1982). Buffering is conceivably most effective in droplets resting on such lesions (Hutchinson et al. 1986; Adams and Hutchinson 1987) and injured plant parts are more subjected to ion loss by leaching than are intact plant parts (Tukey and Morgan 1963; Tukey 1970; Smith 1981).

Less is known about the leaching and uptake of anions (Evans 1984; Evans et al. 1986; Skiba et al. 1986). At a pH greater than 3, the cuticle should theoretically be much less permeable to anions than to cations (Schönherr and Huber 1977; cf. Riederer Chap. 3-A, this Vol.). Experimental results on the influence of rainwater acidity on the diffusivity of anions seem to confirm this hypothesis. While Schönherr and Huber (1977), Evans et al. (1981) and Evans et al. (1986) have found increasing anion diffusion through cuticles of different plant species with decreasing pH, Skiba et al. (1986), Scherbatskoy and Klein (1983), and Wittwer and Teubner (1959) have failed to find any such intercorrelation.

In intact leaves, ion diffusion seems to be more intensive in certain regions, e.g., in the vicinity of guard cells, vascular bundles, and basal cells of trichomes (Schönherr and Bukovac 1970). During senescence, the physicochemical structure of the cuticle changes (Riederer Chap. 3-A, this Vol.) and it becomes more susceptible to leaching (Tukey and Tukey 1969). Further, the degree of leaching depends on the thickness and chemical composition of the cuticle, its wettability in terms of its contact angle (Evans 1984; Hutchinson et al. 1986; Kannan 1986), air humidity (Tukey et al. 1962; Larsen 1986), residence time of droplets on the leaf (Mecklenburg et al. 1965; Tukey 1970; Adams and Hutchinson 1984; Adams and Hutchinson 1987), the nutrient status of the plant (Tukey et al. 1962), and the sulfate to the nitrate ratio of the treating solution (Irving 1985; Skiba et al. 1986). Diffusion of ions through cuticles may also take place under those dry conditions when no rain drops rest on the leaf, but this contribution to total leaching is not well understood (Reiners et al. 1986).

Because trichomes on spruce epidermis are absent and the wettability of the cuticular membrane is poor (Miller 1985), the loss of solutes through healthy cuticles of *Picea abies* should be very low (Evans 1982). Nevertheless, neutralization of acid rain on spruce needles has been reported (Hutchinson et al. 1986).

Little is known about the leaching of ions from the bark of conifers, but Reiners and Olson (1984) suggest that this pathway plays an important role.

In contrast to the uncertainty about the actual processes of leaching at the cuticle or the bark level, balance models on spruce stands indicate intensive leaching (Ulrich 1983; Matzner 1984; Meiwes et al. 1984). However, the estimated rates are not yet comparable with those from laboratory experiments.

Ion uptake should follow the same rules as leaching because the cuticles and the bark are nonliving membranes (Dreyer et al. 1981). There are, however, indications that direction plays a role, but conclusive explanations can not yet be offered (Kannan 1986).

The scope of our studies in the forests of the Fichtelgebirge was to determine leaching rates and pathways for different ions, and uptake of ions through above-ground plant parts. Therefore, we used three different experimental approaches:

1. short-term experiments in which water droplets were applied to needles and the diffusion of selected ions through the cuticular layer was measured;
2. short-term experiments in which twigs and seedlings were sprayed and the efflux and influx of ions were measured;
3. long-term experiments in which the fluxes of ions in two spruce stands were measured and a canopy balance model was applied.

Comparing results will give indications about the role of leaching in the nutrient cycling of the spruce stands in the Fichtelgebirge.

2 Can Needle Surfaces Buffer Acid Rain?

2.1 Direct Measurement of pH Buffering on Spruce Needles

Single droplets of rainwater were applied to spruce needles, and pH and magnesium (one of the most intensively leached ions) were measured. After a reaction time of 5 h, pH and magnesium were again measured. Results indicated that the buffering intensity of the leaf surface was due to deposited aerosol particles and ion exchange through the cuticle, assuming that the latter process was linear with time.

Experiments were carried out under laboratory conditions on needles sampled at the healthy site (see Chap. 0–2, this Vol.) in March 1987. Whole branches were cut and brought into the laboratory for no longer than 2 days before analysis. Three categories of needles were selected for buffering experiments: green needles with no visible injury, yellow needles, and brown, almost dead ones.

All experiments were carried out isothermally. For one measurement, ten artificial rain droplets of 2.5 μl each with an ionic composition typical of the rainwater in the Fichtelgebirge (Klemm et al. 1987; Fig. 1a) were each put on one needle. Due to the geometry of the spruce needles, the wetted leaf area was almost rectangular. Its size was determined by measuring its length and width for typical needles under a microscope. The covered leaf area estimated with this procedure was 1.7 mm^2 for each droplet and the resulting average droplet thickness was 1.5 mm.

The needles were immediately put into a reaction box where 100% air humidity was established by distilled water to avoid evaporation of the droplets. After 15 min, two needles were removed from the box so that magnesium concentration and pH in the droplets could be measured. Similarly, after 5 h, two needles were removed so that magnesium concentration could be measured again. The volume of the droplets decreased as a result of higher vapour pressure H_2O over the curved droplet surface. Therefore, the volume of five droplets was measured by removing each droplet from the needle with a syringe, putting all of them into a 0.5 ml Eppendorf reaction cap, and weighing them. Preliminary tests showed that about 73% of applied liquid could be recovered with this weighing procedure. The average decrease in volume was 45% for the 5-h period, but the wetted area was not reduced.

pH measurements were carried out with a specially constructed mini-pH electrode. The glass electrode was based on the principle of potassium electrodes

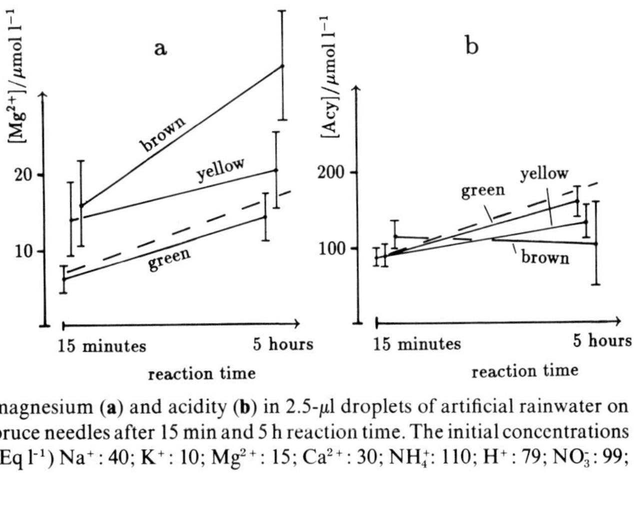

Fig. 1. Concentrations of magnesium (**a**) and acidity (**b**) in 2.5-μl droplets of artificial rainwater on brown, yellow, and green spruce needles after 15 min and 5 h reaction time. The initial concentrations in the artificial rainwater (μEq l^{-1}) Na^+: 40; K^+: 10; Mg^{2+}: 15; Ca^{2+}: 30; NH_4^+: 110; H^+: 79; NO_3^-: 99; SO_4^{2+}: 140; Cl^-: 45

of Thomas (1978). The geometry of the sensitive tip (diameter 0.1 mm) allowed measurements on small wetted surfaces. The liquid junction of the reference electrode consisted of an open tip glass cappillary (tip diameter 0.05 mm) filled with 0.1 mol l^{-1} KCl solution (e.g., Camman 1977). Millivolt readings were carried out with a twin high impedance ($10^{15}\Omega$) potentiometer built by the electronic workshop of the Universität Bayreuth. Taking into consideration the slope of the glass cell (50 mV/pH), its reproducibility ($\pm$ 1.5 mV), and changes in liquid junction potential ($\pm$2 mV), the margin error of the pH measurement was $\pm$0.08 unit. Magnesium concentrations were measured with an AAS. Finally, all procedures were tested on teflon and parafilm, two reference surfaces which do not chemically react with artificial rainwater.

2.2 *pH and Magnesium Concentrations on Wetted Spruce Needles*

Results are summarized in Fig. 1. Each point is the average value of 20 measurements. For magnesium (Fig. 1), great variations within each group lead to high standard errors. Assuming that no reaction took place between the rainwater and the surface of the leaf, 7,5 μmol l^{-1} and 17 μmol l^{-1} Mg^{2+} should have been found at the start and at the end of the reaction time respectively (*dashed line* in Fig. 1a). This seems to have been true for green needles. For yellow needles, no highly significant deviation from the predicted nonbuffering value could be measured. However, the values are above the dashed line in Fig. 1 and therefore indicate the presence of magnesium on the leaf surface. On the brown needles, increased concentrations at the start show that $2.2 \pm 1.4 \times 10^{-11}$ mol Mg^{2+} were introduced into the drop either by solubilization of deposited particles or by rapid leaching from dead plant material. This introduced amount was $1.3 \pm 0.8 \times 10^{-9}$ mol cm^{-2}. While the increase in concentration during the reaction time was not significant, it indicates some leaching occured during the five hours. Due to great standard deviations, Mg^{2+} concentrations in the droplets would have to have increased by more than 2.3×10^{-5} mol per droplet in 5 h to be statistically significant. This corresponds to a leaching rate of 3.9×10^{-11} mol cm^{-2} h^{-1}.

For pH, the results are similar. In Fig. 1b, acidity data are calculated as [Acy] = 10^{-pH}, assuming that no weak acids were present (cf. Hantschel and Klemm 1987). For green and yellow needles, no acids could be found (<20 μ mol l^{-1} or $< 5.8 \times 10^{-10}$ mol cm^{-2}), and no buffering could be determined ($< 7.3 \times 10^{-10}$mol cm^{-2} h^{-1}). For brown needles, the results show higher standard deviations because some needles showed low pH's at the beginning of the experiment and some very high pH's after the reaction time (pH around 4.6). For these cases, the initial acidity of the rainwater (7.9×10^{-5} mol l^{-1}) was almost buffered (buffering: 5.4×10^{-5} mol l^{-2}). During the reaction time of five hours, the proton/cation exchange of these needles was 1.6×10^{-9} mol cm^{-2} h^{-1}.

It was not possible to measure both pH and Mg^{2+} in the same droplet. In some cases, both high pH's and high Mg^{2+} concentrations occured, but with the methods employed, it could not be determined whether these high concentrations coincided within one droplet. For single cases, buffering action on brown spruce needles could be measured. Part of this buffering was due to cation exchange vs. Mg^{2+}, but there should have been at least one more reaction, e.g., cation exchange of K^+ and/or Ca^{2+}. Buffering of acidity on cuticles probably occurs only on needles that are mechanically damaged, so that ion diffusion through leaf injuries is not effectively hindered.

3 Short-Term Studies of Sulfate Uptake by Needles

In order to contribute to knowledge of the fate of sulfate deposited by acid precipitation on conifer leaves, uptake or radioactively-labeled sulfate and its partitioning into the shoot axis and neighboring needles of Norway Spruce were studied under field conditions.

3.1 Experimental

Several approximately 8- to 10-year-old Norway spruce trees growing in a natural stand in the Fichtelgebirge and in the Botanical Garden of Bayreuth were used in the studies. The experiments were carried out on sunny days and the trees were maintained under natural environmental conditions as much as possible. Because of the heterogeneity of the plant material growing in uncontrolled field conditions and with varying weather conditions, each experiment represents a unique set of data.

As the spruce needle surface repels water, and the droplets tend to coalesce into a continuous film from which the liquid drains off, only very small volumes of acid or salt solution could be applied to the needles. Upon termination of an uptake experiment, the needles to which the isotope-labeled solution had been applied were separated from the collected twig sections, rinsed with water and, as was the rest of the collected material, were frozen in liquid N_2, and ground in a mortar. The powder was then extracted with boiling 70% ethanol and the radioactivity of the extracts was measured with a liquid scintillation counter. Since it was unlikely that ^{35}S was substantially incorporated from sulfate into compounds insoluble in boiling 70% ethanol during the short-term experiments, the residue of the extraction was not further investigated.

3.2 Time Kinetics of $^{35}SO_4^{2-}$ Uptake

Evans et al. (1985) have shown that the permeability of chloride and heavy metal ions into bean leaves increased with the time of exposure of the leaf to simulated acid rain. This observation suggests that even small amounts of strong acids (which effectively occur upon drying and absorption of acid rain droplets) could hydrolyze cuticular membrane material and hence facilitate the penetration of hydrophilic substances into the leaf tissue. By contrast, rapid uptake of $^{35}SO_4^{2-}$ from acidic solutions into *Phaseolus* leaves during an initial 20-min period decreased to a rate which remained constant for 3 h (Evans et al. 1981). Hence, different anions obviously react differently with different leaf surfaces; general predictions cannot be made unless the behaviour of anions has been tested for the respective plant material.

In order to study sulfate penetration from sulfuric acid into Norway spruce needles, uptake kinetics were established over a 2-h period. For each investigated time-span, one 5-μl droplet of 50 m mol l^{-1} $H_2^{35}SO_4$ was placed onto each of two separate needles of a current-year sprout. The droplets were prevented from drying by cautiously spraying the tree twice with water. After 15, 30, 60, 90 and 120 min reaction periods, the incorporation of ^{35}S into the needles to which the acid had been applied, and into the current-year and the supporting 1-year-old twigs was determined.

Since the amount of ^{35}S which was extractable from the 1-year-old axis was less than 2% of that extractable from the current-year shoot axis, significant export of labeled sulfur into older twig sections should not have occurred. Uptake of sulfate from the 50 m mol l^{-1} sulfuric acid by the treated needles was more or less linear over the 2 h and resulted in an incorporation of approximately 50% of the applied ^{35}S during that time-span. The linear rate of uptake suggested that even at this high acid concentration, no feed-forward effect on the uptake of sulfate from the applied acid, e.g., via lysis of the epicuticular wax layer or the cuticle, was operative. As shown in Fig. 2, ^{35}S initially accumulated in the treated needles, whereas only a minor proportion of the incorporated anion had entered the shoot axes and migrated further into the attached needles (fraction "other needles"). After 1 h of sulfuric acid application, the fraction of ^{35}S having been incorporated into the axes (predominantly into that of the current-year shoot) increased substantially. One-half hour later, the presence of the isotope in the untreated needles had significantly increased. From the nature of the experiment, the initial accumulation and delayed mobility of ^{35}S in the treated needles is not easily understood. Perhaps a time-dependent switch-over from sulfate transport via the outer free space to that within the symplast is responsible for the observed phenomenon.

3.3 *Influence of pH on the Uptake of Sulfate by Norway Spruce Needles*

Schönherr (1976a) observed a substantial increase in water permeability of isolated cuticles of *Citrus aurantium* with increasing pH. However, the rate of $^{35}SO_4^{2-}$ incorporation into bean leaves has been found to increase with decreasing pH, while Rb^+ (as a model for K^+) has been shown to be taken up less efficiently at low pH (Evans et al. 1981, 1985). Such observations indicate that especially in natural stands buffering substances by the leaf (Hutchinson et al. 1986) or cation exudation by the leaf (Hutchinson et al. 1986; Adams and Hutchinson 1987) may considerably alter the pH of droplets layered onto the leaf surface. In the experiments reported here with Norway spruce, rather high ionic strength and

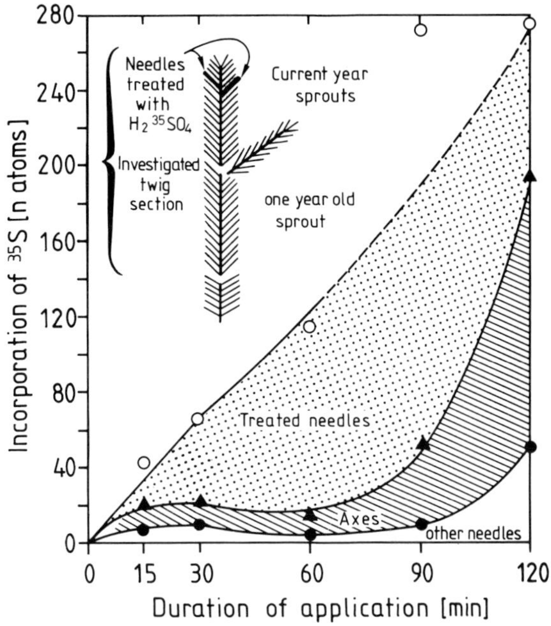

Fig. 2. Time kinetics of ^{35}S-uptake by Norway spruce needles. Two 5-μl-droplets (each 0.25 μmol sulfuric acid) of 50 mmol l^{-1} $H_2{}^{35}SO_4$, each placed on the upper surface of one current-year needle, were applied for the time periods indicated on the abscissa. Incorporation of ^{35}S into the treated needles, the current-year plus 1-year-old axes (see *insert*), and the needles attached to these sprouts ("other needles") was measured

short uptake time spans (10 min) were used to circumvent secondary effects. The results of two relevant experiments are shown in Tables 1 and 2. As long as the droplets remained liquid, ^{35}S-incorporation from a neutral solution was slightly higher than that from an acidic one. In particular, the mobility of sulfate from the treated needle to the rest of the twig was consistently higher when the needle was not simultaneously stressed by a high proton concentration. Tables 1 and 2(A) also indicate that the dependence of sulfate uptake on the concentration of the applied solution was stronger than that on the pH. In summary, spruce needles show an increased uptake of sulfate at higher pH-values, at least for very short treatment periods, and thus appear to react in manner contrary to that of bean leaves.

3.4 Incorporation of ^{35}S upon Drying of the Superficially Applied Solution

Evaporation of a droplet of an aqueous solution on a leaf surface may conceivably result in at least two effects: an increase of solute concentration, and a change of the pH of the solutions. Alkalization has been observed with bean leaves when the applied droplet had a pH-value higher than 4, while acidification

Table 1. ^{35}S-uptake [n atoms] by a spruce needle (center or tip) and the supporting twig from a 5-µl droplet of either 10 m mol l^{-1} $H_2{}^{35}SO_4$ (pH 1.6) or $Na_2{}^{35}SO_4$ (pH 6.5) over a period of 10 min

$^{35}SO_4^{2-}$ Applied to	pH Solution	^{35}S-Content of the treated needle	^{35}S-Content of the the rest of the twig		Total ^{35}S-uptake
			Needles	Axis	
Needle center	1.6	0.3	0.9	0.5	1.7
	6.5	0.3	1.1	1.7	3.1
Needle tip	1.6	0.3	1.1	1.0	2.4
	6.5	1.0	0.8	0.9	2.7

Table 2. ^{35}S-Uptake [n atoms] by a spruce needle and the supporting twig from a 5µl-droplet of either 200 m mol l^{-1} $H_2{}^{35}SO_4$ (pH 1) or $Na_2{}^{35}SO_4$ (pH 7.0) over a 10-min period

Experiment[a]	pH of Solution	^{35}S-Content of needle of application	^{35}S-Content of the rest of the twig	Total ^{35}S-uptake
A	1	9	9	18
	7	0.5	24	24.5
B	1	4	172	176
	7	0.6	180	180.5

[a] A: Droplet kept wet; B: droplet dried up during the application period.

has resulted from treating the leaves with a solution of a pH lower than 3.1 (Evans et al. 1985).

Evaporation to dryness of the droplets applied to spruce needles enhanced sulfate uptake by almost one order of magnitude and concommitantly overshadowed any influence of the initial pH (Table 2). Drying did not simply increase absorption of ^{35}S by the treated needle, but promoted the flux from the needle into the shoot axis, where approximately 90% of the label was found to be localized. Hence "drying up" in this case is not only a matter of solvent evaporation, but rather must be understood as resulting in an enhanced uptake of the liquid.

4 Uptake of Water and Solutes Through Bark and Needle Surfaces

The exchange of water and solutes through bark and needles of spruce twigs was studied as mass flow and diffusive processes, respectively. Pathways for absorption of water and solutes exist in the region of radial rays across the twig bark. It is shown that the contribution of ion exchange via water pathways across the bark may be significant for the whole plant ion balance (Katz et al. 1988). When twigs at low water potential were sprayed with water (Fig. 3), the water potential first increased rapidly and then more slowly. This was caused by the mass flow of water, which appeared to be related to the water potential (hydrostatic pressure) difference between the outside of the twig and the xylem. For twigs carrying a full set of needles and for partially defoliated twigs, a change in water potential from

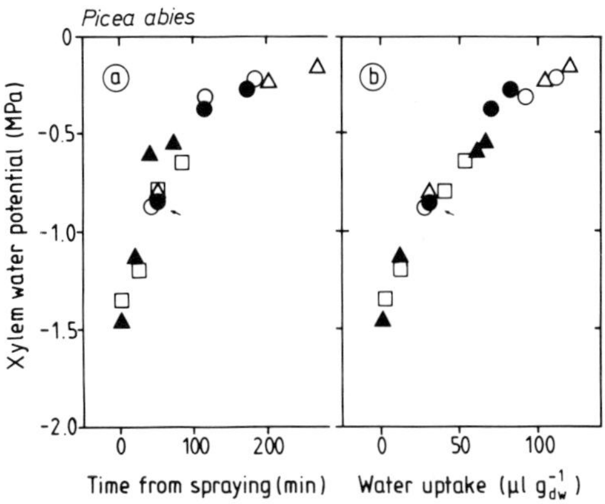

Fig. 3. Xylem water potential of twigs in relation to the time after spraying (**a**) and the resulting change in water uptake (**b**) for twigs with a complete set of needles. Each *symbol* represents a different twig, n = 5 (Katz et al. 1988)

-1.4 MPa to -0.2 MPa was equivalent to a total water uptake of 60 to 80 $\mu l\ g^{-1}$ dry wt (6%–8%, s.e. 1% of relative water content). The water transport across the bark was complicated by the complex structure of the different tissues. Apoplastic and symplasmatic pathways may exist across the bark, the phloem and the cambium and into the xylem. However, the initial slope of the water uptake curve can be used to estimate a lower limit of water permeability, given a simple two-compartment model for water flow from outside into the xylem and across the barrier possibly represented by the cambium. The hydraulic conductivity of needled twigs (Table 3) was $2 \times 10^{-9}\ m\ s^{-1}\ (MPa)^{-1}$, which was two orders of magnitude smaller than values known for cell membranes (Steudle and Zimmermann 1984).

One pathway of water exchange was identified using fluorescein as a polar dye (Fig. 4). During rehydration, the dye stained only the rays and the parenchyma cells in the bark and wood. This suggests that there are water pathways across the bark which allow movement of solutes of high molecular weight if the bark is hydrated. Fluorescent dyes did not penetrate needle surfaces, which does not mean that water and ion movement may not have taken place.

The exchange of magnesium chloride by convective mass flow through the twig surface was investigated using distilled water or solutes with concentrations close to that of the xylem sap. In this experiment 2.2 $mmol\ l^{-1}$ was sprayed onto the bark of twigs having a water potential of -1.0 MPa (Fig. 5). When water saturation was nearly complete, no change in xylem sap concentration could be detected at the cut surface of the twig since the applied solutions had not yet reached it (Fig. 5: 1st spraying cycle). When the twig was again dried to -1.0 MPa by expressing xylem sap with the pressure chamber and sprayed a second time with distilled water or magnesium chloride solution (Fig. 5: 2nd spraying cycle), the expressed sap showed an increased concentration with the magnesium chloride spray and a decreased concentration with the distilled water.

In the field, mass flow will only occur as an uptake mechanism if rain contacts the forest canopy at a low water potential. During the growing season 61 such rain events may occur (see Chap. 1-B). During the initial linear phase of water uptake

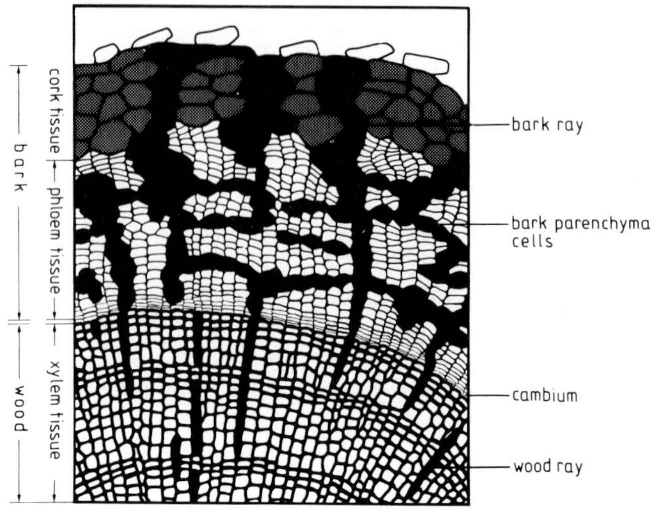

Fig. 4. Distribution of fluorescent dyes (fluorescein, *black cells*) in the bark and xylem rays following uptake of solution through the outer surface

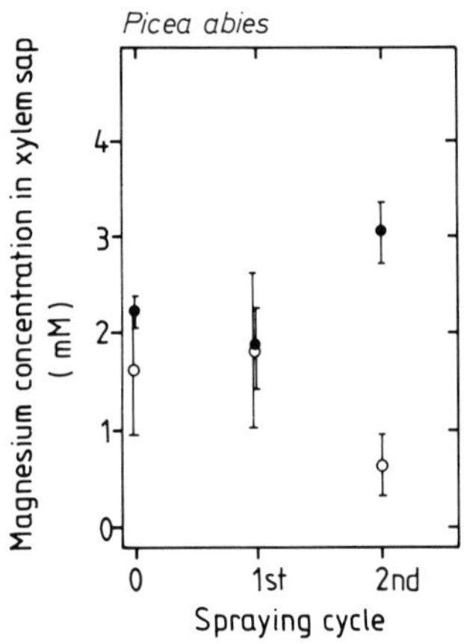

Fig. 5. Magnesium concentration in xylem water of twigs following repeated spraying of 2.2 mmol l^{-1} magnesium chloride solution ($MgCl_2$) (●) or distilled water (○). The solution was sprayed on twigs with a water potential of -1.0 MPa. Following the first spraying after the surface of the twigs was dry, the twigs were again brought to a water potential of -1.0 MPa by expressing xylem sap and then sprayed again with water or with the magnesium chloride solution

from -1.5 MPa to -0.8 MPa a total of 5.1 l m^{-2} ground a^{-1} may occur, but the total gain of magnesium or nitrogen due to convective flow would be small when compared with the requirements for growth (< 1%, Table 3).

Once the twig is hydrated, the pathways between the outside and the xylem are open for the diffusion of ions into or out of the twig. The direction and magnitude of this process should depend on the concentration difference between the two compartments, the permeability of the barrier, the diffusion potential, and the time of exposure. The exchange of magnesium chloride by diffusion was

Table 3. Hydraulic conductivity of water and permeability coefficient of magnesium chloride ($MgCl_2$) across the bark of spruce twigs[a]

Hydraulic conductivity for mass flow of water ($m\ s^{-1}\ MPa^{-1}$)	2×10^{-9}
Permeability coefficient for diffusion of magnesium chloride $MgCl_2$ ($m\ s^{-1}$)	9×10^{-9}

For the calculation, a two-compartment model of solute and xylem separated by the cambium was considered. For the initial 55 min the mass flow of water was calculated from the initial slope of the water uptake curve to be 32 $\mu l\ g^{-1}$ dry wt for a surface area of 3.3 $10^{-3}\ m^2\ g^{-1}$ and an average driving force of 1.17 MPa (Fig. 3). The estimate of the diffusion coefficient for magnesium chloride was based on an uptake of 0.7 mol Mg m^{-3} over a period of 5 h and an average concentration difference between the solution on the bark and the xylem of 1.25 mol Mg m^{-3}. It is assumed that the total twig and needle biomass participates in the uptake by mass flow.

measured by treating twigs for 5 h with a 2 mmol l^{-1} magnesium chloride solution. In this experiment the initial xylem sap concentration was 0.4 mmol l^{-1} (s.e. = 0.1), which increased to 1.1 mmol l^{-1} (s.e. = 0.3). The permeability coefficient of the magnesium of the needled twigs was estimated to be 9×10^{-9} m s^{-1} (Table 3). This was lower than the permeability coefficient of calcium but higher than that of potassium for root cell membranes (Pitman 1976). Under natural conditions the concentration of magnesium is higher in the xylem than in rain or fog (see Chaps. 1-B and 4-F), and therefore, during a precipitation event when the canopy is wet a magnesium transport from the xylem to the surface may occur. The estimated range of magnesium loss (Table 4) during an annual period is in the range of the magnesium requirements for growth. This estimate does not consider saturation effects. It is uncertain as to how much time and what proportion of twigs are moistened by rain and fog. The range may represent an upper limit for leaching.

For nitrogen, the concentration gradient for diffusion is in the opposite direction. There is a high concentration of ammonium and nitrate in rain and fog water (see Chap. 1-B) but a low concentration in the xylem. This should lead to a net uptake. Assuming that the permeability coefficient for nitrogen ions is the same order of magnitude as that estimated for magnesium, the uptake of ammonium and nitrate via aboveground parts (Table 4) may contribute 2–18% of the nitrogen requirement for annual growth. This calculation does not consider losses of amino acids from the xylem to the outside.

5 Leaching of Cations by Acid Rain from Twigs and Single Needles

We investigated leaching by rain of minerals from twigs and detached needles under controlled laboratory conditions. Loss of minerals was evaluated on the basis of the total mineral content of each single twig or needle.

Table 4. Total loss of magnesium ions, calculated from the diffusion coeffcient[a]

		Transport		Measured canopy	Requirement for above-ground growth	
		Diffusion	Mass flow	Leaching/ uptake	Healthy	Declining
	Rain	5 to 42 10^{-3}				
Mg	Fog	4 to 40 10^{-3}				
	Total	9 to 82 10^{-3}	40 10^{-6}	1–3 10^{-3}	21 10^{-3}	12 10^{-3}
NH_4^+-N	Rain	2 to 17 10^{-3}				
and	Fog	12 to 74 10^{-3}				
NO_3^--N	Total	14 to 91 10^{-3}	1 10^{-3}	46–58 10^{-3}	469 10^{-3}	423 10^{-3}

[a] Assuming 91 rain and 33 fog events per year of 18-h wettening time per event as the upper limit, or 61 rain events during the growing season and 33 fog events of 9-h wettening time and 50% wettening of the total twig biomass as the lower limit (see Chap. 1-B). The average xylem sap concentration was 0.4 mol Mg m^{-3} (Osonubi et al. 1988) while the average concentration of rain was 9×10^{-3} mol Mg m^{-3} and 27×10^{-3} mol Mg m^{-3} of fog (Chap. 1-B). The total uptake of nitrogen (N) was calculated assuming the same range of wettening time and the same diffusion coefficient. The concentration of nitrogen (N) was much higher than that of magnesium (Mg) (155×10^{-3} mol N m^{-3} in rain and 1.91 mol N m^{-3} in fog; see Chap. 1-B).

5.1 Materials and Methods

Twigs and needles of *Picea abies* (L. Karst.) were obtained from 25-year-old trees grown in the local Botanical Garden of Würzburg. The trees showed no visible symptoms of disease. Twigs with two needle ages were mounted in perspex cuvettes and exposed to experimental solution under the conditions described in more detail in Fig. 6. As a reference for the amount of solutes leached, the total solute content from stem segments and needles was measured. Stem parts were cut into small pieces dried at 80°C for 48 h. They were then ground to a fine powder with quartz sand. The homogenate was suspended in 4 ml of distilled water and boiled in a water bath for 30 min. After centrifugation, the clear supernatant was used to analyze water soluble compounds. To evaluate the contribution of the needle and the bark separately, experiments were also performed with detached needles. Approximately ten needles were mounted in a styrofoam block fitted into a standard 1-cm cuvette filled with 2 ml of water or leaching solution. Cuvettes were gently shaken during incubation at room temperature. At certain intervals, the 2-ml leaching solution was replaced and used for various analytical procedures.

5.2 Leaching of Cations from Foliated Twigs

Rather large rain variation effects in the leaching rates from twigs with distilled water or with acidic solutions had to be compared by subsequent treatment of individual samples. Results of such an experiment are given in Fig. 7. For

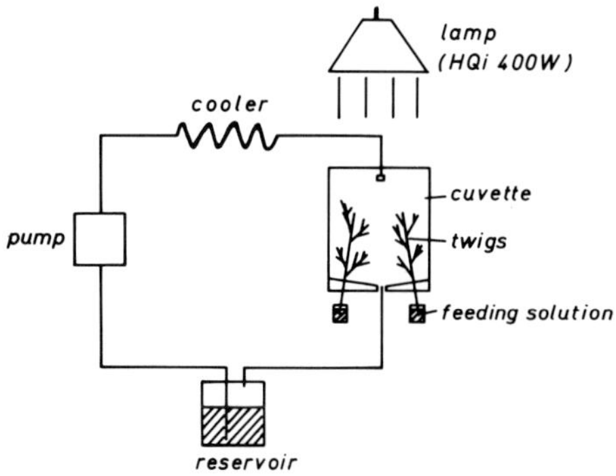

Fig. 6. Schematic presentation of the cuvettes and the system by which whole twigs were sprayed with water or various solutions. On top of the perspex chamber, a nozzle was connected by tubings with a Teflon-coated pump (Procon Standex Int., Krefeld, FRG). At the bottom of the chamber, the water was collected and guided through Teflon tubings to a reservoir (0.5 l vol). From there it flowed back to the pump. The water from the pump first had to pass through a cooler (glass) in order to keep the water temperature constant. Rain was provided only during the day (8 h) in intervals of 2 min rain/4 min break. The pumping speed was 7 l/h. At night (16 h), chambers were opened to allow the twigs to dry up. During the day, twigs were illuminated from the top with a HQi-lamp (400 W, Schreder, Winterbach, FRG) providing a light intensity of about 400 $\mu E\ s^{-1}$ (PAR). Four twigs (usually with 2 needle ages) were mounted in the chambers so that the stem base remained outside the chamber. This allowed feeding of water or other solutes to the stem during the experiments. Aliquots (usually 5 ml) were taken from the reservoir at various times for solute analysis

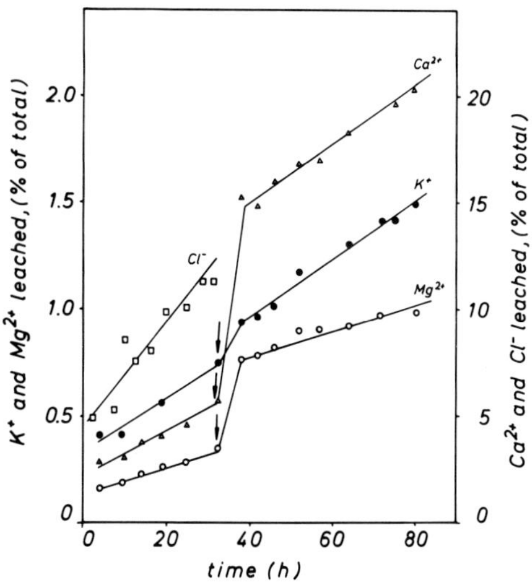

Fig. 7. Leaching of various ions from spruce twigs as a function of time. Twigs were first treated for about 26 h with distilled water. At the times indicated (see *arrows*), the water in the reservoir was replaced by H_2SO_4. A representative experiment from a series of 4 experiments is shown

Table 5. Average ion content of whole twigs (with the two youngest needle ages) and single needles. Values are given in mg g^{-1} fresh weight ± SD (n = 3–7)

Ion	Concentration (mg g^{-1} fresh wt) twigs	needles
K^+	2.60 ± 0.32	3.91 ± 0.10
Mg^{2+}	0.31 ± 0.05	0.46 ± 0.07
Ca^{2+}	0.54 ± 0.12	0.35 ± 0.03

comparison, the total absolute ion content in twigs (with two needle ages) in single 2-year-old needles are given in Table 5. The major inorganic ions washed out of the twigs were the cations Ca^{2+}, Mg^{2+}, and K^+ (Fig. 7, Table 5). The efflux was linear within the time span of the treatment. Relative efflux rates were in the range of 0.007–0.25% of the total cation content per hour. Relative efflux was faster for Ca^{2+} and slower for Mg^{2+} by at least one order of magtitude. With the relative permeabilities of the ions through the cuticles (see introduction to this chapter), the relatively high leaching rates for Ca^{2+} can only be interpreted as the result of high Ca^{2+}/Mg^{2+} and Ca^{2+}/K^+ ratios in the cell wall (cf. Bauch and Schröder 1982; Mengel and Kirkby 1987). After switching from distilled water to dilute acid (H_2SO_4, 1 m mol l^{-1}, pH 2.95), there was an immediately rapid but transient increase in the efflux of cations. Subsequently, rates returned closely to the original values. Anion efflux could not be measured simultaneously since the high sulfate concentrations interfered with the anion chromatography. Stimulation of cation efflux was similar when a solution of Na_2SO_4 (pH 5.8) was used instead of sulfuric acid (Fig. 8). Thus, the rapid and transient increase in the cation efflux was caused by the exchange of cations for H^+ or Na^+ at the twig surfaces. Leaching of cations was also stimulated when salt solutions ($AlCl_3$ or KCl, 20 m mol l^{-1}) were fed through the stems. In most experiments, feeding of salt solutions not only caused a rapid transient increase of the leaching rate, but higher rates were maintained throughout the duration of the experiment (results not shown here).

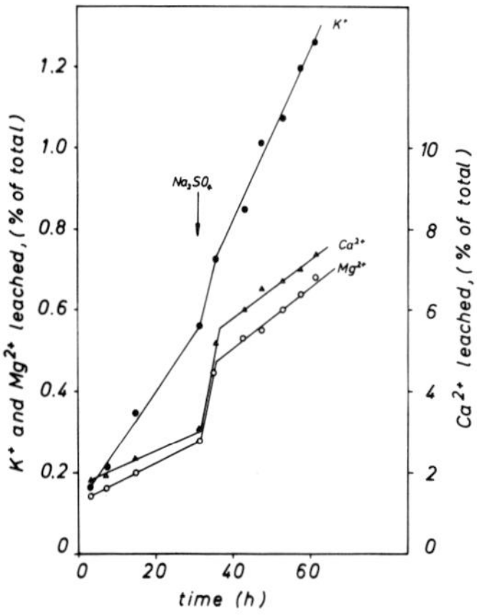

Fig. 8. Leaching of cations from spruce twigs as a function of time. The experiment was similar to the experiment in Fig. 7, but with Na_2SO_4 (1 mmol l^{-1}, pH 5.8) instead of H_2SO_4 was used

5.3 Leaching of Cations from Detached Needles

The experiments with twigs could not separate leaching of cations from the bark and needles. Therefore, efflux from detached needles (without the contribution of the bark) was examined and compared with the efflux from twigs. Because spraying of single needles with water or dilute acid under the same conditions employed with whole twigs was difficult, efflux was measured by immersing needles in water or in H_2SO_4 (1 m mol l^{-1}). The relative efflux rate of K^+ and Mg^{2+} from single needles into water was very similar to those rates obtained with whole twigs (Table 6, cf. Fig. 7). Rates of Ca^{2+} efflux, however, were lower and often below the detection limit (i.e., much less than 0.016% of total per h). This indicates that the higher rates of Ca^{2+} leaching from twigs were due to the exchange of Ca^{2+} from the stem parts and not to the Ca^{2+} exchange from the needles. The efflux of cations from detached needles was not stimulated by treatment with H_2SO_4 (Table 6). With single needles prelabeled with $^{14}CO_2$, leaching assimilates was also measured. Relative leaching rates (as a percentage of total water-soluble labeled compounds) were about one order of magnitude lower than the efflux of inorganic cations (data not shown here). We therefore suggest that inorganic cations are preferentially removed from the needle surface and the apoplast while leached organic compounds originate mainly from the membrane-surrounded symplast.

Table 6. Efflux of cations from single needles into a surrounding medium (distilled water or sulfuric acid 1 mmol l^{-1})[a]

Ion	Spray solution	
	H_2O	H_2SO_4
K^+	0.012	0.012
Mg^{2+}	0.010	0.011
Ca^{2+}	$\gg$0.01	$\gg$0.01

[a] Relative rates are given as percent of total h^{-1}. A representative experiment from a series of 5 similar experiments. Standard deviation of relative efflux was 10–30%, but the ratio of efflux in water and in H_2SO_4 was practically identical in all experiments.

In spite of the rather heavy rain treatment applied in our experiments, leaching of inorganic mineral ions from whole twigs was low and was only transiently increased by acid rain. In our experiments, the acid rain treatment included a 14-h night-phase without rain. During this period, needles were allowed to dry. Drying entailed concentrating acids and salts in the droplets adhering to the twig surfaces so that the effects of acid rain were exaggerated. In contrast to experiments with whole trees in the field, cut twigs are not able to replace mineral solutes lost by leaching. This is presumably the main reason why cation efflux was only transiently increased by acid rain.

6 Leaching of Minerals from Seedlings under Laboratory Conditions

Nutrients which are removed from the tree canopy by leaching can only be replaced by increasing uptake from the soil. If the leaching mechanism is a cation vs. proton exchange in the canopy, acidification of the rhizosphere may result (Ulrich 1981).

Flückiger et al. (1986) qualitatively showed in laboratory experiments that acidification of the rhizosphere of beech seedlings is a function of acid treatment of the leaves. The goals of our experiments were (1) to determine under laboratory conditions whether the quantification of cation leaching out of the spruce canopy was a function of acid deposition and (2) to quantify simultaneously the leaching effects on the acidification of the rhizosphere. Three-year-old Spruce seedlings were cultivated in nutrient solutions (350 ml per tree) and part of the needles were removed prior to the experiments. Acid treatments of trees (10–15 replications) were carried out. Each tree was sprayed with 67 ml per day of either water or sulphuric acid of pH 2.7 or pH 2.4 respectively during a period of 10 days. One treatment contained unaffected control trees. Total throughfall of water and pH 2.4 treatments was sampled and K^+-, Ca^{2+}- ang Mg^{2+}-contents were measured.

At the end of the experiment, acidity (end point titration of pH 8.4) and P-contents of the nutrient solutions were measured. In addition, element contents of the spruce needles were analyzed (For more details see Kaupenjohann et al. 1988). Leaf area was estimated according to Oren et al. (1988). Stem area was determined by measuring the diameter at the middle and the length of each branch with the assumption that the branch was cylindrical.

The amounts of K^+, Ca^{2+} and Mg^{2+} leached from the tree canopies (Table 7) showed no Ca^{2+}- and Mg^{2+}- and relatively low K^+-leaching in the water treatment. Treatment with pH 2.4 clearly increased the leaching of ions, especially that of Mg^{2+}. Table 8 shows the amounts of leached nutrients related to leaf area. As the ratio leaf area:stem area is about 1.45 in our seedlings, the results may be compared with those of older spruces in the conclusions of this chapter.

Figure 9 shows that increased leaching is associated with a significant increase in the acidity of the nutrient solutions.

Table 7. K^+, Ca^{2+}, and Mg^{2+} leaching out of the canopy and acidity of the nutrient solution of spruce for different treatments

Treatment	Cation leaching K^+	Ca^{2+} (μEq/tree)	Mg^{2+}	Acidity of nutrient solution
Water	25	0	0	440
pH 2.4	55	20	65	500
D minus B	30	20	65	60

Table 8. Total K^+. Ca^{2+}, and Mg^{2+} leaching out of the canopy related to the leaf area

	Cation leaching K^+	Ca^{2+} (nEq cm^{-2})	Mg^{2+}
Water	56	0	0
pH 2.4	124	45	146

Determination of the acidity change ΔAcy (acidity plus phosphorus taken up by the plants as $H_2PO_4^-$) was done at the end of the experiment as following (cf. Kaupenjohann et al. 1988):

$$\Delta Acy(\mu Eq\ l^{-1}) = Acy(\mu Eq\ l^{-1}) + \Delta P(\mu Eq\ l^{-1})$$

ΔP is the applied P minus the remaining P in the nutrient solution at the end of the experiment. Difference in N uptake (the nutrient solution contained only NO_3^--N) of the differently treated seedlings was not statistically significant (Kaupenjohann et al. 1988). However, as NO_3^--contents in the nutrient solutions of acid-treated trees seemed lower compared to the water-treated and untreated trees. ΔAcy is more likely to be underestimated than overestimated by the equation.

Comparing the water treatment with the pH 2.4 treatment, the difference in total acidity was about 60 μEq H^+ per tree (Table 7). On the other hand K^+-, Ca^{2+}- and Mg^{2+}-leaching induced by the acidic treatment was a total of 115 μEq per tree (Table 7). The calculations suggest an imbalance between nutrient losses from the canopy and uptake from the nutrient solution of about 55 μEq per tree. Analysis of nutrient contents of the trees needles showed that the expected loss of nutrients could be nearly explained by the loss of Mg^{2+} (Table 9)

Mg^{2+} contents decreased from 1.3 g kg^{-1} in needles of trees sprayed with water to 0.9 g kg^{-1} in the pH 2.4 treatment. The difference between the two was a total of 45 μEq Mg^{2+} per tree (Table 9). Indeed, no acidity effects seemed to occur when

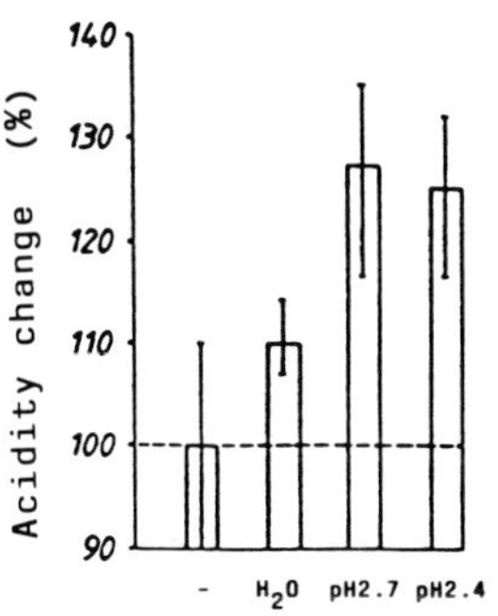

Fig. 9. Acidity changes (% of untreated control = 399 μEq H^+/tree) of the nutrient solutions 10 days after starting the experiments

Table 9. Mineral element contents and total dry matter of 1-year-old spruce needles

Treatment	Needle element contents					
	(g kg⁻¹)				(μEg/tree)	(g needles/tree)
	N	K	Ca	Mg	Mg	
Untreated	17.7	6.8	11.2	1.5	350	2.80
Water	17.7	6.8	9.4	1.3	285	2.64
pH 2.7	16.7	7.1	9.5	1.1	290	3.19
pH 2.4	16.9	6.1	10.3	0.9	240	3.20
pH 2.4 – water					-45	

total Mg amounts in water and pH 2.7-treated seedlings were compared. However, this result has to be attributed to the initially lower dry weight of the trees sprayed with acid of pH 2.7 (Table 9).

7 Leaching of Minerals with a Canopy Balance Model

As there are no methods to directly measure the leaching of ions under field conditions, canopy leaching was calculated with a canopy balance model published by Ulrich (1983). The total deposition of an ion x with the throughfall (TD_x) is divided between precipitation input (PD_x), interception deposition (ID_x), and a sink term (S_x) which contains all fluxes described in the introduction of this chapter:

$$TD_x = PD_x + ID_x + S_x \qquad (1)$$

The following procedure of estimating the sink term is based on general knowledge of the behavior of single ions in ecosystems as well as on long-term measurements of all precipitation components, mainly in the Solling (Matzner 1984). For several macro elements, the most important process of the sink term is needle leaching. Few elements, e.g., Na^+ and Cl^-, play a relatively unimportant role in plant metabolism and show low leaching rates compared to deposition input. Therefore, for Cl^- Eq. (1) simplifies to:

$$TD_{Cl^-} = PD_{Cl^-} + ID_{Cl^-} \qquad (2)$$

Ulrich (1983) assumes that the concentration ratios of the ions are equal in precipitation deposition and interception deposition, respectively ($ID_{Cl^-}/ID_x = PD_{Cl^-}/PD_x$). Now a factor representing the capability of the canopies to adsorb aerosol particles can be calculated (Eq. 3):

$$ID_x = (ID_{Cl^-}/PD_{Cl^-}) \times PD_x \quad x = H^+, K^+ Ca^{2+}, Mg^{2+} SO_4^{2-} \qquad (3)$$

Measuring TD_{Cl^-}, TD_x, PD_{Cl^-}, and PD_x (PD as incident precipitation) over a period of several months and using Eqs. (2) and (3), needle leaching is the only unknown parameter in Eq. (1), sink term S_x, and can therefore be calculated.

The amount of the leached cations (K^+, Ca^{2+}, and Mg^{2+}) depends on the proton buffering rate of the canopy. The applied canopy balance model, which has successfully been used in different forest ecosystems (Matzner 1984; Büttner et al. 1986), assumes that leaching of K^+, Ca^{2+}, and Mg^{2+} is due to proton exchange with the canopy. Hantschel and Klemm (1987) also showed the importance of weak, organic acids to proton buffering in spruce stands.

In Fig. 10, the canopy difference of cation fluxes (Ulrich 1983), particulate deposition (PD_x) and leaching (S_x), is plotted for the 2 years and shows that high canopy differences for potassium were obviously dominated by the leaching component. K^+-leaching rates for the two periods were significantly higher in the damaged spruce stand at Oberwarmensteinach (2.2 g m^{-2} in 1985) than in the healthy stand at Wülfersreuth (1.1 g m^{-2} in 1985). This was only partly caused by higher throughfall amounts at Oberwarmensteinach than at Wülfersreuth (cf. Chap. 4-C, this Vol.) and higher K^+-contents in the spruce needles at Oberwarmensteinach (Kaupenjohann et al. 1987).

Figure 10 also shows the same site differences for the Ca^{2+}-leaching fluxes. In 1985, 0.6 g m^{-2} were leached at Oberwarmensteinach and 0.3 g m^{-2} at Wülfersreuth. The proportion of leaching rates of the canopy difference was smaller at both sites and throughout the experimental period. Mg^{2+} showed a similar response (Fig. 10). In the healthy stand, less Mg^{2+} was leached at Wülfersreuth than at Oberwarmensteinach. Because the proportion of Mg^{2+} leaching fluxes on canopy differences can be less than 50%, it is therefore the smallest of the three compared base cations.

High potassium-leaching rates in damaged forest stands were also measured by Matzner (1984) and Leonardi and Flückiger (1988). The high contribution of K^+-leaching rates to canopy differences reflects the high mobility of this ion in plant tissues (Mengel et al. 1987). Skiba et al. (1986) referred to the dependency

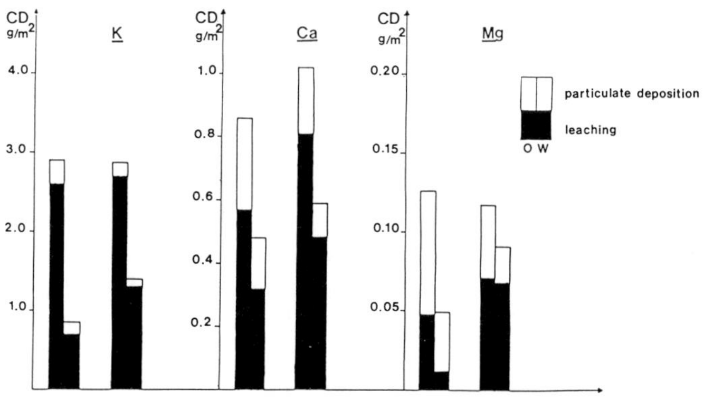

Fig. 10. Canopy differences (*CD*) balanced with the model of Ulrich (1983) for the study sites Oberwarmensteinach (*O*) and Wülfersreuth (*W*)

of K^+-leaching fluxes and proton buffering in canopies. This process probably causes the differences between the two spruce stands. Nutrient status seems to be a factor generally influencing leaching. Spruce stands with less decline phenomena and lower leaching rates, as at Wülfersreuth, are characterized by higher nutrient contents, especially for Ca^{2+} and Mg^{2+} (Kaupenjohann et al. 1987). Changes of leaching fluxes after fertilization confirm this (see Chap. 4-G, Hantschel 1987). Calcium and magnesium are more important to particulate deposition and canopy differences than is potassium.

8 Conclusions

The experiments described in this chapter differ in many respects, as the time scales ranged from minutes to years, different plant materials (from single needles to whole canopies) were used, experimental solutions varied considerably in chemical composition, and different ions were measured.

Because we tried to find a common basis for comparing all the data, all fluxes have been given in flux per area and per time unit. In Table 10a, we present fluxes, as determined from experiments, in relation to the surface area of needles or twigs. In Table 10b, the fluxes for a whole year were calculated on a ground-area basis. We assumed that

1. Norway spruce carried 2.6 kg needles and twigs per m^2 ground area at our study area (Oren et al. 1986),
2. the leaf area index (LAI) in the experimental stand at Wülfersreuth (experiment 6) was 8 (Hantschel 1987),
3. the total leaf area per ground area was LAI $\times$ 2.65 $m^2\ m^{-2}$ (Oren et al. 1986),
4. a 5-μl droplet covered twice the leaf area of a 2.5 μl droplet
5. the ratio of leaf area to bark area in 1- and 2-year-old twigs was 10, and
6. specific leaf weight was 195 g m^{-2} (projected leaf area, Oren et al. 1988).

Uptake of sulfate (Experiment 2) was much slower than the leaching of the cations. This finding supports the theoretical prediction of low penetration rates for anions through intact cuticles. Substantial leaching of the cation K^+, Mg^{2+}, and Ca^{2+}, however, was found, if plant material was treated with dilute acid, acidic rainwater, or salt solutions.

The leaching rates observed in the various experiments covered a range of two orders of magnitude. Leaching rates in the stand of intact trees and under field conditions (Experiment 6) were in the range 0.025, 0.002, an 0.01 mol $m^{-2}\ a^{-1}$ for K^+, Mg^{2+}, and Ca^{2+}, respectively. Under laboratory conditions, leaching rates were at least four times higher. Computations from Experiments 3 and 4 (Table 10b) show that the flux of ions through the bark was probably in the same range as the flux through the cuticles. It is concluded from Experiment 4 that cations can be leached from both needles and stems, and the predominant leaching pathway for calcium is through the bark. The uptake by mass flow is at least one order of magnitude lower than the leaching by molecular diffusion.

Table 10a. A summary of leaching (–) rates and uptake (+) rates per wet needle or twig surface under the conditions defined in the various experiments. All data were converted into 10^{-9} mol m^{-2} s^{-1}

Exp. No.	Authors	K^+	Mg^{2+}	Ca^{2+}	SO_4^{2-}	NH_4^+ + NO_3^-	Surface	Experimental solution	Duration of experiments
1	Klemm (Sect. 2, this Chap.)		< –0.1 –total: –4.4 for single needles–				Needles	Artificial rainwater	5 h
2	Kuhn and Beck (Sect. 3, this Chap.)				$+4 \cdot 10^{-11} \ldots 4 \cdot 10^{-9}$		Needles	Na_2SO_4, H_2SO_4 10...200 mmol l^{-1}	≤ 2 h
3	Katz et al. (1988)		$+2.4 \cdot 10^{-5}$			$+6.4 \cdot 10^{-5}$	Twigs	$MgCl^2$ 2 mmol l^{-1}	1 h
4	Mitterhuber et al.	–0.38	–0.04	≪ –0.3			Needles	Na_2SO_4, H_2SO_4	≤ 80 h
	(Sect. 5, this Chap.)	-3	-0.2	-2.3			Twigs	1 m mol l^{-1}	
5	Kaupenjohann et al. (Sect. 6, this Chap.)	-1.4	-0.84	-0.26			Seedlings	H_2SO_4, pH 2.4	10 days

Table 10b. Leaching (–) and uptake (+) rates per year and ground area (mol m^{-2} a^{-1})

Exp. no.	Authors	K^+	Mg^{2+}	Ca^{2+}	SO_4^{2-}	NH_4^+ + NO_3^-	Surface	Comments
1	Klemm (Sect. 2, this Chap.)		$<$ –0.02				Needles	Diffusion
2	Kuhn and Beck (Sect. 3, this Chap.)				$+6 \cdot 10^{-12} \ldots 6 \cdot 10^{-10}$		Needles	Diffusion
3	Katz et al.		$+6.3 \cdot 10^{-5}$			+0.04	Twigs	Massflow
	(Sect. 4, this Chap.)		–0.04			+0.07	Twigs	Diffusion
4	Mitterhuber et al.	–0.06	–0.006	$\ll$ –0.05			Needles	Diffusion
	(Sect. 5, this Chap.)	–0.41	–0.33	–0.38			Twigs	
5	Kaupenjohann et al. (Sect. 6, this Chap.)	–0.22	–0.14	–0.04			Saplings	Diffusion
6	Hantschel et al. (Sect. 7, this Chap.)	–0.025	–0.002	–0.01			Whole Canopy	Balance model

[a] In experiments 1–5 we assumed 100 rain (see Chap. 1-B) and 33 fog (Trautner and Eiden, submitted) events per year, and the whole plant surface was wetted for 18 h in each event with the experimental solution, respectively.

There seems to be no unique permeability of a spruce needle surface to ions, but we suggest that leaching takes place on distinct leaves or parts of leaves (Experiment 1). Needles with a damaged wax layer, for example, should be more susceptible to leaching.

Leaching of cations by diffusion is a cation exchange reaction (K^+, Mg^{2+}, Ca^{2+} vs. H^+). The cation exchange intensity was presumed to depend linearly on the ion concentrations both in the treatment solutions and in the leaf interior. In Experiment 4, however, leaching rates only transiently increased with increasing hydrogen ion concentrations in the treatment solution. Further, a correlation of the leaching intensity and the ion concentration of the cells could not be observed in Experiment 4. We conclude from these findings that increasing acidity enhances the transfer of cations from the root system to the aboveground plant parts, and the ion flux seems to be directly from the xylem to the leaf surfaces.

This hypothesis is supported by the fact that leaching does not alter the content in the cells in Experiment 4: about 100–200 days of uninterrupted rain treatment and unchanged efflux would have been required to deplete the twigs by about 50% of their total K^+ or Mg^{2+} content in the absence of any replacement of minerals by uptake from the soil. Upon adequate cation supply by the soil, plants can easily compensate such a mineral loss under natural conditions.

Mengel et al. (1987) arrived at the conclusion that leaching rates of cations are too small to accelerate forest decline caused by nutrient deficiency. However, Zöttl and Mies (1983), Zöttl and Hüttl (1985), and Kaupenjohann et al. (1988) claimed that the decline of spruce is intensified by leaching of Mg^{2+} on very Mg^{2+}-poor stands. This hypothesis is supported by the results of Experiment 5, as treatment of seedlings with dilute acid led to lower Mg^{2+} contents in the needles.

In our opinion, the effect of leaching upon the nutritional status of the plant has to be discussed in terms of nutrient *cycling* in the plant/soil system. Because protons enter the plants through the leaf and stem surfaces, the leaf interior is acidified and protons are transported into the roots. Protons are then extruded in a cation exchange reaction (Experiment 5) which leads to an acidification of the rhizosphere. Cations like Mg^{2+} are transported by the same mechanism but in the opposite direction. Once leached, they will find their way back to the soil, but it is not clear if they will be leached through the soil or if they will again be available for the roots (see Chap. 4-I, this Vol.). Thus, if the concentrations of certain nutrient cations in the rhizosphere are low and/or uptake mechanisms of the roots are hindered, leaching may lead to a nutrient deficiency in the plant.

References

Abrahamsen G (1983) Sulphur pollution: Ca, Mg and Al in soil and soil water and possible effects on forest trees. In: Ulrich B, Pankrath J (eds) Effects of accumulation of air pollutants in forest ecosystems. Reidel Publ Co, Dordrecht, pp 207–218

Adams CM, Hutchinson TC (1984) A comparison of the ability of leaf surfaces of three species to neutralize acid rain drops. II. The influence of leaf wettability, leaf age and rain duration on changes in droplet pH and chemistry on leaf surfaces. N Phytol 97:463–478

Adams CM, Hutchinson TC (1987) Comparative abilities of leaf surfaces to neutralize acidic raindrops. New Phytol 106:437–456

Ashenden TW, Bell SA (1987) The effects of simulated acid rain on the growth of three herbaceous species grown on a range of british soils. Environ Pollut 48:295–310

Bauch J, Schröder W (1982) Zellulärer Nachweis einiger Elemente in den Feinwurzeln gesunder und erkrankter Tannen (*Abies alba* Mill.) und Fichten (*Picea abies* [L.] Karst.). Forstwiss Centralbl 101:285–294

Berg VS (1987) Plant cuticle as a barrier for acid rain penetration. In: Hutchinson TC, Meema KM (eds) Effects of atmospheric pollutants on forests, wetlands and agricultural ecosystems. NATO ASI Ser G, Ecol Stud vol 16, Springer, Berlin Heidelberg New York Tokyo, pp 145–154

Büttner G, Lamersdorf N, Schultz R, und Ulrich B (1986) Deposition und Verteilung chemischer Elemente in küstennahen Waldstandorten – Fallstudie Wingst, Abschlußber, Ber Forschungszentrums Waldökosysteme/Waldsterben, Reihe B, 1:1–136

Camman K (1977) Das Arbeiten mit ionenselektiven Elektroden. Springer, Berlin Heidelberg New York, 1–226

Caporn JM, Hutchinson TC (1986) The contrasting presponse to simulated acid rain of leaves and cotyledons of cabbage (*Brassica oleracea* L.). New Phytol 103:311–324

Cole DW, Johnson DW (1977) Atmospheric sulfate additions and cation leaching in a douglas fir ecosystem. Water Resour Res 13:313–317

Cronan CS (1984) Biogeochemical responses of forest canopies to acid precipitation. Acid Precip Ser 5:65–79

Dreyer SA, Seymour V, Cleland RE (1981) Low proton conductance of plant cuticles and its relevance to the acid-growth theory. Plant Physiol 68:664–667

Evans LS (1982) Biological effects of acidity in precipitation on vegetation: a review. Environ Exp Bot 22:155–169

Evans LS (1984) Botanical aspects of acidic precipitation. Bot Rev 50:449–489

Evans LS Curry TM (1979) Differential responses of plant foliage to simulated acid rain. Am J Bot 66:953–962

Evans LS, Gmur NF, Da Costa F (1977a) Leaf surface and histological perturbations of leaves of *Phaseolus vulgaris* and *Helianthus annuus* after exposure to simulated acid rain. Am J Bot 64:903–913

Evans LS, Gmur NF, Kelsch JJ (1977b) Perturbations of upper leaf surface structures by simulated acid rain. Environ Exp Bot 17:145–149

Evans LS, Curry TM, Lewin KF (1981) Responses of leaves of *Phaseolus vulgaris* L. to simulated acid rain. New Phytol 88:403–420

Evans LS, Santucci KA, Patti MJ (1985) Interactions of simulated rainfall solutions and leaves of *Phaseolus vulgaris* L. Environ Exp Bot 25:31–40

Evans LS, Canada DC, Santucci KA (1986) Foliar uptake of ^{15}N from rain. Environ Exp Bot 26:143–146

Flückiger W, Brauns S, Flückiger-Keller H, Leonards S, Asche N, Bühler U, Cier M (1986) Untersuchungen über Waldschäden in festen Buchenbeobachtungsflächen der Kantone Basel-Landschaft, Basel-Stadt, Aargau, Solothurn, Bern Zürich und Zug. Schweiz Z Forstwes 137: 917–1010

Frevert T, Klemm O (1984) Wie ändern sich pH-Werte im Regen-und Nebelwasser beim Abtrocknen auf Pflanzenoberflächen? Arch Meteorol Geophys Bioclimatol B 34:75–81

Hantschel R (1987) Wasser- und Elementbilanz von geschädigten, gedüngten Fichtenökosystemen im Fichtelgebirge unter Berücksichtigung von physikalischer und chemischer Bodenheterogenität. Bayreuther Bodenkundl Ber 3:1–219

Hantschel R, Klemm O (1987) Characterisation of weak acidity in selected precipitation samples from a forest ecosystem. Tellus 39B:354–362 Ser B 34:75–81

Hindawi IJ, Rea JA, Griffis WL (1980) Response of bush bean exposed to acid mist. Am J Bot 67:168–172

Horntvedt R, Dollard GJ, Joranger E (1980) Effects of acid precipitation on soil and forest. 2. Atmosphere – vegetation interactions. In: Drablos D, Tollan A (eds) Proc Int Conf Ecol Impact Acid Precip Norway, SNSF – Project, 192–193

Hutchinson TC, Adams CM (1987) Comparative abilities of leaf surfaces to neutralize acidic raindrops. I. The influence of calcium nutrition and charcoal-filtered air. New Phytol 106:169–183

Hutchinson TC, Adams CM, Gaber BA (1986) Neutralization of acidic raindrops on leaves of agricultural crop and boreal forest species. Water Air Soil Pollut 31:475–484

Irving P (1985) Modeling the response of greenhouse-grown radish plants to acidic rain. Environ Exp Bot 25:327–338

Jacobson JS (1980) Experimental studies on the phytotoxicity of acid precipitation: the United States experimence. In: Hutchinson TC, Havas M (eds) Effects of acid precipitation on terrestical ecosystems. Plenum Press, NY, pp 151–160

Jonhson DW, Turner J, Kelly JM (1982) The effects of acid rain on forest nutrient status. Water Resour Res 18:449–461

Johnston JW, Shriner DS, Kinerley CK (1986) The combined effects of simulated acid rain and ozone on injury, chlorophyll, and growth of radish. Environ Exp Bot 26:107–113

Kannan S (1986) Foliar absorption and transport of inorganic nutrients. Crit Rev Plant Sci 4:341–375

Katz C, Oren R, Schulze ED, Milburn J (1989) Uptake of water and solutes through twigs of *Picea abies* (L.) Karst. Trees 3:33–37

Kaupenjohann M, Zech W, Hantschel R, Horn R (1987) Ergebnisse von Düngungsversuchen mit Magnesium an vermutlich immissionsgeschädigten Fichten (*Picea abies* (L.) Karst.) im Fichtelgebirge, Forstwiss Centralbl 106:78–84

Kaupenjohann M, Schneider BU, Hantschel R, Horn R, Zech W (1988) Sulfuric acid rain treatment of *Picea abies*: Effects on nutrient solution and throughfall chemistry as well as on spruce nutrition. Z Pflanzenernähr Bodenkd 151:123–126

Klemm O, Riederer M, Frevert T (1987) pH-Response on leaves and isolated cuticles of *Hedera helix* L. after wetting with artificial rainwater. Environ Exp Bot 27:349–355

Kreutzer K, Bittersohl J (1986) Stoffauswaschungen aus Fichtenkronen (*Picea abies* (L.) Karst) durch saure Beregnung. Forstwiss Centralbl 105:357–363

Larsen BR (1986) In vivo buffering and concentration of simulated acid rain drops on leaves of selected crops. Water Air Soil Pollut 31:401–407

Leonardi S, Flückiger W (1988) Indirect effects of acid mist upon the rhizosphere and the leaves buffering capacity of beech seedlings. In: Mathy P (ed) Air pollution and ecosystems. D Reidel, Dordrecht, Boston, Lancaster, Tokyo, pp 697–700

Lepp NW, Dickinson NM (1976) The pH of leaf surfaces, and its modification by atmospheric pollution. In: Kaerenlampi L (ed) Proc Kuopio meeting plant damages caused by air pollution, Kiopio, 138–139

Matzner E (1984) Deposition und Umsatz chemischer Elemente im Kronenraum von Waldbeständen. Ber Forschungszentrums Waldökosysteme/Waldsterben 2:61–87

McFarlane JC, Berry WL (1974) Cation penetration through isolated leaf cuticles. Plant Physiol 53:723–727

Mecklenburg RA, Tukey HB Jr, Morgan JV (1965) A mechanism for the leaching of calcium from foliage. Plant Physiol 41:610–613

Meiwes KJ, Hauhs M, Gerke H, Asche N, Matzner E, Lamersdorf N (1984) Die Erfassung des Stoffkreislaufs in Waldökosystemen – Konzept und Methodik –. Ber Forschungszentrums Waldökosysteme/Waldsterben 7:68–142

Mengel K, Kirkby EA (1987) Principles of plant nutrition. Int Potash Inst, Worblaufen-Bern, Switzerland, 4 edn

Mengel K, Lutz HJ, Breininger MT (1987) Auswaschung von Nährstoffen durch sauren Nebel aus jungen intakten Fichten (*Picea abies*). Z Pflanzenernähr Bodenkd 150:61–68

Miller RH (1985) The Prevalence of pores and canals in leaf cuticular membranes. Ann Bot 55:459–471

Norby RJ, Luxmoore RJ (1983) Growth analysis of soybeam exposed to simulated acid rain and gaseous air pollutants. New Phytol 95:272–287

Oertli JJ, Harr J, Guggenheim R (1977) The pH-value as an indicator for the leaf surface microenvironment. J Plant Dis Prot 12:729–73

Oren R, Schulze ED, Matyssek R, Zimmermann R (1986) Estimating photosynthetic rate and annual carbon gain in conifers from specific leaf weight and leaf biomass. Oecologia 70:187–193

Oren R, Schulze ED, Matyssek R, Zimmermann R (1986) Estimating photosynthetic rate and annual carbon gain in conifers from specific leaf weight and leaf biomass. Oecologia 70:187–193

Oren R, Schulze ED, Werk KS, Meyer J, Schneider BU, Heilmeier H (1988) Performance of two *Picea abies* (L.) Karst. stands at different stages of decline. I. Carbon relations and stand growth. Oecologia 75:25–37

Osonubi O, Oren R, Werk KS, Schulze ED, Heilmeier H (1988) Performance of two *Picea abies* (L.) Karst. at different stages of decline. IV. Xylem sap concentrations of magnesium, calcium, potassium and nitrogen. Oecologia 77:1–6

Paparozzi ET, Tukey HB (1984) Characterization of injury to birch and bean leaves by simulated acid precipitation. In: Linthurst RA (ed) Direct and indirect effects of acidic deposition on vegetation. Acid Precip Ser 5:13–18

Pitman MG (1976) Ion uptake by plant roots. Enzycl Plant Physiol NS 2:95–128

Proctor JTA (1983) Effect of simulated sulfuric acid rain on apple tree foliage, nutrient content, yield and fruit quality. Environ Exp Bot 23/2:167–174

Reed DW, Tukey HB (1982) Permeability of brussels sprouts and carnation cuticles from leaves developed in different temperatures and light intensities. In: Cutler KLA, Alvin KL, Price, CE (eds) The plant cuticle. Academic Press, Lond NY, pp 267–278

Reiners WA, Olson RK (1984) Effects of canopy components on throughfall chemistry: an experimental analysis. Oecologia 63:320–330

Reiners WA, Olson RK, Howard L, Schaefer DA (1986) Ion migration from interiors to outer surfaces of balsam fir needles during dry, interstorm periods. Environ Exp Bot 26:227–231

Scherbatskoy T, Klein RM (1983) Response of spruce and birch foliage to leaching by acidic mists. J Environ Qual 12:189–195

Schönherr J (1976a) Water permeability of isolated cuticular membranes: the effect of pH and cations on diffusion, hydrodynamic permeability and size of polar pores in the cutin matrix. Planta 128:113–126

Schönherr J (1976b) Water permeability of isolated cuticular membranes: the effect of cuticular waxes on diffusion of water. Planta 131:159–164

Schönherr J, Bukovac MJ (1970) Preferential polar pathways in the cuticle and their relationship to ectodesmata. Planta 92:189–201

Schönherr J, Huber R (1977) Plant cuticles are polyelectrolytes with isoelectric points around three. Plant Physiol 59:145–150

Skeffington RA, Roberts TM (1985) The effects of ozone and acid mist on Scots pine saplings. Oecologia 65:201–206

Skiba U, Pleirson-Smith TJ, Cresser MS (1986) Effects of simulated precipitation acidified with sulphuric and/or nitric acid on the throughfall chemistry of sitka spruce *Picea sitchensis* and heather *Calluna vulgaris*. Environ Pollut Ser B 11:255–270

Smith WH (1981) Forest vegetation as a sink for gaseous contaminants. Air Pollut For, Springer, Berlin Heidelberg New York, pp 1–379

Steudle E, Zimmermann U (1984) Water relations of plant cells: further development of the pressure probe and of techniques for measuring pressure dependent transport. In: Cram WJ, Janáček K, Rybová R, Sigler K (eds) Membrane transport in plants. Academia, Prague, pp 73–82

Teuchert E, Teuchert I (1983) Möglichkeiten der Anreicherung von Schwefelsäure in Baumkronen. Forstwiss Centralbl 103:181–186

Thomas RC (1978) Ion-sensitive intracellular microelectrodes. Academic Press, Lond NY, pp 1–110

Trautner F, Eiden R (1988) A measuring device to quantify deposition of fogwater and ionic input by fog on small spruce trees. Trees 2:92–95

Tukey HB (1970) The leaching of substances from plants. Ann Rev Plant Physiol 21:305–324

Tukey HB, Morgan JV (1963) Injury to foliage and its effect upon the leaching of nutrients from above-ground plant parts. Physiol Plant 16:557–564

Tukey HB, Tukey HB (1969) The leaching of materials from leaves. In: Scharrer K, Linser H (eds) Handbuch der Pflanzenernaehrung und Duengung. Springer, Berlin Heidelberg New York, S 585–594

Tukey HB, Wittwer SH, Bukovac MJ (1962) The uptake and loss of materials by leaves and other above-ground plant parts with special reference to plant nutrition. Agrochimica 7:1–28

Ulrich B (1981) Theoretische Betrachtungen des Ionenkreislaufs in Waldökosystemen. Z Pflanzenernähr Bodenkd 144:647–659

Ulrich B (1983) Interaction of forest canopies with atmospheric constituents: SO_2, alkali and earth alkali cations and chloride. In: Ulrich B, Pankrath J (eds) Effects of accumulation of air pollutants in forest ecosystems. Reidel Publ Co, Dordrecht, pp 33–45

Unsworth MH (1984) Evaporation from forests in cloud enhances the effects of acid deposition. Nature (Lond) 312:262–264

Wittwer SH, Teubner FG (1959) Foliar absorption of mineral nutrients. Annu Rev Plant Physiol 10:13–32

Wood T, Bormann FH (1975) Increases in foliar leaching caused by acidification of an artificial mist. Ambio 4:169–171

Wood T, Bormann FH (1977) Short term effects of simulated acid rain upon the growth and nutrient relations of *Pinus strobus*, L. Water, Air Soil Pollut 7:479–488

Zöttl HW, Hüttl R (1985) Schadsymptome und Ernährungszustand von Fichtenbeständen im südwestdeutschen Alpenvorland. Allg Forstz 40:197–199

Zöttl HW, Mies E (1983) Nährelementversorgung und Schadstoffbelastung von Fichtenökosystemen im Südschwarzwald unter Immissionseinfluß. Mitt Dtsch Bodenkd Ges 38:429–434

3-D Atmospheric Pollutants and Plant Metabolism

O.L. LANGE, U. HEBER, E.-D. SCHULZE, and H. ZIEGLER

1 Introduction

Large amounts of sulfur dioxide (SO_2) and nitrogen oxides (NO_x) are continuously released into the atmosphere. When they react with water, these gases can form sulfurous acid (H_2SO_3), nitrous acid (HNO_2) and nitric acid (HNO_3). The weak acids are easily oxidized to the strong acids sulfuric acid and nitric acid. Photodissociation of nitrogen dioxide (NO_2) in the atmosphere produces oxygen radicals which give rise to the formation of ozone (O_3).

SO_2, NO_x and O_3 are major gaseous air pollutants. There is no doubt that they are potentially toxic to man, animals, and plants. However, it is still a matter of scientific dispute as to whether these pollutants can contribute to, or be responsible for, the decline of forests at the concentrations normally found in the atmosphere or considered legally safe. Thus, it is important to determine at which concentrations plant cellular reactions are capable of preventing or coping with toxic effects produced by these pollutants. Unfortunately, no general answer can be given because different organisms are differently equipped to prevent or counter damage. Even the tolerance of individual organisms differs depending on age, physiological and nutritional state, and general stress conditions.

In 1982, the annual emission of SO_2 was 3×10^6 tons in the FRG (Kalmbach 1985). The figure for nitrogen oxides was slightly higher (3.1×10^6 tons). Although the FRG is in fact a net exporter of air pollutants, imports balance exports to a considerable extent. For sulfur dioxide, deposition is about 80% of emission (Pankrath 1985). A similar percentage is assumed for nitrogen oxides. From this, one can estimate the average annual sulfur deposition to be about 150 mmol m^{-2} ground area and the deposition of oxidized nitrogen about 200 mmol m^{-2} ground area. For comparison, an average value of the content of sulfur in a spinach leaf is 4 mmol m^{-2} projected leaf area, and for nitrogen 100 mmol m^{-2}.

As SO_2 and NO_x react with water to produce acids which can be oxidized, the maximum number of protons that can dissociate from the reaction products is on average in the FRG about 500 mmol m^{-2} a^{-1} (2 H^+ from the reaction with SO_2 and 1 H^+ from the reaction with NO_2). In the Fichtelgebirge, the total measured deposition of protons was 233 mmol H^+ m^{-2} a^{-1} (see Chap. 4-I). A large proportion of the acids is neutralized by other components of the deposition. 188 mmol m^{-2} a^{-1} was measured in the throughfall of the spruce forest and an estimated 45 mmol H^+ m^{-2} ground area a^{-1} was thought to enter the needles (see Chap. 4-I). This

E.-D. Schulze, O.L. Lange, R. Oren (Eds.)
Ecological Studies Vol. 77

figure can be compared with the buffering capacity of the cytoplasm of green mesophyll cells in spinach, which is about 2 μmol H^+ mg^{-1} chlorophyll per pH unit at the pH of the cytoplasm (Pfanz and Heber 1986). This corresponds to about 0.6 mmol $H^+ \cdot m^{-2}$ leaf area per pH unit. If the pH of the cytoplasm were actually lowered by one pH unit, severe metabolic damage would be unavoidable. Control of the cytoplasmic pH is of paramount importance for the stabilization and maintenance of cellular metabolism (Smith and Raven 1979; Davies 1986). Key chloroplast enzymes such as fructose bisphosphatase have a distinct pH optimum activity (Baier and Latzko 1975; Woodrow et al. 1984). A shift by one pH unit would effectively inactivate the enzyme. At pH 7 of the chloroplast stroma, photosynthesis cannot proceed at a significant rate (Werdan et al. 1975). It requires a stroma pH approaching 8. To maintain pH within its proper control range, stabilizing mechanisms are at work. In the face of acid import, pH stabilization can be achieved only by the production of base or by the export or destruction of acid.

Another form of cellular damage by air pollutants is oxidation (Elstner 1982). In the light, oxygen radicals are involved in the oxidation of the sulfite which has been formed from SO_2. If not trapped by cellular quenching systems, the radicals can also oxidize cellular constituents. In addition to attacking the unsaturated fatty acids of biomembranes, ozone and other photooxidants, rapidly oxidize SH groups. In many enzymes, SH groups form part of the catalytic center. The active conformation of other enzymes depends on the reductive splitting of S-S bridges (Buchanan 1980). Oxidation will inactivate these enzymes. Obviously, cells must be capable of trapping oxidants before they can act, or of reversing oxidation after it has occurred. Oxidized components which have been damaged beyond repair must be replaced. Survival will depend on the capacity of trapping and on repair mechanisms.

These are examples of the direct threat which an organism faces when it absorbs noxious gases. However, indirect effects of air pollutants may actually be even more dangerous than direct effects. SO_2 and NO_x easily dissolve in rain. When stomata are opened, they are very effective gas exchange systems. However, they do not allow the entry of water or solutes dissolved in water (cf. H. Ziegler 1988). The ionized solutes contained in acid rain penetrate only very slowly through the protective tissue of the leaf epidermis (see Chap. 3-C). Direct uptake of acids by leaves is therefore insignificant. Rather, acids enter the soil through rain and interact with soil constituents.

The interactions of protons and cations cause major changes in soil chemistry and the leaching of cations to groundwater. Studies on ecosystem balance show that atmospheric acid inputs and forest practices contribute to the acidification of soils (Chap. 4-I). The implications and impacts from these alterations of soil properties for spruce in the Fichtelgebirge are discussed in Part 4 of this Volume. This chapter analyzes which properties plants must possess in order to cope with the direct threat posed by the absorption of noxious gases. The question as to which factors play a predominant role in the forest decline observed in the Fichtelgebirge is also addressed.

2 Fluxes of Gaseous Air Pollutants into Leaves

For a gas to be transferred from the surrounding atmosphere into a plant, the concentration gradient between the atmosphere (source) and the destination (sink) and the magnitude of barriers on the path towards the sink are important. We restrict discussion here to a quantitative consideration of diffusional transport of gaseous air pollutants into leaves which are structured so as to both facilitate and control gas exchange.

Leaves are protected against excessive loss of water by an epidermis, which is covered by a cuticular layer. In the epidermis, stomatal pores can open and close, facilitate or restrict gas exchange. When opened, they may cover not much more than 1% of the leaf area. Different air pollutants encounter the same obstacles on their path into the plant, but because of different physical properties they respond to these barriers differently. For this reason, we treat the transfer of SO_2, NO_2, and O_3 separately.

2.1 *Uptake of SO_2*

According to the flux equation

$$F = g \times \Delta c, \tag{1}$$

the flux F from the source to the sink is proportional to the concentration or activity gradient of SO_2 between the source and the sink Δc. The conductance or permeability coefficient, g, is inversely related to the magnitude of the barriers opposing flux, or the flux resistance, R. Table 1 compares, on an unit leaf-area basis, the conductance g of the cuticular layer with that of opened stomata of *Citrus* leaves for SO_2 diffusion. Although the cuticle is moderately permeable to SO_2 and O_2 compared to water, it is highly impermeable to both the gases and water when compared to opened stomata. Table 1 shows that when the stomata are closed, penetration is drastically reduced. Stomatal closure can therefore protect leaves against SO_2 concentrations which would otherwise damage them if their stomata were opened. This is important in respect to seasonal variations in atmospheric SO_2 concentrations. In winter, SO_2 concentrations often rise

Table 1. Conductance, *g*, of the cuticle and of open stomates of a *Citrus* leaf for the diffusion of different gases[a]

	Cuticula (g, cm s^{-1})	Stomata (g, cm s^{-1})
SO_2	2.01×10^{-5}	0.12
O_2	3.05×10^{-5}	0.18
CO_2	5.4×10^{-6}	0.14
H_2O	1.09×10^{-7}	0.22

[a] Boundary layer phenomena are included. Data for conductance recalculated from Lendzian (1984).

to dangerous levels during temperature inversions, but stomata are mostly closed.

Because the diffusional conductance of stomata is proportional to stomatal aperture, and as the conductance of the boundary layer of air around a leaf changes depending on wind velocity (Miszalski and Ziegler 1980), g in Eq. (1) must be determined if information on SO_2 fluxes into leaves shall be obtained. This can be done by measuring transpirational water loss from the leaf, which permits the calculation of the total needle conductance for the diffusion of water vapor. Through multiplication of the resulting conductance value with the factor 0.529, conductance for SO_2 is obtained. Actually, the measured conversion factor is larger than the one calculated on the basis of diffusion coefficients (Taylor and Tingey 1983), but we will base calculations on values derived from diffusion data.

The driving force of diffusional flux of SO_2 into a leaf is the concentration gradient between the atmosphere and the leaf interior. To calculate SO_2 fluxes according to Eq. (1), the intercellular SO_2 concentration must be known. It is a function of SO_2 influx through the stomata and SO_2 uptake by the leaf cells. On the path into the cells, the first significant barriers are cell wall and plasmalemma. The cell wall is hydrated so that SO_2 can dissolve in an aqueous phase and react with water to form bisulfite and protons as

$$SO_2 + H_2O \rightarrow HSO_3^- + H^+. \qquad (2)$$

As the equilibrium of this reaction is far on the right side of Eq. (2), diffusional transport of SO_2 through the cell wall to the plasmalemma is supported not only by a gradient of SO_2 but also, much more efficiently, by the larger gradient of bisulfite. Thus, the cell wall is not a significant barrier to SO_2 transfer. The conductance of the plasmalemma for SO_2 has been determined by measuring the uptake of SO_2 into mesophyll protoplasts both radiometrically using ^{35}S (Pfanz et al. 1987a), and potentiometrically (Ghisi et al., unpubl.). Values have been g = 0.0125 cm s^{-1} (unstirred sample, isotope method) and g = 0.04 cm s^{-1} (stirred sample, potentiometry). At first glance, such values appear to indicate a conductance of the plasmalemma which is small compared to that of open stomata. However, SO_2 is a highly soluble gas. At 20°C, its concentration in solution is 32 times higher than its concentration in air. Furthermore, the ratio of the area of the plasmalemma of leaf cells to that of the leaf area is about 11 in young barley leaves (Pfanz et al. 1987a), and may be much higher in many other leaves. If solubility and leaf area are taken into account, g of the plasmalemma can be directly compared with stomatal conductance. It is about 10 cm s^{-1} or more, as opposed to 1 cm s^{-1} when the stomata are widely open. This difference is even greater when stomata are partially closed. Thus, the plasmalemma is an insignificant barrier to SO_2 transfer compared to the stomatal barrier (Pfanz et al. 1987b; Laisk et al. 1987). Bisulfite and sulfite anions do not appear to be transferred across the plasmalemma to any significant extent. The damaging effects of sulfite solutions on cells are directly related to the concentration of dissolved SO_2. No relationship with the concentrations of HSO_3^- or SO_3^{2-} has been observed (Spedding et al. 1980b).

Once SO_2 has entered the cytoplasm, it is rapidly removed by reaction (2) and by

$$HSO_3^- \rightarrow SO_3^{2-} + H^+ \qquad (3)$$

so that its concentration in the cytosol is extremely low (Spedding et al. 1980a). At a cytoplasmic pH of 7.5, the ratio of HSO_3^- to SO_2 is about 4×10^5 and that of SO_3^{2-} to SO_2 is higher than 10^6. Rapid removal of SO_2 by dissociation inside the cells and a very low barrier of the plasmalemma to SO_2 penetration results in a rapid flux of SO_2 into the cells. As a result, the intercellular SO_2 concentration can, on first approximation and for the purpose of calculating SO_2 flux according to Eq. (1), be considered zero (Laisk et al. 1987).

Once atmospheric concentrations and boundary layer and stomatal conductances are known, flux data can be obtained. Calculations made for two levels of atmospheric pollution and for different values of boundary layer and stomatal conductance are shown in Table 2. A level of 0.01 $\mu l\ l^{-1}$ SO_2 (25 $\mu g\ m^{-3}$) corresponds to the annual average of SO_2 immission in the Fichtelgebirge (see Chapter 1-B). Peak values of 0.2 $\mu l\ l^{-1}$ (500 $\mu g\ m^{-3}$) or more are not rare during temperature inversions. The data show that fluxes of SO_2 into a leaf, which are normally slow at low atmospheric pollution levels, can be fast during short-term exposure to peak pollution levels. The burden on cellular pH-stat mechanisms and on other detoxification mechanisms is correspondingly high. Leaves are

Table 2. Calculated SO_2 influx into a leaf from air polluted with 0.01 or 0.05 $\mu l\ l^{-1}$ SO_2 and intracellular proton production from SO_2 in relation to the buffering capacity of the leaf

a	*b*		*c*	*d*
SO_2, $\mu l\ l^{-1}$	g, cm s^{-1}	g, mmol $m^{-2}\ s^{-1}$	$\mu mol\ m^{-2}\ h^{-1}$	Hours
0.01	0.025	10.4	0.37	200
	0.05	20.8	0.75	400
	0.1	41.6	1.5	200
	0.2	83.2	3	100
0.2	0.025	10.4	7.5	40
	0.05	20.8	15	20
	0.1	41.6	30	10
	0.2	83.2	60	5

a: Atmospheric SO_2 in $\mu l\ l^{-1}$;
b: boundary layer and stomatal conductance, g, for SO_2 in cm s^{-1} (and in mmol $m^{-2}\ s^{-1}$ for 20°C);
c: rate of SO_2 entry in $\mu mol\ m^{-2}$ leaf area h^{-1};
d: exposure time in hours required to decrease the cytoplasmic pH by one unit if the pH-stat mechanism is thought to be inoperative (i.e., ratio maximum intracellular H^+ production from SO_2/buffering capacity of the cytoplasm = 1); the buffering capacity of the cytoplasm is considered to be 0.6 mmol H^+ m^{-2} leaf area per pH unit.

Assumptions (1) All SO_2 entering the mesophyll cells is oxidized to sulfuric acid (reality differs from assumption: SO_2 can be detoxified by reduction).
(2) All H^+ produced is retained in the cytoplasm (reality differs from assumption: the pH-stat mechanism tends to stabilize the cytosolic pH).

actually more efficient traps for atmospheric SO_2 than they are for CO_2. Because of the high solubility of SO_2 in aqueous phases, particularly in a canopy with several layers of leaves, SO_2 is filtered from the air. When the stomata are opened, leaves are on a unit-area basis much better traps for SO_2 than dry soil; diffusion gradients are much steeper into the leaves than into dry soil.

Table 2 simplifies matters. In order to have a more realistic view of the amount of SO_2 which is actually taken up by spruce needles in the Fichtelgebirge, first attempts have been made to model SO_2 fluxes (see also Fischer 1967; Winner et al. 1985b). From gas exchange measurements in the field, diurnal courses of stomatal conductance under natural conditions are known (Wedler 1986; Beyschlag et al. 1987). Together with the SO_2 concentrations in the air which have been recorded at the same site (see Chap. 1-B), diurnal courses of stomatal conductance allow the calculation of the SO_2 uptake of the needles (Laisk et al. 1987). These estimates are related to the wind velocity in the gas exchange cuvettes. Such calculations will become different at the level of the total forest stand, where the highly decreased boundary layer conductance in the canopy becomes the dominating parameter (see Jarvis and McNaughton 1986) and where the SO_2 partial pressure of the air decreases from the top of the canopy to the ground, due to SO_2 uptake by the plant biomass. The following calculations, therefore, represent conditions for branches which are freely exposed to the polluted atmosphere at the top of the canopy. Two examples are presented for the site at Oberwarmensteinach.

Figure 1 shows climatic data and net photosynthesis of spruce during a summer day with low maximal SO_2 concentration (9 $\mu g\ m^{-3}$ or 0.0034 $\mu l\ l^{-1}$). SO_2 uptake peaked in the morning at the highest stomatal conductance. It was about 100 pmol $m^{-2}\ s^{-1}$. The total daily SO_2 gain was 2.5 $\mu mol\ m^{-2}$ needle surface area. The ratio of net carbon gain to concurrent sulfur intake was, on a molar basis, about 30 000. As the ratio of carbon to sulfur is less than 1000 in many rapidly growing plants, but as high as 10 000 or more in older trees (Heber et al. 1987), the sulfur burden was slight.

A day in September (Fig. 2) was characterized by a sudden increase in SO_2 concentration for 4 h with an hourly average up to 85 $\mu g\ m^{-3}$ (0.032 $\mu l\ l^{-1}$). This led to a SO_2 flux into the needles of almost 500 pmol $m^{-2}\ s^{-1}$ and resulted in a SO_2 uptake of about 6.5 $\mu mol\ m^{-2}$ during the day. At the same time, net photosynthesis was lower than in Fig. 1 because of lower light intensities. On this particular day, the ratio of carbon gain to sulfur intake from the atmosphere was 3280. For trees, this can be a considerable sulfur burden (Heber et al. 1987).

Available gas exchange data do not yet allow a comprehensive and exact modeling of SO_2 uptake by the spruce needles during the course of the year. Therefore, we do not yet know the real SO_2 impact for the needle mesophyll, (see the respective estimates for the ecosystem model in Chap. 4-I.) Uncertainties also exist in regard to the calculation of SO_2 fluxes from measured transpiration data. Measured SO_2 fluxes have been found to be actually faster than fluxes calculated on the basis of $g_{H_2O} = 0.529\ g_{SO_2}$ (Taylor and Tingey 1983). Further modeling is necessary to calculate the long-term SO_2 influx into single needles in the field, and such calculations must be extended to canopies.

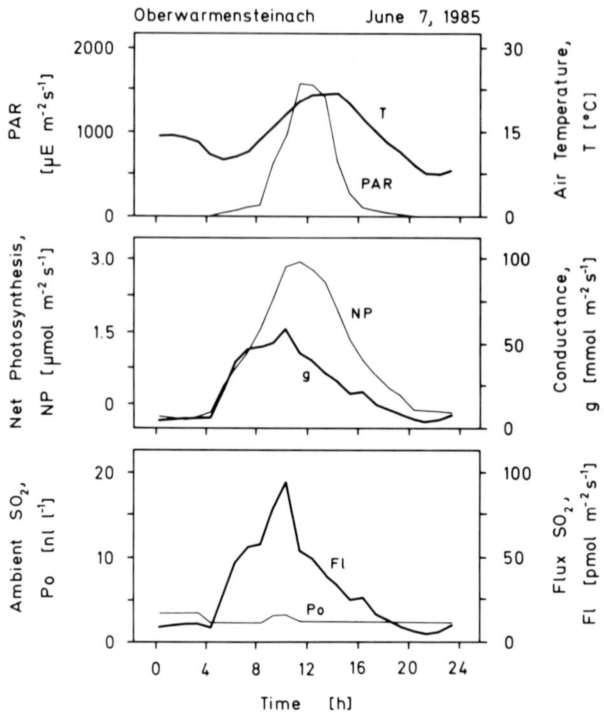

Fig. 1. Diurnal time courses of (*top*) air temperature and photosynthetically active radiation (PAR); (*middle*) net photosynthesis rate and needle conductance for water vapor diffusion; (*bottom*) ambient SO_2 concentration and simulated SO_2 uptake of the needles (fluxes related to total needle surface). 1-year old spruce needles. Oberwarmensteinach (June 7, 1985): data from Wedler (1986)

2.2 Fate of SO_2 Inside the Leaf

It has been mentioned that SO_2 is rapidly converted into HSO_3^- and SO_3^{2-} once it has entered the cytoplasm. The contribution of the large central vacuole of leaf cells to SO_2 trapping is insignificant (Pfanz et al. 1987b; Laisk et al. 1987). In the mesophyll, metabolization of the anions formed from SO_2 occurs predominantly in illuminated chloroplasts (I. Ziegler 1975; Schmidt 1979). Transfer of SO_3^{2-} across the chloroplast envelope is catalyzed by the phosphate translocator (Ziegler and Hampp 1977; Hampp and Ziegler 1977; Hampp et al. 1980). Although the conductivities of the chloroplast envelope and the tonoplast for SO_2 do not differ much from the conductivity of the plasmalemma (Pfanz et al. 1987a), the contribution of SO_2 diffusion through the internal biomembranes to the total transport of sulfur is small compared to the catalyzed transport of the anions. This is because the intracellular SO_2 concentration is too low to support SO_2 gradients capable of driving much diffusional transport (Spedding et al. 1980a).

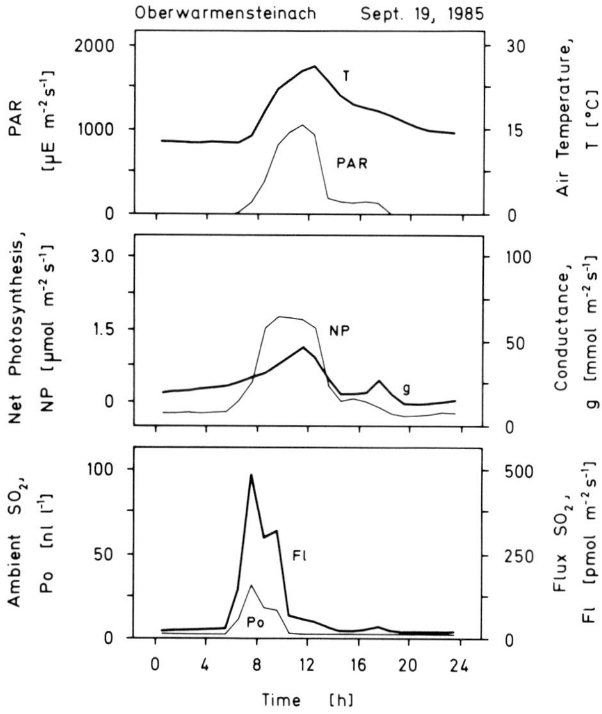

Fig. 2. As for Fig. 1, but September 19, 1985. Note that the ordinate scales of the panel below for both figures differ by the factor of 5

In the chloroplasts, the anions HSO_3^- and SO_3^{2-} are either reduced to the sulfide level or oxidized to sulfate (Weigl and Ziegler 1962; Hällgren 1978; Rennenberg 1984; I. Ziegler 1975). The difference is important.

Oxidation produces SO_4^{2-}. Because sulfuric acid is a stronger acid than the sulfurous acid formed by hydration of SO_2, oxidation increases the danger of cellular acidification. There are several possibilities with which to counter it. H^+ can be removed by cellular buffers according to

$$A^- + H^+ \rightarrow AH. \tag{4}$$

When H^+-generation exceeds the cellular buffering capacity, this possibility is of limited value.

Another means of coping with excess H^+ is to export it from the cytoplasm. Active pumping of H^+ into the vacuole of plant cells is well known (Sze 1985). Transfer of SO_4^{2-} has been shown to be stimulated by ATP (Kaiser et al. 1988). Still, the capacity of the vacuole to accumulate sulfuric acid is genetically limited (Laisk et al. 1987). The pH of the cell sap from differentiated leaf tissue is rather well defined (Smith and Raven 1979). Experimental alterations produce compensatory action by the enzymic pH-stat of the cytosol (W. Kaiser, personal communication). There is no indication so far of a significant influence of potentially acidic gases on the vacuolar pH.

Thirdly, acid can be removed by its reaction with base. OH^- is produced during the reduction of nitrate in the light:

$$NO_3^- + 3\,H_2O \rightarrow NH_4^+ + 2\,OH^- + 2\,O_2 \quad (5)$$

or in the dark

$$C_6H_{12}O_6 + 3\,NO_3^- + 3\,H_2O \rightarrow 3\,NH_4^+ + 6\,OH^- + 6\,CO_2. \quad (6)$$

It should be noted that equations (5) and (6) as well as equations given below integrate a complex set of individual biochemical reactions which cannot be outlined in detail. The equations show that nitrate reduction results in alkalization.

Indeed, levels of NO_3^- decrease rapidly in leaves exposed to a penetrating acid (Pfanz and Heber, unpubl.). Another possibility of coping with the generation of excess H^+ inside the cytoplasm is to ensure the degradation of organic acids (Davies 1986). There is actually a rapid and significant decrease in the concentration of malate and other organic acids on exposure of leaves to an acidic gas. However, degradation of acids is limited by acid availability and by the necessity of maintaining organic acid anions at a level sufficient to sustain metabolism.

Although all potential possibilities of compensating for the SO_2-dependent generation of H^+ appear to be exploited by the cellular pH-stat mechanism, removal of the H^+ generated during the oxidation of SO_2 appears to remain a critical problem. For this reason, excessive accumulation of SO_4^{2-}, which is derived from SO_2, must cause a reduction in cellular viability. When sulfur levels reach a maximum of 2000 to 3000 $\mu g\ g^{-1}$ dry wt, spruce needles are usually shed (Lyr et al. 1967). In the Fichtelgebirge, the total sulfur content of the current flush was about 1100 to 1200 $\mu g\ g^{-1}$ dry wt. At a damaged site close to Oberwarmensteinach, total content increased to an average of 1500 $\mu g\ g^{-1}$ in needles which were 4 years old, while at the undamaged site at Wülfersreuth, 2300 $\mu g\ g^{-1}$ was measured (Table 5). The oldest needles (age classes 7 and 9) of exposed large trees (ca. 100 years) at Oberwarmensteinach contained less than 2100 $\mu g\ g^{-1}$ sulfur. There was no proportionality between sulfur content and damage (see Sect. 3.4).

It appears that translocation of sulfur prevents excessive accumulation at the sites of uptake (Weigl and Ziegler 1962). However, no evidence is available that airborne sulfur can be excreted as sulfate by the root system. Proton excretion, on the other hand, is a well-established phenomenon. Protons are exchanged by the roots for cations such as Ca^{2+}. The significance of proton/cation exchange for relieving proton stress due to SO_2 incorporation in the crown is discussed in Chapter 4-I. However, nothing is known on long distance transport of H^+ from the needles to the root system. Nevertheless, soil scientists consider this means of proton loss to be substantial (see Chap. 4-B). The magnitude of the uptake of Ca^{2+} is such that it could contribute significantly to detoxification of airborne sulfur.

Whereas oxidation of SO_2 to sulfuric acid aggrevates the problem of acidification, reduction alleviates it (Sekiya et al. 1982; Heber et al. 1987). Through hydrogen sulfide as an intermediate, reduction leads to the formation of sulfur-containing compounds such as cysteine, methionine, and glutathione (Weigl and Ziegler 1962; Rennenberg 1984). Light-dependent H_2S formation may be written as

$$SO_3^{2-} + 2\,H^+ \rightarrow H_2S + 1\tfrac{1}{2}\,O_2 \tag{7}$$

and

$$SO_4^{2-} + 2\,H^+ \rightarrow H_2S + 2\,O_2. \tag{8}$$

The reactions show that the protons generated during the hydration of SO_2 are removed from the cytoplasm during reduction. H_2S may be released from the leaves (Hällgren and Fredriksson 1982; Sekiya et al. 1982; Rennenberg 1984), but release accounts for only a low percentage of the SO_2 absorbed. As far as information is available, release does not amount to more than 10 or 15% of SO_2 absorption and cannot, therefore, solve the problem of cytoplasmic acidification.

The synthesis of cysteine from inorganic precursors may be written as

$$H^+ + SO_4^{2-} + NH_4^+ + 3\,CO_2 + H_2O \rightarrow \text{cysteine} + 4\tfrac{1}{2}\,O_2. \tag{9}$$

It can be seen that one of the protons generated during the oxidation of SO_2 to sulfuric acid is consumed during cysteine formation. If nitrate is the amine donor, the reaction is

$$3\,H^+ + SO_4^{2-} + NO_3^- + 3\,CO_2 + 2\,H_2O \rightarrow \text{cysteine} + 6\tfrac{1}{2}\,O_2. \tag{10}$$

As three protons are consumed, it appears at first sight that this reaction is capable of solving the problem of acidification. That this is not fully correct can be demonstrated by an alternative formulation of the reaction (10):

$$SO_4^{2-} + NO_3^- + 3\,CO_2 + 5\,H_2O \rightarrow \text{cysteine} + 6\tfrac{1}{2}\,O_2 + 3\,OH^-. \tag{11}$$

This reaction shows that cysteine synthesis from sulfate and nitrate, which had been taken up from the soil together with cations such as K^+, actually generates OH^- ions (Eq. 11). No such generation takes place when SO_4^{2-} (and NO_3^-) are derived from SO_2 (and NO_2) in the atmosphere (Eq. 10). Thus, according to Eq. (10) reduction of airborne sulfate (and nitrate) is not equivalent to reduction of ions taken up from the soil (Eq. 11). Still, there is no doubt that reduction can ease the problem of SO_2 toxicity (Sekiya et al. 1982).

There is now the question of how effectively reduction can contribute to detoxification. Obviously, effectiveness depends on the capacity for reduction. Unfortunately, no experimental data are available to establish to what extent different organisms differing in their SO_2 sensitivity also differ in their ability to reduce SO_2 or sulfate. Still, as the main products of reduction are used for protein synthesis, it is useful to compare different organisms in regard to protein content and protein production. The sulfur content of plant proteins is about 1% (Ziegler 1983). This permits an estimate of how much SO_2 is detoxified by incorporating its sulfur into protein. Different organisms differ in their protein content. They also differ in growth rate. Growth requires protein production.

It appears that herbaceous plants such as fast-growing, protein-rich field crops should be capable of detoxifying SO_2 more efficiently than trees which have a low percentage of protein in their biomass. It has recently been argued that mature forest trees may be able to completely satisfy their sulfur requirements from air containing 0.004 $\mu l\ l^{-1}$ SO_2 (Heber et al. 1987). Higher SO_2 levels would, over prolonged periods of time, result in sulfate accumulation. On the other hand, protein-rich annuals should be expected to suffer from sulfur deficiency even when exposed to SO_2 levels in air of up to 0.08 $\mu l\ l^{-1}$ if they had no access to sulfate from the soil.

If these deductions could be experimentally substantiated, they would explain why fast-growing grasses prosper in areas where forests have died from overexposure to SO_2. Such forests can be seen in the Erzgebirge close to the border between the GDR and the CSSR (for instance, near Fürstenau). As early as in 1912, Wieler wrote that, in contrast to trees, it was possible to grow field crops and vegetables very near to smelters "*da diese gegen den direkten Angriff der Säure widerstandsfähiger zu sein scheinen als die meisten Holzgewächse*" (since these seem to be more resistant against direct effects of acidity than woody species).

2.3 Uptake of Nitrogen Oxide

Nitrogen reacts with oxygen during the combustion of fossil fuels to form predominantly NO, which is further oxidized to NO_2. As both gases differ in their solubility properties, they encounter barriers of different relative magnitude on their path into cells. Far from the sites of emission, NO_2 concentrations usually exceed NO concentrations considerably. According to the *Sondergutachten* of the *Sachverständigenrat für Umweltfragen 1983* (Drucksache des Deutschen Bundestages Nr. 10/1983, Stuttgart, Mainz), average concentrations of NO_2 are 50–100 $\mu g\ m^{-3}$ (0.025 to 0.05 $\mu l\ l^{-1}$) in industrialized areas and 10–20 $\mu g\ m^{-3}$ (0.005–0.01 $\mu l\ l^{-1}$) in rural areas. At Oberwarmensteinach, monthly NO_2 averages varied in 1986 from between 0.003 and 0.012 $\mu l\ l^{-1}$. In 1987, higher concentrations were generally observed (see Fig. 3 in Chap. 1-B). Corresponding figures for NO were 0.001 to 0.0012 $\mu l\ l^{-1}$ in 1986 and up to 0.006 $\mu l\ l^{-1}$ in 1987. Because the uptake of NO is slower than that of NO_2 even at equivalent concentrations, we will consider only transfer of NO_2 into leaves.

As is the case for SO_2, the cuticle is a far more effective barrier to the entry of NO_2 than open stomata are (see Chap. 3-A). NO_2 diffusion through the stomata is somewhat faster than SO_2 diffusion. Very small gradients of NO_2 facilitate rapid diffusion across the plasmalemma. In aqueous cellular phases, NO_2 is highly soluble. It reacts with water to form nitrous acid and nitric acid.

$$2\,NO_2 + H_2O \rightarrow NO_3^- + NO_2^- + 2\,H^+. \tag{12}$$

These acids are trapped in the slightly alkaline cytoplasm. Rapid uptake of NO_2 by the mesophyll cells depletes the intercellular space of NO_2. Like SO_2, NO_2 fluxes into a leaf can be calculated from the external NO_2 concentrations and the known stomatal and boundary layer resistance to water vapor flux under the simplifying assumption that the intercellular NO_2 concentration is zero. As a result there is a linear relationship between the stomatal and boundary layer conductance for NO_2 and the uptake of NO_2 into leaves.

2.4 *Fate of NO_2 Inside the Leaf*

The leaf mesophyll is capable of nitrate and nitrite reduction. Nitrate reductase is a cytosolic, nitrite reductase, a chloroplastic enzyme (Dalling et al. 1972; Plaut et al. 1977). The capacity of chloroplasts to reduce nitrite is very high compared to NO_2 fluxes from moderately polluted air into leaves. Nitrite formed in the cytosol of leaf cells is rapidly transferred into the chloroplasts. The chloroplast envelope is permeable to both HNO_2 and NO_2^- (Heber and Purczeld 1978). A carrier is probably responsible for transport of the nitrite anion. Light-dependent reduction of nitrate can be written as

$$NO_3^- + 3\ H_2O \rightarrow NH_4^+ + 2\ O_2 + 2\ OH^- \qquad (13)$$

and that of nitrite

$$NO_2^- + 3\ H_2O \rightarrow NH_4^+ + 1\ \tfrac{1}{2}\ O_2 + 2\ OH^-. \qquad (14)$$

It can be seen that the reductive reactions not only consume the protons formed when NO_2 is converted to nitrous and nitric acid [see reaction (10)] but also additional protons. Still, reduction of airborne nitrate (or nitrite) is not equivalent to reduction of the NO_3^- taken up from the soil. If import of the latter is electrically balanced by K^+ import, reduction results in the net formation of two hydroxyl ions, instead of the one hydroxyl ion left if proton production in Eq. (12) is subtracted from the hydroxyl ion formation of Eqs. (13) and (14).

On a unit weight basis, SO_2 and NO_x emissions are comparable in the FRG. On a molar basis, the average deposition of nitrogen per unit ground area should therefore be expected to be only somewhat higher than that of sulfur. In fact measured air pollution by NO_2 was higher than that of SO_2 during the growing season in the Fichtelgebirge (see Chap. 1-B). Therefore, when NO_2 uptake was calculated on a canopy basis, it was found to be higher than the total uptake of SO_2 (18 mmol NO_2 per m^2 ground area and year versus 7.5 mmol SO_2 per m^2 ground area and year, (see Chap. 4-I). The cellular requirements for sulfur and nitrogen differ considerably. Plant proteins contain about 16% nitrogen and only 1% sulfur. The molar ratio of N/S is therefore about 37 in proteins. An analysis of maize leaves has yielded an N/S ratio of 43 and that of maize shoots a ratio of 20 (Salisbury and Ross 1986). The ratio amounts to 30 for *Pinus radiata* (see Chap. 4-A). As the cellular capacity for the reductive detoxification of NO_2 is high, and the fluxes of NO_2 which are calculated from average concentrations of NO_2 as measured in the Fichtelgebirge into leaves with open stomata are low, it is obvious that SO_2 poses a far greater threat to cellular survival than NO_2. As long as there is cellular demand for reduced nitrogen and NO_2 levels in the air are low, NO_2 is a fertilizer rather than a toxic pollutant. However, increased nitrogen availability may accentuate nutritional imbalances, if essential elements such as Mg are in short supply (see Chap. 4-H).

2.5 *Uptake of Ozone*

NO_2 dissociates under the influence of ultraviolet radiation to NO and highly reactive monoatomic oxygen. The latter forms ozone with molecular oxygen. According to the *Umweltbundesamt* (1978) the tolerance limit of 4-h exposure per day is an average concentration of 50 $\mu g\ m^{-3}$ ozone (0.025 $\mu l\ l^{-1}$) for sensitive plants and 300 $\mu g\ m^{-3}$ (0.15 $\mu l\ l^{-1}$) for less sensitive plants. Maximum emission values of the "VDI guideline 2310" of September 1974 are 150 μg ozone m^{-3} (0.08 $\mu l\ l^{-1}$) as a 30 min average and 50 $\mu g\ m^{-3}$ (0.03 $\mu l\ l^{-1}$) for the daily and annual averages.

At Oberwarmensteinach, in the Fichtelgebirge, average monthly ozone concentrations varied in 1986 from between 0.016 and 0.04 $\mu l\ l^{-1}$. They were higher in 1987 (see Chap. 1-B). The highest monthly averages were close to 0.08 $\mu l\ l^{-1}$. Thus, ozone concentrations in the Fichtelgebirge were often close to, or actually higher than, maximum tolerable emission values as defined by the VDI guideline number 2310.

Ozone enters the mesophyll of leaves almost exclusively through the stomata. Although ozone is more soluble in aqueous media than oxygen, it is far less soluble than SO_2 or NO_2. Even though the plasmalemma resistance to the penetration of ozone is comparable to that measured for SO_2 (Heber, Urbach, and Reisberg, unpubl.), reduced gradients reduce transport into the cells. Consequently, intercellular ozone concentrations are significant even when stomata are partially closed. This is exemplified by some representative data in Table 3. Thus, partial closure of stomata is much less effective in protecting leaves against ozone than they are against SO_2 or NO_2.

2.6 *Fate of Ozone Inside the Leaf*

Ozone is a strong oxidant. It can react with many different organic compounds. In a competitive situation, it will be consumed preferentially in reactions which are characterized by significant concentrations of a reaction partner and high second-order velocity constants. Fast reactions will outcompete slow reactions. With double bonds, ozone forms ozonides which decompose in aqueous media. The reaction with double bonds is a fast reaction. In the lipid bilayer of the plasmalemma, unsaturated fatty acid residues dominate over satured residues (Yoshida and Uemura 1986). The concentration of double bonds may reach 2 mol l^{-1} within the lipid core of biomembranes. A computer simulation of diffusion (Kolbowski et al., unpubl.) has shown that the ozone concentration difference across the plasmalemma makes up more than two-thirds of the ozone concentration outside the plasmalemma. At an external concentration of 0.03 $\mu l\ l^{-1}$ ozone, about 1% of the double bonds of fatty acid residues of the plasmalemma may be oxidized per hour when the stomata are open, if ozone degradation in the apoplast is not extensive (little is known so far about such degradation). Lipid destruction and oxidation of membrane proteins may make the plasmalemma the main target of ozone attack. A large part of the ozone entering the cell is already trapped in the plasmalemma. Very little ozone reaches the vacuole. The

Table 3. Simulation of ozone fluxes into a leaf through the stomates, and of the intercellular ozone concentration inside the leaf as a function of the external ozone concentration[a]

Conductance cm s^{-1} (mmol m^{-2} s^{-1})		Flux, fmol cm^{-2} s^{-1}		Intercellular ozone, $\mu l\ l^{-1}$ in the gas phase	
		External ozone, $\mu l\ l^{-1}$			
		0.025	0.15	0.025	0.15
1	(415.0)	49	290	0.023	0.136
0.75	(312.0)	47	282	0.022	0.133
0.25	(104.0)	38	229	0.018	0.108
0.1	(41.6)	27	161	0.0125	0.076
0.05	(20.8)	18	108	0.008	0.050
0.02	(8.3)	9	54	0.004	0.025
0.0075	(3.1)	4	23	0.002	0.011

[a]0.025 $\mu l\ l^{-1}$ is the tolerance limit of sensitive plants during a 4-h exposure period per day and 0.15 $\mu l\ l^{-1}$ is the tolerance limit of less sensitive plants (Umweltbundesamt 1978). Data from a modification of the computer program of Laisk et al. (1987). Indicated is the total needle conductance (stomata plus boundary layer) for water vapor in cm s^{-1} (and in mmol m^{-2} s^{-1} for 20°C).

biomembranes of cell organelles, such as chloroplasts, also contain a high proportion of unsaturated fatty acid residues. However, as the intracellular ozone concentration rapidly decreases with increasing distance from the plasmalemma, they are much less endangered by ozonization than is the plasmalemma. Whenever ozone concentrations in air are high, causing pigment bleaching in chloroplast membranes, damage to the plasmalemma must precede damage to the thylakoids. In the cytoplasm, dominant ozone-consuming reactions are the oxidation of SH groups and of ascorbate. Table 4 lists velocity constants of some oxidation reactions which have been calculated from spectrophotometric ozone measurements. The calculations are based on the assumption of bimolecular reactions, although in some cases deviations from second-order kinetics were observed. Since SH groups play an important role in the catalytic activity of

Table 4. Bimolecular velocity constants k of the reaction of ozone with several solutes[a]

	k (M^{-1} s^{-1})
Linolenic acid	125 000
Fumaric acid	61 000
Succinic acid	10
Cysteine	144 000
Cystine	300
Ascorbate	210 000
Sucrose	240

[a]The measurements were performed at 16°C. (Unpubl. data from Kolbowski, Urbach, Reisberg, and Heber).

enzymes, protein SH-group oxidation interferes with enzyme catalysis. Other oxidative reactions between ozone and proteins contribute to protein alteration (Mudd et al. 1984). Competition between different reactions determines the extent of damage to proteins. Consumption of ozone during the oxidation of cysteine, glutathione or ascorbate protects proteins against oxidative damage; these reductants detoxify ozone. Reduced glutathione is regenerated from oxidized glutathione by glutathione reductase. Metabolic regeneration of cysteine and ascorbate is also not difficult when air is only moderately polluted by ozone (for fluxes see Table 4; at an external concentration of 0.03 $\mu l\ l^{-1}$ ozone, a typical rate of ozone flux into a leaf is 10 nmol mg^{-1} chlorophyll h^{-1} when the stomata are open).

Owing to the presence of considerable levels of antioxidants in the cytosol and of the possibility of reductive repair of oxidized proteins, low levels of ozone are not likely to cause persisting damage in the aqueous phase of the cytoplasm. Damage to membranes is indicated by permeability alterations (Mudd et al. 1984; Pell 1987) and by ethane production (Hock and Elstner 1988). Little is known about the cellular capacity for the repair of biomembranes.

2.7 Pollution Stress in the Fichtelgebirge

If the concentration of air pollutants measured at the station in Oberwarmensteinach is representative of the general situation in the Fichtelgebirge, fluxes of sulfur dioxide into leaves will place a considerable burden on cellular detoxification mechanisms when stomata are open. This might be particularly true for trees suffering from mineral deficiencies which reduce their detoxification capacity. Even when stomata are partially closed, the burden may still be significant at low temperatures, which slow down metabolic detoxification. Since leaf cells are equipped with mechanisms for the vacuolar storage of the excess protons and sulfate (Kaiser et al. 1988) derived from SO_2, acute effects of temporary exposure to elevated concentrations of SO_2 appear to be much less significant in reducing cellular viability than are the long-term effects of chronic exposure to lower concentrations.

Because of the high capacity of leaf cells to detoxify NO_2 and low NO_2 concentrations, NO_2 does not appear to play a direct role as a damaging pollutant in the forest decline observed in the Fichtelgebirge. Still, it should be noted that Mansfield et al. (1988) have reported synergistic effects of SO_2 and NO_2 which make combinations of these gases possibly more harmful than SO_2 and NO_2 individually. Ozone levels in the Fichtelgebirge are close to concentrations considered to be detrimental to sensitive plants.

In the following section we report field studies designed to evaluate any immediate metabolic damage caused by the direct effect of air pollutants on the green spruce biomass at the experimental sites in the Fichtelgebirge.

3 Direct Effects of Atmospheric Pollutants on Metabolic Features of Norway Spruce in the Fichtelgebirge?

There is no doubt that direct exposure of plants to high concentrations of atmospheric pollutants is not only harmful for plant metabolism in vitro and in fumigation experiments, but also causes damage to trees under field conditions (see Guderian 1977, 1985; Winner et al. 1985a; Hutchinson and Meema 1987). In order to explain forest decline, one might hypothesize that the direct impact of pollutants such as SO_2 and ozone on plant metabolism is the major agent responsible for damage. On the other hand, there is clear evidence that soil acidification due to acid deposition and its effects on nutrient availability and root function are important factors in forest damage (Matzner and Ulrich 1987). A combined effect, i.e., pollutant uptake by the green biomass and of soil acidification and depletion is also possible (Prinz et al. 1987; Hock and Elstner 1988).

To determine the causes for forest decline in the Fichtelgebirge, we tried to determine whether indirect effects of soil acidification or direct effects of gaseous air pollutants were responsible for, or contributed to, the observed damage. We are fully aware of the difficulty of such a differentiation. In contrast to short-term effects of high concentrations of pollutants, little is known about the metabolic responses to the long-term exposure of low levels of pollutants experienced by the trees at our experimental sites. Physiological properties of "latent injury" are difficult to define and even more difficult to measure. In addition, interactions between different stress factors must be considered. As photosynthesis is the most important plant function, we will mainly concentrate on gas exchange measurements which have been studied in considerable detail at the experimental sites in the Fichtelgebirge. The general dilemma of studies in the field is that all trees at the experimental sites are exposed to comparable atmospheric pollution stress. Reliable controls are unavailable. Conclusions must therefore be drawn from metabolic responses which correlate with the degree of damage when two sites, or two individual trees, are compared and differ in respect to damage.

3.1 Photosynthesis

Photosynthesis is often considered to be one of the plant metabolic processes most sensitive to SO_2 and O_3 (Hällgren 1984; Lendzian and Unsworth 1983; see I. Ziegler 1975). However, not much information is available about the gas-exchange responses of intact spruce during and after experimental long-term exposure to pollutants under conditions comparable to the field situation found in the Fichtelgebirge. Krause and Prinz (1986) have reported that continuous fumigation with 150 $\mu g\ m^{-3}$ (ca. 0.07 $\mu l\ l^{-1}$) O_3 depressed net photosynthetic CO_2 uptake of the youngest needles significantly after 8 weeks. Keller and Häsler (1987) subjected 12-year-old spruce trees to 100 or 300 $\mu g\ m^{-3}$ (ca. 0.047 or 0.14 $\mu l\ l^{-1}$) O_3 for 45 h of daylight per week during a total growing season. They did not detect any significant effect on CO_2 exchange at the lower ozone concentration, although a significant depression of net photosynthesis was observed after 27 weeks exposure to the higher concentration. An immediate decline of CO_2-assimilation was observed by Atkinson et al. (1988) when leaves of chapparal shrubs were exposed to 0.20 $\mu l\ l^{-1}$ SO_2 in the field. Constant fumigation for 10 weeks of spruce grafts with SO_2 at a concentration of 0.05 $\mu l\ l^{-1}$ resulted in a depression in net photosynthesis (Keller 1978; see Keller 1984).

Measurements of CO_2 gas exchange of damaged and undamaged stands, as well as individual trees and needles of different degrees of damage and different age-classes were compared (see Chap. 4-D). Rates of CO_2 assimilation under controlled conditions (within the range of conditions which occur naturally were similar for green needles of healthy appearing trees which differed with respect to their general degree of tree damage at the two experimental sites, Oberwarmensteinach and Wülfersreuth. Rates were slightly higher at the declining site (Fig. 3; Zimmermann et al. 1988). As far as a comparison is possible, the level of CO_2 assimilation was similar to that reported for spruce earlier, e.g., in 1973 in the Solling mountains (Fuchs et al. 1977; Schulze et al. 1977; see Zimmermann et al. 1988). This was before forest decline had been claimed to occur. Thus, no

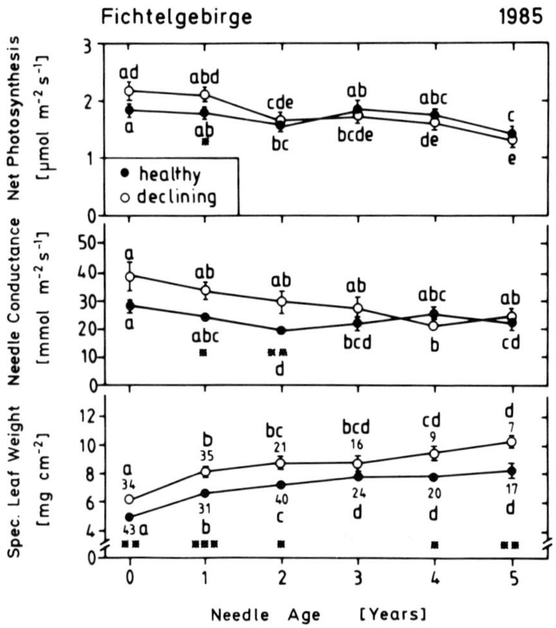

Fig. 3. Comparison of average net photosynthesis rates (*top*), needle conductance (*middle*), and leaf dry wt per total area (*bottom*) for six needle age classes (1980–1985) of green trees at the healthy site at Wülfersreuth and at the declining site at Oberwarmensteinach under similar, controlled climatic conditions (values are based on all-sided needle surface; *vertical bars* indicate ± 1 SE; number of twigs measured in each category is indicated by the *numbers* near the means in the lower figure). Two values within one site not identified by the same letter are significantly different at p 0.05. Differences between sites for each needle age class are marked as: *** p 0.001; ** p 0.01; * p 0.05; no mark: p 0.05 (Zimmermann et al. 1988)

substantial general loss in photosynthetic activity attributable to atmospheric pollution was detected for the healthy appearing trees at the two sites.

However, a depression in photosynthesis was a typical characteristic of damaged trees, and was usually associated with pigment bleaching. This became obvious when natural diurnal time courses of CO_2 exchange were compared (see Chap. 4-D). When expressed on a dry weight, as well as on a needle surface-area basis, net photosynthetic CO_2 assimilation of bleached needles of damaged trees was depressed by 20% to more than 50%, compared to the green needles of the healthy-looking counterparts. The study took place under identical conditions of naturally changing weather. This deviation in performance was due to differences in the activity of the photosynthetic apparatus of the different sets of needles. The damaged needles exhibited an apparently lower quantum use efficiency (Chap. 4-D), and their chlorophyll fluorescence changed characteristically (Wedler, unpublished). Figure 4 shows a set of CO_2 dependencies of photosynthesis for both a damaged, chlorotic, and a healthy appearing, green tree at Oberwarmensteinach. The different needle age-classes differed in their responses; the current flush was not yet fully grown. For comparable pairs of age classes, both the carboxylation efficiency (the primary slope of the response curve) as well as the photosynthetic capacity (maximal rate at saturating light and saturating needle internal CO_2 partial pressure, under standard temperature and humidity conditions) of the damaged needles were dramatically reduced. Such depression of the photosynthetic capacity was a characteristic feature of declining trees in the Fichtelgebirge. As will be discussed later, it was associated with mineral deficiency.

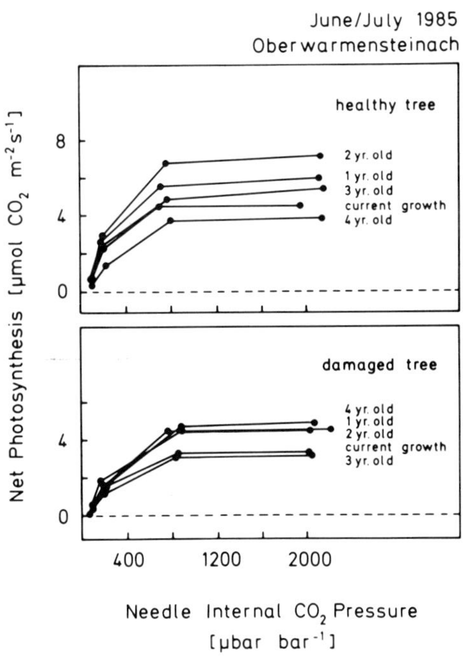

Fig. 4. CO_2 response of net photosynthesis under saturating photosynthetically active radiation flux, standard temperature and air humidity for different needle age classes of a healthy and a damaged spruce tree (current growth not yet fully grown). Oberwarmensteinach (June/July 1985). (After Weikert 1986)

Damaged and healthy appearing trees with different degrees of impairment of their photosynthetic capacity occur at the experimental site Oberwarmensteinach, side by side in the same atmospheric environment. However, such variability was not observed at the healthy site under similar atmospheric conditions. This eliminates the possibility of genetic variation in sensitivity as a reason for the variability in tree appearance. Thus, it seems unlikely that the differences in damage are caused through exposure to different degrees of pollution stress. Typical differences in damage of the photosynthetic apparatus were noted even in different age classes of the same tree. The photosynthetic capacity of the foliage of a healthy tree typically increased during the growing period until the end of the summer. The current year's flush usually exhibited the best performance, usually decreasing slightly with age. This was true when needles of different age classes of one twig were compared concomitantly, as well as when a particular individual's growth was continuously investigated over a period of several years (Weikert, unpublished). With damaged trees, the current year's flush was usually green; bleaching set in only with the aging of the needles. Concomitantly, the photosynthetic activity of the youngest needles was high. It was more depressed in the older age classes than could be expected from aging alone, compared to the healthy trees. Photosynthetic performance of twigs in similar exposure has been studied from 1984–1987. Figure 5 depicts the conditions in the fall for these 4 years. The photosynthetic capacity of the needles of the damaged trees was related to that of the same age-class of the undamaged ones. In all 4 years the current flush of the damaged trees reached rates of 100%, i.e., metabolic damage could not be detected when compared to the current flush of healthy individuals. A pronounced drop in capacity occurred between the current flush and the 1-year-old needles. Subsequently, the classes no longer changed substantially in their photosynthetic capacity with increasing age. This general pattern varied somewhat, depending on the conditions of the individual years.

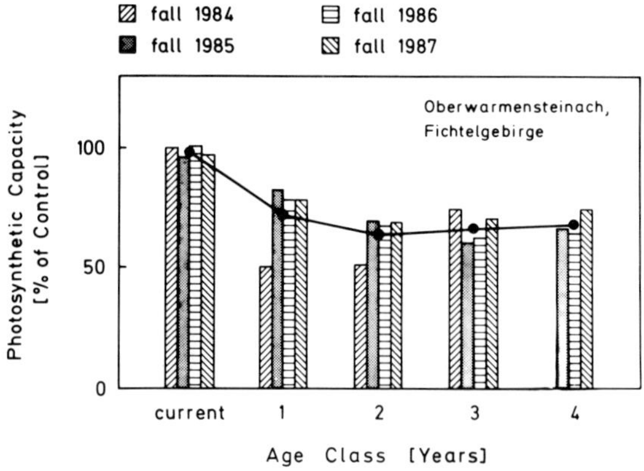

Fig. 5. Photosynthetic capacity (saturating CO_2, saturating photosynthetically active radiation flux, standard temperature and air humidity) for different needle age classes of damaged, chlorotic spruce trees as related to the capacity of healthy control trees (= 100%). Oberwarmensteinach (fall 1984, 1985, 1986, 1987). (After Lange et al. 1989)

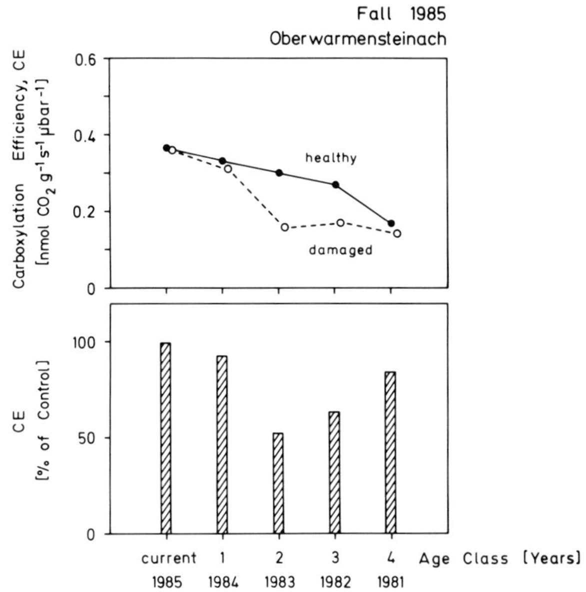

Fig. 6. Carboxylation efficiency (CE, initial slope of CO_2 response curve, saturating photosynthetically active radiation flux, standard temperature and air humidity) for different needle age classes of healthy and damaged spruce trees: *above*, absolute rates; *below*, related to CE of healthy control trees (= 100%). Oberwarmensteinach (end of growing season (1985). (Weikert 1986)

Performance similar to that of photosynthetic capacity was also found for the initial slope of the CO_2 responses of the specific needle age classes. Figure 6 (upper panel) shows a pronounced drop in absolute carboxylation efficiency by almost 50% between the first and the second year of the needle age of damaged trees, but no further decline. When the carboxylation efficiency of the bleached needles was compared with that of the healthy, control trees (Fig. 6, lower panel), a relative recovery for the older age classes was obvious, independent of whether CO_2 exchange was related to needle dry weight, needle carbon content, or needle surface.

According to these findings, there was no progressive impairment of photosynthesis with needle age in damaged trees after an initial substantial depression. Increasing the exposure time of the mature needles of damaged trees to the polluted atmosphere beyond the first (or second) year for further 1, 2, or even 3 years did not increase the damage in needle photosynthesis. This was always true on a relative basis when the same needle age class of damaged and undamaged trees were compared. Sometimes it even occurred in terms of absolute rates of CO_2 exchange. In one case, the same photosynthetic capacity was found in 4- as well as in 8-year-old needles of a large, ca. 100-year-old tree at Oberwarmensteinach. Such performance does not agree with observations made during experiments where plants have been fumigated with atmospheric pollutants. In these cases, plant response usually follows dose-response relationships, i.e., it is not only concentration-dependent but also time-dependent. This has certainly been shown for high pollution stress (Guderian 1977, 1985). However, a similar relationship also seems to hold for long-term impact. Current models, which aim to describe reduction in growth and yield of field crops due to environmental air pollution, usually consider damage in terms of the concentration of toxic substances, as well as the duration of their impact (see Kercher and King 1985). Keller (1978) and Keller and Häsler (1987) found in their long-term fumigation experiments that the time of exposure was an important variable with respect to depression in the photosynthesis of spruce at low concentrations of sulfur dioxide and ozone. Since this was not the case with the depression of photosynthesis of damaged spruce trees in the Fichtelgebirge, it is unlikely that the observed damage can be traced back to the damage caused in the fumigation experiments. Either the impairment of the photosynthetic metabolism in the Fichtelgebirge is not the consequence of a direct impact of atmospheric pollutants on the needle mesophyll, or a long-term uptake of pollutants induces a steady-state situation in which detoxification and repair balance damage. If such damage occurs, it may result in a generally reduced capacity in the primary productivity of the mesophyll. Not much biochemical detail is known about the mechanism of such processes.

The drop in photosynthetic capacity and the bleaching of older needles usually take place during the late spring and early summer months, when the new growth is in flush (see Chap. 4-F). During this period of the year, atmospheric pollution stress is at its minimum (see Chap. 1-B). However, the mineral contents of the foliage, especially in the 1-year-old needle growth of the damaged trees, dramatically change (see Chap. 4-F). These changes are correlated with photosynthetic characteristics. Figure 7 shows the situation at the end of the growing period (1984). Both the 1-year-old needles of the healthy and those of the damaged trees contained much less magnesium than their current flush. However, the Mg-concentration of the damaged needles was less than 200 $\mu g\ g^{-1}$, which is only 50% of that of the healthy counterparts. As discussed in Chapter 4-F, such a deficiency in spruce needles will directly affect photosynthetic metabolism (see Lange et al. 1986). Obviously, this is reflected in the depressed photosynthetic capacity of the damaged, older age classes (see Fig. 5). The direct effects of acid rain and air pollution on mineral nutrition of the trees are discussed further in detail in Part 4 of this Volume.

In order to experimentally demonstrate by field experiment that the direct impact of atmospheric pollutants is apparently not primarily responsible for the depressed photosynthetic capacity and needle-yellowing, terminal buds of single branches of damaged and undamaged trees were removed in the spring (see Lange et al. 1987; Lippert 1988). It is shown in Chapter 4-F that as a consequence of this treatment, the previous year's needles retained minerals that were lost from the needles of intact branches. During the course of the growing season, the performance of the experimental needles, which lacked competition from the new flush, was compared to that of the control needles of the same age class on the branches with the new flush. Throughout the experiment, chloroplast pigments (Fig. 8) of the healthy trees were not affected by the elimination of the new flush

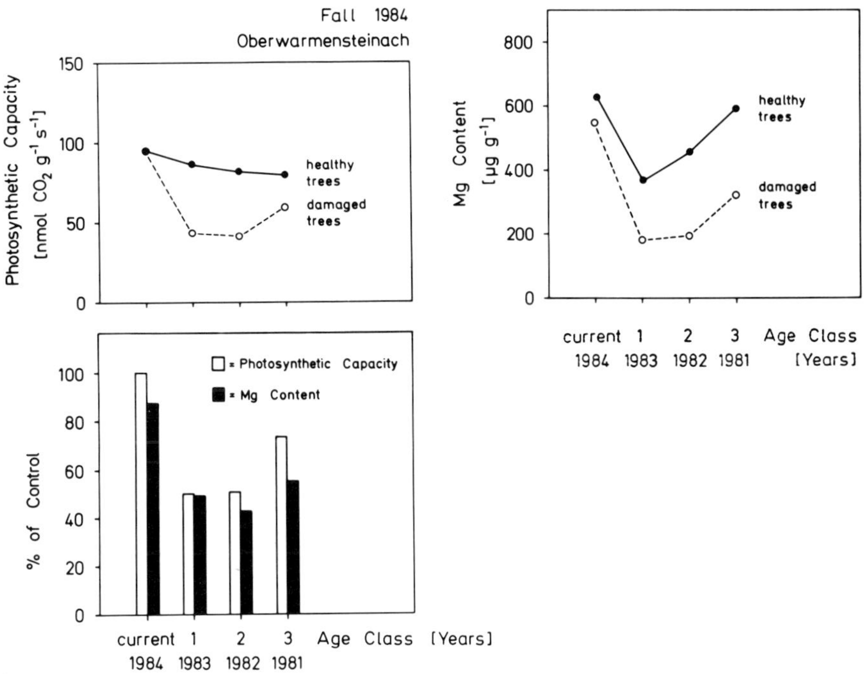

Fig. 7. Photosynthetic capacity (*left, above*, see Fig. 5) and Mg-content (*right*) for different needle age classes of healthy (*drawn lines*) and damaged (*dashed lines*) spruce trees; (*below*) photosynthetic capacity and Mg-content of the damaged trees related to the healthy trees (= 100%). Oberwarmensteinach (end of growing season 1984). (After Lange et al. 1989)

(see Mies and Zöttl 1985). However, a large difference in performance occurred between the two treatments with the heavily damaged trees. Here, the chlorophyll and carotenoid content of the previous year's needles increased substantially on those branches where the new flush had been eliminated. They remained green, while their counterparts on intact branches with competition from the current flush became chlorotic. This response was paralleled by a change in the photosynthetic capacity of the previous year's needles of the different treatments in both of the experimental years, as depicted in Fig. 9 for 1986 (see also Chap. 4-D). The increase in chloroplast pigments, as well as in photosynthetic capacity, for the needles of the damaged tree where the young growth was suppressed was associated with an increase in those minerals which seemed deficient in the control needles with the new flush (see Chap. 4-F). Thus, in contrast to needles of the same age class on intact branches where undisturbed new growth existed side by side in the same atmospheric environment, the experimental needles escaped bleaching and a decrease in photosynthesis. Thus, it seems in this experiment that the bleaching and the loss in photosynthetic capacity resulted from nutrient deficiencies rather than from a direct impact of atmospheric pollution. This conclusion is supported by the fact that fertilization at the Oberwarmensteinach

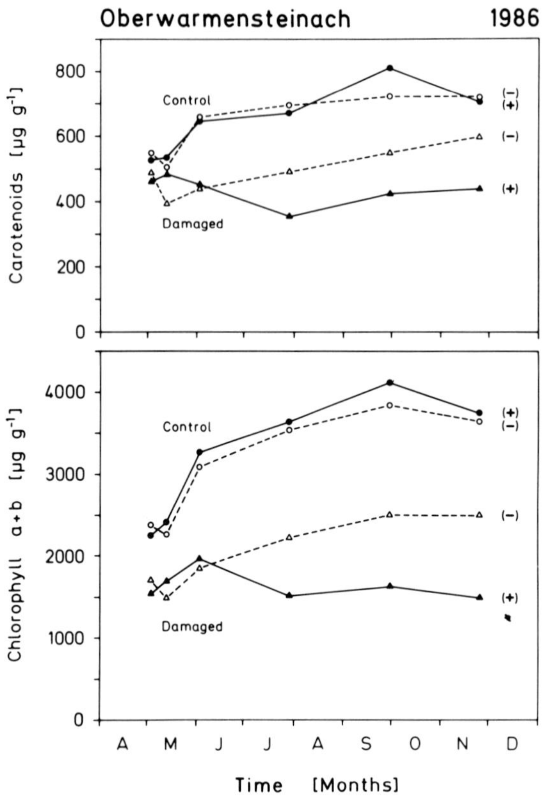

Fig. 8. Seasonal time courses (*abscissa*: months of year 1986) of dry-wt-related chloroplast pigments of the previous year's spruce needles for a healthy control tree and for a damaged tree; branches with intact flush of the current year's needles (+ , *drawn lines*) and experimental branches without the new flush (-, *dashed lines*). *Below*: chlorophyll a + b; *above*: total carotenoids. Oberwarmensteinach, after Lange et al. (1987)

site revitalized formerly damaged trees (Kaupenjohann et al. 1985; see Chap. 4-G). Two years after magnesium had been supplied to their root systems, conditions of atmospheric pollution remaining unchanged, photosynthetic responses of the previously damaged trees were normalized in comparison to nonfertilized, damaged trees exhibiting Mg deficiency (Beyschlag et al. 1987).

3.2 *Stomatal Responses*

According to laboratory experiments, there is no simple pattern of stomatal response to SO_2. The mechanisms of SO_2 and O_3 action on stomatal behavior, and the many interactions with other external and internal influences, are not yet clearly understood (see Mansfield 1976). Stomata have responded sensitively to

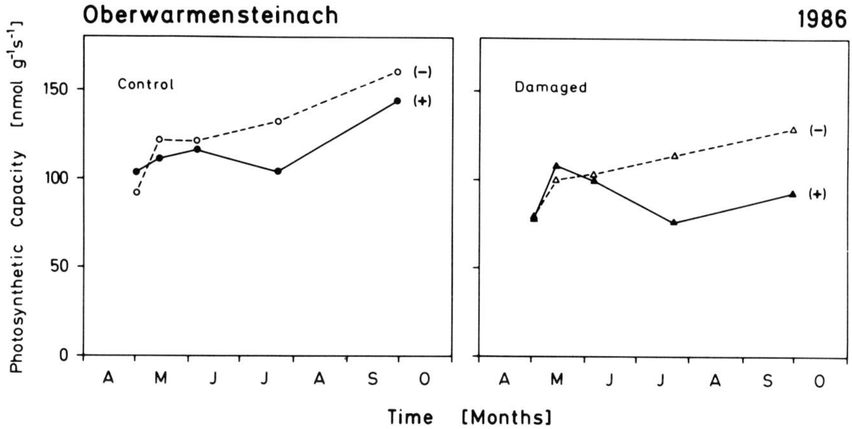

Fig. 9. Seasonal time courses of photosynthetic capacity (see Fig. 5) of the previous year's spruce needles on branches with and without the current year's flush, as described in Fig. 8. After Lange et al. (1987)

pollution stress, and thus the first measurable indications of plant responses to SO_2 stress are often found in transpiration (Black 1985). The stomatal responses of conifers are also altered as a result of exposure to atmospheric pollutants (see Brenninger and Tranquillini 1983). This is not only true for high pollution stress, but it is also the case when trees are fumigated with low concentrations of pollutants. Keller and Häsler (1986) subjected spruce cuttings to 0.025 $\mu l\ l^{-1}$ SO_2 for almost half a year. Stomatal transpiration decreased in light, and stomata reacted sluggishly to changed light conditions. When spruce trees were subjected over the total growing period to 45 h per week of 100 $\mu g\ m^{-3}$ (ca. 0.047 $\mu l\ l^{-1}$) ozone, stomatal conductance in darkness was higher than that of the control. Stomatal responses tended to become insensitive to light/dark changes (Keller and Häsler 1987).

Stomatal performance of damaged and undamaged trees was investigated at both experimental sites in the Fichtelgebirge over four growing periods. Measurements of gas exchange under natural environmental conditions did not show clear differences in stomatal performance between green and damaged, chlorotic needles (Wedler 1986; Pfeiffer 1987). Maximal needle conductance and diurnal response patterns were comparable. The range of needle conductance was similar to that reported for other sites with presumably lower air pollution (Weikert 1986).

Stomatal responses of the spruce trees in the Fichtelgebirge to external factors also did not appear to be impaired. As depicted in Fig. 10 (Zimmermann et al. 1988), in field experiments under steady-state conditions, needles of all relevant age classes and degrees of damage with different Mg-contents show a pronounced and reversible response to changes in air humidity and photon flux density. Wedler (1986) obtained similar results when he compared the humidity responses of chlorotic needles with those of green ones from damaged and healthy

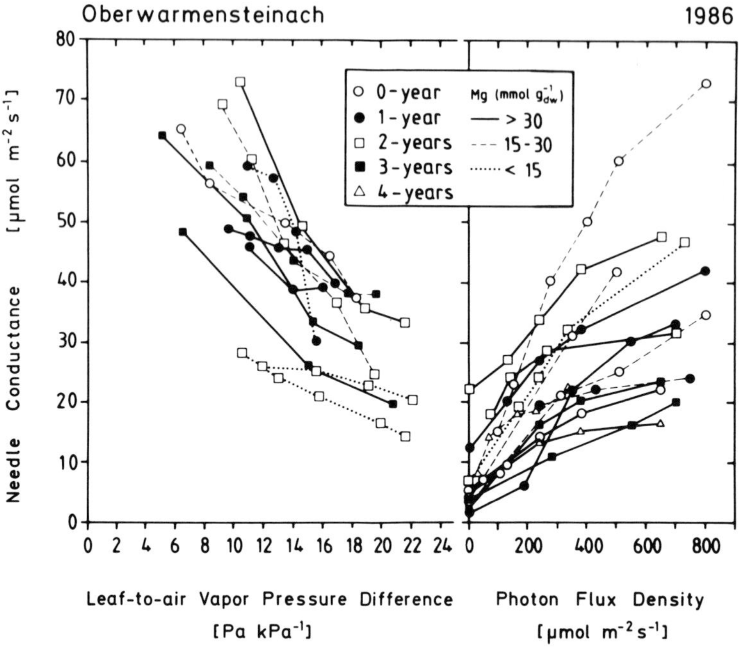

Fig. 10. Steady-state response of conductance for water vapor of different spruce needle age classes at 19 °C needle temperature (*left*) to increasing leaf-to-air water vapor difference (saturating light for photosynthesis) and (*right*) to increasing photosynthetically active radiation flux (constant humidity). Three ranges of needle Mg-content indicated. Oberwarmensteinach (June to September 1986). (After Zimmermann et al. 1988)

trees growing near each other at the Oberwarmensteinach site. Thus, these measurements do not indicate that air pollutants directly influence stomatal responses of spruce needles in the Fichtelgebirge. No evidence has been obtained that observed needle damage is correlated to or even caused through a misfunctioning of the stomatal apparatus which could be attributed to the direct impact of atmospheric pollution.

The same conclusion can be derived from quite a different source of information. According to Farquhar et al. (1982), the ratio $^{13}C/^{12}C$ of the organic material of a plant (the δ ^{13}C-value, see Osmond et al. 1982) reflects the average leaf intercellular CO_2 concentration during the light period. This concentration is decisively determined from the average stomatal leaf conductance, as well as from the photosynthetic activity of the mesophyll. If stomatal performance is altered in damaged spruce needles, changes in their δ ^{13}C-value should be expected. At constant carboxylation efficiency of the mesophyll, increased stomatal conductance should increase the intercellular CO_2 partial pressure, which thus should increase the carbon discrimination and consequently decrease the δ ^{13}C-value (becoming more negative). Carbon isotope ratios have been determined for needles of different age classes sampled from trees of the five fertilization plots (Number I being the control, see Chap. 4-G) from the healthy

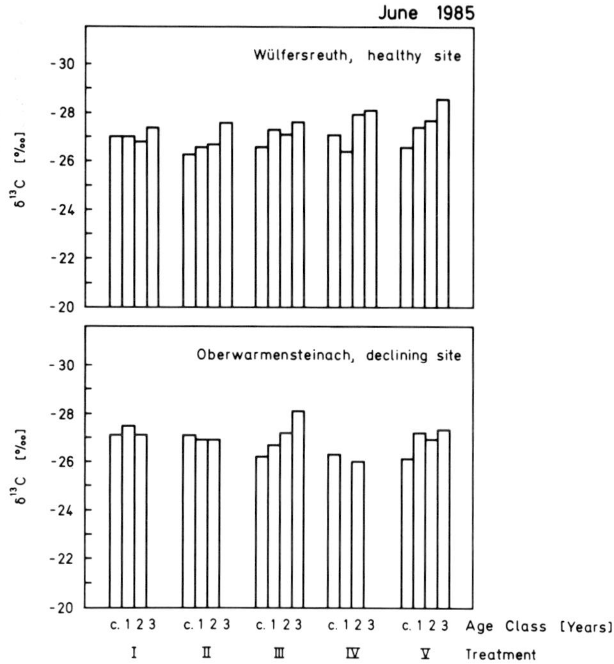

Fig. 11. Carbon isotope discrimination values ($\delta^{13}C$, see text) for needles of different age classes of trees at the healthy (Wülfersreuth) and the declining (Oberwarmensteinach) sites. *Arabic numbers* indicate age of needles (*c.*: current flush (1985); *1:* last year's (1984) growth, etc.); *Roman numbers* indicate the fertilization plots (see Chap. 4-G), *I* being the nontreated control. Sampling date: June 1985. (Unpublished data from Ziegler and Stichler)

site at Wülfersreuth and from the damaged site at Oberwarmensteinach (Fig. 11). There seemed to be a general tendency at both sites that $\delta\ ^{13}C$ decreased with the age of the needles. Possibly this was due to the lower conductances of the younger needles and/or to the age-dependent drop in the photosynthetic capacity of the needles. A lower content of terpene in the youngest needles must also be taken into consideration as a possible reason for the differences in carbon isotope discrimination, because fatty acid and terpene biosynthesis is known to decrease the $\delta\ ^{13}C$-values. However, there was neither a significant difference in the absolute level of the isotope ratios, nor in the age-dependent patterns, between the damaged and healthy trees which would indicate differences in stomatal performance.

3.3 Chloroplast Pigments

Besides needle loss, "bleaching" (or "acute yellowing", see Kandler et al. 1987), i.e., chlorophyll deficiency, is a characteristic feature of tree damage in the Fichtelgebirge. In the typical case, yellowing is confined to the older needle age classes, while the youngest growth remains green (see Introduction, Fig. 3).

Chlorophyll loss starts at the tips of the needles, and is most obvious in the sun crown of the trees. Köstner (1989) investigated chloroplast pigment content and dynamics in damaged and healthy appearing trees at the experimental site at Oberwarmensteinach. For an individual needle age class of the healthy trees, the annual maximum dry weight related chlorophyll content occurred around late summer and fall (Fig. 12). With the age of the needles, this annual maximum increased and reached its highest values for 4-year-old needles, the oldest analyzed (Fig. 13). During the winter months, the pigment content remained remarkably stable. However, a pronounced decrease took place early in the spring when the chlorophyll content dropped to 75–50% of its maximum (see the studies on seasonal changes in chloroplast formation by Senser et al. 1975).

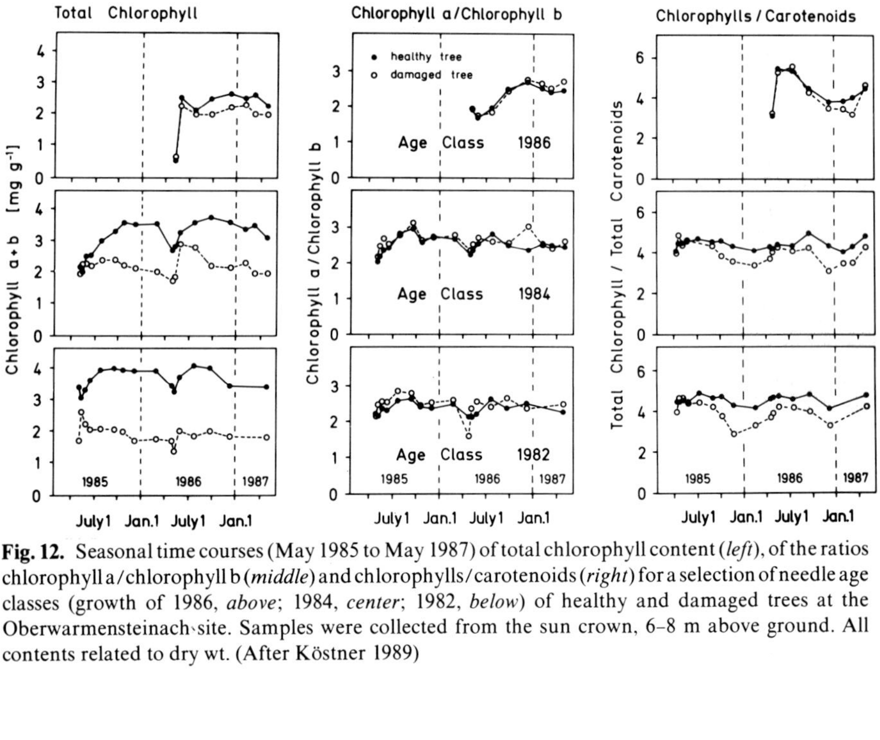

Fig. 12. Seasonal time courses (May 1985 to May 1987) of total chlorophyll content (*left*), of the ratios chlorophyll a/chlorophyll b (*middle*) and chlorophylls/carotenoids (*right*) for a selection of needle age classes (growth of 1986, *above*; 1984, *center*; 1982, *below*) of healthy and damaged trees at the Oberwarmensteinach site. Samples were collected from the sun crown, 6–8 m above ground. All contents related to dry wt. (After Köstner 1989)

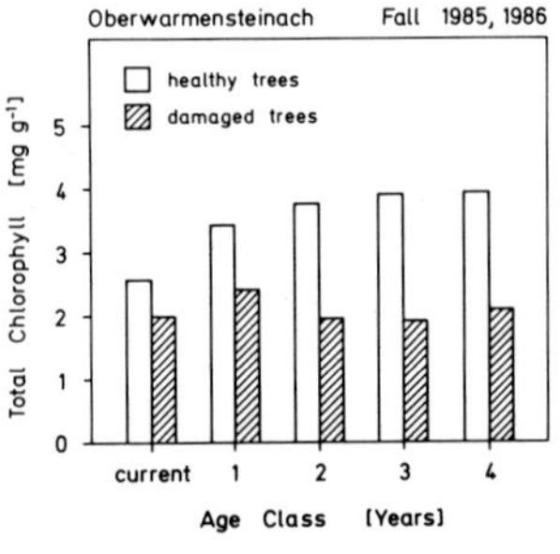

Fig. 13. Annual maxima of total chlorophyll content (dry wt-related) of different needle age classes of healthy and damaged trees. Oberwarmensteinach (end of growing season 1985 and 1986). (After Köstner 1989)

Subsequently, recovery took place in May and in the summer months. For the current flush of the damaged trees, the fall values of chlorophyll content still reached ca. 80% in comparison to the healthy trees (Fig. 13). However, the difference became greater for the older age classes because there was no consistent increase in chlorophyll content from year to year as there was for the healthy needles. With respect to the seasonal dynamics, most of the age classes of the damaged needles showed some increase in chlorophyll content in the spring, which became slowly reversed during the rest of the year. It should be noted that all these changes became apparent not only when pigment contents were compared on a dry-weight basis, but also when they were related to total needle carbon content.

For the damaged trees, annual maxima of total chlorophyll was clearly correlated with the magnesium content of the needles in an almost linear fashion between 100 and 300 $\mu g\ g^{-1}$ Mg (Fig. 14). In contrast, healthy needles contained more Mg, which no longer seemed to control chlorophyll content. It appears that seasonal changes in pigment content of the damaged spruce needles were also decisively influenced by their temporary mineral nutrition status. This can be concluded from the experiment depicted in Fig. 8. For the damaged trees, the pigment content of the mature needles of the intact branches remained at a more or less constant level during summer and fall. The largest difference in chlorophyll content in comparison to their undamaged counterparts developed during late spring and early summer, when the new flush was sprouting (and the SO_2 pollution at the site was minimal). Thus, compared with green needles, the deficit in chlorophyll as well as in carotenoids of the chlorotic needles was not so much due to a net "loss" in pigments, but rather due to their reduced capacity for pigment formation after the spring depression. This appears to be a consequence of nutrient deficiency. Needle pigment-content also increased in the last year's

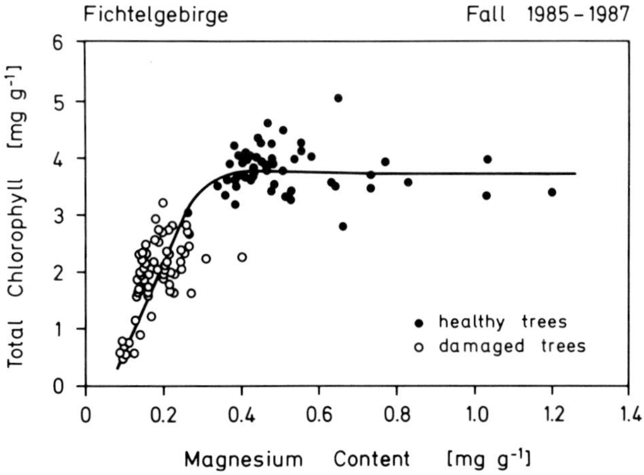

Fig. 14. Total chlorophyll content in relation to magnesium content (dry wt basis) if 1- to 4-year-old needles of 5 healthy and 5 damaged trees. Fichtelgebirge (end of growing season 1985, 1986, 1987). (After Köstner 1989)

needles of the damaged tree, when their mineral supply was improved with the elimination of the competition of the current flush (see Chap. 4-F). It is well known that a Mg-deficiency causes a reduction in chloroplast pigment-formation of plants, which impacts chlorophyll as well as carotenoids (Michael 1941; see Mothes and Baudisch 1958).

In spite of the large differences in total chlorophyll content between the needles of the healthy and damaged trees, the ratios of the different pigment components remained surprisingly stable. This is shown for chl. a/chl. b and for total chlorophyll/total carotenoids in Fig. 12. In an extreme situation, the tips of the chlorotic needles contained only 14% chlorophyll a + b in relation to the green needles of undamaged trees. However, even then the chlorophyll a:b ratio was not significantly changed. The total carotenoids were usually less reduced than the chlorophylls. In very strongly bleached needles there was still 30% of the carotenoid content of green needles present (ca. 700 $\mu g\ g^{-1}$). Amongst the carotenoids, carotene is often more depressed than are the epoxi-carotenoids and lutein.

Fumigation experiments have shown, at least at low concentrations, that SO_2 reduces the different pigment components almost proportionally (Schenk 1971). However, as noted by Kandler et al. (1987), with photooxidative pigment destruction from ozone, one would expect a predominant decrease in carotene compared with the chlorophylls. This was not the case for the damaged needles in the Fichtelgebirge.

3.4 Needle Sulfur Content

If the amount of SO_2 uptake of the spruce needles were involved in mesophyll damage, one would expect a certain correlation between the degree of injury and needle sulfur content. During long-term fumigation experiments, total sulfur in the needles has increased with SO_2 concentration applied, and with injury reached (e.g., Keller and Häsler 1986; see Guderian 1977). However, such a relation did not exist for the trees at the two experimental sites in the Fichtelgebirge. Figure 15 shows the photosynthetic capacity of 3-year-old needles in relation to total sulfur content. There was no correlation between the degree of damage and sulfur content. This is even clearer in Table 5. According to the average of many samples collected during the period 1984–1986 for the site at Oberwarmensteinach, the total sulfur content of green needles from healthy trees tended to be even somewhat higher than that for needles of the same age class from damaged trees. However, the difference was only significant for age class 1. Much fewer sulfur determinations are available for the site at Wülfersreuth (1985), so that a comparison is possible only with caution. Surprisingly, for each of the age classes, needle sulfur content at the healthy site at Wülfersreuth was higher than at the Oberwarmensteinach site, where symptoms of forest decline are present. In most of the data, a clear increase in total sulfur with needle age is recognizable. The relative as well as the absolute increase from one year to the next is higher at Wülfersreuth than at Oberwarmensteinach. If the dry weight-related needle sulfur content reflects at least in a certain way the extent of

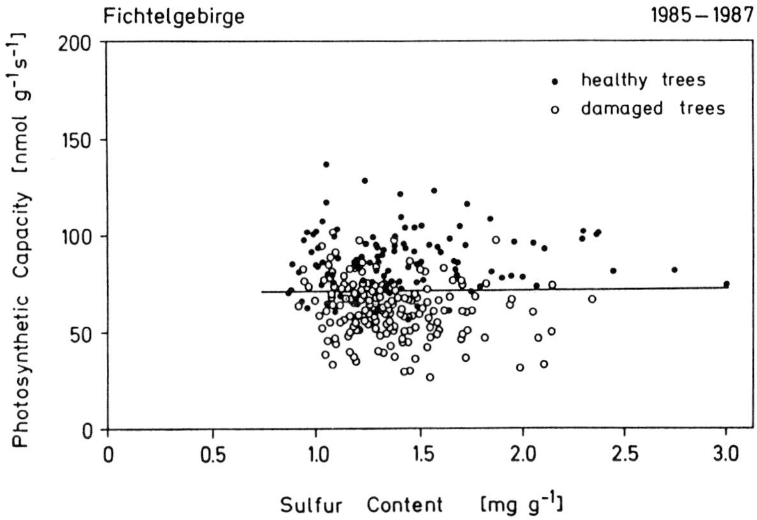

Fig. 15. Photosynthetic capacity (see Fig. 5) as related to total sulfur content (dry wt basis) of 3-year-old needles for damaged (*open symbols*) and for healthy (*solid symbols*) spruce trees; regression line indicated. Oberwarmensteinach and Wülfersreuth (1985, 1986, 1987). (After Langes et al. 1989)

Table 5. Average sulfur content and SE for different needle age classes from healthy appearing and damaged trees at Oberwarmensteinach (1984–1986) and from healthy trees at Wülfersreuth (1985)[a]

Needle age class	Oberwarmensteinach		Wülfersreuth
	Healthy	Damaged	
Current year	1174 ± 67 (54)	1191 ± 45 (95)	1340 (5)
1 year	1224 ± 25 (107)	1415 ± 42 (182)	1554 (5)
2 years	1287 ± 31 (94)	1323 ± 22 (147)	1736 (5)
3 years	1393 ± 36 (101)	1440 ± 23 (142)	2071 (5)
4 years	1488 ± 39 (81)	1572 ± 28 (116)	2307 (4)

[a] in $\mu g\ g^{-1}$, related to dry wt, number of samples in *brackets*

SO_2-incorporation, it appears that the trees take up more of this pollutant in the healthy stand, possibly because stomata tend to be open over longer periods of time under better growth conditions. However, we do not have quantitative figures about sulfur translocation from the needles to the other compartments of the trees, which would be needed for a sound sulfur balance sheet. Nevertheless, there is no indication that damaged needles have incorporated more SO_2 and/or have retained more sulfur than the healthy ones. This would further corroborate the conclusion that the direct impact of SO_2 on the needle mesophyll does not explain the differentiation of the actual patterns of tree damage.

4 Discussion and Conclusions

Pollution levels of SO_2 and ozone were significant in the Fichtelgebirge during the period of study (see Chap. 1-B). However, the field experiments on photosynthetic performance, stomatal behavior, and chloroplast pigment dynamics failed to reveal changes which could be attributed to a direct detrimental impact of atmospheric pollutants. Koch and Lautenschlager (1988) also failed to find significant differences in needle gas exchange when spruce twigs exposed to a "normally" polluted atmosphere were compared to twigs gassed with highly purified air for longer than 17 months during field experiments in the Bayerische Wald area. These results are in remarkable contrast to the estimates of SO_2 and O_3 fluxes into exposed leaves and needles (see Tables 2 and 3), which suggest that the uptake of pollutants through the stomata as calculated from measured atmospheric pollution levels should cause considerable acidification and oxidation over prolonged periods of exposure, if cellular detoxification was not achieved. It must therefore be concluded that either cellular defenses are fully adequate to cope with the atmospheric impact, or that the impact is overshadowed by other damaging events to such an extent that it can not be revealed by our analyses. Indeed, the evidence presented strongly suggests that such damage to spruce trees in the Fichtelgebirge as bleaching and loss of photosynthetic activity is related to mineral deficiencies or imbalances, which must be attributed to either changes in the ability of the roots to absorb ions such as Mg^{++}, or to mineral depletion of the soil. Soil depletion has been considered responsible for tree damage and present forest decline by soil scientists such as Ulrich (1980; Ulrich et al. 1979), Rehfuess (Bosch et al. 1983), Zech (Zech and Popp 1983) and Zöttl (Zöttl and Mies 1983). A similar hypothesis was suggested and experimentally tested already in 1912 by Wieler, for forests directly impacted through fumes from nearby smelters. Part 4 of this Volume is devoted to soil responses to acid rain, and to associated effects on the spruce stands in the Fichtelgebirge.

Although mineral deficiencies can easily explain a reduction in primary productivity, they are scarcely adequate to explain a decrease in the quantum yield of photosynthesis, and are possibly insufficient for interpreting sudden tree collapse as reported in the Fichtelgebirge and in other regions of middle Europe. Soil depletion through the cumulative effect of acid deposition develops as a slow process over prolonged periods of time; mineral deficiencies caused by soil depletion do not suddenly appear.

There is no doubt that plants possess the capability of detoxifying pollutants such as SO_2 and ozone. The pH-stat mechanism of cells can compensate for acidification. Glutathione and ascorbate can scavenge oxygen radicals; they also reduce incorporated O_3. Once cellular constituents such as proteins have been damaged, they can be degraded and replaced by newly-synthesized components. Efficient detoxification and repair underlie the capability of cells to survive a hostile chemical environment. However, damage can be prevented only as long as the necessary enzymic equipment is available in the necessary working condition. In damaged trees, carboxylation efficiency has been shown to be reduced. This suggests a reduction in the concentration of ribulose bisphosphate carboxylase, a major chloroplast protein. A reduction in protein content is to be expected in a situation of mineral deficiency. If it extends to the enzymes responsible for detoxification of pollutants and the repair of damaged cellular constituents, the rates at which these processes work will decrease. It should be noted that Mg^{++}, which has been shown to be deficient in damaged plants, is particularly important for protein synthesis. A shift in the balance between damage and repair might result in a new steady-state situation in which the plant is weakened and productivity decreases. Once the current year's flush of a damaged spruce tree in the Fichtelgebirge deprives last year's needles of part of their Mg^{++}, photosynthetic capacity decreases, but progressive damage to the photosynthetic apparatus does not develop as the needles age further. This may either be interpreted to mean that atmospheric pollutants do not play a significant role in the observed damage, or that they decrease viability by shifting the balance between damage and repair to a decreased level of productivity, which would upset the stability of the biotic system. An oxidative shift in the ratio of reduced to oxidized sulfur has been observed by Mitterhuber et al. (see Chap. 3-C) in damaged needles of spruce in the Fichtelgebirge. It might be suggested that the general burden of the direct effects of atmospheric pollutants in the Fichtelgebirge weakens the trees and makes them susceptible to impairment caused by soil acidification, mineral deficiencies, or probably by secondary biotic effects such as an attack by parasites (see Chap. 2-B). However, so far we are unable to quantify a general, pollution-induced, and "latent" reduction in tree viability. A direct correlation between metabolic characteristics of damaged trees and mineral deficiency appears then to be established. These factors are discussed in detail in Part 4 of this Volume.

References

Atkinson CJ, Winner WE, Mooney HA (1988) Gas exchange and SO_2 fumigation studies with irrigated and unirrigated field grown *Diplatus aurantiacus* and *Heteromeles arbutifolia.* Oecologia 75:386–393

Baier D, Latzko E (1975) Properties and regulation of C-1-fructose-1,6-diphosphatase from spinach chloroplasts. Biochim Biophys Acta 396:141–147

Beyschlag W, Wedler M, Lange OL, Heber U (1987) Einfluß einer Magnesiumdüngung auf Photosynthese und Transpiration von Fichten an einem Magnesium-Mangelstandort im Fichtelgebirge. Allg Forstz 27/28/29:738–741

Black VJ (1985) SO_2 effects on stomatal behavior. In: Winner WE, Mooney HA, Goldstein RA (eds) Sulfur dioxide and vegetation. Stanford Univ Press, Stanford, pp 96–117

Bosch C, Pfannkuch E, Baum U, Rehfuess KE (1983) Über die Erkrankung der Fichte (*Picea abies* Karst.) in den Hochlagen des Bayerischen Waldes. Forstwiss Centralbl 102:167–181

Brenninger C, Tranquillini W (1983) Photosynthese, Transpiration und Spaltöffnungsverhalten verschiedener Holzarten nach Begasung mit SO_2. Eur J For Pathol 13:228–238

Buchanan BB (1980) Role of light in the regulation of chloroplast enzymes. Annu Rev Plant Physiol 31:341–374

Dalling MJ, Tolbert NE, Hageman RH (1972) Intracellular location of nitrate reductase and nitrite reductase. I. Spinach and tobacco leaves. Biochim Biophys Acta 283:505–512

Davies DD (1986) The fine control of cytosolic pH. Physiol Plant 67:702–706

Elstner EF (1982) Oxygen activation and oxygen toxicity. Annu Rev Plant Physiol 33:73–96

Farquhar GD, O'Leary MH, Berry JA (1982) On the relationship between carbon isotope discrimination and the intercellular carbon dioxide concentration in leaves. Aust J Plant Physiol 9:121–137

Fischer K (1967) Cytologische und physiologische Wirkungen von Schwefeldioxid auf höhere Pflanzen. Dissertation, TH Darmstadt

Fuchs M, Schulze ED, Fuchs MI (1977) Spacial distribution of photosynthetic capacity and performance in a mountain spruce forest in northern Germany. II. Climatic control of carbon dioxide uptake. Oecologia 29:329–340

Guderian R (1977) Air pollution. Ecol Stud 22. Springer, Berlin Heidelberg New York

Guderian R (1985) Air pollution by photochemical oxidants. Ecol Stud 52. Springer, Berlin Heidelberg New York

Hällgren JE (1978) Physiological and biochemical effects of sulfur dioxide on plants. In: Nriagu JO (ed) Sulfur in the environment II. Wiley, N Y, pp 163–209

Hällgren JE (1984) Photosynthetic gas exchange in leaves as affected by air pollutants. In: Koziol MJ, Watley FR (eds) Gaseous air pollutants and plant metabolism. Butterworths, Lond, pp 147–159

Hällgren JE, Fredriksson SA (1982) Emission of hydrogen sulfide from sulfur dioxide-fumigated pine trees. Plant Physiol 70:456–459

Hampp R, Ziegler I (1977) Sulfate and sulfite translocation via the phosphate translocator of the inner membrane of chloroplasts. Planta 137:309–312

Hampp R, Spedding DJ, Ziegler I, Ziegler H (1980) The efflux of inorganic sulfur from spinach chloroplasts. Z Pflanzenphysiol 99:113–119

Heber U, Purczeld P (1978) Substrate and product fluxes across the chloroplast envelope during bicarbonate and nitrite reduction. In: Hall DO, Coombs J, Goodwin TW (eds) Photosynthesis 77, Proc IVth Int Congr Photosynthesis, Biochem Soc, Lond, pp 107–118

Heber U, Laisk A, Pfanz H, Lange OL (1987) Wann ist SO_2 Nährstoff und wann Schadstoff? Ein Beitrag zum Waldschadensproblem. AFZ 27/28/29:701–705

Hock P, Elstner EF (eds) (1988) Schadwirkungen auf Pflanzen. Bibliogr Inst Mannheim Wien Zürich

Hutchinson TC, Meema KM (1987) Effects of atmospheric pollutants on forests, wetlands and agricultural ecosystems. Springer, Berlin Heidelberg New York Tokyo

Jarvis PG, McNaughton KG (1986) Stomatal control of transpiration: scaling up from leaf to region. Adv Ecol Res 15:1–49

Kaiser G, Martinoia E, Schröppel-Meier G, Heber U (1989) Active transport of sulfate into the vacuole of plant cells provides halotolerance and can detoxify SO_2. J Plant Physiol 133:756–763

Kalmbach S (1985) Emissionsentwicklung bei den Hauptemittenten. In: Bauer F (ed) Die Sache mit dem Wald. Aktuelles Waldwissen der Allg Forstz, BLV-Verlagsgesellschaft München, pp 162–163

Kandler O, Miller W, Ostner R (1987) Dynamik der "akuten Vergilbung" der Fichte. Allg Forstz 27/28/29:715–723

Kaupenjohann M, Hantschel R, Horn W, Zech W (1985) Nährstoffversorgung gedüngter, unterschiedlich geschädigter Fichten auf immissionsbelasteten Standorten in NO-Bayern. Mitt Dtsch Bodenkundl Ges 43:969–974

Keller T (1978) Einfluß niedriger SO_2-Konzentrationen auf die CO_2-Aufnahme von Fichte und Tanne. Photosynthetica 12:316–320

Keller T (1984) The influence of SO_2 on CO_2 uptake and peroxidase activity. Eur J For Pathol 14:354–359

Keller T, Häsler R (1986) The influence of a prolonged SO_2 fumigation on the stomatal reaction of spruce. Eur J For Pathol 16:110–115

Keller T, Häsler R (1987) Some effects of long-term ozone fumigation on Norway spruce. I. Gas exchange and stomatal response. Trees 1:129–133

Kercher JR, King DA (1985) Modeling effects of SO_2 on the productivity and growth of plants. In: Winner WE, Mooney HA, Goldstein RA (eds) Sulfur dioxide and vegetation. Stanford Univ Press, Stanford, pp 356–372

Koch W, Lautenschlager K (1988) Photosynthesis and transpiration of a spruce in purified and ambient atmosphere. Trees 2:213–222

Köstner BMM (1989) Jahresverlauf der Chloroplastenpigmente von ungeschädigten und chlorotischen Fichten an einem Waldschadenstandort im Fichtelgebirge in Abhängigkeit von Alter und Mineralstoffgehalt der Nadeln. Dissertation Universität Würzburg

Krauze GHM, Prinz B (1986) Zur Wirkung von Ozon und saurem Nebel (einzeln und Kombination) auf phänomenologische und physiologische Parameter an Nadel- und Laubgehölzen im kombinierten Begasungsexperiment. In: Führ F, Ganser S, Kloster G, Prinz B, Stüttgen E (eds) Statussemin 1985 Arbeitsgruppe "Waldschäden/Luftverunreinigungen". Kernforschungsanlage Jülich, pp 208–221

Laisk A, Pfanz H, Heber U (1987) SO_2 fluxes into different cellular compartments of leaves photosynthesizing in a polluted atmosphere. II. Consequences of SO_2 uptake as revealed by computer analysis. Planta 173:241–252

Lange OL, Gebel J, Zellner H, Schramel P (1986) Photosynthesekapazität und Magnesiumgehalte verschiedener Nadeljahrgänge bei der Fichte in Waldschadensgebieten des Fichtelgebirges. In: Führ F, Ganser S, Kloster G, Prinz B, Stüttgen E (eds) Statussemin 1985 Arbeitsgruppe "Waldschäden/Luftverunreinigungen". Kernforschungsanlage Jülich, pp 127–147

Lange OL, Weikert RM, Wedler M, Gebel J, Heber U (1989) Photosynthese und Nährstoffversorgung von Fichten aus einem Waldschadensgebiet auf basenarmem Untergrund. Allg Forst Zeitschr 3/1989:55–64

Lange OL, Zellner H, Gebel J, Schramel P, Köstner B, Czygan FC (1987) Photosynthetic capacity, chloroplast pigments, and mineral content of the previous year's spruce needles with and without the new flush: analysis of the forest-decline phenomenon of needle bleaching. Oecologia 73:351–357

Lendzian KJ (1984) Permeability of plant cuticles to gaseous air pollutants. In: Koziol MJ, Whatley FR (eds) Gaseous air pollutants and plant metabolism. Butterworths, Lond , pp 77–81

Lendzian KJ, Unsworth MH (1983) Ecophysiological effects of atmospheric pollutants. In: Lange OL, Nobel PS, Osmond CB, Ziegler H (eds) Encycl plant physiol, vol 12C Physiological plant ecology, vol IV. Springer, Berlin Heidelberg New York, pp 465–502

Lippert M (1988) Photosynthesekapazität, Chlorophyllgehalt und Nährstoffgehalte von Nadeln geschädigter und ungeschädigter Fichten mit und ohne Konkurrenz des diesjährigen Triebs: Experimenteller Ansatz zur Analyse der Waldschäden im Fichtelgebirge. Diplomarbeit Univ Würzburg

Lyr H, Polster H, Fiedler HJ (1967) Gehölzphysiologie. Fischer, Jena

Mansfield TA (1976) Effects of air pollution on plants. Cambridge Univ Press, NY

Mansfield TA, Lucas PW, Wright EA (1988) Interactions between air pollutants and other limiting factors: In: Mathy P (ed) Air pollution and ecosystems (Proc Int Symp held in Grenoble, France, 18–22 May 1987). Reidel Publ Co, Dordrecht, pp 123–141

Matzner E, Ulrich B (1987) Results of studies on forest decline in northwest Germany. In: Hutchinson TM, Meema KM (eds) Effects of atmospheric pollutants on forest, wetlands and agricultural ecosystems. Springer, Berlin Heidelberg New York Tokyo, pp 25–42

Michael G (1941) Uber die Aufnahme und Verteilung des Magnesiums und dessen Rolle in der höheren grünen Pflanze. Bodenkd Pflanzenernähr 25:65–120

Mies E, Zöttl HW (1985) Zeitliche Änderung der Chlorophyll- und Elementgehalte in den Nadeln eines gelb-chlorotischen Fichtenbestandes. Forstwiss Centralbl 194:1–8

Miszalski Z, Ziegler H (1980) "Available SO_2" – a parameter for SO_2 toxicity. Phytopathol Z 97:144–147

Mothes K, Baudisch W (1958) Untersuchungen über die Reversibilität der Ausbleichung grüner Blätter. Flora 146:521–531

Mudd JB, Banerjee SK, Dooley MM, Knight KL (1984) Pollutants and plant cells: effects on membranes. In: Koziol MJ, Whatley FR (eds) Gaseous air pollutants and plant metabolism. Butterworths, Lond, pp 105–116

Osmond CB, Winter K, Ziegler H (1982) Functional significance of different pathways of CO_2 fixation in photosynthesis. In: Lange OL, Nobel PS, Osmond CB, Ziegler H (eds) Encyclopedia of plant physiol. New ser 12 B, Springer, Berlin Heidelberg New York, pp 479–548

Pankrath J (1985) Grenzüberschreitende Luftverunreinigungen. In: Bauer F (ed) Die Sache mit dem Wald. Aktuelles Waldwissen der Allg Forstz, BLV Verlagsgesellschaft München, pp 170–171

Pell E (1987) Ozone toxicity – is there more than one mechanism of action? In: Hutchinson TC, Meema KM (eds) Effects of atmospheric pollutants on forests, wetlands and agricultural ecosystems. Springer, Berlin Heidelberg New York Tokyo, pp 229–240

Pfanz H, Heber U (1986) Buffer capacities of leaves, leaf cells and leaf cell organelles in relation to fluxes of potentially acidic gases. Plant Physiol 81:597–602

Pfanz H, Martinoia E, Lange OL, Heber U (1987a) Mesophyll resistance to SO_2 fluxes into leaves. Plant Physiol 85:922–927

Pfanz H, Martinoia E, Lange OL, Heber U (1987b) Flux of SO_2 into leaves and cellular acidification by SO_2. Plant Physiol 85:928–933

Pfeiffer S (1987) Netto-Photosynthese und Transpiration einer gedüngten und einer geschädigten Fichte bei natürlicher externer CO_2-Konzentration am Freilandstandort im Fichtelgebirge. Diplomarbeit, Univ Würzburg

Plaut Z, Lendzian K, Bassham JA (1977) Nitrite reduction in reconstituted and whole spinach chloroplasts during carbon dioxide reduction. Plant Physiol 59:184–188

Prinz B, Krause GHM, Jung KD (1987) Development and causes of novel forest decline in Germany. In: Hutchinson TM, Meema KM (eds) Effects of atmospheric pollutants on forest, wetlands and agricultural ecosystems. Springer, Berlin Heidelberg New York Tokyo, pp 1–24

Rennenberg H (1984) The fate of excess sulfur in higher plants. Annu Rev Plant Physiol 35:121–153

Salisbury FB, Ross CW (1986) Plant physiology. Wadsworth Publ Co, Belmont

Schenk J (1971) Untersuchungen über den Einfluß von Schwefeldioxid auf Chloroplastenpigmente. Dissertation, Universität Dresden

Schmidt A (1979) Photosynthetic assimilation of sulfur compounds. In: Gibbs M, Latzko E (eds) Photosynthesis II. Photosynthetic carbon metabolism and related processes. Encyclopedia of plant physiol. New ser vol 6, Springer, Berlin Heidelberg New York, pp 481–496

Schulze E-D, Fuchs MI, Fuchs M (1977) Spacial distribution of photosynthetic capacity and performance in a mountain spruce forest of northern Germany. I. Biomass distribution and daily CO_2-uptake in different crown layers. Oecologia 29:43–61

Sekiya J, Wilson LG, Filner P (1982) Resistance to injury by sulfur dioxide. Correlation with its reduction to, and emission of hydrogen sulfide. Plant Physiol 70:437–441

Senser M, Schötz F, Beck E (1975) Seasonal changes in structure and function of spruce chloroplasts. Planta 126:1–10

Smith FA, Raven JA (1979) Intracellular pH and its regulation. Annu Rev Plant Physiol 30:289–311

Spedding DJ, Ziegler I, Hampp R, Ziegler H (1980a) Effect of pH on the uptake of ^{35}S sulfur from sulfate, sulfite, and sulfide by isolated spinach chloroplasts. Z Pflanzenphysiol 96:351–364

Spedding DJ, Ziegler I, Hampp R, Ziegler H (1980b) Effect of pH on the uptake of ^{35}S sulfur from sulfate, sulfite, and sulfide by *Chlorella vulgaris.* Z Pflanzenphysiol 97:205–214

Sze H (1985) H^+ translocating ATPases: advances using membrane vesicles. Annu Rev Plant Physiol 36:175–182

Taylor GE Jr, Tingey DT (1983) Sulfur dioxide flux into leaves of *Geranium carolinianum* L. Plant Physiol 72:237–244

Ulrich B (1980) Die Wälder in Mitteleuropa: Meßergebnisse ihrer Umweltbelastung, Theorie ihrer Gefährdung, Prognose ihrer Entwicklung. Allg Forstz 35:1198–1202

Ulrich B, Mayer R, Khanna PU (1979) Die Deposition von Luftverunreinigungen und ihre Auswirkungen im Waldökosystem im Solling. Schr Forstl Fak Univ Göttingen 58:1–291

Umweltbundesamt (1978) Medizinische, biologische und ökologische Grundlagen zur Bewertung schädlicher Luftverunreinigungen. Sachverständigenanhörung, Berlin

Wedler M (1986) Einfluß einer Magnesiumdüngung auf Photosynthese und Wasserhaushalt geschädigter Fichten an einem Magnesium-Mangelstandort im Fichtelgebirge. Diplomarbeit Univ Würzburg

Weigl J, Ziegler H (1962) Die räumliche Verteilung von ^{35}S und die Art der markierten Verbindungen in Spinatblättern nach Begasung mit $^{35}SO_2$. Planta 58:435–447

Weikert R (1986) Die photosynthetische Leistungsfähigkeit verschieden stark geschädigter Fichten – Vergleich zwischen einem "belasteten" und "unbelasteten" Standort. Diplomarbeit Univ Würzburg

Werdan K, Heldt HW, Milovancev M (1975) The role of pH in the regulation of carbon fixation in the chloroplast stroma. Studies on CO_2 fixation in the light and dark. Biochim Biophys Acta 396:276–292

Wieler A (1912) Pflanzenwachstum und Kalkmangel im Boden. Bornträger, Berl

Winner WE, Mooney HA, Goldstein RA (1985a) Sulfur dioxide and vegetation. Stanford Univ Press, Stanford

Winner WE, Mooney HA, Kimberlyn W, von Caemmerer S (1985b) Measuring and assessing SO_2 effects on photosynthesis and plant growth. In: Winner WE, Mooney HA, Goldstein RA (eds) Sulfur dioxide and vegetation. Stanford Univ Press, Stanford, pp 118–132

Woodrow IE, Murphy DJ, Latzko E (1984) Regulation of stromal sedoheptulose-1,7-bisphosphatase activity by pH and Mg^{2+} concentration. J Biol Chem 259:3791–3795

Yoshida S, Uemura M (1986) Lipid composition of plasma membranes and tonoplasts isolated from etiolated seedlings of mung bean (*Vigna vadiata* L.). Plant Physiol 82:807–812

Zech W, Popp E (1983) Magnesiummangel, einer der Gründe für das Fichten- und Tannensterben in NO-Bayern. Forstwiss Centralbl 102:50–55

Ziegler H (1983) Physiologie. In: Von Denffer D, Ziegler H, Ehrendorfer F, Bresinsky A Lehrbuch der Botanik für Hochschulen. Fischer, Stuttgart NY, pp 215–483

Ziegler H (1988) Weg der Schadstoffe in die Pflanze. In: Hock B, Elstner EF (eds) Schadwirkungen auf Pflanzen. Bibliogr Inst Mannheim Wien Zürich, pp 35–46

Ziegler I (1975) The effect of SO_2 pollution on plant metabolish. Residue Rev 56:79–105

Ziegler I, Hampp R (1977) Control of $^{35}SO_4^{2-}$ and $^{35}SO_3^{2-}$. Incorporation to spinach chloroplasts during photosynthetic CO_2 fixation. Planta 137:303–307

Zimmermann R, Oren R, Schulze ED, Werk KS (1988) Performance of *Picea abies* (L.) Karst. at different stages of decline. II. Photosynthesis and leaf conductance. Oecologia 76:513–518

Zöttl HW, Mies E (1983) Die Fichtenerkrankung in Höhenlagen des Südschwarzwaldes. AFZ 154:11–114

Part 4. Soil Responses to Acid Rain and Associated Effects on Plants

Preface

In this part of the book, the role of acidic deposition in forest decline by modifying the soil chemical environment is discussed. Forest decline in the Fichtelgebirge is not confined to the declining stand which was analyzed in preceding chapters (see preface); it is a widespread phenomenon with various expressions. At high elevations (700–1000 m) the soils in large areas are phyllitic or granitic acid spodosols, which are low in calcium and magnesium saturation. Here, older needles of young *P. abies* (L.) Karst. trees first turn yellow and are then shed, while older trees seem to shed green needles (Chap. 4-A). This behavior is associated with a low magnesium supply to the trees. At lower elevations, old trees appear to lose their needles even on base-rich basaltic soils because of restricted potassium uptake. This phenomenon is not only restricted to *P. abies* but occurs in other conifers and several broad-leaf species, all of which also suffer chlorosis and premature foliage loss (Chap. 4-A).

Lower soil acidity is associated with higher nutrient cation availability (Chap. 4-B). Soil acidity is affected by the acid input to the soil, the acid production in it, and proton buffering processes. In the organic layer, the main proton consumption is a rapid exchange of easily adsorbed calcium and magnesium, followed by the release of aluminum. The more acidic nature of the more decomposed (O) horizon in comparison to the overlaying, less decomposed (L) horizon results in a release of aluminum from clay minerals in this horizon. A further, slower proton buffering reaction is attributed to microbial oxidation of organically bound metal cations such as calcium, but this process is very limited in phyllitic, compared to basaltic, soils.

Mineral horizons of soils originating from phyllitic or granitic parent material have higher acid neutralizing capacity than the overlaying organic layer, but lower capacity than mineral horizons in basaltic soils (Chap. 4-B). Phyllitic and granitic soils may have little exchangeable calcium and magnesium; therefore, the aluminum released in proton buffering remains in the soil solution. A high aluminum concentration in the solution reduces the competitiveness of calcium and magnesium in the solution for exchange sites, so that they are leached as soil acidification proceeds. This reduces the storage of these elements in the mineral horizons of these highly acidic soils and leaves the organic layers as the main source of nutrient cations.

Soil structure, particularly the effect of the gravel content on the weathering rate, and the effect of aggregation characteristics on the flux of elements in the soil, affects the nutrient supply to trees appreciably. Indeed, the main proton buffering

E.-D. Schulze, O.L. Lange, R. Oren (Eds.)
Ecological Studies Vol. 77

reactions occur at the surface of aggregates so that an analysis of homogenized samples results in overestimating the buffering rate and capacity of the soil (Chap. 4-B). The difference in the estimation of these soil properties is apparent when the results of the two methods are compared for the healthy and declining sites discussed throughout this book. The soils of the two sites developed to somewhat different stages from a phyllitic parent material. The buffering by homogenized soil samples was similar at the two sites. When undisturbed samples were used, the buffering rate and the capacity of the soil at the healthy site were only slightly reduced, but in the soil of the declining site were greatly depressed. The magnesium and calcium mass flow out of aggregates in the soil at the declining site was about half that at the healthy site. It appears that with similar acidic input, ion release occurs faster but for less time at the declining site. Because of a smaller concentration gradient between the aggregate surface and the soil solution at the declining site, an equilibrium is approached fairly quickly and the ion flux from the aggregate to its surface is reduced (Chap. 4-B).

Roots are unable to penetrate into aggregates, and the frequency of mycorrhiza was significantly lower at the declining site (Chap. 4-E). This probably resulted in less efficient nutrient uptake from the aggregates. For roots, the main source of elements is the surface of the aggregates but, due to the leaching of the surface by acidic deposition and the low flux from inside, this source was limited (Chap. 4-B). Moreover, root penetration into the mineral horizons at the declining site was impaired by the conditions created by the acidification processes (Chap. 4-E). Most of the proton buffering in the mineral horizon at the declining site was due to aluminum, while at the healthy site calcium and aluminum contributed equally to the buffering (Chap. 4-B). As a result, the calcium-to-aluminum ratio at both sites was within the range that affects root growth but was higher at the healthy site (Chap. 4-E). However, soil acidification results in a change in several variables, all of which were associated with root development (e.g., calcium, ammonium, and proton concentrations; Chap. 4-E). Therefore, it is difficult to isolate a single variable as the cause of reduced root growth in the mineral horizon, reduced root tip number, and mycorrhizal frequency. Whatever is the causal factor, root growth into the mineral horizons at the declining site was impaired, particularly during the summer.

The main source of nutrient cations at both sites was the organic layer. However, this layer experiences many short-term droughts during the summer when water and element uptake from this layer is reduced (Chaps. 4-C, F). In such periods, the uptake of water and elements from the mineral soil becomes important for maintenance of plant function and growth. The declining site received higher precipitation, but trees there seemed to experience more difficulties in water uptake, as demonstrated by their lower water potential at similar transpiration rate and stomatal conductance in comparison to the healthy site (Chaps. 4-C, D).

This difficulty in maintaining water uptake occurred for three reasons. (1) Pore distribution in both the organic layer and the mineral soil at the declining site enhanced water flux and made it more difficult for roots there to absorb water. (2) A pore discontinuity such as that found in the mineral soil at the healthy site was absent at the declining site. Thus, in contrast to the healthy site, water could flow out of the rooting zone and not accumulate as plant-available water (Chap. 4-B). (3) Uneven root distribution at the declining site restricted water uptake from deeper horizons.

Nutrient concentrations in the xylem sap were related to the water potential of the trees and probably indicated a reduction in the supply of nutrients during soil drying and greater reliance on the internally cycling element pool (Chap. 4-F). Not all nutrients were affected similarly. The supply of magnesium and calcium seemed to become limited at higher water potentials than that of nitrogen. This may have resulted in a nutritional imbalance which could have been stronger at the declining site, where trees experienced moderate water potentials more often than trees at the healthy site. Such an imbalance might have been strengthened by the aerial uptake of nitrogen in different forms by the canopy, accompanied by canopy leaching of the metal cations (Chaps. 4-B, I).

In general, the concentration of elements in the needles showed site differences which were related to the element concentration in the soil solution. Thus, needles at the declining site had lower calcium and magnesium, higher aluminum and similar nitrogen and potassium concentrations in comparison to those at the healthy site (Chap. 4-F). Moreover, needle magnesium concentration decreased with age at the declining site, but increased at the healthy site (Chaps. 4-B,F). This indicated a magnesium deficiency at the declining site. When the plots at the two sites were analyzed as independent data points, no relation between element concentration in the soil solution and the needles was found. However, other associations were found between root tips, mycorrhizal root tips, and calcium and magnesium concentrations in the xylem sap and the concentrations of these elements in needles (Chaps. 4-E,F). In a broader site survey, magnesium concentrations was determined in the solution collected by percolating a solution with a pH value commonly found in precipitation in this region through undisturbed soil samples. These magnesium concentrations were highly correlated to the magnesium concentrations in the needles from these sites (Chap. 4-B).

Two metabolic processes were associated with the magnesium and calcium concentrations in needles. Photosynthetic rates were associated with magnesium concentration in needles deficient in this element (Chap. 4-D) and growth was related to both magnesium and calcium concentrations in needles of several age classes (Chaps. 4-F, H). Photosynthesis, however, did not seem to be the limiting factor, as indicated by amply stored carbohydrates at both sites. Indeed, too small a proportion of the leaf area at the declining site was yellow enough to affect the carbon balance appreciably. In addition, the leaf-area index was similar and relatively high at both sites. A large reduction in either canopy leaf area or the efficiency of the foliage would be needed to explain the 35% lower growth at the declining site. It was concluded that a reduction in the sink for carbohydrates due to magnesium limitation was responsible for the growth reduction (Chaps. 4-D, F, H).

The role of nitrogen and nitrogen input has been assessed in several chapters. Nitrogen appears to limit the production of canopy components (Chaps. 4-A, F,

H). However, this seems to be true only when the magnesium supply is not limiting. Once the magnesium supply becomes limiting, canopies tend to accumulate nitrogen rather than produce more foliage and twigs (Chaps. 4-F, H). Both increased (Chap. 4-A) and decreased (Chap. 4-H) magnesium concentrations were found with increasing nitrogen concentration. This apparent contradiction is related to the time of sampling and, more importantly, to the initial concentration of magnesium in the foliage prior to bud-break (Chap. 4-H). The effect of the interaction among nutrient supplies on growth and maintenance processes becomes apparent only if the seasonal dynamics of the elements is evaluated with respect to each other.

A seasonal analysis revealed that nitrogen can promote production of crown components as long as the supply of magnesium by retranslocation from mature foliage is sufficient to maintain the newly formed foliage above deficiency. As long as nitrogen is limiting the canopy production, magnesium retranslocation from mature foliage will not be great enough to cause deficiency in mature foliage. However, when nitrogen is ample and the magnesium concentration in mature foliage is low, retranslocation will result in chlorosis in mature needles (Chap. 4-H). At the declining site the recharge of magnesium in the mature foliage is limited because it is less available in the soil. Because most of the retranslocatable magnesium is used for canopy production, stemwood production is dependent on the uptake of magnesium from the soil during the stemwood period of growth, which follows canopy production and occurs mostly in the summer. Magnesium uptake from the soil, as shown above, is limited at the declining site. This model of forest decline from the nutritional disharmony in the timing and relative supplies of magnesium and nitrogen is probably applicable to other situations where a plant retranslocatable element is in short supply in the soil (Chap. 4-H).

The supply of magnesium may have limited stemwood growth at the declining site due to a possible pre-decline difference in the soil characteristics between the sites (Chap. 3). The supply of magnesium, however, seems to have become more limited recently as apparent from the chlorosis, which is a relatively new phenomenon. Indeed, an ecosystem balance analysis revealed that as the stands develop, a large proportion of the available magnesium becomes inaccessible in the stem (and will eventually be lost from the site as the stems are harvested), and that appreciable amounts of magnesium are lost via soil leaching, particularly at the healthy site (Chap. 4-I). Because the declining site appears to operate at a conservative but lower rate of magnesium cycling, a magnesium limitation is indicated.

Needle chlorosis from magnesium deficiency can be alleviated by fertilizing or by preventing canopy production through pruning the buds. Two foliar fertilizer applications reduced chlorosis as needle magnesium concentration increased (Chap. 4-G). Soil fertilization with $MgSO_4$ eliminated chlorosis, but reduced foliar nitrogen concentration associated with the lower nitrate and potassium in the soil solution, perhaps due to leaching. Fertilization with dolomite also eliminated chlorosis, but caused an increased nitrogen concentration in the foliage, perhaps because of faster mineralization and a subsequent increase in the nitrate concentration in the soil solution. In addition to the possible complications in tree nutrition and soil processes from the drastic changes associated with fertilization and the possible consequences to streamwater draining a fertilized watershed, it is uncertain how long the fertilizer effect will last. So far, in the 4 years after fertilization, chlorosis has not returned. However, the magnesium concentration in needles from the fertilized plots at the declining site still declines with needle age (Chap. 4-G) as was found in unfertilized plots at that site, while the concentration in needles from unfertilized plots at the healthy site remained constant with age (Chap. 4-F). This indicates that the fertilizer application reduced, but did not eliminate the magnesium deficiency at the declining site. Indeed, growth at the declining site seemed to increase only slightly as a result of the fertilizer application and remained lower than the growth in unfertilized plots at the healthy site at a similar leaf-area index (Fig. 10 in Chap. 4-D).

Bud-pruning simulates the effects of a reduction in nitrogen input by reducing canopy production (Chap. 4-F). An acidification model showed that nitrate input, more than sulfate, is likely to cause further reductions in soil calcium and magnesium saturation (Chap. 4-B). In addition, calculated critical loads of nitrogen indicate that any input of nitrogen to these acid soils would reduce the exchangeable base-cation pools. It is no wonder, therefore, that comparisons of current data on tree nutrition with data collected a decade ago have shown that magnesium and calcium nutrition of trees growing in acid soils has worsened (Chap. 4-A). Moreover, based on the ecosystem balance analysis for the healthy site, it was estimated that conditions there will become similar to those at the declining site in the foreseeable future, if current pollution rates are maintained (Chap. 4-I).

In conjunction with the research reported in other sections in this book, we conclude that forest decline in the Fichtelgebirge stems from nutrient stress originating in magnesium- and calcium-deficient soils. Nitrogen input further depleted the naturally poor soils of these nutrient elements while concurrently creating greater demand for them by the trees.

4-A Mineral Nutrition of Forest Trees: A Regional Survey

M. KAUPENJOHANN, W. ZECH, R. HANTSCHEL, R. HORN, and B.U. SCHNEIDER

1 Introduction

Previous foliar mineral analysis has shown that at high elevation sites in the Fichtelgebirge (700–1000 m a.s.l.), those *Picea abies* (L.) Karst. with yellow tips and even necrosis of older needles exhibit very low Mg, Low Ca and Zn, and rather high S concentrations (Zech and Popp 1983). Several authors have reported similar results from other forest areas of Central Europe (Bosch et al. 1983; Zöttl and Mies 1983; Hüttl 1985; Van Praag and Weissen 1986; Nebe et al. 1987). Another widespread decline phenomenon is the loss of needles, which mainly develops in older spruce at lower elevations in the Fichtelgebirge. Mineral nutrition studies have found that K but not Mg is often insufficient in trees showing needle loss (Zech et al. 1983; Zech et al. 1985; Hüttl 1985; Zöttl and Hüttl 1986; Mengel et al. 1987). Potassium deficiency symptoms, however, have not been observed in the Fichtelgebirge. Other hypotheses to account for needle loss have included water stress (Ulrich 1981), fungal attack (Rehfuess and Rodenkirchen 1984) and frost damage (Rehfuess and Bosch 1986; Friedlanda et al. 1984), all of which may be influenced by the K nutrition of trees. In addition, Klein (1981) has shown that the resistance of K-deficient *P. abies* to SO_2 is reduced, which may be a factor in causing the forest decline in the SO_2-polluted Fichtelgebirge.

Causes for forest decline are probably restricted to special site conditions. To help evaluate the extent to which these causes may account for decline, the nutrient status of different healthy and declining tree species in southern Germany was determined. Trees at different geological sites, which varied from large scale calcareous soils derived from dolomite in the Bavarian Alps to very acid soils derived from granite and phyllite in the Bavarian forest and in the Fichtelgebirge were studied (Fig. 1). For comparison, we also studied *P. abies* at a site near Pressath, which has low acid deposition and has been described as naturally Mg deficient (Küppers et al. 1985).

Because *Fagus sylvatica* (L.), *Acer platanoides* (L.), *Abies alba* (Mill.), *Larix decidua* (Mill.) and *Pinus sylvestris* (L.) have also been found to display symptoms of insufficient mineral supply on acid soils, these species were also included in this study.

E.-D. Schulze, O.L. Lange, R. Oren (Eds.)
Ecological Studies Vol. 77

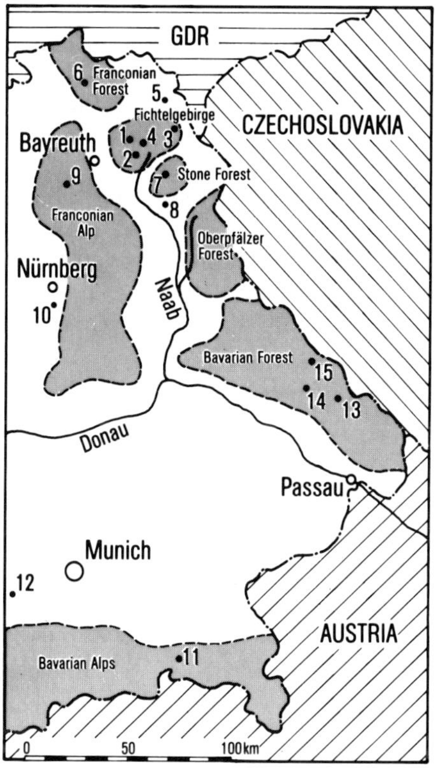

Nr.	Site	Tree species	Altitude	Bedrock
1	healthy	Picea abies	680	Granite/Phyllite
2	declining	Picea abies	700-760	Phyllite
3	Selb	Picea abies	540-580	Basalt
4	Schneeberg	Picea abies, Fagus sylvat., Abies alba, Larix decidua, Acer platan.	900-1000	Granite
5	Rehau	Picea abies	560	Slate
6	Schnappen-hammer	Picea abies	640	Slate
7	Platte	Picea abies	940	Granite
8	Pressath	Picea abies	480	Tertiary sandy deposits
9	Hollfeld	Pinus sylvest.	540	Calcaric deposits
10	Schwabach	Pinus sylvest.	380	Tertiary sandy deposits
11	Kloaschau	Picea abies	950	Dolomite
12	Högelwald[1)]	Picea abies	520-540	Tertiary deposits/Loss
13	Grueben[2)]	Picea abies	790	Granite
14	Hoch-gfeichtet[2)]	Picea abies	1250	Granite
15	Bavarian-Forest[3)]	Picea abies	630	Granite

1) KREUTZER and BITTERSOHL, 1986
2) GIETL, 1988
3) REHFUESS and RODENKIRCHEN, 1984

Fig. 1. Map of southern Germany: the investigated forest sites are mainly located in northeastern Bavaria

2 Sampling Procedure

Foliar mineral analysis is a suitable method for characterizing tree nutrition (Lambert et al. 1983; Bergmann 1986). Höhne (1964) has recommended that conifer samples be taken from trees comparable in age at the end of the vegetation period in late autumn or winter. Since the canopy position of the needles may also influence their nutrient contents, Knabe (1984) has proposed standardizing spruce foliage sampling by taking material from the seventh whorl. Data presented in this chapter, however, mainly refer to element concentrations in spruce needles from the third to fourth whorl because in younger stands, branches of the seventh whorl are often low in the canopy where there is deep shade, severely damaged trees rarely have needles older than 4 years, especially in the lower part of the crown, and effects of fertilization can be determined more easily by analyzing the top part of the crown.

Samples from *P. abies* and *P. sylvestris* were collected in the late autumn for each year from 1983–1986. Each element determination was made on a mixed sample of 3 15 trees. In general, current-, 1-, and 2-year-old needles were analyzed. Additionally, at the healthy site near Wülfersreuth, the declining site near Oberwarmensteinach, and the sites near Schnappenhammer, Pressath and Schneeberg, needles as old as 6 years were also studied. Needle samples of *A. alba* were collected at the end of winter. Current-year foliage of *L. decidua, F. sylvatica* and *A. platanoides* was collected at the end of August from the upper part of the crown (Zech et al. 1985).

Before drying, the plant tissue was cleaned with deionized water. Digestion of plant tissue and analysis of element concentrations were carried out as described by Kaupenjohann et al. (1985). Nutrient concentrations were related to dry weight (dw) of the foliage.

3 Nutrient Status of Trees on Forest Sites in Bavaria

3.1 Nitrogen

The average N concentration in current-year needles of *P. abies* amounted to about 14.0 mg N g_{dw}^{-1} (Table 1) which according to Bergmann (1986) is within the range (13.5–17.0 mg N g_{dw}^{-1}) sufficient for *P. abies*. However, Ingestad (1962) and Tamm (1977) have found that 17–25 mg N g_{dw}^{-1} is necessary for optimum growth of *P. abies*. Such amounts were not observed in our studies. Declining spruce in the Bavarian Alps actually suffers from a N deficiency (9.8 mg N g_{dw}^{-1}). Relatively low N concentrations (12.0–13.4 mg N g_{dw}^{-1}) were also been found at the Schneeberg site, where *P. abies* grows on Spodosols derived from granite. This area is one of the most severely damaged regions in the Fichtelgebirge, as *P. abies* shows distinct needle-tip chlorosis at this site. The N concentrations of *P. abies* needles at the declining and healthy sites near Oberwarmensteinach and Wülfersreuth did not differ and varied from between ca. 13 and 16 mg g_{dw}^{-1} in the years 1983–1986. The N nutrition of older trees was slightly better than that of younger trees at these sites as well as at the site near Selb. The N concentrations of older needles were generally less than those of current-year ones (Fig. 2).

To test possible N-induced dilution of other nutrients, we determined the N and Mg concentration of needles. On a site basis, generally positive correlations indicated better Mg nutrition as the N supply increased (Fig. 2). The growth response of needles to increased N supply was not determined, however. It is possible that Mg dilution did not occur because in contrast to what was found in 1985 at two of the sites (Oren et al. 1988; Chap. 4-H), needle growth did not respond to N. Similarly, declining deciduous trees like *F. sylvatica* and *A. platanoides* also contained less foliar N than their healthy counterparts and the N levels of 11.5 mg N g_{dw}^{-1} in chlorotic leaves of *F. sylvatica* indicated a N deficiency (Table 2).

Nitrogen deposition may directly affect Mg nutrition at sites close to emission sources (Kaupenjohann et al. 1989). Both healthy and greatly damaged *P. sylvestris* showed very high N concentrations in the needles at the Schwabach site close to a poultry farm. Nitrogen concentrations of the needles were strongly correlated with the plant-available N (N_{min}) of the soils (Tables 3 and 4, $r = 0.84$).

Table 1. Mineral concentrations of needles of *Picea abies* at the investigated sites in Bavaria (Fig. 1) (Mean values of n samples, each sample collected from 3–15 trees)

Site	Number	n	Year of sampling	Age of trees (a)	Decline symptoms	N	P	S	K	Ca	Mg	Mn	Zn	Fe	Cu
						mg · g^{-1} dm						µg · g^{-1} dm			
Healthy	1	1	1983	30–40	–	13.8	1.2	1.8	4.5	2.6	0.90	360	20	40	3
		1	1984	30–40	–	15.2	2.3	1.4	7.3	2.7	0.95	320	21	50	4
		1	1985	30–40	–	16.3	2.1	1.9	9.5	4.4	0.99	600	25	50	5
		1	1986	30–40	–	13.8	2.1	1.4	8.4	2.6	1.06	410	20	85	3
		5	1986	20–40	–	12.6 ± 0.84	1.6 ± 0.1	–	7.3 ± 0.24	3.1 ± 0.36	0.72 ± 0.03	–	–	–	–
		5	1984	100	Needle loss	13.2 ± 0.5	1.8 ± 0.3	1.5 ± 0.3	6.5 ± 1.2	2.3 ± 0.3	0.80 ± 0.26	245 ± 125	26 ± 8	62 ± 15	6 ± 1.5
Declining	2	1	1983	30–40	Yellow tip disease	14.0	1.9	1.8	6.1	1.7	0.64	730	17	50	4
		1	1984	30–40	"	14.6	2.4	1.5	9.3	2.2	0.45	1300	23	63	3
		1	1985	30–40	"	16.4	2.6	2.2	13.7	2.6	0.43	970	26	70	4
		1	1986	30–40	"	15.4	2.7	1.9	9.4	2.6	0.55	1075	24	85	5
		5	1986	20–40	"	14.7 ± 1.4	1.8 ± 0.1	–	5.7 ± 0.3	2.1 ± 0.6	0.43 ± 0.05	–	–	–	–
		5	1984	100	Needle loss	14.1 ± 1.2	1.8 ± 0.3	1.4 ± 0.2	5.0 ± 0.9	2.1 ± 0.8	0.6 ± 0.2	1020 ± 300	23 ± 9	59 ± 5	7 ± 2
		18	1984	10–12	Yellow tip disease	13.1 ± 1.9	1.8 ± 0.5	–	5.5 ± 1.5	3.2 ± 1.5	0.51 ± 0.14	–	–	–	–
Selb	3	1	1983	35–40	–	13.9	1.2	2.0	4.9	6.4	2.0	1920	27	57	2
		1	1984	35–40	–	13.5	2.2	1.5	7.1	5.9	1.2	1020	26	62	3
		5	1984	100	Severe needle loss	13.2 ± 1.0	1.5 ± 0.2	1.9 ± 0.4	4.2 ± 0.5	5.0 ± 1.1	1.22 ± 0.15	1060 ± 710	30 ± 8	84 ± 42	6 ± 0.5
Schneeberg	4	10	1984/85	10–12	Severe yellow tip disease	13.4 ± 2.2	1.7 ± 0.3	1.7 ± 0.4	5.6 ± 1.4	1.4 ± 0.3	0.26 ± 0.06	280 ± 140	29 ± 8	84 ± 31	5 ± 2
		1	1985	9	Yellow tip disease	12.0	1.6	–	5.0	2.6	0.42	380	24	44	4

Table 1. *Continued*

Site	Number	n	Year of sampling	Age of trees (a)	Decline symptoms	N	P	S	K	Ca	Mg	Mn	Zn	Fe	Cu
						mg · g_{dw}^{-1}							µg · g_{dw}^{-1}		
Rehau	5	2	1983	40–60	–	13.8	1.8	–	6.3	2.6	0.83	410	16	105	–
Schnappenhammer	6	2	1985	100	Needle loss	16.0	1.4	–	4.1	3.9	1.10	2300	30	78	–
Platte	7	1	1985	20–30	Yellow tip disease	14.8	1.9	1.6	5.9	1.6	0.39	320	19	56	3
Pressath	8	1	1984	15	–	12.1	1.1	1.2	5.3	1.4	0.50	670	38	83	5
Kloaschau	11	6	1985	120	Severe needle loss	9.8 ± 0.9	1.1 ± 0.2	0.8 ± 0.2	3.1 ± 0.7	5.1 ± 0.5	1.9 ± 0.4	170 ± 75	25 ± 14	37 ± 5	4 ± 1
Högelwald	12	1	1985	79	–	16.9	1.7	–	3.7	3.1	1.20	–	–	–	–
		1	1985	14	–	13.3	1.4	–	5.3	3.2	0.90	–	–	–	–
Grueben	13	1	1984			12.9	1.5	0.9	6.5	2.1	0.71	710	22	30	3
		1	1985			13.9	1.7	1.0	4.9	2.3	0.89	800	27	28	3
		1	1986			13.4	1.7	1.0	5.9	2.2	0.81	870	24	45	2
Hochgfeichtet	14	1	1984			12.6	1.8	1.0	6.0	2.0	0.68	290	25	41	3
		1	1985			13.8	1.8	1.0	4.9	2.2	0.62	290	29	48	3
		1	1986			13.1	1.5	0.9	6.1	2.1	0.57	270	23	71	3
Inner Bavarian Forest	16	1	1983	70–80	Needle loss	14.6	1.7	–	4.0	3.4	1.01	–	–	–	–
		1	1983	70–80	Severe needle loss	14.5	1.7	–	5.3	3.3	1.02	–	–	–	–
Average						13.9 ± 2.0	1.7 ± 0.4	1.6 ± 0.4	6.2 ± 2.1	3.2 ± 1.4	0.86 ± 0.46	726 ± 583	26 ± 8	69 ± 26	4.5 ± 2

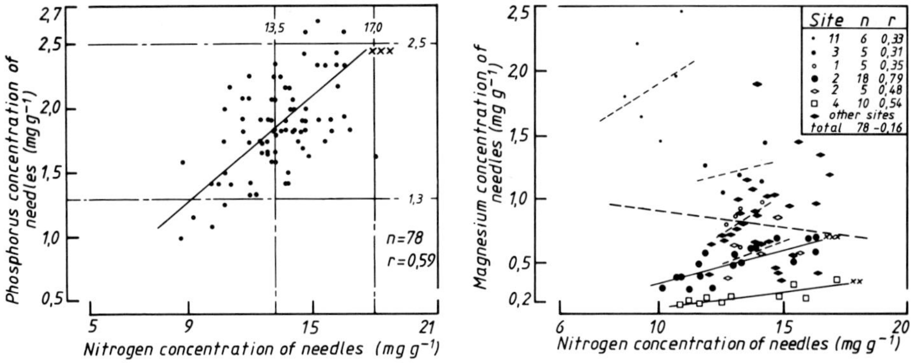

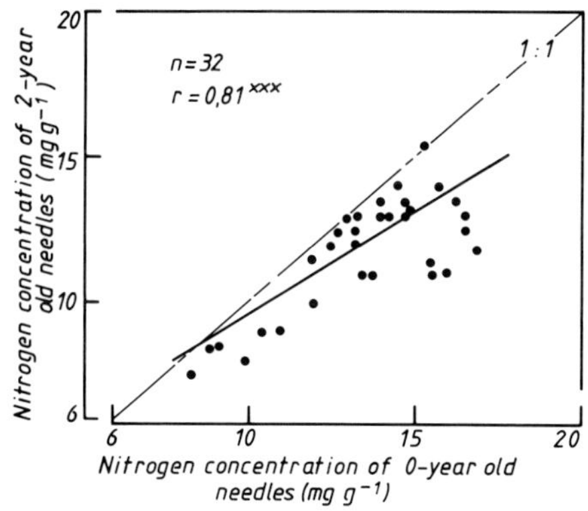

Fig. 2. Nitrogen in spruce needles: relations between concentrations in current (0-year-old needles) and 2-year-old needles and P and Mg concentrations in current-year needles

Table 2. Mineral concentrations of foliage of Silver fir [*Abies alba* (L.)], Larch [*Larix decidua* (Mill.)], Beech [*Fagus sylvatica* (L.)] and Maple [*Acer platanoides* (L.)] growing on Spodosols derived from granite in the Fichtelgebirge (Site 4, Schneeberg)

	N	P	K	Ca	Mg	Al	Fe	Mn	Cu	Zn
	------- mg $\cdot g_{dw}^{-1}$ --------					-------- µg $\cdot g_{dw}^{-1}$ ------------				
Silver Fir Declining, needles with yellow tips	15.1.	1.5	4.7	5.8	0.34	157	70	780	4	26
Larch Declining, pronounced chlorotic needle tips	26.0	1.8	9.5	2.1	0.31	238	119	547	7	16
Beech Declining, chlorosis of the intercostal space	11.5	1.1	7.1	1.7	0.38	112	127	164	6	16
Beech Healthy	22.0	1.4	7.0	4.9	0.51	133	129	940	8	30
Maple Declining, chlorosis of the intercostal space	18.7	0.8	6.0	4.0	0.37	70	145	91	18	26
Maple Healthy	20.3	1.3	8.0	4.7	0.49	70	160	177	12	23

Table 3. Average N-, P-, K-, Ca- and Mg-contents in dry matter of recent needles as well as vitality classes of *Pinus sylvestris* at the experimental plots A – D

Plot	Vitality class	Needle element content (mg g^{-1})				
		N	P	K	Ca	Mg
A	1 (Healthy)	27	1.6	6.1	3.4	0.92
B_1	2	20	1.5	4.9	3.4	0.85
B_2	1	14	1.4	5.2	3.3	0.95
C	4	17	1.3	5.4	3.2	0.91
D	6 (Greatly damaged)	24	1.4	6.7	3.5	0.66

Table 4. pH, Mg-, K- and Ca-concentrations in the soil (0–20 cm soil depth, 1:2 – soil:water extract) and plant-available N (N_{Min}, 0–80 cm soil depth) at the experimental plots A-D

Plot	pH	N_{min}	= $NH_4 - N$	+ $NO_3 - N$	Mg	K	Ca
		kg ha^{-1}			mg l^{-1}		
A	3.91	41.8	36.6	5.2	0.19	13.6	27
B_1	3.93	39.8	32.0	7.8	0.29	10.1	44
B_2	4.20	9.9	19.1	0.8	0.15	6.3	15
C	4.03	20.5	17.4	3.1	0.35	8.6	31
D	3.99	39.2	35.7	3.5	0.29	8.4	27

The generally high N contents of the soils at plots A and D could not be explained without N emissions. Phosphorus, K, and Ca concentrations of the needles did not relate to the health status of the trees. Healthy trees, however, contained sufficient amounts of Mg in the needles, while greatly damaged trees showed significantly lower Mg levels. A strong correlation was found between the Mg concentrations of the needles and the vitality of the trees ($r = 0.84$). In contrast to the findings for N, the correlation between Mg in the soil and needles was negative. Soil chemical conditions (Table 4) and a K/Mg antagonism could not account for the reduced Mg uptake of the declining trees. Low Mg concentrations in the damaged trees might have resulted from the direct effects of NH_4 which was emitted from the farm and which may have increased the leaching of the cations from the canopy of trees downwind.

3.2 Phosphorus

On a regional scale, the average P concentration in *P. abies* needles at all study sites (1.7 mg P g_{dw}^{-1}) was sufficient according to Bergmann's values of 1.3–2.5 mg P g_{dw}^{-1}. The highest values were measured at the declining site near Oberwarmensteinach and varied from between 1.9 and 2.6 mg P g_{dw}^{-1} in the years 1983–1986. The P concentrations at the other sites were lower. The P concentration of the needles at the site near Kloaschau did not even reach the value which Bergmann indicated was sufficient (Table 1, Fig. 2). The P nutrition of *A. alba* (1.5 mg P g_{dw}^{-1}) and *L. decidua* (1.8 mg P g_{dw}^{-1}) seemed adequate, while *F. sylvatica* and *A. platanoides* leaves contained only 0.8–1.4 mg P g_{dw}^{-1}, which may indicate a deficiency for these species (Table 2).

3.3 Sulfur

The S concentrations of *P. abies* needles varied from the low values of 0.8 mg S g_{dw}^{-1} at the Kloaschau site to the very high values of 2.2 mg S g_{dw}^{-1} at the declining site near Oberwarmensteinach. At the other sites in the Fichtelgebirge, the S concentrations of the needles were also high, which may reflect the emissions recorded in the Fichtelgebirge (Chap. 1-B). Trees with a high supply of S accumulated S in older needles (Fig. 3). Because protein synthesis requires a relatively constant ratio of S to N, a determination of the ratio is required if S nutrient status is to be interpreted. Kelly and Lambert (1972) found the following S to N ratio for *Pinus radiata*:

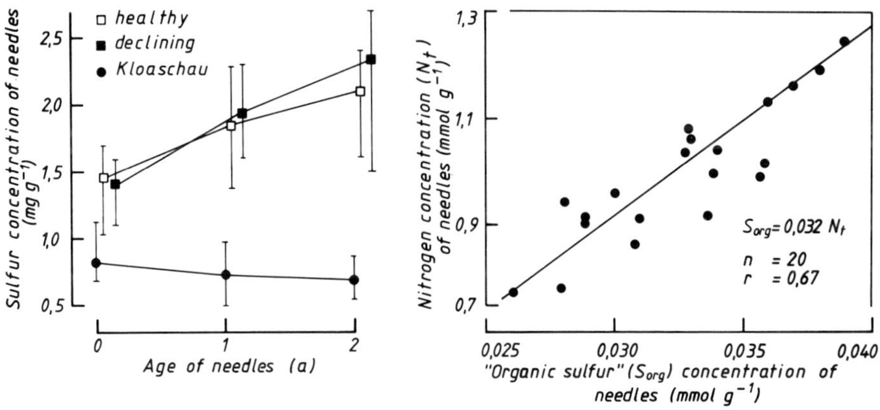

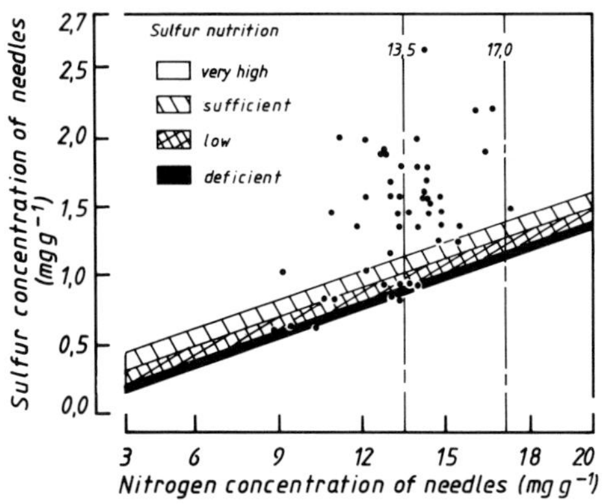

Fig. 3. Sulfur in spruce needles: concentration in needles differing in age (0-year-old needles are current-year needles) and in relation to N nutrition

$$S_{org} = 0.03 \times N_t,$$

where S_{org} is S_t-SO_4-S, and S_t is the total S concentration (mmol g_{dw}^{-1}), SO_4-S is the sulfate S concentration (mmol g_{dw}^{-1}), and N_t is the total N concentration (mmol g_{dw}^{-1}) of the needles. This ratio may also be used for *P. abies*. It seems independent of soil properties and fertilization (Clement 1985). Spruce stands in the SO_2-polluted Fichtelgebirge showed (Fig. 3):

$$S_{org} = 0.032 \times N_t.$$

According to Lambert and Turner (1977), the S nutrient status of *P. radiata* can be characterized as S deficient (S_{org} + < 80 mg S kg_{dw}^{-1}), low to sufficient S status (S_{org} + 80 to 200 mg S kg_{dw}^{-1}), sufficient to high S status (S_{org} + 200 to 400 mg S kg_{dw}^{-1}), and very high S status (S_{org} + > 400 mg S kg_{dw}^{-1}). Applying these values to *P. abies*, the S status of most stands was judged to be generally high, except for the stand in the Bavarian Alps (Fig. 3). In addition, most stands seemed to suffer from an N deficiency relative to S nutrition.

3.4 *Potassium, Calcium and Magnesium*

Great variations in the K, Ca, and Mg values of *P. abies* needles were also found at the different sites (Table 1). On such highly base-saturated soils as those at Kloaschau and Selb, needles showed high Ca and Mg concentrations but low K levels. In contrast, the K concentrations of needles on the more acid soils were often higher. However, Ca and Mg concentrations at sites with acid soils were significantly lower than those on calcareous soils, and in some cases were extremely deficient. In soils with low base saturation (< 10%), Mg and K concentrations of the needles increased proportionally, while in soils with high base saturation, K concentrations dropped while Mg values increased (Fig. 4). Thus, K deficiency may occur where Mg availability is high (due to ion antagonism), while Mg deficiency may result in soils generally poor in cations.

Values for optimum Mg nutrition vary from 1.0 mg Mg g_{dw}^{-1} – 2.5 mg Mg g_{dw}^{-1} (Reemtsma 1986; Knabe 1984; Bergmann 1986). An optimum Mg supply requires no drop in Mg concentration with needle age (Reemtsma 1986). In needles of old *P. abies* at the Schnappenhammer site, Mg concentrations decreased in older needles although the Mg supply of current-year needles seemed adequate. At the healthy site near Wülfersreuth, Mg concentrations even tended to increase slightly in 4-year-old needles. In the declining stand near Oberwarmensteinach, the Mg concentrations decreased in the current-year needles to 3-year-old needles and then increased in older needles (Fig. 4). A similar trend was found at Schneeberg, where *P. abies* showed the same decline phenomena as at Oberwarmensteinach. Low Mg concentrations were strongly correlated with needle-yellowing; the Mg values reflected the field observation that older, shaded needles close to the stem were often greener than younger ones. Only at the Pressath site, where the stand was characterized by strongly Mg deficient soils, did the Mg concentrations decline with the age range investigated.

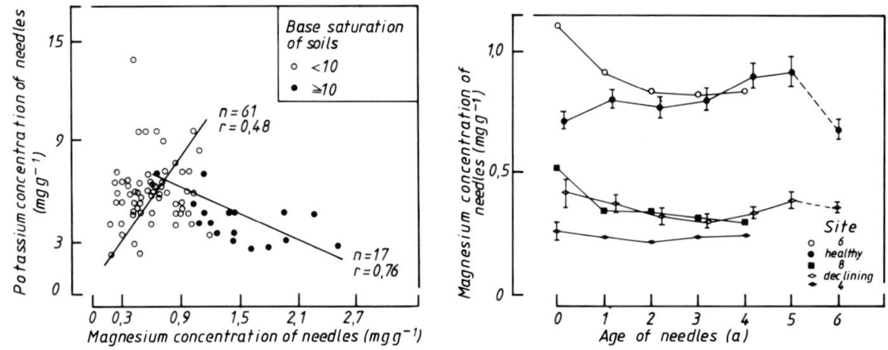

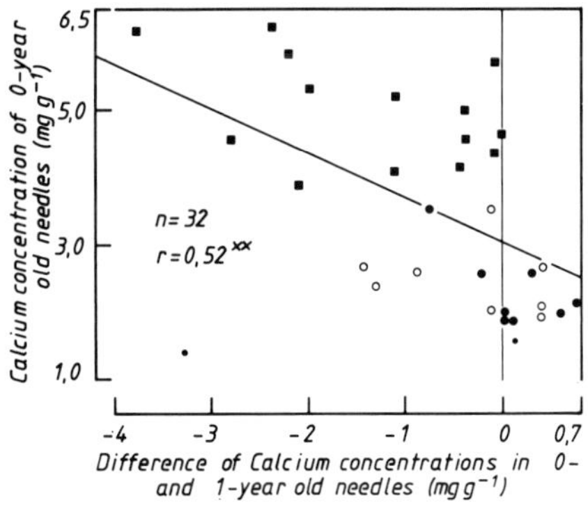

Fig. 4. Alkali and earth alkali metals in spruce needles: relations between Mg and K concentrations in current-year needles, Mg concentrations in needles differing in age, and plot of Ca concentrations in current-year needles (0-year-old needles) over Ca concentrations in current-year minus concentrations in 2-year-old needles

Trees which were 100 years old and growing on the same soils showed significantly higher Mg concentrations at the declining site than did the younger stands (0.62 mg Mg g_{dw}^{-1} in needles of older trees and 0.40 mg Mg g_{dw}^{-1} in younger ones).

The K nutrition of old *P. abies* was worse than that of 40-year-old trees at all sites. Values for old trees (Table 1) compared with those of Bergmann (1986; 5 to 10 mg K g_{dw}^{-1}) indicated insufficient K, particularly at Kloaschau where the K concentration was considered very deficient. The K nutrition of younger stands seemed generally adequate.

The Ca concentrations of *P. abies* needles generally correlated with the Mg levels. The lowest Ca values were found at Schneeberg and in the declining stand near Oberwarmensteinach. According to Bergmann (1986), these values (Table 1) are deficient as 3.5 to 8.0 mg Ca g_{dw}^{-1} is required for optimum growth. Trees on soils with high base saturation were sufficiently supplied with Ca.

In plants Ca is not as mobile as Mg, and therefore should generally accumulate in older plant tissue. However, on acid soils with high emissions, even lower Ca concentrations have been measured in older needles than in current-year ones (Fig. 4). In contrast, at the Pressath site the Ca concentration increased with needle age, although current-year needles had lower concentrations compared with those from other stands.

The other tree species investigated here were very poor in foliar Mg and Ca (Table 2). Declining trees were particularly Mg-deficient.

3.5 Trace Element Nutrition

No correlations between foliar trace-nutrient concentrations and site properties were found. Manganese and Fe concentrations in needles and leaves varied from between 170 and 2300 mg Mn Kg_{dw}^{-1} and 28–105 mg Fe kg_{dw}^{-1} (Table 1). The study trees did not suffer from a lack of either element. It is known, however, that conifers on calcareous soils often develop yellowing of older needles (Zech 1968).

At several sites, the Zn nutrition of spruce was rather poor, especially at the Kloaschau site where severely damaged trees contained only 15 mg Zn kg_{dw}^{-1}. In needles of visibly more healthy trees at the same site, an average of 33 mg Zn kg_{dw}^{-1} was measured. The Zn concentrations of trees at the other sites ranged from between 16 and 38 mg Zn kg_{dw}^{-1} (Table 1).

Copper nutrition of *P. abies* was low at several sites. According to Bergmann (1986), the mean values of about 3.3 mg Cu kg_{dw}^{-1} at the declining and healthy sites indicate insufficient Cu nutrition. Bergmann (1986) has suggested that Cu levels should range from between 9 and 10 mg Cu kg_{dw}^{-1} for optimum growth.

4 Discussion

Results of the regional survey of tree nutrition showed that a strong correlation existed between cation supply and forest decline. The data generally support the results of earlier studies (Zech and Popp 1983; Bosch et al. 1983; Zöttl and Mies 1983; Hüttl 1985; Zech et al. 1985) but on a larger scale.

Nitrogen nutrition of the stands was generally sufficient except for the stand in the Bavarian Alps. Relative to tree growth demands, S and P concentrations were higher than that of N, especially in the declining stands in the Fichtelgebirge which led to an excess of SO_4-S in the needles (Fig. 3). From this observation, a relative N deficiency may be deduced; trees should have been able to assimilate more S into amino acids with higher N levels (Heber et al. 1987). Nitrogen fertilization in combination with the nutrients Mg, K, P and S, however, did not increase needle weight in the declining stand (Chap. 4-G). This might be because either another factor is limiting growth, or because NO_3 uptake by trees is restricted. The latter is indicated by results from a fertilization trial which showed that about 100 kg NO_3-N ha^{-1} were exported from the soil 2 years after fertilization, while only 37 kg N was leached during the same period in the control stand (Hantschel 1987). The difference between the two amounts was nearly 100% of the applied N fertilizer (65 kg NO_3-N ha^{-1}).

Nitrogen and Mg concentrations in needles correlated on a dry weight-basis. Thus, no indications of a Mg dilution by N-induced growth was found, but the relations between needle N and needle growth were not investigated. Internal translocation of nutrients, which must be considered to interpret the contribution of external N supply to tree nutrient relations, however, can not be calculated from these data. This aspect of N nutrition is discussed in Chaps. 4-F and 4-H. A reduction of Mg concentrations in needles may also relate to the direct effects of NH_4 on the canopy in areas where the NH_4 concentration is high (Roelofs et al. 1985).

High P concentrations in the declining *P. abies* stands in the Fichtelgebirge seem to be a recent phenomenon. Our results and data from Zech (1968) and Zöttl and Hüttl (1986) show that P nutrition has improved over the last decades. Zöttl and Hüttl (1986) have concluded that this increase in P may result from the increasing mineralization rates of the forest floor from N deposition. In contrast, Bienkowski et al. (1986) have shown that acid deposition reduces mineralization of organic material. Recently Kaupenjohann et al. (1988) have found that *P. abies* seedlings stressed by H_2SO_4 increased P uptake from nutrient solutions. Because acid stress on the tree canopies might increase the P demand for such metabolic processes as energy transfer during ion uptake, a higher P uptake might be required.

High S concentrations in needles of *P. abies* in Northeast Bavaria reflect the high SO_2 pollution in this region. In the most highly polluted area close to the Czechoslovakian border, S concentrations may reach toxic levels.

Magnesium deficiency is a widespread decline phenomenon affecting mainly younger trees on acid soils at high elevation sites in Bavaria. Magnesium deficiency symptoms exist at about 0.35 mg Mg g_{dw}^{-1} (Zech 1968); however, slight needle-tip yellowing occurs even at needle values of about 0.4 mg Mg g_{dw}^{-1}. In damaged trees, Mg concentrations were lower in older needles than in current-year needles (Fig. 4). However, at the high elevation sites of the Fichtelgebirge (Schneeberg and Oberwarmensteinach) with large acid inputs (Chap. 1-B), the Mg levels tended to increase in the oldest needles. A classical soil-induced Mg deficiency should have resulted in a permanent reduction of Mg concentrations with needle age due to translocation and impaired recharge. This could be seen at the Pressath site (Fig. 4). These findings may indicate that the recently observed Mg deficiency at Schneeberg and Oberwarmensteinach is a new phenomenon.

Calcium concentrations, which were generally low in declining *P. abies*, tended to be lower in 2-year-old needles than in the current-year needles of declining trees, although Ca is considered to be quite immobile in plants (Mengel and Kirkby 1987), and therefore should accumulate in older tissue even on poor soils. This was in fact observed at the Pressath site, which receives low acidic deposition (Fig. 4).

The Mg and Ca concentration distribution in the trees indicate that leaching of these nutrients may be involved in forest decline. Increasing rates of cation leaching out of tree canopies as a result of acid deposition have been shown (Chap. 3-D), which in the case of very acid fog may decrease the Mg foliar concentrations (Kaupenjohann et al. 1988). Leaching rates from older needles may be reduced due to the protection of the younger twigs. Indeed, it has been reported that shaded parts of spruce often look green, while the exposed needles exhibit severe yellowing of needle tips (Zöttl and Mies 1983).

Older *P. abies* at low elevation sites more frequently exhibit lower K than Mg concentrations. In general, soil chemical properties correlate with Mg and K nutrition of trees which supports Hüttl's (1985) findings for forest sites in SW-Germany.

Micronutrient analysis indicated a low Zn supply in some stands. Zinc has been reported as easily leached from the canopy (Mengel et al. 1987). Analysis of beech tree rings from sites near Schneeberg in the Fichtelgebirge has shown increasing Zn concentrations in the wood over the last three decades, but Zn concentrations in the beech leaves have been very low (Schneider and Zech 1989). This result might be explained by a leaching-induced decrease in the Zn concentration in the foliage, while the concentration in the wood reflects only the uptake from the soil.

Copper concentrations of the analyzed forest stands were also close to the deficiency levels indicated by Bergmann (1986). However, Cu application had no effect on tree growth and even lower values than our data have been found in healthy trees in other regions. In a stand of Douglas fir [*Pseudotsuga menziesii* (Mirb. Franco)], for example, a positive reaction to Cu fertilization was not found when foliar Cu concentrations were above 1.3 mg Cu kg_{dw}^{-1} (Oldenkamp and Smilde 1966). Therefore, Cu nutrition is not considered to be an important factor in the forest decline processes in the Fichtelgebirge.

5 Conclusions

The results indicate that the forest decline phenomena in southern Germany are closely associated with impaired tree mineral nutrition. The most important features seem to be a pronounced Mg deficiency on acid soils, low K levels on sites with high base saturation, and usually sufficient N levels and high supplies of P and S.

The recent nutrient status of trees was compared with those values from earlier studies; the results indicate that Mg, Ca, K, and Zn nutrition has become markedly insufficient (Reemtsma 1986; Hüttl and Zöttl 1985) while N, P and S concentrations in foliage have often increased over the last 3 decades.

Sulfuric acid and N deposition are probably the main reasons for the changes in the nutritional balance. These inputs have probably caused leaching of Mg, K, Ca, and Zn from the tree canopy as well as from the soil (Chap. 4-B) and may have diluted tree nutrient status with increased tree growth (Chap. 4-H).

References

Bergmann W (1986) Farbatlas Ernährungsstörungen bei Kulturpflanzen. Fischer, Jena

Bienkowski P, Fischer S, Gozdziewicz J (1986) Reaction of soil system to acid rain under conditions of a laboratory experiment. Ekol pol 34:75–86

Bosch C, Pfannkuch E, Baum U, Rehfuess KE (1983) Über die Erkrankung der Fichte (*Picea abies* Karst.) in den Hochlagen des Bayerischen Waldes. Forstwiss Centralbl 102:167–181

Clement A (1985) N, S, P status and protein synthesis in the foilage of Norway spruce (*Picea abies* (L.) Karst.) and Austrian Slade pine (*Pinus nigra* Arnold var. nigra). Plant Soil 85:345–359

Friedlanda AJ, Gregory RA, Kärencamp C, Johnson AH (1984) Winter damage to foliage as a factor in red spruce decline. Can J For Res 14:963–965

Hantschel R (1987) Wasser- und Elementbilanz von geschädigten, gedüngten Fichtenökosystemen im Fichtelgebirge unter Berücksichtigung von physikalischer und chemischer Bodenheterogenität. Bayreuther Bodenkd Ber 3:1–219

Heber U, Laisk A, Pfanz H, Lange O (1987) Wann ist SO_2 Nährstoff und wann Schadstoff? Ein Beitrag zum Waldschadensproblem. Allg Fortz 42:700–705

Höhne H (1964) Untersuchungen über die jahreszeitlichen Veränderungen des Gewichtes und Elementgehaltes von Fichtennadeln in jüngeren Beständen des Osterzgebirges. Arch Forstwes 13:747–774

Hüttl RF (1985) "Neuartige" Waldschäden und Nährelementversorgung von Fichtenbeständen (*Picea abies* Karst.) in Südwestdeutschland. Freiburger Bodenkd Abh, Freiburg, 16:1–195

Hüttl RF, Zöttl HW (1985) Ernährungszustand von Tannenbeständen in Süddeutschland- ein historischer Vergleich. Allg Forstz 40:1011–1013

Ingestad T (1962) Macroelement nutrition of pine, spruce and birch seedlings in nutrient solutions. Medd Statens Skogsforskningsinst 51:1–150

Kaupenjohann M, Hantschel R, Zech W, Horn R (1985) Nährstoffversorgung gedüngter, unterschiedlich geschädigter Fichten auf immissionsbelasteten Standorten in NO-Bayern. Mitt Dtsch Bodenkd Ges 43/II:969–979

Kaupenjohann M, Schneider BU, Hantschel R, Zech W, Horn R (1988) Sulfuric acid rain treatment of *Picea abies* (Karst. L): Effects on nutrient solution, throughfall chemistry, and tree nutrition. Z Pflanzenernähr Bodenkd 151:123–126

Kaupenjohann M, Döhler H, Bauer M (1989) Effects of N-immissions on nutrient status and vitality of *Pinus sylvestris* near a henhouse. Plant Soil 113:279–282

Kelly J, Lambert M (1972) The relationship between sulphur and nitrogen in the foilage of *Pinus radiata*. Plant Soil 37:395–407

Klein B (1981) Trockenresistenz und Immissionshärte der Fichte. Dissertation, TU München

Knabe W (1984) Merkblatt zur Entnahme von Blatt- und Nadelproben für chemische Analysen. Allg Forstz 39:847–848

Küppers M, Zech W, Schulze ED, Beck E (1985) CO_2-Assimilation, Transpiration und Wachstum von *Pinus sylvestris* L. bei unterschiedlicher Magnesiumversorgung. Forstwiss Centralbl 104:23–36

Lambert MJ, Turner J (1977) Dieback in high site quality *pinus radiata* stands – the role of sulphur and boron deficiencies. N Z J For Sci 7:333–348

Lambert MJ, Hons BS, Hons MS (1983) The use of foliar analysis in fertilizer research. In: Proc Symp on site and productivity of fast growing plantations, May 1984, Pretoria, I:269–291

Mengel K, Kirkby EA (1987) Principles of plant nutrition. Int Potash Inst Bern, 465 pp

Mengel K, Lutz HJ, Breininger MT (1987) Auswaschung von Nährstoffen durch sauren Nebel aus jungen intakten Fichten (*Picea abies*). Z Pflanzenernähr Bodenkd 150:61–68

Nebe W, Fiedler HJ, Ilgen G, Hofmann W (1987) Nutritional disturbances of spruce (*Picea abies* (L.) Karst.) due to immissions in low mountain areas. Flora 179:453–469

Oldenkamp L, Smilde KW (1966) Copper Deficiency in Douglas Fir (*Pseudotsuga menziesii)* (Mirb. Franco). Plant Soil 25:150–152

Oren R, Schulze ED, Werk KS, Meyer J (1988) Performance of two *Picea abies* (L.) Karst. stands at different stages of decline. VII. Nutrient relations and growth. Oecologia 77:163–173

Reemtsma JB (1986) Der Magnesium-Gehalt von Nadeln niedersächsischer Fichtenbestände und seine Beurteilung. Niedersächsische Forstliche Versuchsanst Göttingen, 1–16

Rehfuess KE, Bosch C (1986) Experimentelle Untersuchungen zur Erkrankung der Fichte [*Picea abies* (L.) Karst.] auf sauren Böden der Hochlagen: Arbeitshypothese und Versuchsplan. Forstwiss Centralbl 105:201–206

Rehfuess KE, Rodenkirchen H (1984) Über die Nadelröte-Erkrankung der Fichte [*Picea abies* (L.) Karst.] in Süddeutschland. Forstwiss Centralbl 103:248–262

Roelofs JGM, Kempers AJ, Houdijk ALFM, Jansen J, (1985) The effect of air-borne ammonium sulphate on *Pinus nigra* var. *maritima* in the Netherlands. Plant Soil 84:45–56

Schneider BU, Zech W (1989) Zuwachs- und Elementgehaltsentwicklung in zwei Buchenbeständen im Fichtelgebirge. (in preparation)

Tamm CO (1977) Skogsekesystements reaktion pa växtnäringsstill för sel. Produktionseffekter od miljöförändringar. Skogs Cantbr Akad Tidskr Suppl 11:7–15

Ulrich B (1981) Eine ökosystemare Hypothese über die Ursachen des Tannensterbens (*Abies alba* Mill.). Forstwiss Centralbl 100:228–236

Van Praag HJ, Weissen F (1986) Foliar mineral composition, fertilization and dieback of Norway spruce in the Belgian Ardennes. Tree Physiol 1:169–176

Zech W (1968) Kalkhaltige Böden als Nährsubstrat für Koniferen. Dissertation, TU München, 1–165

Zech W, Popp E (1983) Magnesiummangel, einer der Gründe für das Fichten- und Tannensterben in NO-Bayern. Forstwiss Centralbl 102:50–55

Zech W, Suttner T, Kotschenreuther R (1983) Mineralstoffversorgung vermutlich immissionsgeschädigter Bäume in NO-Bayern. Kali-Briefe, Büntehof 16:565–571

Zech W, Suttner T, Popp E (1985) Elemental analyses and physiological responses to forest trees in SO_2-polluted areas of NE-Bavaria. Water Air Soil Pollut 25:175–183

Zöttl HW, Hüttl RF (1986) Nutrient supply and forest decline in Southwest-Germany. Water Air Soil Pollut 31:449–462

Zöttl HW, Mies E (1983) Die Fichtenerkrankung in den Hochlagen des Südschwarzwaldes. Allg Forst Jagdz 154:110–114

4-B Effects of Acid Rain on Soil Chemistry and Nutrient Availability in the Soil

M. KAUPENJOHANN (Coordinator)
with contributions by L. NÄTSCHER, U. SCHWERTMANN (Sect. 3.1);
P. SÜSSER, U.SCHWERTMANN (Sect. 3.2);
R. HORN, H. TAUBNER, and R. HANTSCHEL (Sect. 4.1);
M. KAUPENJOHANN, R. HORN, R. HANTSCHEL, W. ZECH, and
B.U. SCHNEIDER (Sect. 4.2)

1 Introduction

Nutrient availability in soils is strongly related to soil pH. Plant uptake of Ca and Mg increases with pH while the availability of most trace nutrients increases with decreasing soil pH (Finck 1976). On the one hand, pH determines the solubility of many nutrients, on the other hand pH of the soil solution indicates the concentration of alkali and earth alkali ions.

The pH of the soil solution results from total proton load, which is the sum of acid input plus acid production of the soil and proton buffering processes. Seasonal soil pH fluctuations may be attributed to such various processes as the concentration of the soil solution, the production of nitric acid, or acid deposition (Matzner and Ulrich 1984; Kaupenjohann and Hantschel 1987).

Total proton load of the soil results from either natural processes or anthropogenic activities (Table 1). Conjugate bases corresponding to proton load of the soil may be subdivided into conservative and nonconservative anions (Schnoor and Stumm 1985; Ulrich 1986). Nonconservative anions include organic or carbonate anions, which may associate with protons in the ecologically relevant pH range of soils and may also be transferred into an anhydride and water. In contrast, conservative anions like SO_4^{2-} or NO_3^- remain anions in soils in the same pH range.

According to the mobile anion concept of Reuss (1980), the leaching of cations out of the soil profile requires an equivalent anion transport. Therefore, conservative anions are considered to be potentially the most effective cause of cation leaching, especially from acid soils, into the aquifer. Protons from atmospheric input are mainly accompanied by conservative anions (Chap. 4-I). If these anions are not retained in the ecosystem and are leached out of the soil profile, they induce an equivalent loss of cations. In acid soils, the contribution of acidic deposition to the total proton load may amount to as much as 40%–90% (Matzner and Ulrich 1984; De Vries and Breeuwsma 1985; Bredemeier 1987).

The proton can induce soil responses in three ways: (1) protons may remain in the soil solution and appear in the seepage; (2) protons may be stored in the soil matrix; (3) they may be transferred to the conjugate base of a weaker acid and be either stored or leached. Cases 2 and 3 may be defined as proton buffering, for which several soil reactions are known (Table 2).

E.-D. Schulze, O.L. Lange, R. Oren (Eds.)
Ecological Studies Vol. 77

Table 1. Cause and significance of acid load of soils

Cause	Reactions	Main pH range	Amount (mol H^+ m^{-2} a^{-1})
	Natural sources		
1. Dissociation of carbonic acid	$H_2CO_3 = HCO_3^- + H^+$	6.5–8	10–20
2. Dissociation of organic acids	$R - COOH = R - COO^- + H^+$, $R - NH_3^+ = R - NH_2 + H^+$, etc.	3–6, (1.2, 4.3[a]), 3.8–4.2[b]	
3. Accumulation of cation surplus in the organic matter (decoupling of cation uptake and cation mineralisation)	$R - COOH + M^+ = R - COOM + H^+$		0.03–0.13[c]
4. De-coupling of N-mineralisation and N-uptake (leaching of NO_3^- from mineralisation)	$R - NH_2 + 2O_2 = R - OH + NO_3^- + H^+$		0.3[c]
5. Uptake of NH_4^+	$R - OH + NH_4^+ = R - NH_2 + H_2O + H^+$		0.1[c]
6. Natural emission of SO_2 and NO_x			0.01[c]
7. Oxidation of sulfides	$S + 3/2\ O_2 + H_2O = SO_4^{--} + 2H^+$		
	Anthropogenic Sources		
8. Acidic deposition	H_2SO_4, HNO_3, etc.		–0.8[c]
9. Removal of organic substances			0.01–0.1[c]

[a] pK_{a1}-, pK_{a2}-values of oxalic acid (Weast 1978).
[b] pK_a-values of fulvic acids (Lee 1985; Cronan and Alken 1985).
[c] (Ulrich 1986).

Table 2. pH-Buffer compounds, -reactions, pH ranges of buffering, buffer rate, chemical alterations in soil/solution (Schwertmann et al. 1987) and buffer ranges according to Ulrich (1981)

Buffer compounds[a]	Reaction[a] (examples)	pH range[a]	Buffer intensity[b] ($molH^+/m^{-2}\ a^{-1}$)	Chemical alterations[b] in soil/soil solution	Buffer range[c]
1. Carbonate	$CaCO_3 + 2H^+ = Ca^{2+} + CO_2 + H_2O$	8–4.5	Fast ($>$ 0.2)	Loss of carbonates, Ca, Mg dominates the cation composition of the soil solution	Carbonic acid/carbonate (ph 8.6–6.2)
2. Variable change	$Clay - OH\ M + H^+ = Clay - OH_2 + M^+$ $R - (COO)M + H^+ = RCOOH + M^+$	8–5 6–3	Very fast	Loss of exchangeable cations, decrease of effective cation exchange capacity	Carbonic acid/silicate (pH 6.2–5.0)
3. Silicates	$-(SiO)M + H^+ = -(SiOH) + M^+$ $-AlO_2M + 4H^+ = Al^{3+} + M^+ + 2H_2O$	$<$ 7 $<$ 4.5	Slow (0.02–0.2)	Silicate and clay mineral destruction, decrease of total cation exchange capacity increase of Al-saturation	Exchange buffer range (pH 5.0–4.2)
4. Oxy-/Hydroxy-compounds	$Al(OH)_3 + 3H^+ = Al^{3+} + 3H_2O$	4.8–3	Fast ($>$ 0.2)	increase of Al-concentration in the soil/solution	Aluminium buffer range (pH 4.2–2.8)
	$AlOHSO_4 + H^+ = Al^{3+} + SO_4^{2-} + H_2O$	4.5–3	"	Increase of Al- and SO_4^{2-}-concentrations in the soil/solution	
	$FeOOH + 3H^+ = Fe^{3+} + 2H_2O$	$<$ 3	" ($>$ 0.2)	increase of Fe-saturation	Iron buffer range

[a] Schwertmann et al. (1987).
[b] Blümel (1986).
[c] Ulrich (1981).

Generally, proton buffering by the soil leads to soil acidification, which is defined as a decrease in acid neutralizing capacity (ANC; Van Breemen et al. 1983). How much soil pH decreases with a given proton load depends on the amount of buffering compounds (capacity) and buffering kinetics (intensity). Capacity factors are determined by the quantity of the system, while intensity factors are controlled by the chemical properties of the soil.

The mechanisms and kinetics of proton buffering by the healthy and the declining stands under study in this volume are described in Section 3 of this chapter. The kinetics of proton buffering have been mainly studied through homogenized soil samples. In addition, there is a considerable effect of soil structure on the pH buffering of soils which was investigated through naturally structured soil samples (Sect. 4). Based on all available information, a soil chemistry model was developed to determine to what extent acid deposition causes chemical alterations in the soils at the study sites (Sect. 5). The model integrates the most important buffering mechanisms described in this chapter and predicts the earth alkali ion saturation as a function of acid inputs.

2 Chemical and Physical Properties of the Study Soils

In this chapter, major chemical and physical properties of the soils at the healthy and declining sites are summarized. In addition, data from a base-rich soil derived from basalt are compared with the data from the study soils, which are more acid. At the base-rich site east of the Fichtelgebirge near Selb, 40-year-old spruce stands are healthy, while 100-year-old spruce trees show severe needle loss. According to Soil Taxonomy (1975), the profiles at the healthy and declining sites are classified as Entic Haplorthod and Spodic Fragiumbrept derived from phyllite. The soil derived from basalt is a Dystric Eutrochrept.

The medium-textured, gravelly, and well-structured soils at both the healthy and declining sites are very porous (Table 3). The bulk density values range from between 0.62 and 1.49 g cm^{-3} and increase with depth. Apart from the A-horizon's bulk density values of 0.62 g cm^{-3} at the healthy site and 1.2 g cm^{-3} at the declining site, all the other data range in the same order of magnitude for the various horizons at both sites. Consequently, the two soils are well aerated and have about 10–20% plant-available water at field capacity. The macroscopic continuity, particularly at the declining site, is very high throughout the profile. Thus, hydraulic conductivity at the declining site remains high. At the healthy site, on the other hand, a sharp decrease from more than 140 cm d^{-1} to less than 40 cm d^{-1} has been found at a depth of 85 cm. The basaltic soil near Selb is also fine-textured and well-structured. The total pore volume ranges between 35 and 45%. The air capacity is smaller, but the amount of plant-available water is similar to that at the two study sites. Due to higher bulk density, the water infiltration rate is reduced compared to the study sites (for more details see Hantschel 1987).

The soils differed very little in total C, N, and P contents throughout the profiles (Table 3). However, differences between the soils were distinct for total K, Ca, and Mg stores (see also Chap. 4-I). Total K concentrations were highest at

Table 3. Physical and chemical properties of mineral soils at the declining and healthy sites as well as near Selb (Hantschel 1987; Süsser 1987; Kaupenjohann 1989)

Site	Horizon	Depth (cm)	Texture	Structure	Bulk density (g cm^{-3})	Total element contents C (g kg^{-1})	N	K	Ca (eq kg^{-1})	Mg	pH ($CaCl_2$)	CEC (mmol (+) kg^{-1})	Alkali plus earth-alkali (AES) saturation (%)
Declining	Ahe	0– 10	slU	Fine subang[a]. bl.[b]	1.20	17.0	0.91	0.927	0.011	0.155	3.17	86	4.3
	Bsv1	– 30	lU	Medium subang. bl.	1.15	15.0	0.79	0.965	0.010	0.177	3.79	66	6.2
	Bsv2	– 55	slU	Medium subang. bl.	1.44	6.0	0.56	0.529	0.018	0.186	4.23	43	11.2
	Bv	– 80	slU	Medium subang. bl.	1.49	2.0	0.94	1.687	0.032	0.177	4.36	43	13.6
	BvCv	–100	slU	Medium subang. bl.	1.47	3.0	0.61	0.876	0.046	0.098	4.53	41	20.0
Healthy	Ah	– 10	l'U	Fine subang. bl.	0.62	104.0	3.00	0.517	0.054	0.194	2.76	138	9.6
	Bv1	– 30	l'U	Thin platy, fine subang.	1.37	7.0	0.60	0.767	0.039	0.243	3.70	68	6.3
	Bv2	– 60	l'U	Medium subang. bl.	1.37	4.0	0.55	1.423	0.047	0.286	4.11	53	12.3
	BvCv	–100	slU	Corse subang. bl.	1.43	2.0	0.43	1.215	0.019	0.393	4.27	58	18.6
Selb	Ah	– 30	lT	Medium granular bl.	1.05	18.8	1.45	0.410	0.321	0.521	3.82	198	86.3
	Bv1	– 55	lT	Medium subang. bl.	1.54	5.0	0.40	0.168	0.666	0.719	4.57	263	90.2
	Bv2	– 75	lT	Medium subang. bl.	1.59	4.2	0.35	0.107	0.385	0.719	4.81	289	92.0
	Bvg	– 95	lT	Medium subang. bl.	1.60	3.0	0.20	0.046	0.470	0.792	4.62	253	90.0
	Bg	–100	lT	Medium subang. bl.	1.61	6.0	0.50	0.019	0.531	0.813	4.59	280	93.0

[a] Subangular.
[b] Blocky.

the declining site and rather low at Selb, while the inverse was true for Ca and Mg. The pH values measured in 0.01 n $CaCl_2$ solution were generally low, even near Selb, although alkali plus earth alkali ion saturation (AES) was higher than 90% at the Selb site (Table 3). This contrast may be attributed to the high proportion of variable charge (Süsser 1987). However, base saturation of the soils at the declining and healthy sites was very low. In particular, the Mg concentration was low in the soil of the declining site (Table 4). Soil water extracts contain the most easily available soil nutrient fractions. Data of ion concentrations in 1/2-soil/water-extracts are shown in Table 4. The pH-values were generally higher compared to those in $CaCl_2$ extracts (Table 3), which indicates the presence of exchangeable acidity. Moreover, Mg and Ca concentrations in the soil solution at the declining site were lower and the K availability was higher than in the soil at the healthy site. Aluminum concentrations were low at Selb but higher at both the declining and healthy sites, while the Ca to Al ratio was higher at the healthy site compared to the declining site.

3 Mechanisms and Kinetics of Buffering in Acid Soils

3.1 Organic Horizons

The atmospheric input of strong mineral acids has led to an accelerated consumption of acid neutralization capacity (ANC) of soils in industrialized areas and has turned ANC into equivalent amounts of base neutralizing capacity (BNC; van Breemen et al. 1983). Informations on the nature and magnitude of ANC and BNC in various soils and informations on the rates of H^+ and OH^- consumption of the soils (buffering) are essential if the rates of further acidification and "base" cation and Al release are to be predicted. To quantify these processes, titrations at constant pH (pH-stat titrations) and percolation techniques have been widely used (e.g., Stahlberg 1984; Cronan 1985; James and Riha 1986; Brown 1987).

This study analyzed seven soils in order to expand our current understanding of the size and nature of ANC. Two soils were at the healthy and declining sites, two at the site near Selb, and one at the site at Schneeberg (an acid soil derived from granite). Because the ANC in the organic C-rich surface horizons was expected to be essentially different from that in the mineral horizons underneath and information on the ANC in organic horizons is sparse (James and Riha 1986), results are described in two parts with Section 3.2 reporting on the mineral horizons. Proton reactions with the material of the L, O, and some Ah horizons were studied by three different methods:

1. pH stat titration, in which 1.25 g of an air-dried sample was titrated in 10 ml distilled water with 0.005 or 0.020 M HCl in an automatic titrator (Radiometer TTT 80) at 20° C set to pH 3.00. The amount of acid was recorded as a function of time.

Table 4. Exchangeable cations [mmol (+) kg^{-1}] and ion concentrations in water extracts ($\mu mol\ l^{-1}$) of mineral soil material at the declining and healthy sites and at the Selb site (Kaupenjohann 1989)

Site	Horizon	K		Ca		Mg		Al		pH
		ex mmol (+) kg^{-1}	wa $\mu mol\ l^{-1}$	ex	wa	ex	wa	ex	wa	wa
Declining	Ahe	1.1	92	1.2	12	0.26	10	63	59	4.09
	Bsv1	1.1	150	1.1	15	0.10	8	60	82	4.35
	Bsv2	1.0	90	1.0	19	0.10	9	35	85	4.52
	Bv	1.0	92	1.0	14	0.10	7	35	126	4.42
	BvCv	0.8	212	0.9	14	0.02	10	31	67	4.87
Healthy	Ah	1.5	100	11.0	344	0.76	75	126	56	3.58
	Bv1	0.7	95	2.8	297	0.25	58	60	78	4.19
	Bv2	0.8	120	2.5	226	0.10	43	46	70	4.40
	BvCv	0.8	95	1.2	109	0.04	37	45	70	4.87
Selb	Ah	3.8	74	160	279	2.9	76	15	30	4.75
	Bv1	2.1	26	202	242	50.3	140	n.d.[a]	11	6.33
	Bv2	2.0	61	163	352	110.0	226	n.d.	n.d.	7.22
	Bvg	2.9	29	190	302	55.0	185	n.d.	n.d.	6.29
	Bg	3.1	35	205	257	70.3	165	n.d.	n.d.	6.15

[a] n.d. = not detectable.

2. Batch titration, in which 20 g of an air-dried sample was kept in 1 liter of water for a few days, then brought to pH 3.0 with 3 M H_2SO_4 and stored in a dark room at 24°C. One ml toluene was added at the beginning and twice again thereafter. Solutions were collected and filtered through a 50 nm membrane filter.
3. Acid percolation in which a glass tube (28 cm long, 2.4 cm wide) was filled to a height of 12 cm with field-moist soil (between 10 and 40 g) and gently compressed.

The percolation took place under aerobic conditions. Soluble salts were removed by prepercolation with water for 4 days followed by a 90 ml/d (2–3x pore volume) percolation with 10^{-3} M H_2SO_4 for 5 weeks at 24°C. This amount of acid corresponded to a 35-fold annual acid input (= 2 kmol ha^{-1}). Magnesium, Ca, Al, Fe_{total} and Mn were determined in the various reaction solutions by atomic absorption spectrometry (AAS), Fe^{2+} colorimetrically after Childs (1981) and NH_4 with an ion sensitive electrode.

For the potentiometric titration of one sample (healthy site, 1), organic matter was purified by two water extractions followed by sedimentation. This removed most of the inorganic constituents and concentrated organic C (C_{org}) from 27.2 to 40.2%. The purified organic material was H-saturated with 2M HCl and leached with water until the Cl concentration was < 0.05 mmol l^{-1}. A separate lot was Al-saturated in the same way with 0.11 M $AlCl_3$ at pH 3.0. A 1-g sample was titrated at 20°C in 10 ml 0.1 M KCl with HCl and KOH, respectively, by pH-stat titration in steps of 0.3 to 1.0 pH units until the titrant had been consumed for each step. Proton consumption or consumed H^+ (H_{cons}) was calculated from these titrations with a blank for the acid titration.

The equilibrium soil solution (ESS) was produced by equilibrating air-dried soil with distilled water at 4°C for 8 days. The ESS was then separated by centrifugation at 2200 g, filtered through a 50-nm membrane filter, and frozen until analyzed. Organic C was determined by dichromate oxidation and total acidity by automatically titrating 1 ml with 5×10^{-3} M NaOH.

3.1.1 Chemical Properties and Actual Acidification Status of the Organic Layers

The organic nature of the L and O horizons was documented by a range of 21–44% C_{org} (Table 5). Because of its strong acid character (pH < 4), the effective cation exchange capacity (CEC_{eff}, i.e., the CEC at field pH) of these horizons is low (93–217 mmole(+) kg^{-1}). The pH and the percent alkali plus earth alkali saturation (AES) are considerably higher in the L than in the O horizons, although the acid input comes into contact with the L horizon first. Both the higher pH and AES indicate a distinct litter input of cations which originates from root uptake from the lower soil horizons.

The pH is lower and the CEC_{eff} is higher in the O horizons than in the L horizons. This may result from organic acid production during litter decomposition and humification and from the formation of functional groups with lower pK values in the O horizon. The more acid nature of the functional groups may also have led to Al-release from clay minerals as indicated by a much higher

Table 5. Chemical data of the organic surface horizons

Profile No.[a]	Parent material	Horizon	Thick-ness cm	pH ESS[b]	pH[c] soil	Ct -%-	CEC[eff] mmol (+) kg^{-1}	AES[d] %	Al[d] %
OW 1	Phyllite	O	4	3.87	3.20	40.2	93	76	13
OW 1		Ah	4	3.57	3.02	10.3	73	27	47
OW 4		L	4	3.64	3.00	40.6	116	67	20
OW 4		Oh	6	2.95	2.65	30.8	157	25	52
WÜ 1	Phyllite	L	4	4.03	3.30	44.2	156	90	4
WÜ 1		Oh	3	3.40	2.95	27.2	190	56	35
WÜ 3		L	5	3.63	3.00	42.7	161	76	17
WÜ 3		Oh	2	2.98	2.64	39.2	209	36	49
SC	Granite	L	1	3.30	2.83	37.3	116	50	34
SC		Oh	4	2.98	2.62	38.5	170	38	44
SC		O	5	3.06	2.83	20.8	134	21	64
ST 1	Basalt	O	5	4.13	3.62	33.8	217	83	9
ST 1		Ah	5	4.19	3.62	5.4	154	44	44
ST 3		O	5	4.20	3.65	33.0	202	77	13
ST 3		Ah	5	4.17	3.60	6.6	168	44	48

[a] OW = declining; WÜ = healthy; SC = Schneeberg; ST = Selb.
[b] Equilibrium soil solution.
[c] In 0.01 $MCaCl_2$.
[d] Alkali + earth alkali saturation (Ca + Mg + Na + K)/CEC_{eff} and Al saturation, respectively.

Al-saturation in the O horizon. A significant linear correlation existed between AES and the pH (2.95–4.20) of the ESS (AES = –138 + 57 pH; r = 0.95; n = 10) for the phyllitic and granitic soils. This contrasts with the correlation for the mineral horizons where AES is low (< 20%) and fairly constant in the pH range of 2.9–4.8 (Fig. 1). The main source for Na, K, Mg, and Ca in these highly acidic soils is thus the organic upper part of the soil profile; the permanent charge sites of clay minerals are occupied predominately by Al.

3.1.2 Proton Consumption During pH 3 Stat Titration

Where H_{cons} reached an endpoint within 20 min or less in the phyllitic and granitic soils, the two basaltic soils still consumed H^+ at 10^3 min (Fig. 2). The time curves for the initially fast reaction in the first group did not fit a first-order equation. In the basaltic soils, however, the fast reaction was followed by a slower one which obeyed first-order kinetics. The basaltic soil reaction rate was 0.12 h^{-1}. For these soils, final values for H_{cons} were obtained by iteration through a first-order reaction equation. Between the different samples, the final values of H_{cons} varied by one order of magnitude (Table 6). These values were positively correlated with exchangeable Ca + Mg + K and with $pH_{initial}$–3.0 (r = 0.75 and 0.83 respectively; n = 12). However, even after the values were normalized as H_{cons} kg^{-1} pH^{-1}, the variation was still large (Table 6). The range was similar to that found by Federer and Hornbeck (1985) but considerably lower than that reported by James and Riha (1986) for organic horizons, where the values were estimated from the slope of pH-vs-HNO_3-added curves.

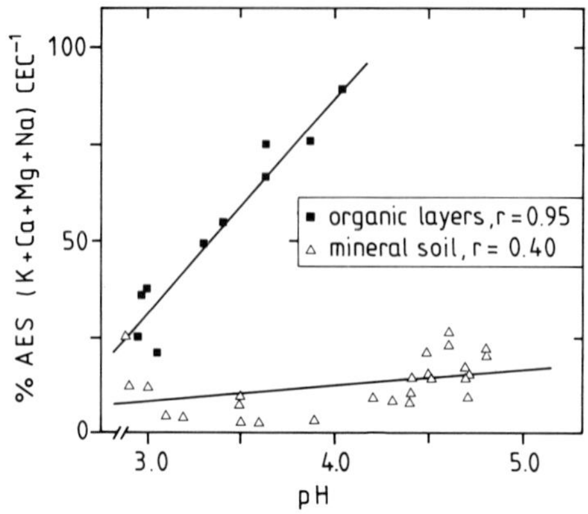

Fig. 1. Alkali plus earth alkali-cation-saturation (*AES*) of organic and mineral soil horizons vs pH of the equilibrium soil solution

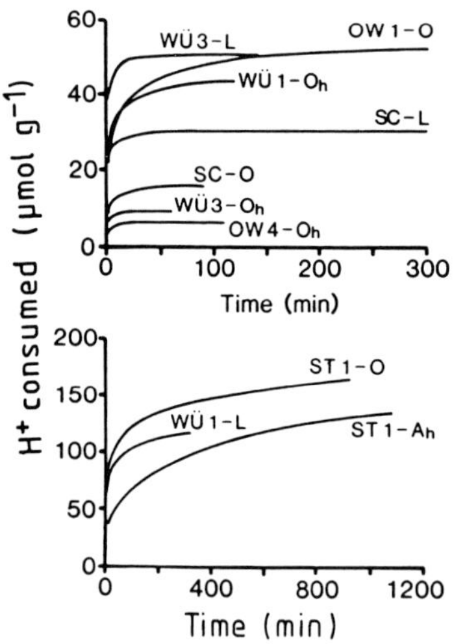

Fig. 2. Proton consumption vs time of various L, O, and A_h horizons (see Table 1)

3.1.3 Cation Exchange as a Buffering Reaction

As H^+ was consumed cations were released into solution. Their magnitude followed the order Ca $\gg$ Al $\geq$ Mg $>$ Mn^{2+} $\approx$ Fe^{2+} (Table 6). This sequence follows that of Brown (1987) and James and Riha (1986) except for the magnitude order of Al which was considerably higher in our samples. The sum of cations (ΣM) approximates H_{cons} in most of the samples ($\Sigma M\ H^{-1}_{cons} \approx 1$) which indicates that H is predominately consumed by H-metal cation exchange. This contrasts with James and Riha (1986) where $\Sigma M\ H^{-1}_{cons}$ ranged from between only 0.14–0.63. Some of our samples, particularly the basaltic ones, even had ratios > 1 (Table 6). These extra cations may have been derived from dissolution of organic salts during titration without protons being consuming. All soils released some Mn^{2+} as well as essentially divalent Fe as determined from an essential

Table 6. Protons consumed (H_{cons}) and cations released into solution during pH-stat titration at pH 3.0

Profile	Horizon	Initial pH	Ca	Mg	Al^{3+}	Fe^{2+a}	Mn^{2+}	ΣM	H^{b}_{cons}	H_{cons} pH^{-1}
			mmol (+)/kg							
OW 1	O	3.97	34.3	7.2	5.5	0.9	6.0	53.9	52.3	54
OW 1	A_h	4.02	6.1	2.5	3.3	2.3	0.8	15.0	16.4	16
WÜ 1	L	4.22	90.7	12.5	9.2	1.2	5.7	119.3	116.0	95
WÜ 1	O_h	3.64	34.5	6.9	8.1	1.7	1.0	52.2	43.4	68
SC	O_h	3.49	11.0	2.9	5.9	1.0	0.3	21.1	17.2	35
SC	O	3.60	9.5	2.4	9.0	1.4	0.2	22.5	16.0	27
ST 1	O	4.19	136.2	18.6	36.2	2.3	19.7	213.0	164.1	138
ST 1	A_h	4.46	91.7	10.4	36.1	0.5	22.7	161.4	134.1	92

[a] Determined by α, α'dipyridyl.
[b] H_{cons}: consumed protons = ANC (pH 3.0).

identity of Fe by AAS and by α,α'-dipyridyl in a NH_4Cl extract of these soils: ($Fe_{dipy} = 0.03 + 0.93\ Fe_{AAS}$; r = 0.991; n = 22).

The concentrations of NH_4Cl exchangeable cations after pH-stat titration as well as changes due to titration are given in Table 7. Generally, Ca and Mg dropped, while Al and H increased after titration. In the organic horizons, CEC_{eff} remained either constant or decreased, but in the two A_h horizons CEC_{eff} increased. The CEC_{eff} decrease probably resulted from an increased protonation of organic functional groups in which H was no longer NH_4-exchangeable. The CEC_{eff} increase in the A_h horizon could have been caused by interlayer-Al removal at pH 3.

In conclusion, the main proton consumption of the organic horizons occurred because of very rapid exchange of easily accessible adsorbed cations. The additional, somewhat slower H-consuming process in the basaltic soils was accompanied by an appearance of Al in solution and at exchange sites.

Table 7. Exchangeable cations and CEC_{eff} after pH-stat titration at pH3[a]

Profile	Horizon	Ca	Mg	Al	H	CEC_{eff}
				mmol (+) kg		
OW 1	O	40.6(−13.6)	4.2(− 3.4)	18.2(+ 6.1)	18.4(+16.2)	90.6(− 2.6)
OW 1	A_h	7.8(− 4.6)	1.6(− 2.2)	42.8(+ 8.6)	23.7(+11.4)	85.5(+12.6)
WÜ	L	92.4(−25.0)	6.2(− 6.0)	19.8(+13.2)	14.9(+12.7)	143.1(−13.1)
WÜ 1	O_h	59.2(−29.8)	5.2(− 6.0)	63.5(− 2.1)	21.6(+ 9.3)	158.8(−31.0)
SC	O_h	46.6(− 5.8)	5.4(− 1.6)	74.8(0.0)	29.9(+ 5.6)	170.1(+ 0.1)
SC	O	24.6(+ 4.2)	2.6(− 1.6)	80.4(− 5.5)	10.6(− 4.6)	127.0(− 7.0)
ST 1	O	85.0(−67.2)	7.2(−12.2)	33.4(+13.8)	12.0(+10.1)	153.8(−63.4)
ST	A_h	44.8(− 7.6)	6.0(− 7.0)	108.4(+41.0)	6.8(+ 5.4)	190.8(+36.8)

[a] Figures in parenthesis indicate change after titration.

3.1.4 Reduction of Fe and Mn as Buffering Reactions

In the long-term (300-day) experiment, ANC was much higher in all the samples than in the short-term experiment (compare Table 8 with Table 6). This increased ANC was not accompanied by an enhanced release of Ca and Mg but by the release of Mn^{2+}, and even more so, of Fe^{2+} into solution (Table 8). Fe^{2+} and Mn^{2+} must have been formed by microbial reduction, probably because the toluene which had been added to suppress bacterial activity decomposed over time.

The reduction of Fe(III)oxides and Mn(IV)oxides consumes 2 mole H^+ per mole of metal:

$$4FeOOH + CH_2O + 8H^+ \rightarrow 4Fe^{2+} + CO_2 + 7H_2O; \quad (1)$$
$$2MnO_2 + CH_2O + 4H^+ \rightarrow 2Mn^{2+} + CO_2 + 3H_2O. \quad (2)$$

These reactions consume ANC (hydroxide buffering after Schwertmann et al. 1987) which is lost from the pedon only if Fe^{2+} is leached. ANC will be restored if $Fe^{2+}$is reoxidized to Fe(III)oxides. This was demonstrated when air was introduced into the system. The concentration of Fe decreased because the dissolved Fe^{2+} became oxidized and precipitated, while the concentration of Al increased as expected from the decrease in pH (Fig. 3).

3.1.5 Mineralization of Organic Matter as a Buffering Reaction

The slight increase in Ca (and Mg) over time during the long-term experiment (Fig. 3) was probably due to microbial oxidation of organic salts (e.g., Ca-oxalate) which consumes protons:

$$Ca(COO)_2 + \tfrac{1}{2}O_2 + 2H^+ \rightarrow 2CO_2 + H_2O + Ca^{2+}. \quad (3)$$

The experimental conditions were favorable to microbial activity with respect to temperature and water supply. In the overall proton balance of pedons, protons consumed by reaction (3) balance those protons released by the roots during cation uptake.

From Figure 3, H_{cons} approximated ΣM (Ca + Mg + Al + Fe^{2+} + Mn^{2+}) at all sampling dates. The regression for all samples was $H_{cons} = -4.6 + 1.09\,\Sigma M$; $r = 0.97$, $n = 78$). The slight average excess (9%) of H_{cons} over ΣM may be partly explained by H adsorption of anions (not measured).

Table 8. ANC and dissolved Fe^{2+} and Mn^{2+} in the long-term batch experiment

Site	Horizon	Fe^{2+}	Mn^{2+}	ANC
			mmol (+)/kg	
OW 1	O	42.7	13.8	109.5
OW 1	A_h	108.6	5.6	169.5
WÜ 1	L	57.4	10.6	315.0
WÜ 1	O_h	136.6	1.7	255.0
SC	O_h	56.1	6.3	112.5
SC	O	72.9	2.2	150.0
ST 1	O	278.4	37.4	729.0
ST 1	A_h	95.5	99.8	427.0

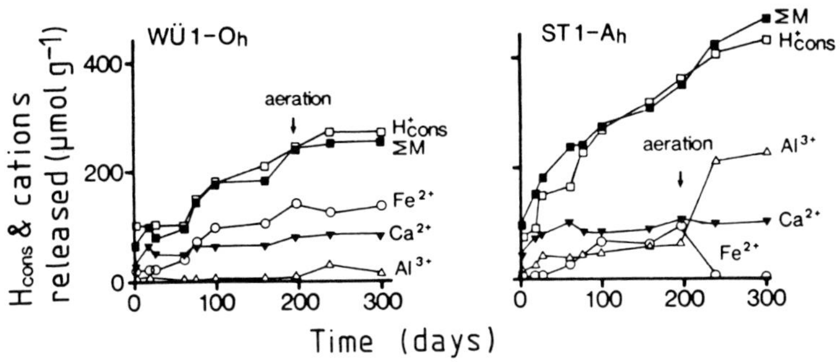

Fig. 3. Proton consumption and cation release as a function of time in the pH 3 batch experiment for a phyllitic (WÜ1-O_h) and a basaltic soil (ST1-A_h)

3.1.6 pH Changes and Cation Release in the Percolation Experiment

The pH of the eluate decreased from 4.04 and 3.73 to ca. 3.2 in the L and O_h horizons of the phyllitic soil within 35 days. In the 2 basaltic soils, pH remained constant at 4.4 in the A_h horizon but increased to ca. 6.0 in the O horizon (data not shown). In the 2 phyllitic soils, only 66% and 50% of H^+ added was buffered and pH decreased throughout the whole column. In contrast, exchangeable Al increased slightly in the L horizon but decreased in the O_h horizon (Table 9). In the O and A_h horizon of the basaltic soil, where 100% of H^+ added was neutralized, pH dropped slightly in the upper part of the column but remained the same or increased in the lower part. However, exchangeable Al increased in the upper part and decreased in the lower part of the column (Table 9).

Buffering can again be essentially explained by the sum of cations (ΣM) released into solution (Table 10). NH_4 was included in ΣM because it also appeared in the eluate. It may have been derived from ammonification of organic N:

$$\text{-R-NH}_2 + H^+ + H_2O \rightarrow \text{-R-OH} + NH_4^+, \quad (4)$$

which is a H-consuming process. The pH increase above the initial pH in the basaltic soil and its eluate can be explained by microbial oxidation of organically bound metal cations, especially Ca. This is supported by a high and constant Ca flux throughout the experiment in the O horizon where the pH increase was highest. In the phyllitic L horizon, Ca dropped sharply after about 15 days (Fig. 4). According to Eq. (3), the base HCO_3^- would be released when the organic salts were metabolized so that the pH would rise (Ulrich et al. 1979; van Breemen et al. 1983).

3.1.7 BNC and ANC of Humics

As shown in the previous sections, humics appear to carry most of the proton-accepting sites of the organic soil horizons. To determine the apparent acid strength of these sites, material in its natural state (OM) from the O_h horizon at the healthy site was exhaustively leached with HCl (H-OM) and $AlCl_3$ (Al-OM) and then titrated. The titration curves were practically linear (not shown) as found

Table 9. pH and exchangeable Al (both in *M* NH_4 Cl) in the upper, central and lower parts of the column before and after 10^{-3} *M* H_2SO_4 percolation

Profile	Horizon		Column position[a]	pH	Al^{3+} mmol (+)/kg
WÜ 3	L	Before		2.97	23.3
			u	2.58	33.8
		After	c	2.63	30.3
			l	2.66	28.6
WÜ 3	Oh	Before		2.62	63.7
			u	2.54	49.6
		After	c	2.58	49.6
			l	2.61	58.4
ST 1	O	Before		3.56	12.9
			u	2.91	53.1
		After	c	3.06	35.6
			l	4.00	6.0
ST 1	Ah	Before		3.50	70.8
			u	3.27	101.1
		After	c	3.43	65.5
			l	3.49	53.1

[a] *u, c, l*: *upper, central* and *lower* parts of the column.

Table 10. Protons consumed (H_{cons}) and sum of cations released (ΣM) in the percolation experiment after 35 days

Profile	Horizon	Initial pH	H^{a}_{cons}	Ca	Mg	Al^{3+}	Mn^{2+}	NH_4	ΣM
			mmol (+)/kg						
WÜ 3	L	4.04	220.1	150.8	14.8	0.0	5.1	22.2	192.9
WÜ 3	O_h	3.73	109.5	58.9	12.4	4.8	1.7	17.3	95.1
ST O	O	4.19	250.5	51.2	11.4	7.8	9.7	24.7	253.0
ST 1	A_h	4.37	75.9	187.2	20.7	1.7	18.7	1.8	81.9

[a] H_{cons}: consumed protons.

by Bache (1984) and Federer and Hornbeck (1985). An expanded Henderson-Hasselbalch plot was produced (Fig. 5) from

$$pH = pK_a + \log[\alpha(1 - \alpha)^{-1}] \quad (5)$$

in which 100% base consumption was arbitrarily set as pH 7 (α = fraction dissociated). The samples OM, H-OM and Al-OM had initial pHs of 3.03; 2.20, and 3.00 respectively and their BNC between pH 3.0 and 7.0 were 2.01, 1.82, and 1.56 mmol g^{-1} C_{org}, respectively. Thus Al saturation reduced the BNC of H-OM by 15% and that of OM by 22%. A reduction in BNC was also found by

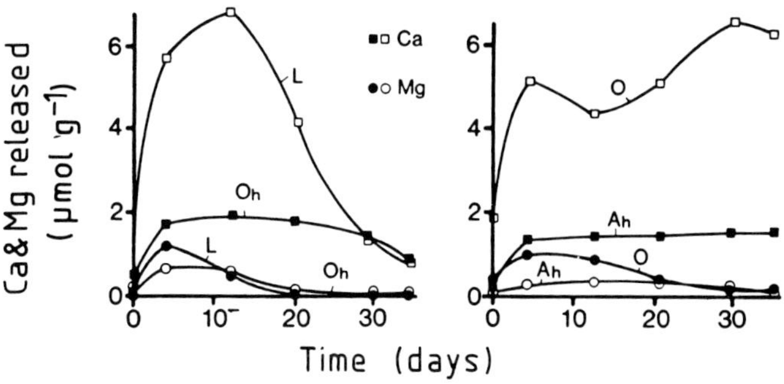

Fig. 4. Ca and Mg released over time during 10^{-3} M H_2SO_4 percolation from 2 horizons of phyllitic (*left*) and basaltic (*right*) soils

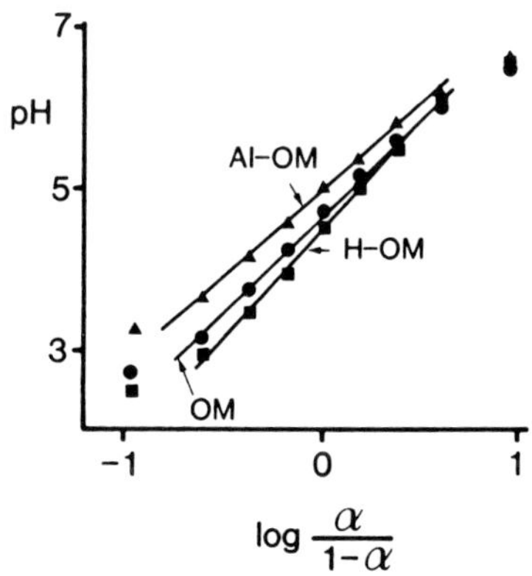

Fig. 5. Henderson-Hasselbalch plot of titration curves of WÜ1-O_h material before and after H and Al saturation

Martin and Reeve (1958) and Hargrove and Thomas (1982). The pK_a values (Eq. 5) for the OM, H-OM and Al-OM, samples were 4.49, 4.65, and 4.96, respectively. These indicated reduced acid strength by Al saturation. Apparently, the organically bound Al was no longer deprotonized to the same extent as the protonized sites of OM and H-OM, probably because part of the Al was complexed by humics. However, full equilibrium may not have been attained during the time span of up to 1 day. Some BNC may also be expected from the components of the EES because of dissolved Al and dissolved organic C (DOC; James and Riha 1986). Total Al ranged from between 0.065 and 0.56 mmol l^{-1}, labile Al (reactable with 8-hydroxychinolin in 15 s) ranged from between 0.023 and 0.260 mmol l^{-1}, and DOC ranged from between 233 and 764 mg l^{-1} (to be published elsewhere). Titration curves of up to pH 9 were generally sigmoidal, although no definite buffering range for Al^{3+} could be ascertained (Fig. 6). The BNC between initial pH and pH 7 was between 0.96 and 3.15 mmol OH l^{-1}. In a multiple linear regression analysis, extinction at 465 nm (as a measure of humics) yielded an adjusted r^2 of 0.65. Labile Al increased r^2 to 0.80 (CV: 15%) while total Al, pH, and DOC did not improve r^2.

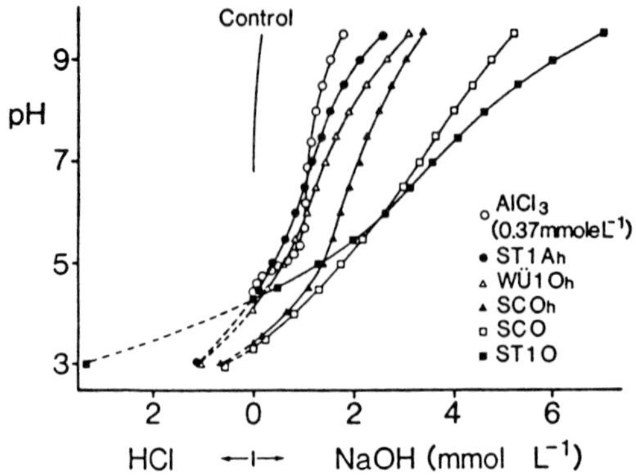

Fig. 6. Titration curves of equilibrium soil solutions (ESS)

3.2 Mineral Horizons

In Section 3.1 of this chapter, we showed that different proton- consuming reactions occurred at different rates and with different ANCs when protons were added to the organic horizons of acid forest soils formed from granite, basalt, and phyllite. Fast reactions consisted of an exchange of Ca, Mg, and Al by H^+. Part of the H^+ was NH_4Cl extractable. In addition to the fast reactions, much slower ones (several weeks) occurred under limited O_2 supply in a batch experiment. Substantial amounts of Fe (and some Mn) oxides were dissolved by microbial reduction, and 2 mol of protons were consumed per mole of Fe and Mn reduced. Another slow proton-consuming reaction under aerobic conditions was the oxidation of Ca (and Mg) salts of organic acids.

In this Section, we report on the mechanisms and kinetics of proton reactions in the mineral horizons thought to react differently from those of horizons dominated by humics. Of particular interest was whether such horizons would consume protons not only by simple cation exchange but also by reactions with hydroxides, oxides, and/or silicates during the experiment.

The study concentrated on four separate soils at the declining and healthy sites and at the Selb and Schneeberg sites. Proton addition methods were described in Section 3.1. ANC predictions of proton buffering under natural conditions were limited by the use of sieved fine earth as compared to a structured soil in the field. Recently the phyllitic soil was found to buffer fewer protons in its natural structure than the basaltic soil (Kaupenjohann and Hantschel 1987). This may be attributed to the difference in structure (crumbs vs. subpolyhedric) and/or to the dominating type of ANC.

All cations and Si were determined by AAS and sulphate was determined by turbidimetry (Bardslay and Lancaster 1965) and ion chromatography. Clay minerals were identified by X-ray diffraction (XRD). Aluminum was extracted by acid NH_4-oxalate at pH 3 (Schwertmann 1964). The clay fraction was collected by sedimentation after suspension in distilled water without any added dispergent. The clay was equilibrated for 4 days over a range of pH values from between 1 and

12 with HCl and NaOH, washed and freeze-dried, and analyzed in a Linseis instrument by thermogravimetry (TGA) at a heating rate of 10°C min^{-1}. The digitized weight-loss curves were differentiated.

3.2.1 pH 3 Stat Titrations: ANC, Kinetics and Cation Balance

Figure 7 shows the experimental data of a typical proton consumption curve which could be described by two pseudo-first-order reactions (Lasaga 1981; Aringhieri and Pardini 1983):

$$ANC_{tot} = ANC_1(1-e^{-k1t}) + ANC_2(1-e^{-k2t}) \quad (6)$$

by which the amount of H consumed at any time t (ANC_{tot}) is divided into two ANCs and their corresponding reaction rates (k). Both ANC and k were found by iteration. For the curve in Fig. 7 ANC_1 was 80.7 and ANC_2 was 130.8 mmol kg^{-1}, and k_1 and k_2 were 1.5 and 0.04 h^{-1}, respectively. Table 11 summarizes the kinetic parameters of the soils studied. ANCs generally increased with depth of the soil profile. With the exception of the B_{v1b} horizon of the profile at the healthy site, ANCs were low for the strongly acidic phyllitic and granitic soils but much higher

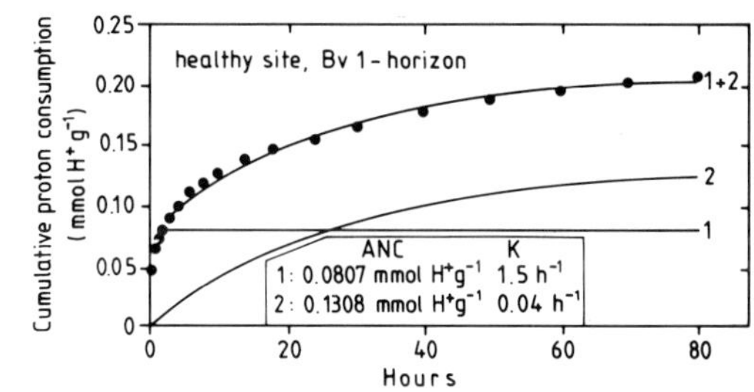

Fig. 7. Kinetic curves of proton consumption of sample WÜ Bv1b during pH 3 stat titration. Experimental data and 2 calculated pseudo-first-order curves are given

Table 11. Acid neutralizing capacities (ANC) and reaction rates (k) of protons during pH-stat titration at pH 3

Profile	Horizon	ANC_1 mmol(+)/kg	k_1 h^{-1}	ANC_2 mmol(+)/kg	k_2 h^{-1}	ANC_{tot} mmol(+)/kg
ST1	Ah1	85.8	3.44	103.4	0.05	189
	Bv1	127.9	2.79	80.2	0.07	208
	Bv2b	123.5	3.08	90.8	0.09	214
OW4	Ah	3.1	21.03	2.3	5.30	5
	Ae2	13.2	4.29	4.7	0.45	18
	C	35.7	1.86	48.3	0.05	84
WÜ3	Ah	8.6	13.94	0.5	0.64	9
	Bv1b	80.7	1.46	130.8	0.04	211
	Bv1b[a]	22.4	5.55	22.2	0.04	46
	Cvb	67.0	3.15	52.7	0.13	120
SC	Ae	2.4	5.17	2.2	0.03	4
	Bhs	12.8	7.26	12.0	0.20	24
	Cv	47.0	1.38	25.3	0.04	72

[a] After preextraction with 0.2 M NH_4-oxalate pH 3 and Ca saturation.

for the less acidic basaltic soil. The high ANC_{tot} of the B_{v1b} horizon was drastically reduced ca. 80% by preextraction with acid NH_4-oxalate. Horizons richer in C_{org}, such as the A_h horizons of the two phyllitic profiles, showed high k values but very low ANCs. This agrees with the values obtained for the organic horizons described in Section 3.1 of this chapter. In contrast, in the mineral horizons the k values were much lower and ANCs were usually higher than in the organic horizons.

Similarly, the study soils differed in the amount of cations released into solution (Table 12) and the resulting change in induced exchangeable cations (Table 13). An approximate equivalency between H_{cons} and ΣM in the basaltic soil indicated that protons induced an equivalent exchange of these cations. Essentially, no exchangeable H^+ was detectable. As determined by AAS, ex-

Table 12. Cations and sulfate released into solution and protons consumed (H_{cons}) during pH-stat titration at pH 3.

Profile	Horizon	Al	Ca	Mg	K	ΣM	SO_4	H_{cons}
		mmol(+)/kg						
ST1	Ah1	36.5	139.4	13.2	12.9	202.0	–	200.0
	Bv1	18.7	175.0	17.0	12.9	223.6	3.0	222.0
	Bv2b	14.6	163.4	15.2	9.4	202.6	0	224.0
OW4	Ah	3.1	1.3	0.9	3.6	8.9	0	5.5
	Ae2	16.2	0	0.3	13.8	30.3	0	18.0
	C	92.8	0.4	0.4	18.0	111.6	1.2	86.5
WÜ3	Ah	8.0	1.0	0.9	5.2	15.1	0.4	9.0
	Bv1b	250.9	0.4	0.4	18.5	270.2	17.8	226.5
	Cvb	124.8	0.1	0.4	11.1	136.4	2.5	120.5
SC	Ae	4.3	2.1	0.7	14.9	22.0	2.9	4.4
	Bhs	19.6	0.6	0.9	16.0	37.1	1.0	28.0
	Cv	121.4	0	0.3	16.5	138.2	1.1	74.5

Table 13. Exchangeable cations and acidity before (*b*) and after (*a*) pH-stat titration at pH 3

Profile	Horizon	Al b	Al a	Ca b	Ca a	Mg b	Mg a	K b	K a	Exchangeable acidity b	Exchangeable acidity a
		mmol(+)/kg									
ST1	Ah1	0.6	80.8	107.3	43.1	23.0	8.6	2.1	3.9	21.1	82.0
	Bv1	7.9	102.8	243.6	131.2	100.2	58.1	3.8	4.7	5.8	110.5
	Bv2b	6.4	108.4	277.2	160.3	116.6	73.5	1.9	6.1	4.2	117.0
OW4	Ah	62.4	56.1	6.0	1.6	1.2	1.1	2.1	1.9	74.6	79.5
	Ae2	87.4	73.7	2.0	0.8	0	0.4	0.9	1.1	87.6	77.0
	C	12.1	26.8	3.4	0.8	0.3	0.1	0.8	1.3	15.3	29.5
WÜ3	Ah	104.9	105.2	3.0	1.4	1.0	1.5	1.3	1.7	126.8	121.0
	Bv1b	22.1	49.0	2.4	0.5	0.2	0.9	0.9	1.7	25.6	46.5
	Cvb	40.4	55.3	2.7	0.5	0.2	0.3	1.5	1.9	36.2	57.0
SC	Ae	53.2	47.9	5.7	2.4	0.8	0.9	1.1	3.6	59.6	56.5
	Bhs	89.8	72.6	6.3	1.2	0.7	0.7	–	1.5	78.1	75.5
	Cv	15.6	22.9	3.4	0.8	0.2	0.2	0.7	1.1	16.1	28.0

changeable Al agreed with total exchangeable acidity (TEA), which was estimated by potentiometric titration as: Al(AAS)(mmol(+) kg^{-1}) = 1.02 TEA(mmol OH kg^{-1})-0.865; r = 0.997; n = 28. Among the exchangeable cations, Ca and Mg decreased and Al increased (Table 13). Within 80 h, H^+ freed Al from nonexchangeable sources. With a much higher replacing power than H_3O^+, Al in turn replaced Ca and Mg.

As the more acidic phyllitic and granitic soils were almost devoid of exchangeable Ca and Mg and had an Al saturation of 85–95% of CEC_{eff}, the Al released by the protons remained almost completely in solution. Again, TEA was equal to exchangeable Al so no exchangeable H could be measured except in the humiferous A_h horizons. This contrasts to the organic horizons (see Sect. 3.1) where higher amounts of NH_4Cl exchangeable H^+ occurred.

In addition to the ANC between initial pH and pH 3 (see Table 11), ANC was also determined as a function of pH (mmol kg^{-1} pH^{-1}) (Table 14). In the pH range of 3–5, ANC increased with decreasing pH. In contrast, in horizons rich in humics ANC per pH was essentially constant over a wide pH range.

3.2.2 Long-Term Batch Titration at pH 3

Table 15 presents the cation-proton balance of a 160-day batch titration at pH3. In the B and C horizons, $H_{cons} \simeq \Sigma M$, whereas in the more humiferous A horizons, $H_{cons} > \Sigma M$. This may be due to the microbial formation of some Fe^{2+} and Mn^{2+} which, as shown in Section 3.1 of this Chapter, is related to the presence of organic matter. The generally well-balanced B and C horizons indicated that microbial reduction did not play a substantial role in these horizons, which are poor in C_{org}.

3.2.3 Sulfuric Acid Percolation

Within 56 days, about 100 mmol H^+ (as 10^{-3} M H_2SO_4) percolated at a rate of 100 ml d^{-1} through soil columns after soluble salts were removed by preextraction with water. The pH of the percolate (Fig. 8) was buffered at pH 5.5 and 4.4 in the two soils with high ANC but it dropped to about pH 3.3 in those with low ANC. Initial and terminal values of exchangeable Al (Al_{ex}), oxalate soluble Al (Al_o) and the pH of the soil in the column are given in Fig. 9. Al_o decreased and pH and Al_{ex}

Table 14. Acid neutralizing capacity per pH unit of two soil samples between their initial pH and pH 4.0, 3.5, and 3.0, respectively, as determined by pH-stat titration within 80 h

Profile Horizon Initial pH	pH of pH-stat titration	ANC between initial pH and pH-stat	ANC/pH
		mmol(+)/kg	
ST1	4.0	53	67
Bv1	3.5	114	87
4.8	3.0	208	116
WÜ4	4.0	25	35
Bv1b	3.5	103	86
4.7	3.0	212	124

Table 15. Protons consumed and Al, Ca, and Mg released in a 160-day batch titration at pH 3

Profile	Horizon	Al	Ca	Mg	ΣM	H_{cons}
		mmol(+)/kg				
ST1	Ah1	43.3	67.4	17.4	128	222
	Bv1	10.6	104.8	41.6	157	173
	Bv2b	11.1	108.5	45.6	165	174
SC	Ah	2.8	2.8	0.8	6	12
	Ae	5.8	1.7	0.4	8	14
	Cv	98.2	0.1	0	98	70
WÜ3	Ah	8.6	1.1	1.0	11	26
	Bv1b	144.2	0.1	0	144	122
	Cvb	95.2	0.1	0	95	113

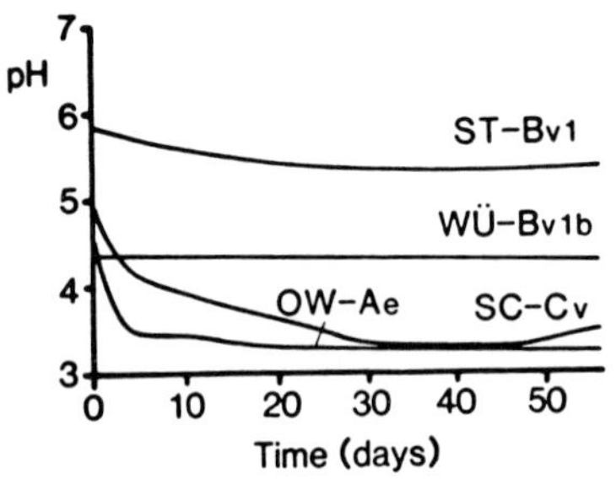

Fig. 8. pH vs time curves of the leachate during 10^{-3} M H_2SO_4 percolation through 4 soils (Samples ST-Bv1; WÜ-Bv1b; SC-Cv; and OW-Ae, see Tables 5 and 11)

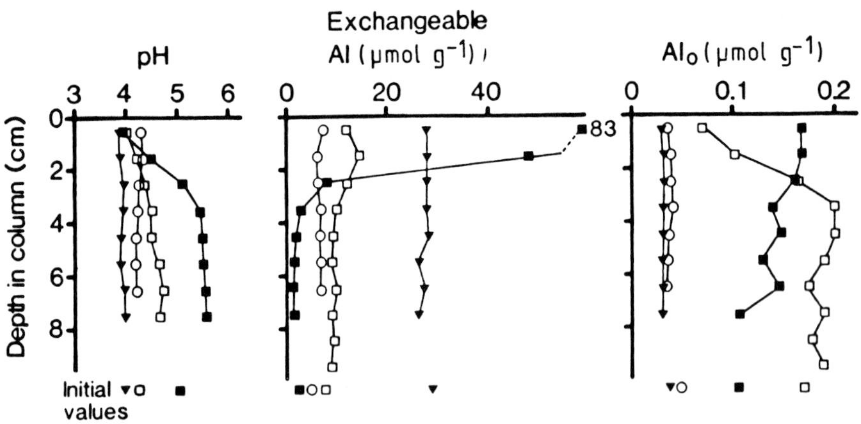

Fig. 9. pH, exchangeable and oxalate-extractable Al (Al_o) in the column of 4 soils after 10^{-3} M H_2SO_4 percolation. Samples SC-Cv(*o*); OW4-Ae2 (▼); WÜ3-Bv1b (□); ST1-Bv1 (■)

Table 16. Column balance of Al_o (mmol/column)

	Sc Cv	OW4 Ae2	ST1 Bv1	WÜ3 Bv1b
Before	2.41	1.85	5.37	8.52
After	1.20	1.24	6.87	8.30
Loss/gain	-1.21	-0.61	1.50	-0.22
Al leached	0.90	0.83	0.01	1.66

remained essentially the same in the two soils with low ANC. This loss can be reasonably accounted for by the Al found in the eluate (Table 16). In those soils with a high ANC, pH decreased and Al_{ex} increased in the upper part of the column while pH increased and Al_{ex} decreased in the lower part of the column. Al_o dropped in the upper part of the phyllitic soil column but increased in the whole basaltic soil column.

The differentiation within the column is of particular interest. In the basaltic soil, the increase in Al_{ex} by 0.65 mmol in the upper 0–3 cm was accompanied by an Al_o increase of 0.85 mmol. The Al_o increase below 3 cm (0.83 mmol), however, indicated the neoformation of Al_o from Al, which was released in the upper part of the column at low pH and which precipitated in the lower part of the column at higher pH (Fig. 9). For this reason, essentially no Al left the column. In the phyllitic sample with high ANC, 1.66 mmol Al was leached out. Al_o decreased in the upper 0–3 cm section by 0.83 mmol and increased in the 4–10 cm section by 0.63 mmol, whereas the change in Al_{ex} was rather small. Again, Al appeared to have been dissolved in the upper part of the column (loss in ANC) and re-precipitated in the lower part of the column (gain in ANC).

3.2.4 Nature of the Reactive Al Compound

To characterize the solid phase beyond primary and clay silicates which would actively accept protons, release Al, and which may have been neoformed in the column experiments, further experiments were conducted. This solid phase could have been an Al hydroxide, hydroxy-sulfate, or hydroxy-silicate. The most likely candidate for the Al hydroxy-sulfates was jurbanite, $AlOHSO_4$ (pK_{sp} = 17.8; Nordstrom 1982), which on the basis of solubility product principles has been claimed to exist in numerous acidic forest soils with high sulfate input (Ulrich et al. 1979; Prenzel 1982; Nilson and Bergkvist 1983; Khanna et al. 1987; Evans and Zelazny 1987). Mulder et al. (1987a), however, have criticized the use of saturation indices ($\log Q_p K_p^{-1}$; Q_p:ion product; K_p:solubility product of solid phase p), claiming that Q_p may be fortuitously similar to K_p. So far, jurbanite has not been positively identified as a solid phase in acidic forest soils.

Figure 10 shows the plot of soil equilibrium solution activities in a phase stability diagram for amorphous $Al(OH)_3$ (pK_s = 32.8), gibbsite (pK_s = 34.1) and jurbanite (pK_s = 17.8). These activities were calculated from Prenzel's model (1982). The data of the less acid solutions (subsoils) indeed concentrated around the jurbanite line, but they were also close to that of amorphous $Al(OH)_3$. Thus, precipitation of either phase may have been possible in the B horizons. As usual,

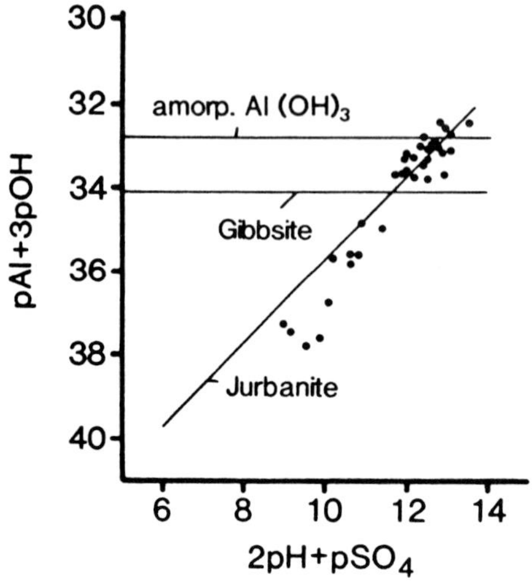

Fig. 10. Equilibrium soil solution (ESS) composition in a stability diagram of gibbsite, jurbanite, and amorphous $Al(OH)_3$

however, considerable scattering of the data was noticeable. The more acid top soil solutions appeared highly undersaturated with respect to the $Al(OH)_3$ phases but also with respect to jurbanite.

In an attempt to synthesize jurbanite, 21 10^{-4}–10^{-2} M $Al_2(SO_4)_3$ solutions containing either no additional electrolytes or alternatively, 0.5, 0.25, and 0.3 mmol l^{-1} Ca, K, and Na, were brought to a pH of between 2.8 and 4.8 with NH_3 and stored at 25°C for 6 months. In some solutions, the small amounts of precipitate formed were X-ray amorphous. Two of these had Al/SO_4 ratios of 2.6 (pH 4.0) and 3.8 (pH 4.8), which were much higher than that of jurbanite. In addition, the solutions were supersaturated with respect to the solubility product of jurbanite, but were close to the solubility product of amorphous Al hydroxide (data not shown). Amorphous Al hydroxy-sulfate with a non-stoechiometric Al/SO_4 ratio, which may have been formed, would have had an OH/Al ratio $<$ 3 if the compound had been dissolved by acids as:

$$Al(OH)_{3-2m}(SO_4)_m + (3-2m)H^+ \rightarrow Al^{3+} + (3-2m)H_2O + mSO_4^{2-}. \qquad (7)$$

The same value should have been found for the ratio of H_{cons}/Al_{rel}. Table 17 shows that the H_{cons}/Al_{rel} ratio was around 2.5 at pH 3 for the B and C horizons of the phyllitic and granitic soils. However, the amount of SO_4 released was much too low to account for the proton deficiency.

To study sulfate further, selected soil samples were equilibrated at various pH with HCl. The sulfate released (SO_{4rel}) was measured. In all samples, SO_{4rel} increased with increasing pH (Fig. 11 *left* and *center*) either because SO_4 was desorbed or because an Al hydroxy-sulfate was deprotonated:

$$Al(OH)_{3-2m}(SO_4)_m + 2mOH^- \rightarrow Al(OH)_3 + mSO_4^{2-}. \qquad (8)$$

On lowering the pH, however, SO_{4rel} remained constant in some samples (Fig. 11, center), but increased in others (Fig. 11, left). These increases could not have

Table 17. Protons consumed and Al and SO_4 released after equilibrating various samples at different pH

Profile	Horizon	Equilibrium pH	H_{cons} mmol/kg	Al_{rel} soln. mmol (+)/kg	Al_{rel} exch. mmol (+)/kg	$\frac{H_{cons}}{Al_{rel}}$	SO_{4rel} mmol (−)/kg
WÜ3	Bv1b	3.0	226.5	250.9	26.9	2.45	17.8
	Bv1b	3.5	108.0	121.4	17.5	2.33	6.8
	Bv1b	4.0	25.7	39.3	8.9	1.60	3.8
	Cvb	3.0	120.5	124.8	14.9	2.59	2.5
SC	Cv	3.0	74.5	121.4	7.3	1.74	1.1
OW4	Cv	3.0	86.5	92.8	14.7	2.41	1.2

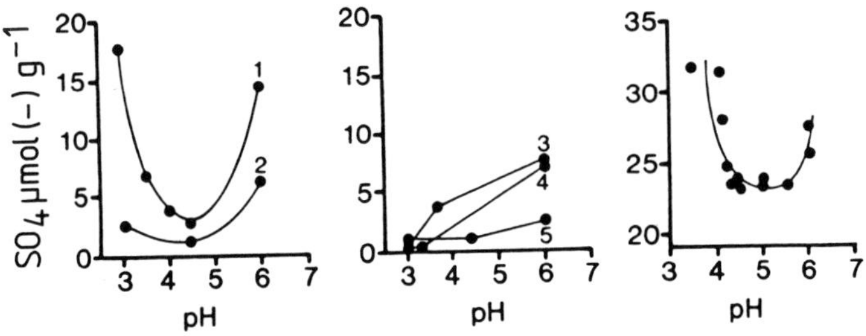

Fig. 11. Sulfate solution concentration after equilibrating 5 soils (*left* and *center*) and a 0.03 M $Al_2(SO_4)_3$ solution (*right*) between pH 3 and 6. The soils are *1* WÜ3-Bv1b; *2* WÜ3-Cvb; *3* SC-Bhs; *4* WÜ3-Ah; *5* ST1-C (see Table 1)

resulted from SO_4 desorption but rather may have been due to dissolution of an SO_4-containing Al hydroxy compound by protonation (Eq. 7). A similar minimum of concentration of SO_4 around pH 5 was also obtained when a 0.03 molar solution of $Al_2(SO_4)_3$ was brought to various pH values between 3.5 and 6 and analyzed immediately after precipitation (Fig. 11, right). The amorphous precipitate formed had a molar Al/SO_4 ratio of about 2.5.

Differential TGA curves (Fig. 12) were prepared from the clay fraction from the B_{v1b} horizon of the phyllitic soil which had the highest Al release at pH 3 (Table 17). This fraction showed weight loss maxima at about 120, 300, and 500°C. The first was due to loss of adsorbed water and the third to clay mineral dehydroxylation. After equilibrating for 40 h at various pH values, the weight loss between 200 and 350°C (2nd peak) decreased both with decreasing pH in the acid range and with increasing pH in the alkaline range from 59 (untreated) to a minimum of 19 g kg^{-1} (pH 12) (Table 18). Although there was also a weight loss between 200 and 350°C of 25 to 29 g kg^{-1}, the basaltic soil showed no peak and the shape of the curves was essentially independent of pretreatment at various pH values (not shown). A lack of or only a weak (at pH 1) treatment effect was noticed for the clay mineral peak at 500°C. Quantitative data in Table 18 show a

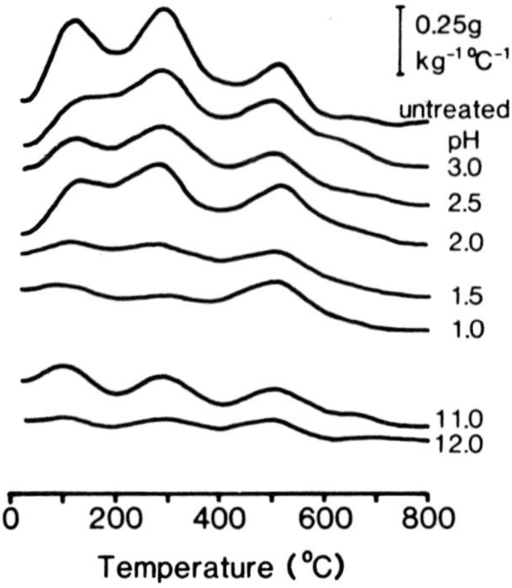

Fig. 12. Differentiated TGA cures of soil clay WÜ Bv1b after equilibration between pH 1 and 12

Table 18. Reduction in weight loss; Al and Fe released and H consumed in the Bv1b horizon of a phyllitic soil after equilibration (40 h) at various pH

Treatment (pH)	Decrease in weight loss between 200 & 300°C[a] g kg^{-1}	Al	Fe	H
		released mol (+) kg^{-1}		consumed
3.0	15.6	0.74	0.02	0.78
2.5	20.2	0.89	0.06	1.12
2.0	20.7	1.18	0.10	1.26
1.5	31.9	1.41	0.45	2.50
1.0	35.3	2.01	0.82	4.20
11.0	36.3	1.57	0.22	2.94
12.0	40.2	2.41	0.37	3.86

[a] Weight loss of untreated sample was 58.8 g kg^{-1}.

significant relationship between the reduction in weight loss between 200 and 350°C and the amount of Al (or Al + Fe) dissolved during the treatment at various pH values. At pH 2–3, the amount of H_{cons} was similar to $(Al + Fe)_{rel}$, but rose at pH < 2, and again at pH 11 and 12. This indicated other sources of proton consumption at very low and very high pH. The amounts of SO_4 and Si released (mol kg^{-1}) were, respectively 2.7×10^{-3} and 0.064 at pH 2, and 2.4×10^{-3} and 0.050 at pH 3.

In an attempt to synthesize the compound causing the weight loss peak between 200 and 350°C, three samples (Al1, Al4, Al5) were synthesized at 25°C following Johansson (1962). Aluminum metal was dissolved in concentrated H_2SO_4 (2.35 ML^{-1} Al and 2.93 ML^{-1} SO_4). The precipitate was separated without (Al5) and with water washing (Al4) to enhance hydrolysis, and dried at 40°C. Al1 was prepared from a $10^{-2}ML^{-1}$ $Al_2(SO_4)_3$ solution by quickly titrating it with NaOH to pH 4. Al2 was prepared by titrating a $10^{-3}ML^{-1}$ $Al_2(SO_4)_3$ solution slowly with gaseous NH_3 to pH 7 and again washing the precipitate and drying it at 40°C. Bulk composition of these preparations showed a range of the SO_4/Al mole ratio from between 0.38 and 1.18 (corresponding to an OH/Al range of 2.24–0.65).

X-ray diffraction of Al1 and Al5 identified crystalline Al hydroxy-sulfates (jurbanite and possibly meta-aluminite) and amorphous Al hydroxides (Al2 and Al3), respectively, by a very broad band at ca. 4.1 A. Al4 was transitional. Thermal gravimetric analysis exhibited a peak at ca. 150°C for water loss. This weakened and broadened as OH/Al increased, and a shoulder, which computer simulation showed to have a maximum at 228°C, developed on the high-temperature side of the peak (Fig. 13). Thus, as SO_4 was replaced by OH, a dehydroxylation peak developed which was similar to that identified in the clay fractions of the soils. Together with the small amount of sulfate and silicate released into solution with acidification and the activity product in the equilibrium soil solution, this dehydroxylation result indicated that the subsoil compound producing the TGA peak may at least be partly caused by a poorly-ordered Al hydroxide rather than jurbanite or amorphous Al silicate. Further work is under way.

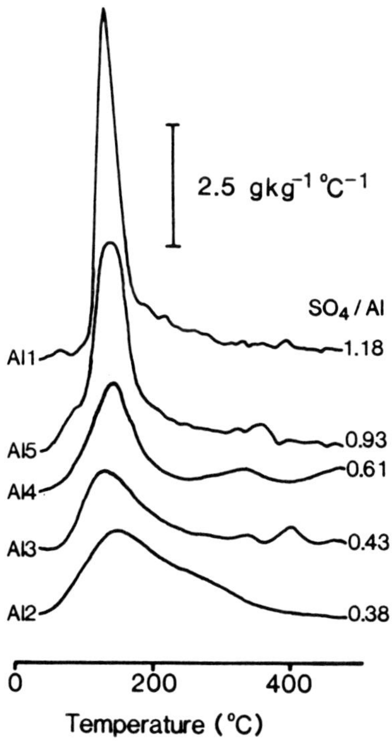

Fig. 13. Differentiated TGA curves of some synthetic Al hydroxy sulfate compounds

3.2.5 Soil Acidification and Nutrient Availability

As shown earlier, proton buffering releases mostly Al-ions into the soil solution. This effect by itself may decrease plant uptake of Mg, K and Ca ions, due to ion antagonism. Moreover, increasing Al saturation increases the exchange coefficient for the exchange reactions of Mg and Ca versus Al (Fig. 14). As a result, Mg and Ca ions can compete for exchange sites to a lesser extent when soil acidification takes place, and therefore become more easily washed out of the soil profile. This may reduce the storage of Mg and Ca ions released by silicate weathering.

To estimate the Mg release by weathering of silicates, batch titrations (pH 3, 120 days) were carried out with soil samples previously saturated with Ba (Süsser 1987). The results were used to calculate the yearly Mg release of different soil horizons (Table 19) based on the soil bulk density shown in Table 3. Both the

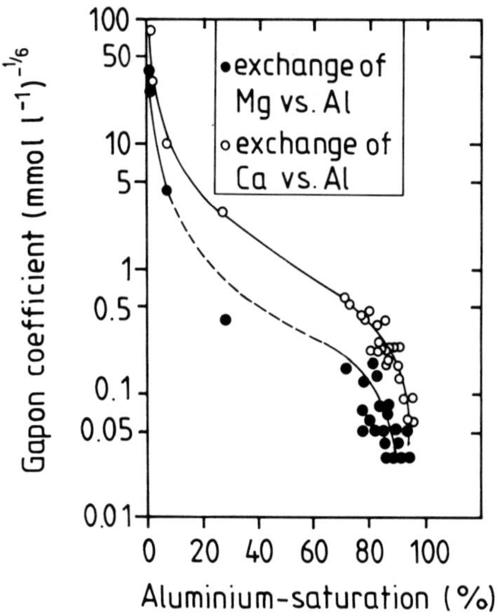

Fig. 14. Gapon coefficients for Mg and Ca vs. Al exchange as a function of Al-saturation of soil samples

Table 19. Mg release as a result of silicate weathering

Site	Horizon	Mg (mmol m^{-2} dm^{-1})
Selb	Ah1	20.4
	Bv1	25.2
	Bv2	26.0
Healthy	Ah	11.8
	Bv1b	0.8
	Cvb	3.0
Declining	Ae	5.4
	Bv1	0.7
	Cv	0.4

healthy and declining stands received only very low amounts of Mg from weathering, compared to the stand on the basaltic soil. The rates of Mg release were higher in the A horizons, in part due to mineralization of organic matter. Therefore, only part of the Mg could be considered to be a net supply from rock weathering.

4 Effects of Soil Structure on Water Transport, Proton Buffering, and Nutrient Release

Ecosystem balances quantify the processes which occur under in situ conditions and which mainly depend on physical and chemical properties of soils. Theoretically, in fine-textured and well-structured soils, the physical and chemical properties of single aggregates and interaggregate pore systems, as well as the continuity of the inter- and intraaggregate voids, affect water transport, nutrient fluxes, and the process of proton buffering. However, the effect of soil structure on water transport is only included in water budget models on a macroscopic scale (= mobile water), while the effect of the intraaggregate voids (= immobile water) is not included. Beven and Germann 1982; Becher 1985). In contrast, models on ion transport in soils are generally based on data obtained for homogenized soils, and assume that the whole solid phase of soils contributes to exchange processes (Nye and Ameloko 1986). Consequently, the potential capacity of soils is quantified although the actual properties of soils in the field may vary greatly.

In this chapter, the process of aggregate formation, the theory of water and ion fluxes in structured soils, and some physical and chemical properties of aggregates are described in order to understand the differences in mass flux, ion diffusion, and proton buffering determined from single aggregates, from structured bulk soil, and from homogenized soil samples at both sites in the Fichtelgebirge.

4.1 Soil Aggregate Formation and Its Possible Role in Water and Nutrient Transport

4.1.1 The Process of Soil Aggregate Formation

In soils with high clay content under field conditions, the mineral particles (sand, silt, and clay) tend to form structured units known as aggregates. This process commonly occurs when soils dry and swell and is enhanced by normal biological activity (Hillel 1980). The aggregates can be very small (e.g., crumbs < 2 mm) or range to a size of more than 10 cm for polyhedra, subangular blocks, prisms or columns. While swelling may lead to an expansion of contracted particles, which is followed by a rewetting of aggregated soils, a complete disaggregation is not possible (Horn 1976). During consecutive swelling and drying cycles, harder and smaller aggregates develop. Their compaction depends on water surface tension, mineral particle mobility, and aggregated particle tensile strength. Aggregation is further enhanced by such biological and chemical processes as flocculation, and cementation by organo-mineralic bonding (Dexter et al. 1988). Because the particles contract, single aggregates become denser than bulk soil as the bulk soil develops interaggregate pores. While interaggregate pores may be very wide, intra-aggregate capillaries are not only smaller in diameter, but also fewer in number than interaggregate pores.

Water surface tension may also cause particle transport and particle size distribution differences in aggregates. Repeated wetting and drying cycles, as well as a higher mobility of clay particles in the pores, results in a heterogeneity of grain size distribution in the aggregate. The outer skin can be clay-enriched as the center of the aggregate becomes coarser (Horn 1987). The arrangement of voids can be quantified by the tortuosity of the pore system. This is expressed as the ratio of average length of pore passages to the geometric length of the soil specimen.

4.1.2 Soil Water Fluxes

The water fluxes (v) in soils can be described and quantified by the Darcy law, given laminar flow and a homogeneous pore system:

$$v = k\, dw\, dx^{-1}, \tag{9}$$

where k is the hydraulic conductivity (cm s^{-1}), w is the matrix potential (hPa), and x is the flow distance (cm). The term dw dx^{-1} represents the hydraulic gradient.

Generally, the values of the hydraulic gradient vary only by half an order of magnitude, depending on water potential, grain, and pore size distribution (Hartge and Horn 1977). The values of the hydraulic conductivity, however, range from between 10^{-2} cm s^{-1} and 10^{-11} cm s^{-1} depending on water potential, texture, and structure. Under saturated conditions, hydraulic conductivity may range from between 10^{-2} and 10^{-3} cm s^{-1} in a sandy soil and between 10^{-4} and 10^{-7} cm s^{-1} in a clay. As hydraulic conductivity is also affected by structure and texture, it is greater if the soil is highly porous, fractured, or aggregated than if it is tightly compacted and dense. The value of hydraulic conductivity not only depends on the pore volume, but also on the continuity of conducting pores. Thus, in structured soils with large cracks, the k-value increases while flow velocity is

Table 20. Saturated hydraulic conductivity kf (cm s^{-1}) of structured bulk soil horizons and single aggregates of B horizons in a gleyic Vertisol *A* and Flusivol *B*, (according to Gunzelmann and Horn 1987)

	Structure	Texture	Saturated hydraulic conductivity kf (cm s^{-1}) Bulk soil	Aggregates
A	Subangular blocky	Loamy clay	1.1×10^{-2}	$3.5 \times 10^{-6} \pm 2.2 \times 10^{-6}$
A	Blocky	Loamy clay	1.6×10^{-3}	$4.8 \times 10^{-6} \pm 2.7 \times 10^{-6}$
B	Prismatic-blocky	Loamy clay	2.6×10^{-5}	$6.0 \times 10^{-6} \pm 2.5 \times 10^{-6}$
B	Prismatic	Sandy loam	3.8×10^{-3}	$3.4 \times 10^{-3} \pm 2.4 \times 10^{-3}$

strongly reduced inside the aggregate due to shrinkage (Table 20). With increasing aggregte density, the k-value decreases up to four orders of magnitude for single aggregates as compared to bulk soil. If, however, the same kind of aggregate contains more sand than silt and clay, there is no difference in the k-value as compared to bulk soil.

The effect of structure on hydraulic conductivity persists under unsaturated conditions (Fig. 15). At high water potentials, the unsaturated hydraulic conductivity in single aggregates decreases with the compaction of the structural elements (prisms less than polyhedrons or subangular blocks). Only after exceeding the cross-over suction values at very negative potentials (Hillel 1980) are hydraulic conductivity values higher in the aggregates compared to the bulk soil.

4.1.3 Soil Aggregation, Ion Mass Flow and Diffusion

Mass flow and diffusion within the heterogenous composition of soils affect ion transport because the accessibility of solid surfaces for solutions is reduced, and because different flow velocities in the dual pore system cause imbalances in the exchange processes during the passage of the liquid (Schulin et al. 1986; Palma et al. 1984). These imbalances can occur more often and more intensely in aggregates adjacent to coarse pores, because equilibration time depends on (1) the ion transport out of the liquid to the aggregate surface; (2) the ion diffusion from the outer surface into intraaggregate pores; (3) the adsorption at the soil particles; (4) the chemical reaction with soil surfaces; (5) the desorption from the particle surface; (6) the transfer of chemical reactants from the inner part of aggregates to the surface; and (7) the transport in the interaggregate pore system.

Ion transport rate J (mol m^{-2} d^{-1}) in soils due to mass flow and diffusion can be calculated as:

$$J = vc + (-D\ dc\ dx^{-1}) + a, \tag{10}$$

where v is the mass flow of water, c is the ion concentration, D is the effective molecular or ionic diffusion coefficient in the soil solution ($cm^2\ s^{-1}$), dc dx^{-1} is the

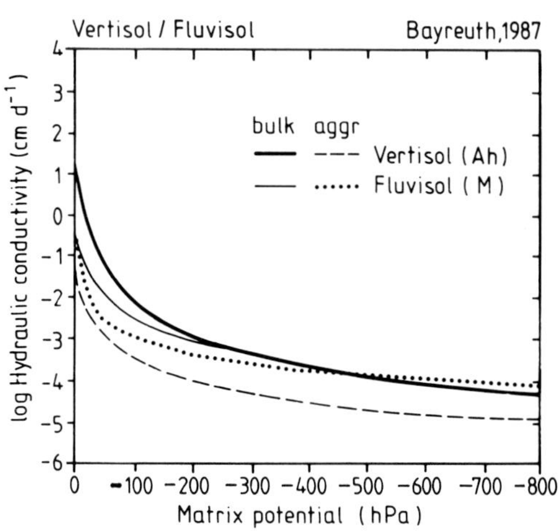

Fig. 15. Hydraulic conductivity of bulk soil and soil aggregates as a function of soil water potential

concentration gradient along the distance, and a represents a sink or source term (Barber 1962).

Laboratory experiments showed that Ca and Mg mass flow by gravity through single aggregates (polyhedrons; texture: clayey loam) was much smaller than through the undisturbed structured soil or through the same homogenized soil material for a constant amount of water (1 liter) (Table 21).

The ion diffusion coefficient *D* (Eq. 10) describes diffusion both in water and at soil surfaces. Although ion mobility at the soil surface is reduced, ion diffusion close to soil surfaces contributes to ion fluxes to some extent and therefore should not be completely neglected:

$$-D\ dc\ dx^{-1} = -D_l\ dc_l\ dx^{-1} - D_s\ dc_s\ dx^{-1}, \tag{11}$$

where l represents the liquid phase and s the solid phase. The water flow near the particle surface is less than in the pore center because the soil solution becomes more viscous with decreasing distance to particle surfaces and because the soil solution near the soil surface (Stern-layer) is more concentrated than in the pores. Consequently, the osmotic component must be considered in the calculation of the effective diffusion coefficient:

$$D = D_l w_l f_l\ dc_l\ dx^{-1} + D_s w_s f_s\ dc_s\ dx^{-1}, \tag{12}$$

where w is the water volume at the surface of soil particles. An impedance factor, f, includes the ratio of the geometric dimensions of a soil sample to the actual path length through water-filled pores for a diffusing molecule or ion. This parameter depends on both the volume and the geometry of the water pathway. Thus, with decreasing water content, the value for f decreases (Beese and Wierenga 1983). In unsaturated soils with a high clay content, the f-factor is determined by the reduced thickness of the water-films coating the particles, the increasing density of exchangeable cations adsorbed to clay surfaces and the corresponding exclusion of anions, and an increase in the viscosity of the liquid.

Because the bulk density of single aggregates is greater than that of bulk soil, and because the average pore size and pore continuity in single aggregates is much

Table 21. Calcium and magnesium mass flow (mmol $\times$ m^{-2} $\times$ s^{-1}) at water saturation through single aggregates, a structured bulk soil sample, and homogenized material by gravity (B_t horizon, Luvisol derived from loess, clayey loam, polyhedrons)

	Mass flow (mmol $\times$ m^{-2} $\times$ s^{-1})		
	Aggregate	Bulk soil	Homogenized material
Saturated hydraulic conductivity (cm $\times$ s^{-1})	$4.8 \times 10^{-6} \pm 2.7 \times 10^{-6}$	1.6×10^{-3}	2×10^{-5}
Ca	$2.28 \times 10^{-6} \pm 2.7 \times 10^{-6}$	2.58×10^{-4}	4.8×10^{-5}
Mg	$1.21 \times 10^{-6} \pm 2.7 \times 10^{-6}$	4.03×10^{-4}	1.9×10^{-5}

smaller due to the higher clay content at the outer skin, the K diffusion out of single aggregates is reduced compared to that out of bulk soil at a given time. However, the accessibility of the particle surfaces inside the aggregates as well as the concentration gradient between the chemical composition of the soil bathing solution and the ion concentration of the soil itself all further affect ion diffusion.

When K-release rates per mass unit of soil out single aggregates (polyhedrons; texture: loamy clay), structured bulk soil, and homogenized soil material (< 2 mm) (P-horizon, Vertisol derived from Amaltheenton) were determined under saturated conditions over a period of 128 h, the K-release rates were always highest for the completely homogenized material (Fig. 16). The larger the aggregates were, the smaller were the release rates at a given bathing solution. However, when the bathing solution, e.g., the Calcium acetate/Calcium

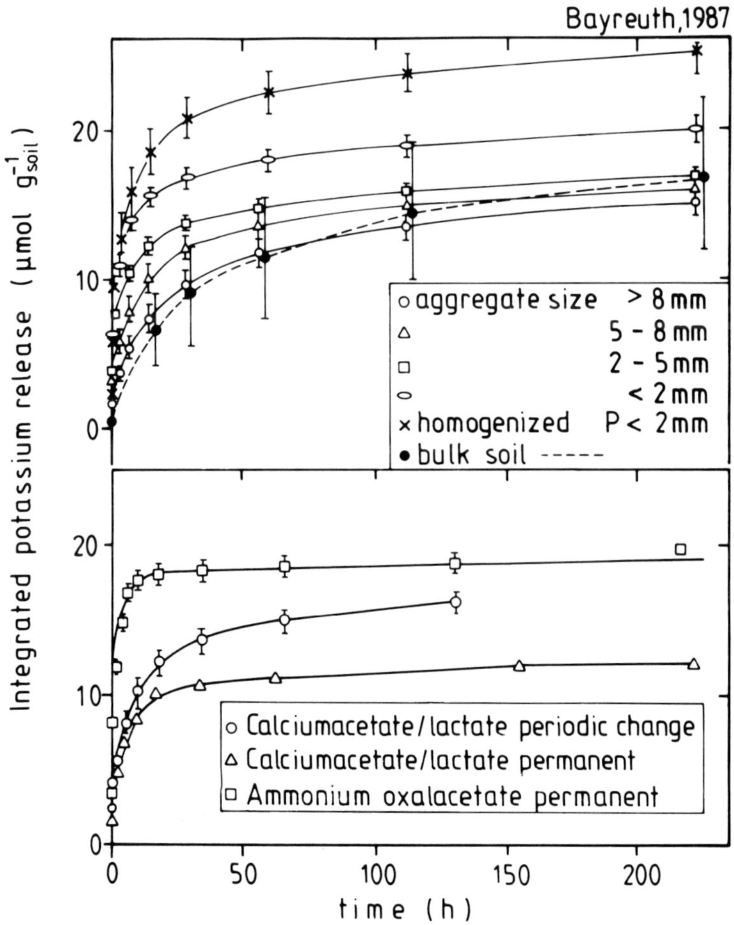

Fig. 16. K release of different soil aggregates and bulk soil into different bathing solutions as a function of time

lactate in which the soil had been kept during the experiments, was changed periodically in order to keep the concentration gradient constant throughout the whole experiment, the K-release out of the same size aggregates was higher compared to that out of soil exposed permanently to the same solution. When the concentration gradient between the soil and the bathing solution was increased (e.g., to 1 mol l^{-1} Ammonium acetate), a doubling of the ion release was observed just after 30 mins and a steady state was reached after 8 h. However, ion release did not reach steady-state in either Calcium acetate/Calcium lactate treatments even after 120 h. Magnesium and Ca release rates out of single aggregates (2–5 mm, 5–8 mm, > 8 mm) per soil mass (Fig. 17) were similar to K release rates.

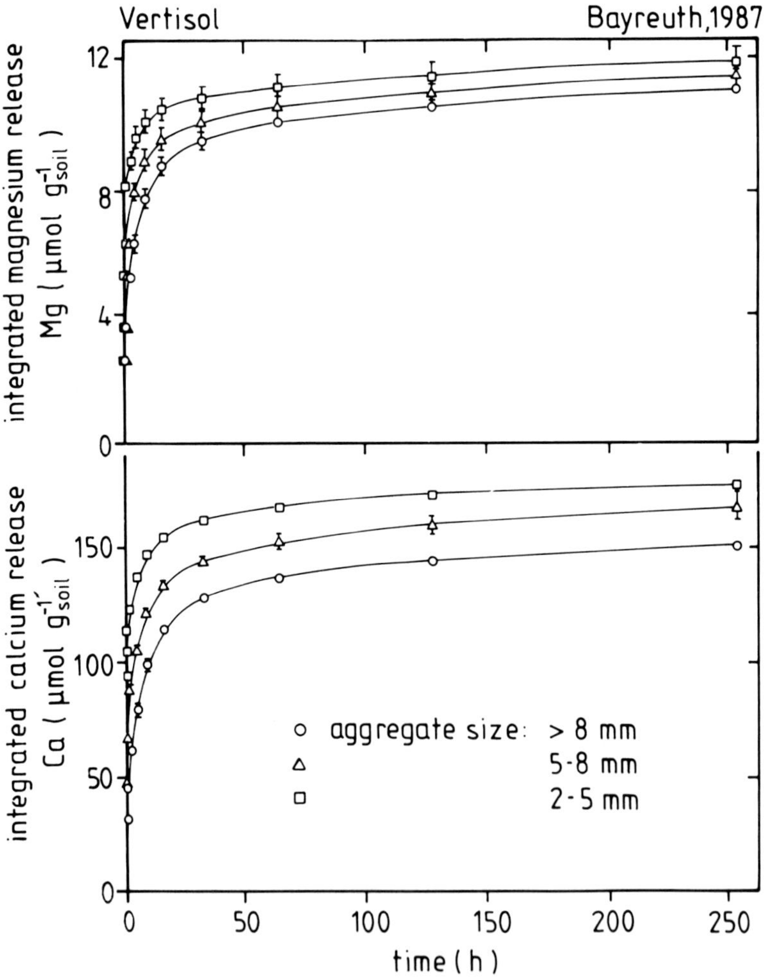

Fig. 17. Ca and Mg release of soil aggregates differing in size as a function of time

With increasing soil mass (i.e., larger aggregates), K per surface area increased (Fig. 18). The amount of K extracted from larger aggregates was much higher than that from the smallest aggregates (< 2 mm) because of greater soil volume or soil mass per unit of aggregate surface area. Again, with the stronger extraction solution, the ion concentration in the bathing solution increased (not shown).

4.1.4 Soil Aggregation and Root Growth

Equations (9) and (10) show that ion transport by mass flow and diffusion depends on flow length at a given hydraulic conductivity and matrix potential. In structured soils, roots grow in soil cracks or coarse pores surrounding single aggregates and do not penetrate the aggregates in order to reduce the length of the flow pathway. Aggregates are penetrated by smaller roots up to a depth of 1–2 mm (Altemüller and Haag 1983) to increase nutrient uptake (Marschner et al. 1986; Jungk and Claassen 1986). Root penetration into single aggregates may be prevented by a higher proportion of finer pores at the outer skin of the aggregates. These finer pores usually prevent aeration of the inner aggregate. Aeration occurs only with lower water potentials, when even the fine pores are dry. In addition, the resistance to root penetration is greater due to the higher bulk density values of the aggregates compared to bulk soil (Horn et al. 1987; Gunzelmann and Horn 1987). Thus, root length or root surface density (cm cm^{-3} or cm^2 cm^{-3} soil) is very high at the aggregate surface.

Nutrient uptake mainly occurs from the aggregate surface where roots may exude additional protons as a result of proton buffering by the tree canopy (see Chap. 3-D). Thus, as long as ion transport by mass flow or diffusion in the

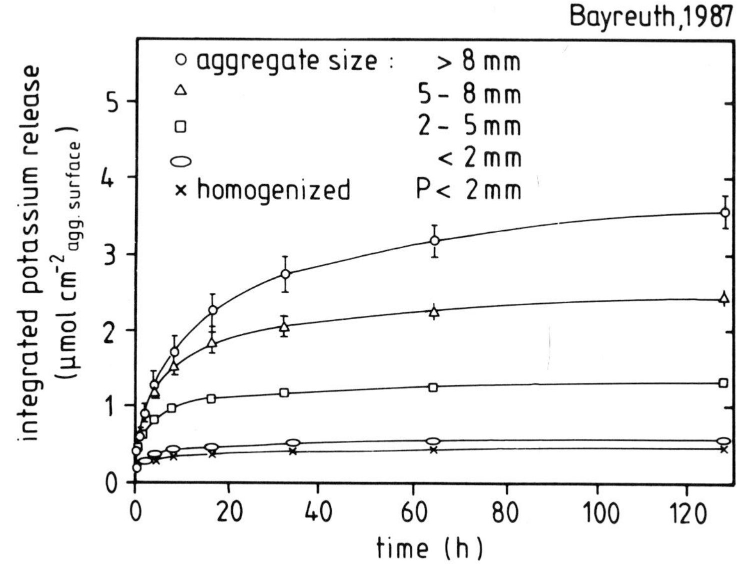

Fig. 18. K release of different soil aggregates related to aggregate surface (cm^{-2}) as a function of time

intra-aggregate pore system is small, an imbalance between the chemical composition of the outer skin and the inner part of the aggregate will persist (Horn 1987). This imbalance results in reduced accessibility of plant available nutrients which may be modified only by mycorrhizal ion transport from the inner part of aggregates.

Precipitation mainly percolates through interaggregate coarse pores. Thus, rain water, e.g., acid rain, will primarily leach cations from the aggregate surface. The reduced accessibility of the soil nutrients and increased root surface in the larger pores of the bulk soil near depleated aggregate surfaces may therefore enhance plant nutrient stress.

4.2 Water Flux, Proton Buffering and Nutrient Release in Structured Forest Soils

In Section 4.1 of this chapter, it was shown that soil structure formation principally results in (1) separation of the total amount of percolating water in a mobile (= percolation in interaggregate voids) and immobile fraction (= percolation in intra-aggregate pores); (2) reduced accessibility of particle surfaces inside single aggregates, which affects ion mass flow and ion diffusion; and (3) restriction of root growth to the outer surfaces of single aggregates.

In this section, we report on results which verify the ecological relevance of soil aggregates in water and ion transport for the soils at both sites in the Fichtelgebirge. Nutrient release characteristics of saturated and undisturbed (percolation experiments) or sieved (batch experiments) mineral and organic soil samples from the healthy and the declining sites were extracted partly with sulfuric acid (Kaupenjohann and Hantschel 1987) or distilled water. Additionally, soil samples from the A-horizons of a Dystric Haplochrept (from a site near Selb, see also Sect. 2) and a Lithic Haploboroll derived from dolomite with carbonates (from the Bavarian Alps) were also included in order to present the complete pH range and the effect of different parent material on nutrient release and proton buffering.

The proton buffering rate of the soil (mmol H^+ m^{-2} h^{-1}) is defined as the difference between proton load and proton content in the soil extracts. The time factor (h) is given by the hydraulic conductivity kf (cm s^{-1}), while the pH buffering rates in the batch experiments were defined by the difference between proton load and proton content in the extract after 60 min of shaking.

4.2.1 Water Transport

At the healthy and declining sites in the Fichtelgebirge, the influence of soil aggregation on water transport could be derived by the hydraulic conductivity and pore size distribution characteristics of the bulk soil as a first approximation. While the values of hydraulic conductivity were extremely high under saturated conditions, a sharp decrease occurred when water potential was less than −100 hPa (Hantschel 1987). Furthermore, the pore size distribution showed the dominance of coarse pores, perhaps because the fine textured soil horizons contained 20–60% gravel. However, up to 15% of these coarse pores represented interaggregate pores. The bulk density values of aggregates in the upper B-horizons at both sites were 1.38 g cm^{-3} as compared to 1.15 g cm^{-3} for the bulk soil at the healthy site, and 1.15 g cm^{-3} to 1.35 g cm^{-3} for the bulk soil at the declining site. Finally, aggregation affects water transport by influencing the maximum water saturation under in situ conditions (Hantschel et al. 1987) and the change in water potential following precipitation events.

4.2.2 *Proton Buffering*

The proton buffering rate of sieved samples (0–5 cm depth) was smallest for the soils at the declining site. In contrast, with increasing amount of AES (healthy site, and Dystric Eutrochrept), the proton buffering rate increased to the 1:1 line for the soil samples with free carbonates (Fig. 20). Nevertheless, no change in the slopes with increasing proton load could be detected because of the complete accessibility of the particle surfaces in homogenized samples to bathing solutions. These results corresponded to those reported in Section 3.2.

In experiments with undisturbed, saturated soil samples a different trend emerged for the declining site. The proton buffering rate of the soil at the declining site reached a maximum value of about 19 mmol H^+ m^{-2} h^{-1} at an acidic load of nearly 40 mmol H^+ m^{-2} h^{-1}. The slopes of the other curves, on the other hand, did not show any significant decline. Theoretically, the maximum proton load of 40 mmol H^+ m^{-2} h^{-1} is equivalent to a rain intensity of 400 mm h^{-1} (pH 4), but such intensity has not been recorded. Proton concentration in fog (pH 2.3; Chap. 4-B) is much higher than in rain. However fog deposition is unlikely to have the intensities of 8 mm h^{-1} (pH 2.3) which are required to exceed the proton buffering rate of the soil. In Chap. 3-D, it is shown that part of the atmospheric acid input is buffered by tree canopies and enters the rhizosphere by additional cation uptake. Calculations for the declining stand showed that on average about 50% of the incoming acids were buffered during pasage through the canopy (Hantschel 1987). This acidity may load the soil immediately around the roots of trees. As the roots occupy 1–10% of the total pore volume of the soils, precipitation intensities of 2.6–26 mm h^{-1} with pH of 3.5 exceed the pH buffering rates of the top soil at the declining stand. Such values are quite common at the declining site (Trautner 1988).

4.2.3 *Nutrient Release*

Undisturbed soil samples were repeatedly percolated (e.g., 4 times replacement of the total water volume = 4 pore volumes) with distilled water and a sharp decrease in nutrient release was found with increasing amount of percolated water (Fig. 19). Both the slope of the curves and the final ion concentration in the soil solution were always smaller at the declining site than at the healthy site. These differences can be explained by the smaller aggregates at the declining site as compared to the healthy site. It is tempting to speculate that for the same acidic input, ion release out of smaller aggregates at the declining site was faster for a shorter period of time than that out of aggregates at the healthy site. Consequently, the reduction of the ion concentration gradient for a given volume of

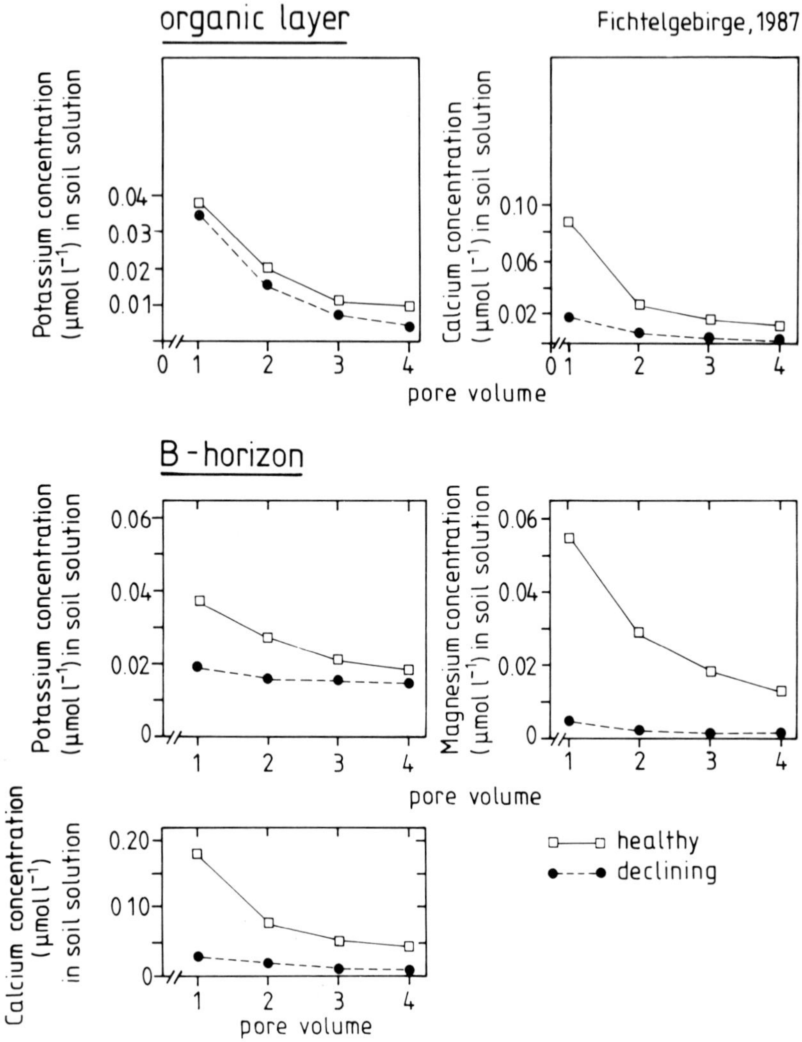

Fig. 19. Cation concentrations in water percolates of different soils as a function of quantity of percolated water (pore volumes)

percolating solution occurred earlier in the smaller aggregates at the declining site than it did in the larger aggregates at the healthy site. As a result of a smaller concentration gradient between the solution and the aggregate surface, a greater reduction of the ion fluxes towards equilibrium must have occurred. Thus, the outer surface of aggregates relative to the center was more deprived of cations (Horn 1987).

As a percolating solution for extracting cations from undisturbed soil samples, sulfuric acid had only slightly affected K-release. The release of bi- and

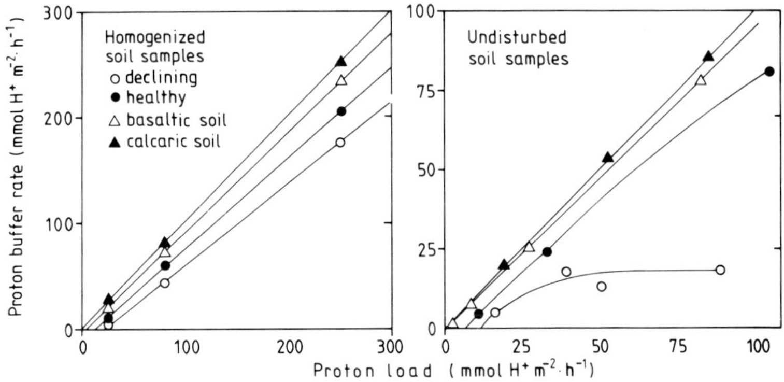

Fig. 20. Proton buffering rate of sieved and undisturbed soil samples as a function of proton load

trivalent cations, however, generally increased with the acidity of the percolating acid (Fig. 21). Consequently, the K-equivalent fraction in the soil solution decreased sharply with the increasing acidity (Fig. 22). This result could explain the low foliar K-concentrations of plant tissue in the old spruce stands, which received higher acidic depositions than the younger ones (Hantschel 1987) at the declining site (Kaupenjohann et al. 1987a). It is quite remarkable that very small amounts of protons caused a reduction in K availability in soils containing free carbonates. Recent studies of spruce nutrition on calcareous soils have shown that K nutrition has declined over the last two decades (Hüttl and Zöttl 1985). This might be explained by the slight increase in the atmospheric acidic load over the last two decades even in the so-called "clean air" areas.

The Mg supply at the declining site was very low at low rates of acidic loads. However, Mg release increased with the increasing proton load of the soils so that as a result of the acidic load, this soil chemical alteration was reflected in the actual nutrient status of the trees. The young spruce stands, in which acidic inputs were relatively low, suffered from a Mg deficiency while the older trees had higher concentrations of Mg in the needles (see Chap. 4-B). The consequence of a higher proton input in the older stands would, however, be more soil leaching. The equivalent fraction of Mg in the percolation soil extracts was determined for an acidic load corresponding to the average calculated proton buffering by the canopy of trees at seven forest sites. This Mg-equivalent portion of the soil extracts correlated with the corresponding Mg concentrations of the spruce needles. The correlation was very strong (Fig. 23). Similar relationships existed between equivalent fractions of K and Ca in the percolation extracts and nutrient status of the spruce stands.

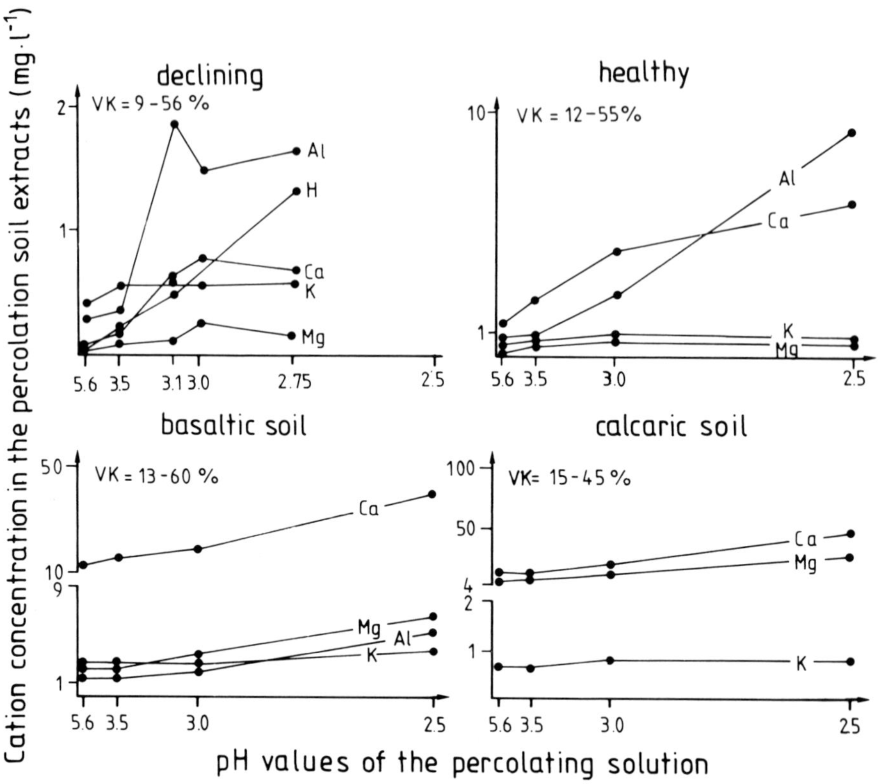

Fig. 21. Cation concentrations in H_2SO_4 percolates of soils as a function of pH of the percolating solution

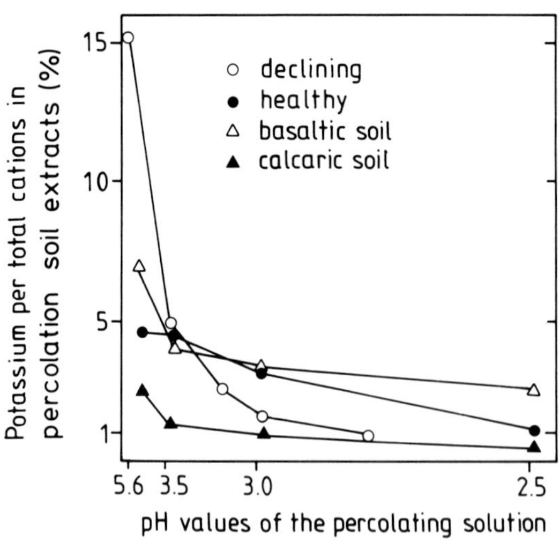

Fig. 22. K per total cations (%) in H_2SO_4 percolates of soils as a function of pH of the percolating solution

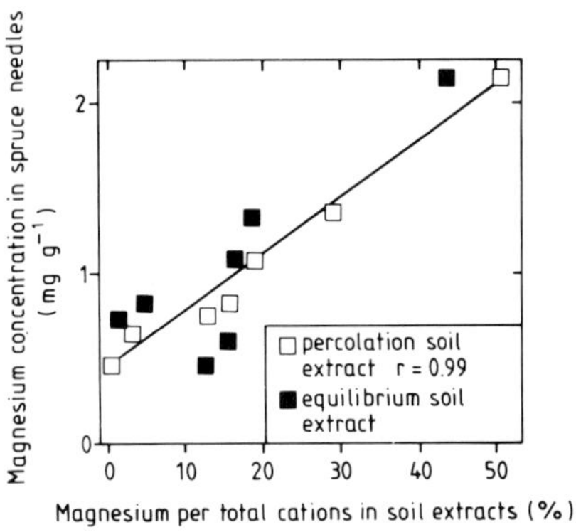

Fig. 23. Relationship between Mg per total cations (%) in soil extracts and Mg concentrations in current-year spruce needles

5 Conclusions

The chemical composition of soils has at least a predisposing effect on forest decline processes (Zech and Popp 1983; Hüttl 1985) as seen in the relationship between the Mg concentration in needles and homogenized soil samples. However, such a relationship becomes even more significant when the same studies are conducted with undisturbed structured soils.

At both sites, pH buffering in mineral soils mainly occurs by dissolution of reactive, poorly ordered Al-hydroxy-compounds. This buffering mechanism leads to increasing concentrations of Al in the soil solution at both the healthy and declining sites. Nevertheless, due to higher Ca concentrations in the mineral soil solution at the healthy site, the Ca/Al ratio is still higher than that at the declining site. These soil chemical properties may be one of the reasons why tree roots do not exploit the mineral soil horizon at the declining site as much as they do at the healthy site (see Chap. 4-E). However, as a result of the more favorable chemical composition of the organic soil layers, tree roots have been mainly concentrated in the organic soil layers at both sites. Proton buffering at variable charge humic sites have mainly released Ca and Mg ions, while exchange of protons with Al has occurred to a lesser extent. Such proton buffering has resulted in relatively low Al concentrations at both sites. Similar findings have also been reported from studies of organic soil layers in several forest stands elsewhere (James and Riha 1986; Murach 1984).

Soil structure affects nutrient release and proton buffering to a great extent. Batch experiments and pH stat titrations with sieved soils from organic and mineral soil horizons have shown that proton buffering reactions are very fast (Schaller and Fischer 1985). Thus, the buffering capacity should exceed the acid input. However, in structured soils, the heterogeneity of the pore size distribution in inter- and intraaggregate voids and the differences in pore continuity can result in reduced accessibility of buffering compounds. Thus, acidic load may exceed the soil buffering rate and cause a biological stress. At the declining site, a significant relationship was found between acid load, nutrient release of soils, and nutrient concentrations of spruce needles. This indicates that there was a strong link between recent acid inputs and tree nutrition.

The question remains, however, whether acid input has significantly influenced soil chemical composition over the last decades and, if so, whether acid input is responsible for the actual chemical soil status.

In comparing historical and recent soil data, several authors have concluded that acid deposition has markedly reduced base saturation, pH, and ANC (a review of available data is given by Berden et al. 1987). As historical soil data was not available for our study, this approach could not be used to assess changes in the soils of the Fichtelgebirge. Therefore, a soil chemical equilibrium model was developed which considered the most important proton buffering reactions of the soils. These were (1) cation exchange (Ca, Mg, Al); (2) solution of Al-hydroxy-compounds ($Al(OH)_3$, $Al(OH)SO_4$); and (3) rock weathering. The model estimates Ca and Mg saturation of the exchange complex, ANC changes, and H, Ca, Mg, Al and SO_4 concentrations in the soil solution as a function of acid inputs into the mineral soil. Acidic inputs have been estimated from the measured mineral soil input (Hantschel 1987) plus those proton equivalents which enter the soil compartment via the rhizosphere through pH buffering of the canopy (see Chap. 3-C). Validation of the model has been carried out with chemical data of soil solutions which were sampled over a 2-year period at the study sites. Because the model omits the effects of soil structure, the ecologically relevant acidification of the soils may be underestimated.

The effects of acid rain on soil chemical status were tested by varying the originally calculated proton input by two orders of magnitude (range of 0.1–10). Calculations of Ca plus Mg saturation of the Ae-horizon of the soil at the declining site are presented as an example (Fig. 24). Ca plus Mg saturation slightly decreased during the 2- year modeling period with measured actual acid input data. Reducing the acid inputs by a factor of 0.1 led to a slight increase in the base saturation, which may be attributed to base inputs (which had not been adequately reduced) and rock weathering. Base saturation did not further decrease when acid inputs were increased by a factor of 10 when the conjugate

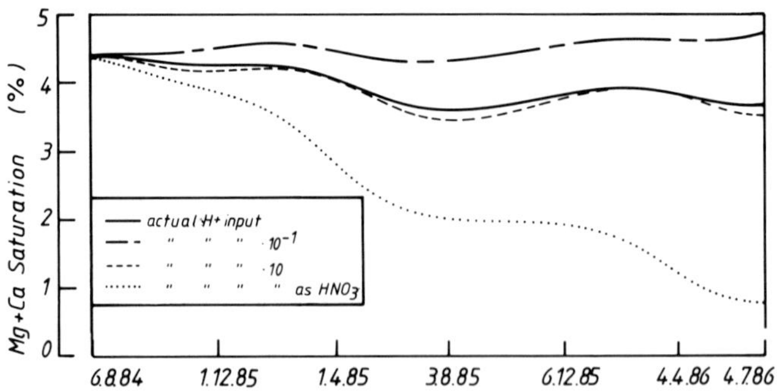

Fig. 24. Model of Ca + Mg saturation of the Ae horizon at the declining site between 06.08.84 and 04.07.86 as a function of different proton loads

base was sulfate, but the ANC of the soil did. This effect may be attributed to the storage of $Al(OH)SO_4$ in the soil. As long as formation of this salt is not restricted, either kinetically or by organic compounds, it may be considered to be a perfect sink for sulfuric acid (Prenzel 1983). The mobile anion precipitates with Al and therefore no further leaching of cations occurs from the soil profile.

Nitrate is also considered to be a mobile anion which can leach cations effectively out of the soil because there is hardly any storage mechanism for this anion in soils. In Chapter 4-I, for example, it has been shown that more than 50% of cation leaching is due to nitrate. Therefore, in another model run, the conjugate base of the input acid was assumed to be nitrate. This time the base saturation strongly decreased when the actual acidic input was increased by a factor of 10. The effect of nitrate leaching on soil acidification based on these calculations seems very strong. Recently, critical loads for N have been calculated (Nilsson and Grennfelt 1988) to maintain a constant exchangeable base cation pool. The authors assume that 3–15 kg N ha^{-1} yr^{-1} is the maximum load for coniferous forests, depending on the productivity of the sites. However, declining stands with deficiencies of Mg, Ca, K or other nutrients may not be able to tolerate any N inputs. Based on deposition rates of 15–25 kg N ha^{-1} yr^{-1} at the study sites (see Chap. 4-I), increased N and sulfuric acid loads may promote soil acidification.

The modeling of acidification processes indicates that there is a considerable effect of acidic inputs on depletion of nutrients from the soil, as has been demonstrated for other stands (Süsser 1987). Current soil acidification from acid inputs is only slightly compensated by weathering of mineral soil, because weathering rates are hardly enhanced by increased acid loads (April and Newton 1985; Fölster 1985; Van Breemen et al. 1986; Mulder et al. 1987b). In terms of Mg nutrition, rock weathering in both the healthy and declining stands is also negligible.

References

Altemüller HJ, Haag T (1983) Mikroskopische Untersuchungen an Maiswurzeln im ungestörten Bodenverband. Kalibriefe 16:349–363

April R, Newton R (1985) Influence of geology on lake acidification in the ILWAS watersheds. Water Air Soil Pollut 26:373–386

Aringhieri R, Pardini G (1983) Interactions between OH^- ions and soil surfaces: a cinetic study. Can J Soil Sci 63:741–748

Bache BW (1984) The role of calcium in buffering soils. Plant Cell Environ 7:391–396

Barber M (1962) A diffusion and mass flow concept of soil nutrient availability. Soil Sci 93:39–49

Bardslay CE, Lancaster JD (1965) Chemical and microbiological properties. In: Black CA, Evans DD, White JL, Emsinger LE, Clark FE (eds) Methods of soil analysis. ASA. Madison, USA, part 2:1102–1116

Becher HH (1985) Mögliche Auswirkungen einer schnellen Wasserbewegung in Böden mit Makroporen auf den Stofftransport. Z Dtsch Geol Ges 136:303–309

Beese F (1982) Gesetzmäβigkeiten beim Transport gelöster Stoffe im Boden. Beitr Hydrol Sonderheft 4:267–300

Beese F, Wierenga PJ (1983) The variability of the apparent diffusion coefficient in unsaturated soils. Z Pflanzenernähr Bodenkd 146:760–771

Berden M, Nilsson SJ, Rosen K, Tyler G (1987) Soil acidification extent, causes and consequences. Nat Swed Environ Prot Board, Rep 3292:1–164
Beven K, Germann P (1982) Macropores and water flow in soils. Water Res 18:1311–1325
Blümel WD (1986) Waldbodenversauerung, Gefährdung eines ökologischen Puffers und Reglers. Geogr Rundschau 38:312–320
Bredemeier M (1987) Quantification of ecosystem-internal proton production from the ion balance of the soil. Plant Soil 101:273–280
Brown KA (1987) Chemical effects of pH 3 sulphuric acid on a soil profile. Water Air Soil Pollut 32:201–218
Childs, CW (1981) Field tests for ferrous iron and ferric organic complexes (on exchange sites or in water-soluble forms) in soils. Aust J Soil Res 19:175–180
Cronan CS (1985) Chemical weathering and solution chemistry in acid forest soils: Differential influence of soil type, biotic processes and H^+ deposition. In: Drever JI (ed) The Chemistry of Weathering. Reidel Publ Co Dordrecht, pp 175–195
Cronan CS, Alken GR (1985) Chemistry and transport of soluble humic substances in forest watersheds of the Adirondack Park, New York. Geochim Cosmochim Acta 49:1697–1705
Dexter AR, Horn R, Kemper W (1988) Two mechanisms of agehardening of aggregates. J Soil Sci 39:163–175
De Vries W, Breeuwsma A (1985) Acidification of Dutch forest soils as influenced by natural soil processes, extraction of biomass and acid rain. Ned Barbouwijdschr 57:1111–117
Driscol CT, Van Breemen N, Mulder J (1985) Aluminum chemistry in a forested spodosol. Soil Sci Soc Am J 49:437–444
Evans A Jr, Zelazny IW (1987) Effects of sulfate additions on the status of exchangeable aluminum in a Cecil soil. Soil Sci 143:410–417
Federer CA, Hornbeck JW (1985) The buffer capacity of forest soils in New England. Water Air Soil Pollut 26:163–173
Finck A (1976) Pflanzenernährung in Stichworten. Hirt, Kiel, 1–200
Fölster H (1985) Proton consumption rates in holocene and present day weathering of acid forest soils. In: Brever JJ (ed) The chemistry of weathering. Reidel Publ Co, Dordrecht, pp 197–209
Gunzelmann M, Horn R (1987) Wasserhaushaltsuntersuchungen an natürlich gelagerten Bodenaggregaten. Mitt Dtsch Bodenkd Ges 53:395–401
Hantschel R (1987) Wasser- und Elementbilanz von geschädigten, gedüngten Fichtenökosystemen unter Berücksichtigung von physikalischer und chemischer Bodenheterogenität. Bayreuther Bodenkd Ber 3:1–219
Hantschel R, Kaupenjohann M, Horn R, Zech W (1986) Kationenkonzentration in der Gleichgewichts- und Perkolationsbodenlösung (GBL und PBL) -ein Methodenvergleich. Z Pflanzenernähr Bodenkd 149:136–139
Hantschel R, Durner W, Horn R (1987) Die Bedeutung von Porenheterogenitäten für die Erstellung von pF/WG – und K/O Kurven. Mitt Dtsch Bodenkd Ges 53:403–409
Hargrove WL, Thomas GW (1982) Titration properties of Al-organic matter. Soil Sci 134:216–225
Hartge KH, Horn R (1977) Spannungen und Spannungsverteilungen als Entstehungsbedingungen von Aggregaten. Mitt Dtsch Bodenkd Ges 25:25–33
Hillel D (1980) Fundamentals of Soil Physics. Academic Press, Lond NY
Horn R (1976) Festigkeitsänderungen infolge von Aggregierungsprozessen eines mesozoischen Tones. Diss TU Hannover
Horn R (1987) The role of structure for nutrient sorptivity of soils. Z Pflanzenernähr Bodenkd 150:13–16
Horn R, Stock J, Dexter AR (1987) The influence of soil structure on penetration resistance. Z Pflanzenernähr Bodenkd 150:342–347
Hüttl RF (1985) "Neuartige" Waldschden und Nährelemantversorgung von Fichtenbeständen (*Picea abies* Karst.) in Südwestdeutschland. Freiburger Bodenkundliche Abhandlungen, Freiburg, 16:1–195
Hüttl RF, Zöttl HW (1985) Ernährungszustand von Tannenbeständen in Süddeutschland- Ein historischer Vergleich. Allg Forstz 40:1011–1013
James BR, Riha SR (1986) pH buffering in forest soil organic horizons: Relevance to acid precipitation. J Environ Qual 15:229–234

Johansson G (1962) The crystal structures of $(Al_2(OH)_2(H_2O)_8)(SO_4)$ $2H_2O$ and $(Al_2(OH)_2$-$(H_2O)_8)(SeO_4)_2$ $2H_2O$. Acta Chim Scand 16:403–420

Jungk A, Claassen M (1986) Availability of phosphate and potassium as the results of interactions between root and soil in the rhizosphere. Z Pflanzenernähr Bodenkd 149:411–427

Khanna PK, Prenzel J, Meiwes KJ, Ulrich B, Matzner E (1987) Dynamics of sulfate retention by acid forest soils in an acidic deposition environment. Soil Sci Soc Am J 51:446–452

Kaupenjohann M (1989) Chemischer Bodenzustand und Nährelementversorgung immissionsbelasteter Fichtenstandorte in NO-Bayern. Bayreuther Bodenkundl Ber 11:1–202

Kaupenjohann M, Hantschel R (1987) Short-timed pH-buffering of disturbed and undisturbed forest soil samples. Z Pflanzenernähr Bodenkd 150:156–160

Kaupenjohann M, Hantschel R, Zech W, Horn R (1987a) Mögliche Auswirkungen "Sauren Regens" auf die Nährstoffversorgung von Wäldern. Kalibriefe, Büntehof 18:631–638

Kaupenjohann M, Hantschel R, Zech W, Horn R (1987b) Bodenextrakte zur chemischen Kennzeichnung der Nährstoffversorgung säurebelasteter Fichtenstandorte. Mitt Dtsch Bodenkd Ges 55/II:607–612

Lasaga AC (1981) Rate laws of chemical reactions. In: Lasaga AC, Kirkpatrick RJ (eds) Reviews in mineralogy, Vol 8: Mineralogical Society of America. Washington DC, Book Crafters Inc. Chelsea, Michigan, 1–68

Lee YH (1985) Aluminium speciation in different water types. In: Andersson F, Olsson B (eds) Lake Gardsjön – an acid forest lake and its catchment. Ecol Bull 37:109–119

Marschner H, Römheld V, Horst WJ, Martin P (1986) Root induced changes in the rhizosphere: importance of the mineral nutrition of plants. Z Pflanzenernähr Bodenkd 149:441–456

Martin AE, Reeve R (1958) Chemical studies of podzolic illuvial horizons. III. Titration curves of organic-matter suspensions. J Soil Sci 9:89–100

Matzner E, Ulrich B (1984) Rates of deposition, of soil internal production and of turnover of protons in two forest ecosystems. Z Pflanzenernähr Bodenkd 147:290–308

Mulder J, Van Dobben HF, De Visser PHB, Booltink HWG, Van Breemen N (1987a) Effect of vegetation cover on atmospheric deposition and soil acidification. Proc Int Symp "Acidification and water pathways". Bolkesjo I:79–90

Mulder J, Van Grinsven JJM, Van Breemen N (1987b) Impacts of acid atmospheric deposition on woodland soils in the Nederlands. III: Aluminum chemistry. Soil Sci Soc Am J 51:1640–1646

Murach D (1984) Die Reaktion von Feinwurzeln auf zunehmende Bodenversauerung. Gött Bodenkd Ber 72:1–126

Nätscher L (1987) Art, Menge und Wirkungsweise von Puffersubstanzen in Auflagehorizonten forstlich genutzter Böden des Fichtelgebirges. PhD Diss, TU München, 132 pp

Nätscher L, Schhwertmann U (1989) Mechanism and kinetics of proton buffering in acid soils. I. Organic horizons. Soil Sci Soc Am J (submitted)

Nilson SI, Bergkvist B (1983) Aluminum chemistry and acidification processes in a shallow podzol on the Swedish west coast. Water Air Soil Pollut 20:311–329

Nilsson J, Grennfelt P (1988) Reprint of the workshop report on critical loads for sulfur and nitrogen. Miljorapport, Stockholm 16:1–31

Nordstrom DK (1982) The effect of sulfate on aluminum concentrations in natural waters: Some stability relations in the system Al_2O_3-SO_3-H_2O at 298 K Geochim Cosmochim Acta 46:681–692

Nye PH, Ameloko A (1986) A comparison of measured and theoretical soil acidity diffusion coefficients over a wide range of pH. J Soil Sci 37:191–196

Palma RM, Arrigo NN, Conti ME (1984) Chemical characteristics of soil aggregates. Sci del Suelo II:93–97

Prenzel J (1982) Ein bodenchemisches Gleichgewichtsmodell mit Kationenaustausch und Aluminiumhydroxosulfat. Gött Bodenkd Ber 72:1–113

Prenzel J (1983) A mechanism for storage and retrieval of acid in acid soils. In: Ulrich B, Pankrath J (eds) Effects of accumulation of air pollutants in forest ecosystems. Reidel Pub Co, Dordrecht, 157–170

Reuss O (1980) Simulation of soil nutrient losses resulting from rainfall acidity. Ecol Model 11:15–38

Schaller G, Fischer WR (1985) Short-time pH buffering of soils. Z Pflanzenernähr Bodenkd 148:471–480

Schnoor JL, Stumm W (1985) Acidification of aquatic and terrestrical systems. In: Stumm W (ed) Chemical processes in lakes. Wiley, NY, 311–338

Schulin R, Flühler H, Mansell RS, Selim HR (1986) Miscible displacement of ions in aggregated soils. Geoderma 38:311–322

Schwertmann U (1964) Differenzierung der Eisenoxide des Bodens mit Ammoniumoxalatlösung. Z Pflanzenernähr Düng Bodenkd 105:194–202

Schwertmann U, Süsser P, Nätscher C (1987) Proton buffer compounds in soils. Z Pflanzenernähr Bodenkd 150:174–178

Soil Taxonomy (1975) Agricultural handbook No. 436. Soil Conserv Serv, US Dep Agric, Washington DC

Stahlberg S (1984) Rapid acid and base titration of soil for determination of exchangeable cations and CEC. Acta Agric Scand 34:71–83

Süsser P (1987) Art, Menge und Wirkungsweise von Puffersubstanzen in Mineralbodenhorizonten forstlich genutzter Böden des Fichtelgebirges. Dissertation, TU München, 1–35

Trautner F (1988) Entwicklung und Anwendung von Meßsystemen zur Untersuchung der chemischen und physikalischen Eigenschaften von Nebelwasser und dessen Deposition auf Fichten. Dissertation, University of Bayreuth, 1–60

Ulrich B (1981) Ökologische Gruppierungen von Böden nach ihrem chemischen Bodenzustand. Z Pflanzenernähr Bodenkd 144:289–305

Ulrich B (1986) Natural and anthropogenic components of soil acidification. Z Pflanzenernähr Bodenkd 149:702–717

Ulrich B, Mayer R, Khanna PK (1979) Deposition von Luftverunreinigung und ihre Auswirkungen in Waldökosystemen im Solling. Schr Forstl Fak Univ Gött Niedersächs Forstl Versuchsanst 58:291 S

Van Breemen N, Mulder J, Driscoll CT (1983) Acidification and alkalinization of soils. Plant Soil 75:283–308

Van Breemen N, De Visser PHB, Van Grinsven JJM (1986) Nutrient and proton budgets in four soil-vegetation systems underlain by Pleistocene alluvial deposits. J Geol Soc 143:659–666

Weast RC (1978) Handbook of chemistry and physics. 58th edn, CRC Press, West Palm Beach, Florida D-150

Zech W, Popp E (1983) Magnesiummangel, einer der Gründe für das Fichten- und Tannensterben in NO-Bayern. Forstwiss Centralbl 102:50–55

4-C Water Relations of Two Norway Spruce Stands at Different Stages of Decline

E.-D. Schulze, R. Hantschel, K.S. Werk, and R. Horn

1 Introduction

Understanding water relations is necessary to assess forest vigor, since numerous plant processes are influenced by the water status of the soil and the plant, e.g., growth, nutrient uptake, the production and release of growth substances and leaf gas exchange (Bradford and Hsiao 1982; Schulze 1986). However, the availability of water for sustaining these processes is not only dependent on control mechanisms at the individual plant level, but is also determined by the canopy structure and boundary conditions at the stand level (Jarvis and MacNaughton 1985; Jarvis 1987). The balance between rainfall and evapotranspiration at the stand level determines to a large extent the transport of substances into and out of the ecosystem. Water percolating through the soil profile carries nutrients which may be leached below the root zone into the groundwater (Ulrich 1987). The physical properties of the soil will in turn affect the availability of water and nutrients for plant uptake.

The water balance at the stand level and the water relations of trees at two forest sites was calculated to determine if it correlates with the observed features of forest decline, namely, needle-yellowing and reduced growth. One experimental site included trees at various stages of decline (Oberwarmensteinach), while trees at the other (Wülfersreuth) consisted of apparently healthy trees. Determining the water balance also provides the basis for estimations of the forest nutrient fluxes (see Chap. 4-I).

2 Assessment of the Forest Water Balance

The methods used to determine both the water relations of the trees and the water balance at the stand level are described in detail by Hantschel (1987), Werk et al. (1988), and Osonubi et al. (1988). Measurements refer to either five control plots per site (control plots, 80 m^2) or to five treatment plots per site, all of which were used for soil studies including fertilization (referred to as soil plots, 500 m^2). These soil plots contained one additional untreated control plot per site (soil control plot, 500 m^2).

The *water balance* of a defined ecosystem is determined by precipitation input, evapotranspiration output, surface runoff and seepage, and storage change

E.-D. Schulze, O.L. Lange, R. Oren (Eds.)
Ecological Studies Vol. 77

(Landsberg 1986, Benecke and Van Der Ploeg 1981). In the study reported here, surface runoff was negligible in both stands. Subsurface flow of water occurred on some soil plots with fertilizer treatments (Hantschel 1987), the results of which are not included in this chapter. The water content of the soil was close to field capacity most of the time because of a surplus of rain (Hantschel 1987). Therefore, changes in storage were small. Evaporation, transpiration, and seepage of intercepted water from the wet canopies were the most important processes in the seasonal and annual water budget of the stands at both sites. Calculations of the water balance made from direct measurements of whole tree transpiration, hydraulic properties of the soil horizons, soil water status, precipitation, and throughfall were supplemented by modeling.

Precipitation was measured in the open on each site with six rain and five snow gauges, each 0.5 m above ground, with an opening of 100 cm^2 and 918 cm^2, respectively. At the declining site, the set of rain collectors was positioned above the canopy on a 50 × 50 m square patch of a spruce plantation bordering the experimental plots. At the healthy site, the rain gauges were located in a large forest clearing (100 × 200 m) at a distance of about 100 m from the experimental plots and separated from the plots by a rim of old trees. The precipitation values of the rain or snow gauges were averaged to provide a single input value.

Canopy throughfall was measured with six rain gauges, each having a 100 cm^2 opening, and six snow gauges, each having a 918 cm^2 opening. These were positioned under the canopies of the five soil plots per site (Hantschel 1987).

Interception loss (evaporation from the wet canopy) was calculated as the difference between precipitation and canopy throughfall. Because only the variation in throughfall was quantified, the variations in interception loss was not determined. Local variations in precipitation may account for some of the variation in throughfall.

Litter evaporation was estimated to be less than 0.01 $l\,m^{-2}\,d^{-1}$ for a dry summer day (Hantschel 1987) and was therefore not incorporated into the water budget (Hantschel 1987). This process may, however, cause drying of the top layer of the litter under the canopy opening or at a low leaf-area index.

Tree transpiration was measured directly from selected green trees by a modified heat-flux method on the five control plots per site (Čermák et al. 1976, Schulze et al. 1985, Werk et al. 1988). Transpiration was also modeled from meteorological data according to Running (1984). This model was adjusted to match the measured water potential in the soil profile (Hantschel 1987). The model uses a single data set of meteorological conditions (see Chap. 1-B) for both sites; it was matched with the soil water status of a single control plot from the soil studies on each site. The model does not include the effects of soil water potential on transpiration and, therefore, overestimates the canopy water loss during dry periods. To remedy these problems, the model was used to estimate transpiration from meteorological data only when the soil was saturated at field capacity and during periods when heat-flux measurements were not available. During periods of no water stress, the model results were similar to direct measurement values. Each heat-flux value consists of an average of five control plots at each site standardized to a leaf area index of 10 (Werk et al. 1988).

Soil water tension and content was measured at five depths with five mercury tensiometers at each depth and at five locations per plot with a neutron scattering probe (Wallingford, USA) at the five soil plots per site. Tensiometer measurements were also made at two depths at five locations per plot on the five control plots per site (Hantschel 1987).

Soil hydraulic properties were derived from (1) water content/water potential curves, which were corrected to the effective pore volume (Hantschel et al. 1987) determined in the field on the soil plots; and (2) hydraulic conductivity/water potential curves, which were determined in the laboratory and under in situ situations with the values of the effective pore volume.

Soil drainage was calculated from the difference between monthly canopy throughfall and transpiration assuming constant storage capacity.

3 The Water Budget of the Healthy and the Declining Forest Sites

At the declining site, total rainfall and canopy throughfall were 17 and 11% higher respectively than at the healthy site (Table 1; Fig. 1). About 20% of precipitation was lost by evaporation from the wet canopy after rain or fog. Although there was no significant difference in transpiration between the sites when canopy transpiration was measured ($p < 0.05$) or modeled, seepage, which was calculated as the difference between throughfall and transpiration, was 30% higher at the declining site.

All components of the water budget varied greatly in seasons and years. About 40% of the precipitation fell during winter (November–April), and 60% of the total seepage to groundwater occurred then. The largest seasonal precipitation difference between the two years was observed in summer. The period from July–October 1985 was drier than in the previous year. At both sites, it was the only season in which months with a negative water balance were found. However, this period appeared shorter at the healthy site despite lower rainfall.

Based on the summer water balance, when precipitation is low the declining site is more drought-prone than the healthy one (it is interesting to note that the local name of the declining site is *Dürrberg*, which translates into "dry hill" and gives a historic reference to the droughty nature of this site). This drought sensitivity can be explained by the differences in soil texture and root distribution at the two sites. The mineral soil at the healthy site is a loamy silt to silty loam while the soil at the declining site is sandy, loamy silt. In addition, the organic layer at the healthy site is more finely decomposed (mor) compared to the more coarse humus (raw humus) at the declining site. The mineral soil at the two sites has a similar capacity to supply water to the roots if the water content is near field capacity (soil water potential of ca. –60 hPa). However, between that level and a

Table 1. The water balance of the healthy and declining forest site[a] (l m^{-2} ground area)

Time	Healthy site					Declining site				
Yr/M	Prec.	Throughfall		Trans.	Seep.	Prec.	Throughfall		Trans.	Seep.
84/7	140.8	128.2	(9.2)	75.4	53.3	167.7	151.9	(4.3)	75.4	76.5
8	144.8	125.3	(8.0)	75.4	49.9	154.7	143.2	(11.4)	75.4	67.8
9	211.0	189.3	(10.2)	68.4	120.9	251.1	217.0	(7.8)	51.0	166.0
10	76.4	52.0	(5.2)	0	52.0	79.9	60.7	(2.0)	0	60.7
11	79.1	53.1	(8.1)	0	53.1	96.2	85.4	(6.9)	0	85.4
12	33.2	18.3	(2.6)	0	18.3	40.0	40.7	(6.0)	0	40.7
85/1	112.1	93.4	(13.2)	0	93.4	140.2	114.9	(20.7)	0	114.9
2	49.7	41.9	(11.7)	0	41.9	47.4	38.9	(6.6)	0	38.9
3	29.0	28.0	(3.8)	0	28.0	46.8	26.5	(3.4)	0	26.5
4	56.2	43.1	(3.9)	0	43.1	75.3	44.6	(8.1)	0	44.6
5	96.2	85.9	(6.5)	46.6	39.3	101.2	89.2	(5.0)	48.2	41.0
6	148.0	94.2	(11.8)	55.6	38.6	202.8	161.5	(8.3)	67.9	93.6
7	87.8	70.7	(8.7)	85.2	−14.5	100.4	84.1	(4.5)	103.4	−19.3
8	91.9	74.3	(6.1)	40.3	34.0	66.1	41.2	(3.5)	51.0	−9.8
9	98.1	69.9	(8.3)	14.8	55.1	106.7	75.2	(5.8)	16.4	58.8
10	63.0	47.5	(4.7)	3.9	43.1	21.9	9.8	(2.3)	4.5	5.3
11	99.5	58.2	(11.1)	0	58.2	105.7	90.0	(4.5)	0	90.0
12	66.1	63.5	(5.5)	0	63.5	94.1	74.5	(7.8)	0	74.5
86/1	93.2	128.7	(18.3)	0	128.7	160.6	154.3	(11.8)	0	154.3
2	17.2	16.4	(1.1)	0	16.4	27.6	25.0	(5.5)	0	25.0
3	95.3	73.0	(6.6)	0	73.0	106.3	84.4	(12.1)	0	84.4
4	42.7	26.4	(2.9)	0	26.4	44.8	24.4	(10.3)	0	24.4
5	136.2	104.9	(14.3)	68.6	36.3	143.2	113.4	(6.1)	64.4	49.0
6	149.4	109.1	(16.1)	62.3	46.8	154.2	127.1	(2.6)	43.6	83.5
Averages for seasons										
1984										
7–9	496.6	446.3	(16.1)	219.2	224.1	573.5	512.1	(20.2)	201.8	310.2
10–3	379.5	288.7	(30.4)	0	288.7	450.5	346.5	(28.0)	0	346.5
1985										
4–6	300.4	223.2	(20.8)	102.2	121.0	379.3	295.3	(14.6)	116.1	179.2
7–9	277.8	214.9	(16.2)	140.3	74.3	273.2	200.6	(8.3)	170.8	29.7
10–3	434.3	375.5	(40.0)	3.9	371.6	516.2	438.0	(26.6)	4.5	434.4
1986										
4–6	328.3	240.4	(32.6)	130.9	109.5	342.2	268.9	(9.8)	108.0	137.5
Annual averages of 2 years										
4–9	701.6	562.4		298.3	264.5	784.1	638.5		300.6	328.3
10–3	406.9	332.1		0	330.2	483.4	392.2		0	390.5
Sum	1108.5	894.5		298.3	594.7	1267.5	1030.7		300.6	718.8

[a] Precipitation was measured once at each site, Throughfall values are an average of 5 plots per site (SE). Transpiration was measured as mass flow in 85/7 until 85/10 (Werk et al. 1988). All other transpiration data are calculated from an energy balance model (Hantschel 1987). Seepage is calculated as the difference between throughfall and transpiration. The input data for 84/7 and 84/8 were incomplete and have been partially extrapolated.

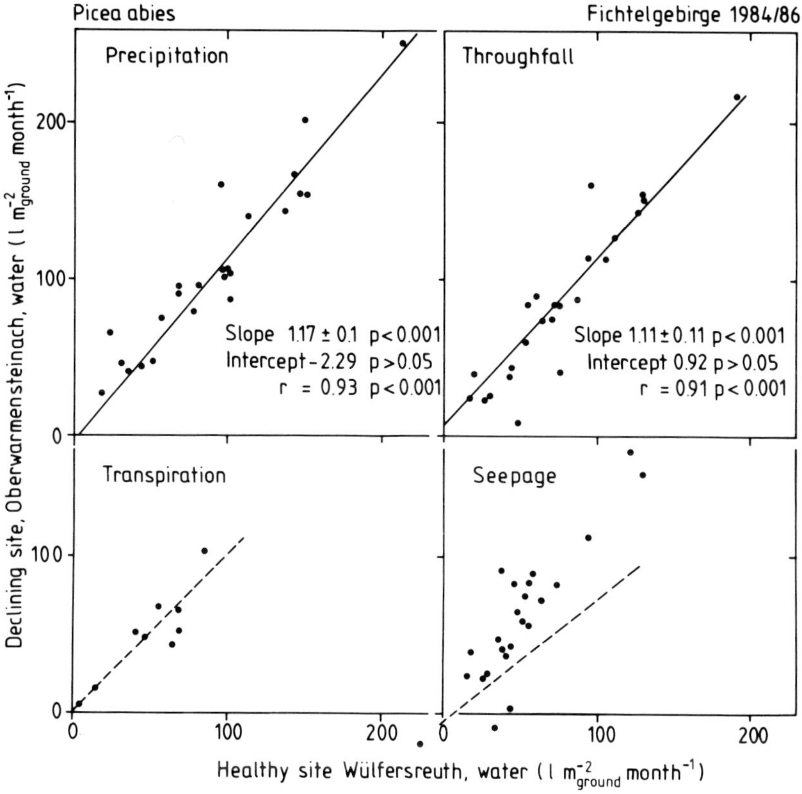

Fig. 1. Precipitation, canopy throughfall, transpiration and seepage of water at the healthy (Wülfersreuth) as compared to the declining (Oberwarmensteinach) forest site for the period from July 1984 to June 1986 (data from Hantschel 1987 and from the German Weather Bureau)

water potential of –300 hPa, the range where the soil water potential fluctuates most of the year (Fig. 6), water flux is higher and water is absorbed by the roots at the healthy site with greater ease than at the declining site. This is indicated from the shallower slopes of K with ψ at that part of the relationship in the healthy site (Fig. 2). Although data on the organic layer are not available, similar differences should occur between the sites as a result of the differences in the organic layer texture. During the dry season, the trees at the healthy site grew more roots into the mineral horizon while trees at the declining site did not (Fig. 5; see also Meyer et al. 1988). Thus, as the organic layer dried and water supply from this layer was curtailed at both sites, the larger root system at greater soil depth at the healthy site was able to more easily absorb extractable water; in comparison, at the declining site, the smaller root system in greater soil depth made water more difficult to extract.

In addition, pore discontinuity was high at the healthy site. A zone of high hydraulic conductance (5–50 cm depth in the mineral horizon) is enclosed

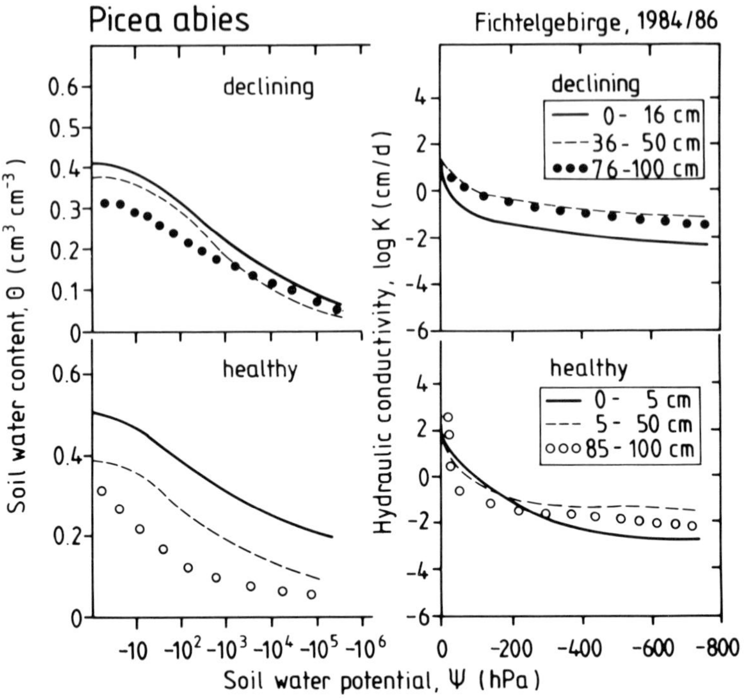

Fig. 2. Soil water content and hydraulic conductivity in relation to soil water potential of different soil horizons at the healthy and declining sites (data from Hantschel 1987)

between two zones of lower conductance (Fig. 2), which probably results in a higher water content there and potentially increases plant-available water. Such conditions were not found at the declining site where the hydraulic conductance first increased with depth and then remained constant, thus permitting uninterrupted water flow to a depth far below the restricted root system at that site.

Canopy transpiration similarities between both sites were derived from modeling and confirmed by direct measurements of water flow through the stem (Werk et al. 1988). On cloudless days, the water loss from the canopy was linearly correlated with the vapor pressure deficit in the atmosphere (Fig. 3). This correlation indicates a strong coupling of the canopy to the atmosphere (Jarvis and McNaughton 1985). Although the declining site has a southeast exposure and the healthy site has a northwest exposure, the declining site receives more clouds and rain than the healthy site. This difference in precipitation reduces transpiration on a seasonal basis and compensates for the differences in exposure between the sites. Canopy conductance (slope of the transpiration/vapor deficit relationship) is not significantly different between the sites (Werk et al. 1988). Temperature and vapor pressure deficit changed, whenever light changed (Fig. 4), but light may have been more important in regulating transpiration than the vapor pressure deficit during dawn and dusk.

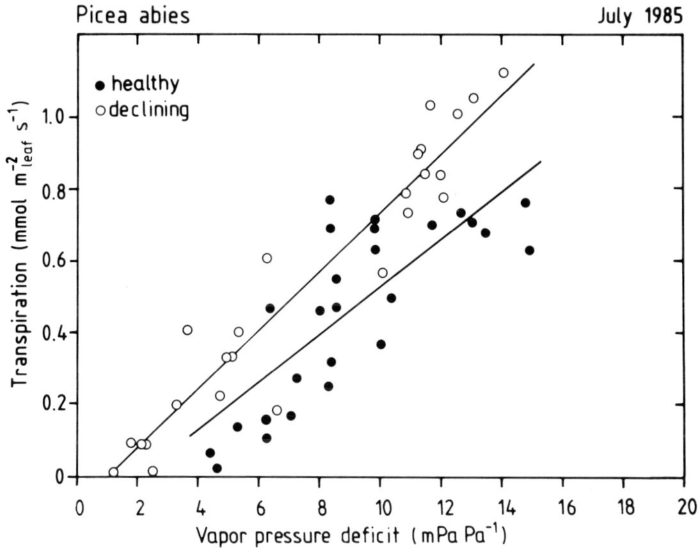

Fig. 3. Canopy transpiration as measured by xylem flow increased with the vapor pressure deficit of the atmosphere at the healthy and declining site (from Werk et al. 1988)

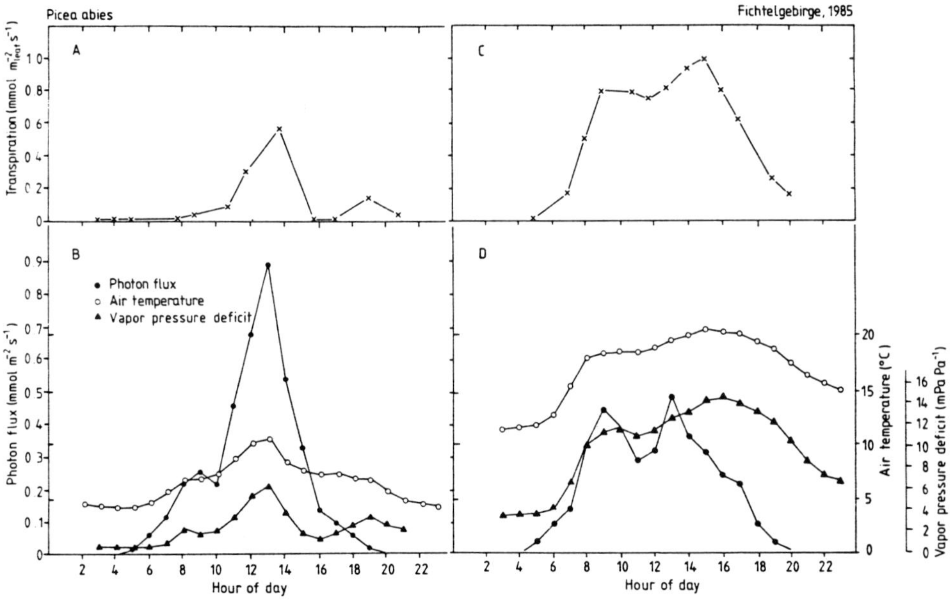

Fig. 4. Daily course of canopy transpiration and climatic factors for an overcast, humid (**A,B**) and a bright, dry (**C,D**) day (Werk et al. 1988)

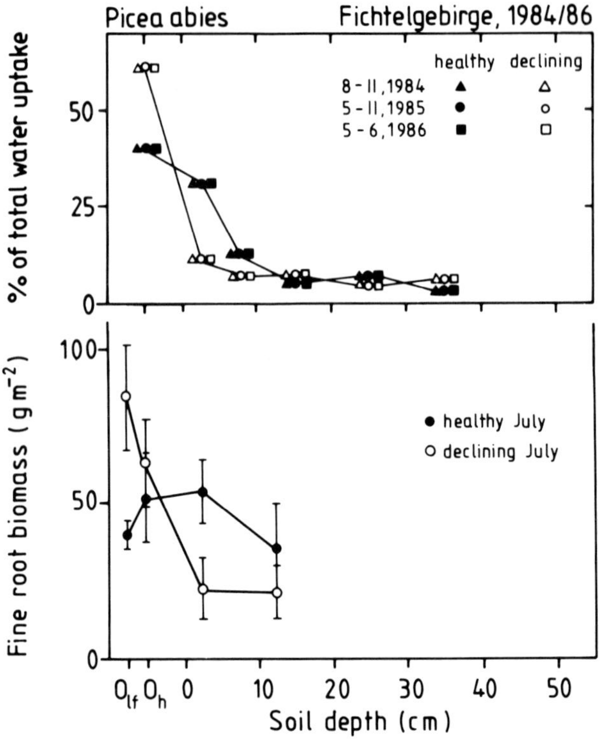

Fig. 5. Water uptake in % of the total (recalculated from Hantschel 1987) and distribution of fine roots in different soil horizons (Chap. 4-E)

4 Water Uptake and Soil Water Status

The distribution of water uptake by trees was derived from the soil water budget on the soil plots (recalculated from Hantschel 1987). A large difference in water extraction from the organic and mineral soil horizons was found for both sites (Fig. 5). The declining site absorbed more than 60% of its water from the litter horizon and an additional 10% from the 0–5 cm soil horizon. In contrast, the trees at the healthy site absorbed only 40% of their water from the litter horizon and an additional 30% from the 0–5 cm soil horizon. This difference between the healthy and declining sites resulted from differences in root distribution. The declining site had more roots in the litter and a distinct reduction in fine root mass in the mineral soil horizon (Meyer et al. 1988), particularly during the summer when most of the transpiration occurred. It is shown in Chapter 4-E that these differences in root distribution are correlated with the different soil calcium to aluminum ratios of the two sites. These different ratios are related to the different long-term proton inputs (see Chap. 4-B).

The soil water potential in the mineral horizon was generally very high at both sites (Fig. 6). However, during periods of low rainfall, water potentials at 15 cm depth (below the main root horizon) decreased sharply at both sites. A low

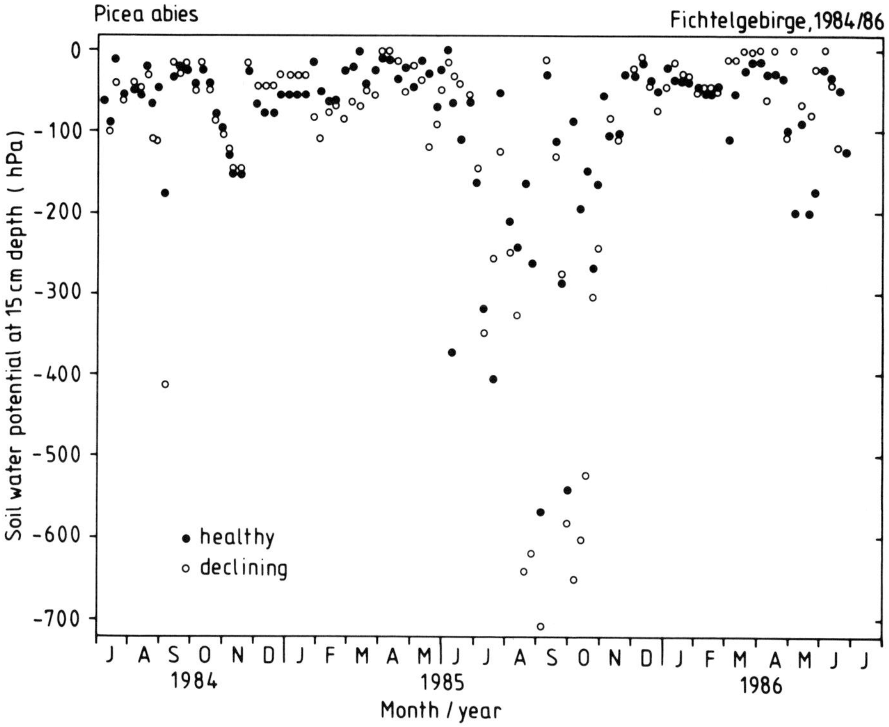

Fig. 6. Soil water potentials at 15 cm soil depth of the healthy and declining sites for the periods from July 1984 to June 1986. The measurements were taken once per week (Hantschel 1987)

water potential in the mineral soil indicates that the litter layer may have become very dry, at least in times of low precipitation. Because the litter layer also has low water-holding capacity and high hydraulic conductivity, the root system conducts water rapidly so that the availability of water to trees in dry periods is strongly reduced.

The decrease in plant water potential in dry periods affected the nutrient uptake of the trees. The magnesium concentration in the xylem sap decreased with decreasing water potential, and this reduction occurred at a higher potential than the equivalent drop in the nitrogen xylem sap concentration (see Chap. 4-F, Osonubi et al. 1988).

5 Conclusions

The water balance shows that the declining site received significantly more rain. This was the major variable that determined the total water budget at each site, since evaporation from the wet canopy and canopy transpiration were similar for both sites. Despite higher total rainfall, the declining site was more prone to short-term drought in periods of little rain than the healthy site because of differences in soil properties. Soil chemical properties affect root distribution and therefore the horizons for water uptake. At the declining site, tree root growth is restricted primarily to the litter horizon and trees depend more on water uptake from the litter than at the healthy site. It is shown in Chapter 4-E that this difference in root distribution results from a lower calcium to aluminum ratio in the mineral soil at the declining site. Increased root growth in the humus layer at the declining site is an indication of maximizing uptake of nutrients and water from this layer. This increase in root growth in the organic layer results in a smaller volume of soil which roots can effectively exploit, and causes lower water potentials at the declining site, thus restricting nutrient uptake (Osonubi et al. 1988; Chap. 4-F). Because low water potentials reduce the magnesium uptake more strongly than nitrogen uptake (Chap. 4-F). Repeated periods of weak drought could reduce the uptake of magnesium at the declining site while not affecting the uptake of nitrogen. In Chapter 4-H, such nutrient imbalance and nutrient stress is linked to forest decline.

Acknowledgement. R. Oren contributed ideas to this chapter and helped in its preparation.

References

Benecke P, Van Der Ploeg RR (1981) The soil environment. Encycl Plant Physiol NS vol 12A:539–560

Bradford KG, Hsiao TC (1982) Physiological responses to water stress. Physiological Plant Ecology II. In: Lange OL, Nobel PS, Osmond CB, Ziegler H (eds) Encyclopedia Plant Physiol. Springer, Berlin Heidelberg New York, NS vol 12B:263–324

Čermák J, Kučera J, Penka M (1976) Improvement of the method of sap flow rate determination in full-grown trees based on heat balance with direct electric heating of the xylem. Biol Plant 18:105–110

Hantschel R (1987) Wasser- und Elementbilanz von geschädigten, gedüngten Fichtenökosystemen im Fichtelgebirge unter Berücksichtigung von physikalischer und chemischer Bodenheterogenität. Bayreuther Bodenkd Ber 3: 1–219

Hantschel R, Durner W, Horn R (1987) Die Bedeutung von Porenheterogenität für die Erstellung der pF/WG- und K/psi-Kurven. Mitt Dtsch Bodenkd Ges 53:403–409

Haverkamp R, Vauclin M, Touma J, Wierenga PJ, Vachaud G (1977) A comparison of numerical simulation models for one-dimensional infiltration. Soil Sci Soc J 41:285–294

Jarvis PG (1987) Water and carbon fluxes in ecosystems. In: Schulze ED, Zwölfer H (eds) Potentials and limitations of ecosystem analysis. Ecol Stud vol 61:50–67. Springer, Berlin Heidelberg New York Tokyo

Jarvis PG, McNaughton KG (1985) Stomatal control of transpiration: scaling up from leaf to region. Adv Ecol Res 15:1–49

Landsberg JJ (1986) Physiological ecology of forest production. Academic Press, Lond NY, 191 pp

Meyer J, Werk KS, Oren R, Zimmermann R, Schulze ED (1988) Performance of *Picea abies* (L.) Karst. at different stages of decline V. Root tips and ectomycorrhiza development and their relations to above ground and soil nutrients. Oecologia 77:7–13

Osonubi R, Oren R, Werk KS, Schulze ED (1988) Performance of *Picea abies* (L.) Karst. stands at different stages of decline. IV. Xylem sap concentrations of magnesium, calcium, potassium, and nitrogen. Oecologia 77:1–6

Running S (1984) Microclimate control of forest productivity: analysis by computer simulation of annual photosynthesis/transpiration balance in different environments. Agric For Meteorol 32:2–22

Schulze ED (1986) Carbon dioxide and water exchange in response to drought in the atmosphere and in the soil. Annu Rev Plant Physiol 37:247–274

Schulze ED, Čermák J, Matyssek R, Penka M, Zimmermann R, Vasicek F, Gries W, Kučera J (1985) Canopy transpiration and water fluxes in the xylem of the trunk of *Larix* and *Picea* trees – a comparison of xylem flow, porometer and cuvette measurements. Oecologia 66:475–483

Ulrich B (1987) Stability, elasticity, and resiliance of terrestrial ecosystems with respect to matter balance. In: Schulze ED, Zwölfer H (eds) Potentials and limitations of ecosystem analysis. Ecol Stud 61:11–49, Springer, Berlin Heidelberg New York Tokyo

Werk KS, Oren R, Schulze ED, Zimmermann R, Meyer J (1988) Performance of *Picea abies* (L.) Karst. stands at different stages of decline. III. Canopy transpiration of green trees. Oecologia 76:519–524

4-D CO_2 Assimilation and the Carbon Balance of Healthy and Declining Norway Spruce Stands

R. OREN and R. ZIMMERMANN

1 Introduction

The declining Norway spruce (*Picea abies* L. Karst.) stand's lower wood production and needle chlorosis could have been caused by many factors (Oren et al. 1988a; Chap. 3-D). Climatic conditions (Fig. 1; Chap. 1-B) and stand characteristics affect site occupation and maintenance respiration (Table 1; Chap. 2). However, climatic and stand differences between the stands, as well as the water relations (Werk et al. 1988; Chap. 4-C), were excluded as possible reasons for decline, because they were either similar in both stands or more favorable for tree growth at the declining site. Both direct damage and nutrient stress to needles could have produced the apparent symptoms of decline and caused a reduction in CO_2 assimilation so as to reduce growth. No direct damage to needles was detected when green needles at both sites and needles ranging from green to yellow at the declining site were studied (Lange et al. 1985; Zimmermann et al. 1988; Chap. 3-D). In this chapter, some results are presented which indicate that only nutrient stress was involved in reducing photosynthesis at the declining site. This reduction in CO_2 assimilation is assessed as to whether a reduction in CO_2 uptake could explain the lower growth found at the declining site.

Needle-yellowing may be associated, in part, with several changes in needle structure and function related to decline, but not causing it (see Chap. 2). Thus, a reduction in foliage function due to primary causes can only be identified by comparing apparently healthy foliage at both healthy and declining sites. By including foliage that does not appear healthy, the range of data is broadened; correlations can be made to establish findings based on green needles and to generate hypotheses about possible reasons for a decline in apparently healthy needles (Chap. 3).

Thus, a healthy-looking tree was chosen from each plot at both sites. The information generated from these trees (i.e., photosynthesis, transpiration, and seasonal changes in carbohydrates and nutrients) represented the conditions of healthy trees with no decline symptoms. In contrast, biomass, growth, root distribution and characteristics, and soil parameters all represented stand level conditions. In two of the plots at the declining site, these conditions included trees with yellow, thin crowns at the start of the study and some trees in an additional plot became chlorotic during the growing season. To support the conclusion generated from studying green trees at the two sites, measurements were also made on yellow trees at the declining site.

E.-D. Schulze, O.L. Lange, R. Oren (Eds.)
Ecological Studies Vol. 77

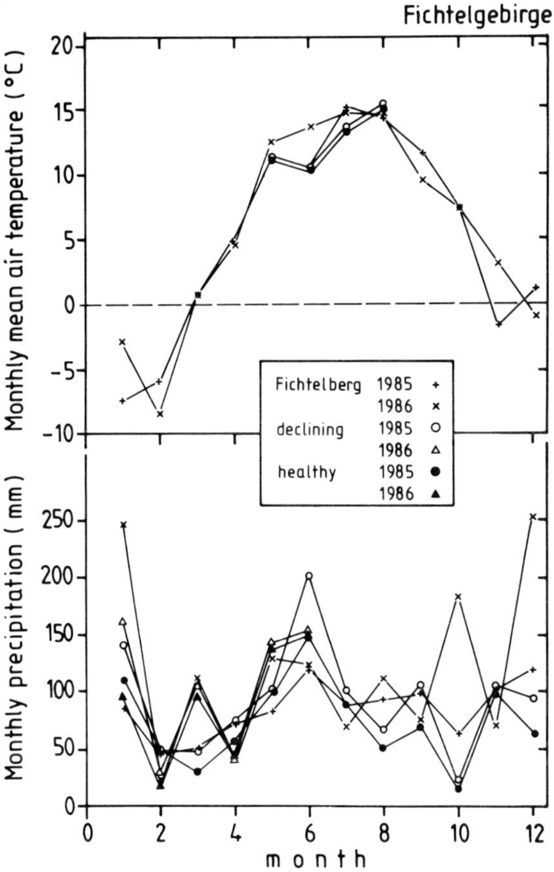

Fig. 1. Seasonal course of air temperature [after Deutsches Wetteramt 1987 and Trautner (pers. comm.)] and precipitation [after Hantschel 1987 and the Deutsches Wetteramt 1987] at healthy and declining sites

The seasonal pattern of carbohydrate concentration in all tree parts, save the stem, showed that carbohydrates accumulated faster in trees at the healthy site before bud-break and that carbohydrates were used faster after bud-break (Fig. 2; Oren et al. 1988a). Figure 3 (top, after Höll 1985) presents the seasonal pattern in the stem of a Norway spruce tree in a different location to indicate the annual trend in carbohydrates accumulation and use. The difference between the sites is even more pronounced when the total difference in carbohydrate accumulation and use is presented on an area basis (Fig. 3, *bottom*; Oren et al. 1988a). This was calculated by multiplying the concentrations of carbohydrates in each plant component and the biomass per unit of area of the respective component. The mean concentration or biomass of all 10 plots was used, unless site means were significantly different ($P \leq 0.05$; Oren et al. 1988a), in which case the site means were used in the calculations.

Table 1. Stand description, biomass, and growth of healthy and declining Norway spruce stands located in the Fichtelgebirge, northeast Bavaria, FRG (Oren et al. 1988a)[a]

Site	Healthy		Declining		t-test significance
Stand parameter					
Age (yrs)	30	(1)	30	(2)	n.s.
Tree density per m^2	0.31	(0.02)	0.44	(0.14)	n.s.
Basal area at 1.3 m (cm^2 m^{-2})	32.3	(1.9)	26.8	(2.7)	n.s.
Quadratic mean diameter at 1.3 m (cm)	11.5	(0.4)	9.7	(1.1)	n.s.
Height (m)	10.04	(0.51)	7.93	(0.90)	n.s.
Height to base of live crown (m)	3.62	(0.40)	1.59	(2.5)	0.004
Leaf-area index (m^2 m^{-2})	10.6	(0.6)	11.9	(1.1)	n.s.
Biomass (g m^{-2} ground)					
Needle	1767	(103)	2009	(263)	n.s.
Twig	637	(88)	729	(233)	n.s.
Branch	734	(38)	810	(106)	n.s.
Stem					
Wood	13251	(1424)	7288	(944)	0.010
Bark	1135	(151)	693	(34)	0.041
Roots (to 20 cm in mineral soil)					
Fine (< 2 mm)	196	(30)	165	(30)	n.s.
Large (> 2–10 mm)	156	(39)	204	(31)	n.s.
Growth (g m^{-2} ground a^{-1})					
Needle 1. from equation	318	(41)	313	(30)	n.s.
2. from litter fall	216	(35)	182	(15)	n.s.
Twig	99	(9)	98	(9)	n.s.
Branch	69	(4)	77	(10)	n.s.
Stem					
Wood	1073	(131)	693	(75)	0.043
Bark	41	(15)	44	(9)	n.s.
Roots					
Large (2–10 mm)	0		0		–
Fine (< 2 mm)	18	(3)	12	(1)	n.s.
Quadratic diameter increment at 1.3 m (cm)	0.35	(0.01)	0.38	(0.05)	n.s.
Basal area increment at 1.3 cm (cm^2 m^{-2})	2.0	(0.1)	2.2	(0.3)	n.s.
Above ground annual relative growth rate (g g^{-1} above ground weight)					
Stemwood only	0.06	(0.01)	0.06	(0.01)	n.s.
Total above ground	0.09	(0.01)	0.10	(0.01)	n.s.
Above ground annual growth efficiency (g m^{-2} leaf area)					
Stemwood only	100.8	(9.9)	58.3	(3.5)	0.010
Total above ground	140.9	(11.9)	92.1	(2.4)	0.013

[a]One SE is given when appropriate, n = 5. Root biomass and production were estimated from Schneider et al. (Chap. 4-E), according to Vogt et al. (1986).

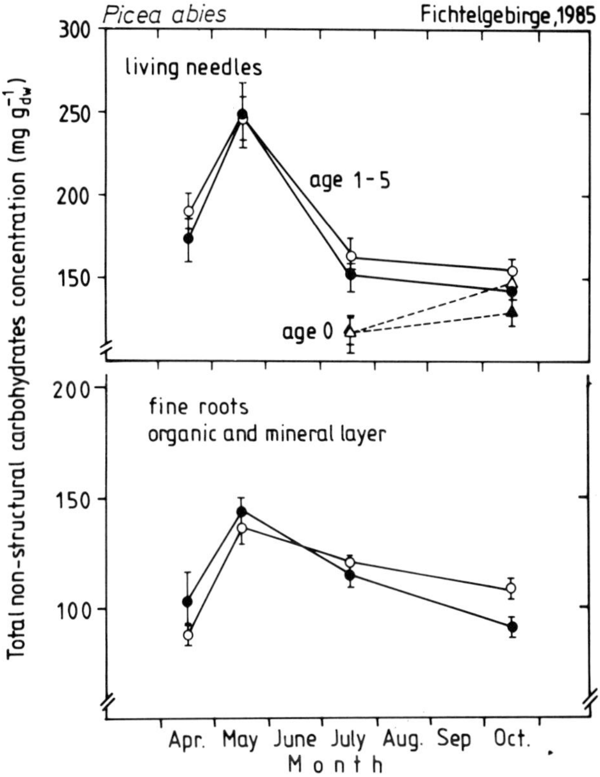

Fig. 2. Seasonal course of total nonstructural carbohydrate (X ± SE) concentration in 1-to-5-year old (*circles, full line*) and current-year (*triangles, broken line*) needles (*top*) of green trees and in fine roots (< 2 mm, *bottom*) in plots in healthy and declining Norway spruce stands (Oren et al. 1988a)

Both photosynthesis and photosynthate demand for growth were impaired at the declining site. However, it is impossible to determine which of the two mechanisms was responsible for the decline. It seems that growth was limited less by carbohydrates than by nutrient supply. Despite a similar level of stored carbohydrates at both sites at the time of bud-break, demand for carbohydrates by trees at the declining site was lower than that at the healthy site.

Measurements of photosynthetic properties helped to reject direct damage as an explanation of the decline (Chap. 3-D) and supported the hypothesis that a nutritional disharmony is the cause of both a lower photosynthetic rate and lower growth.

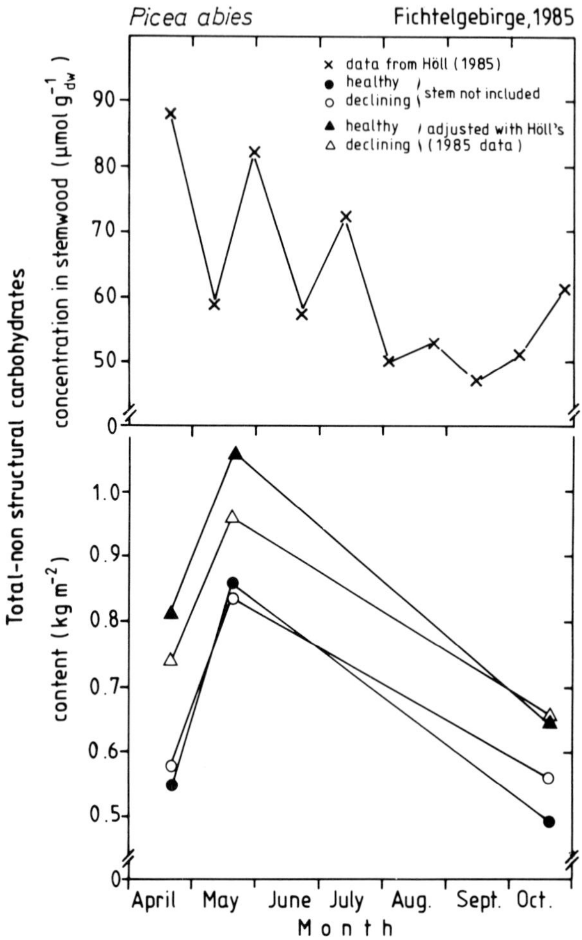

Fig. 3. Seasonal course of total nonstructural carbohydrate in stemwood of Norway spruce trees near Munich, FRG (*top*, after Höll 1985), and the sum of all components expressed by ground area (*bottom*; after Oren et al. 1988a)

2 Photosynthesis and Stomatal Response

The evergreen habit of Norway spruce implies that the photosynthetically active tissue of needles contributes to the total carbon gain of spruce for at least 5 years (Fuchs et al. 1977; Schulze et al. 1977). During this time, the needles are permanently exposed to varying climatic conditions and pollution. Several authors have suggested that forest decline in Central Europe is caused by an exposure to toxic substances which damages needles directly and leads to changes in either the chemical composition and structure of the cuticle or membranes (Heber et al. 1987; Chap. 3-A), the reactivity of stomata to climatic changes, or to malfunctions and destruction of the photosynthetic apparatus (Chap. 3-D). Others have proposed that the loss of essential nutrients from the needle by leaching leads to dysfunctioning enzymatic processes.

Possible pathways of damage can be isolated by studying carbon uptake at the needle level. This allows short-term processes, including stomatal regulation of gas exchange, to be observed and the annual carbon budget of stands to be assessed.

Comparisons based on green needle measurements at both the declining and healthy sites were used to evaluate the possible effects of air pollution before visible symptoms had occurred. Differences between sites should have increased with needle age as needles had been exposed to a larger dose of pollutants. In addition, a study of needles ranging from green to yellow were used to assess how chlorotic needles must be before photosynthesis is appreciably reduced.

Measurements of apparent photosynthesis, transpiration, and leaf conductance were carried out at both sites from 1984–1988 during the growing season. Site comparisons of assimilation rates were carried out with two CO_2/H_2O-porometers (Schulze et al. 1982; Zimmermann et al. 1988). Diurnal courses of photosynthetic uptake and photosynthetic response to changes of single climatic factors were measured by four climatized gas exchange chambers (Schulze et al. 1972). Maximum assimilation rates under nonlimiting climatic conditions were measured by a minicuvette system (Lange et al. 1986). All gas exchange rates were related to total leaf area unless stated otherwise.

2.1 Gas Exchange Rates in Relation to Site and Needle Age

If atmospheric pollutants at the declining site affected photosynthesis directly, then green needles there should have shown lower rates of photosynthesis in comparison to the healthy stand. And since older needles were exposed to emissions over several growing seasons, older needles should have received a greater pollution dose. Thus, a decrease in assimilation rates with needle age should have been greater at the declining site. Both hypotheses were rejected by observations (Zimmermann et al. 1988) as reported by Lange et al. (Chap. 3-D). These observations are briefly discussed here.

Measurements of gas exchange rates of green needles under controlled conditions, within the range of conditions which occur naturally, were either similar at both sites or higher at the declining site (Zimmermann et al. 1988; see Fig. 3, Chap. 3-D). The similarity was maintained when the effect of stomatal conductance on CO_2 uptake was excluded (Fig. 4). The rates of photosynthesis were similar to those reported for healthy spruce elsewhere (Pelkonen and Luukkanen 1974; Schulze et al. 1977). In addition, leaf conductance values were similar to rates reported by Matyssek (1985) and Keller and Häsler (1984). It appears that the gas exchange rates of green needles at the two sites in the Fichtelgebirge have not been affected by anthropogenic pollution and are not related to forest decline.

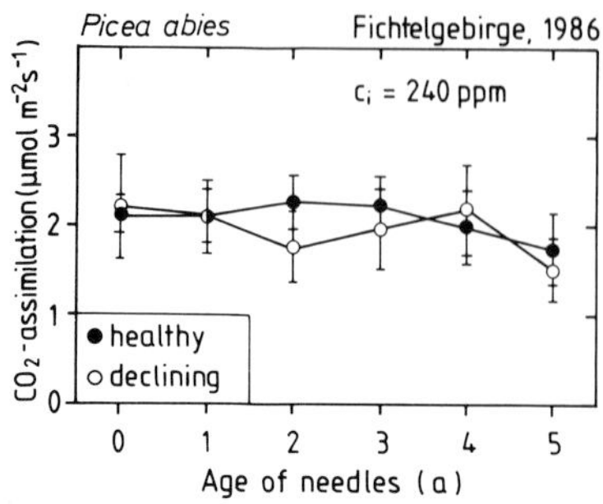

Fig. 4. CO_2-assimilation rates of Norway spruce needles of different age classes in the healthy (Wülfersreuth) and the declining (Oberwarmensteinach) Norway spruce stands, calculated for a leaf conductance (g) of 30 mmol m^{-2} s^{-1} from the relationships of assimilation and leaf conductance. Values are on an all-sided needle-area basis; *vertical bars* indicate ± 1 SE of the estimate (Zimmermann et al. 1988)

Green needles at both sites showed a reduction in photosynthetic rate and stomatal conductance with needle age (Zimmermann et al. 1988; see Fig. 3 in Chap. 3-D). This relationship is commonly observed in conifers (Fuchs et al. 1977). To exclude the age-dependent effect of stomatal conductance on CO_2-assimilation rates, comparisons were made at a constant level of leaf conductance. CO_2-assimilation rates were similar at both sites and did not change with needle age, and so there was no indication of reduced photosynthesis as a dose response to atmospheric pollution. Gas exchange properties of needles of all ages at the two sites may have been similar because high levels of SO_2 occur only in winter when assimilation is very low. In addition, high winter SO_2 concentration may have no effect on the photosynthesis in the following growing season (MacLean et al. 1986). Finally, summer ozone levels were not high at the declining site (Oren et al. 1988a).

2.2 Stomatal Conductance in Relation to Changes in Climatic Factors

Stomatal closure, which limits CO_2 uptake directly by increasing the diffusive resistance between atmosphere and mesophyll, may occur during exposure to such atmospheric pollutants as SO_2 (Guderian 1985), ozone (Keller and Häsler 1984), or a combination of both (Olszyk and Tibbitts 1981a,b; Gross 1987). Stomatal opening leads to high transpiration water loss and a reduction of assimilation due to pollutants' effects in the mesophyll (Guehl and Aussenac 1987). Stomatal opening has been observed during fumigation experiments with SO_2 and has been implicated in the forest decline of areas highly polluted with SO_2.

In the Fichtelgebirge, green and partly yellow needles of all age classes in the declining stand responded similarly to changes in air humidity and photon flux density (Zimmermann et al. 1988; see Fig. 11 in Chap. 3-D). In addition, the absolute level of leaf conductance did not change consistently with age. Thus, atmospheric pollutants at the declining site did not affect the ability of stomata to respond to changes in light or humidity, or cause permanent stomatal paralysis even in older needles exposed to the pollutants for many years.

2.3 Diurnal Course of Assimilation

Total uptake of CO_2 is limited by climatic factors as well as soil parameters, i.e., the water supply of trees. A low water supply can lead to stomatal closure or mesophyll effects because of a reduction in water potential during the dry season. Several authors have discussed the impact of drought on assimilation rate (Rehfuess 1987) or proposed that assimilation becomes limited because of midday stomatal closure; this midday depression has been observed in other species (Lange et al. 1987). The diurnal photosynthesis courses of *P. abies* needles in the Fichtelgebirge over an entire growing season showed no depression in assimilation (Fig. 5; Zimmermann et al. 1988). The decreasing water potential

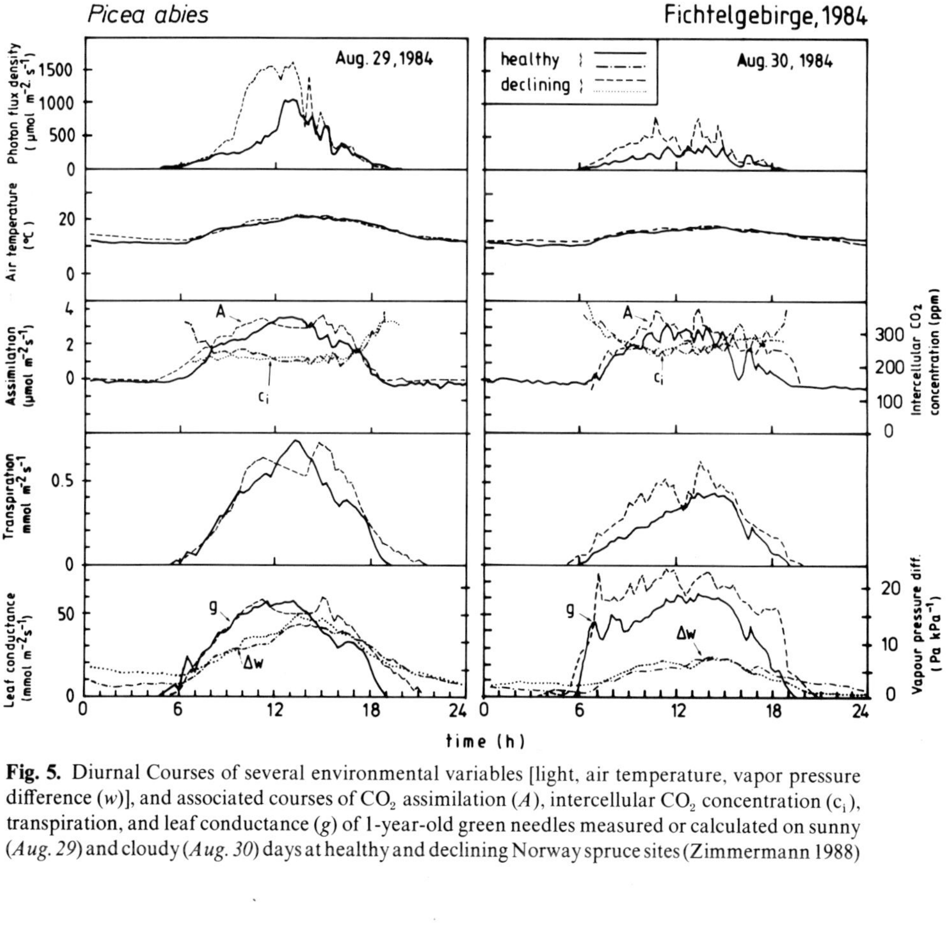

Fig. 5. Diurnal Courses of several environmental variables [light, air temperature, vapor pressure difference (w)], and associated courses of CO_2 assimilation (A), intercellular CO_2 concentration (c_i), transpiration, and leaf conductance (g) of 1-year-old green needles measured or calculated on sunny (*Aug. 29*) and cloudy (*Aug. 30*) days at healthy and declining Norway spruce sites (Zimmermann 1988)

during an extended drought period was not associated with a depression in the photosynthetic rate (Zimmermann 1985).

The model of CO_2 uptake, described in the next section, assumes that the sites have similar daily uptake rates. While the diurnal course of photosynthesis was not identical at the two sites, the total daily carbon uptake of both sites (i.e., the area under the curves) was similar in both sunny and cloudy days (Fig. 5; Zimmermann 1985).

3 Seasonal Carbon Balance

To link the estimation of growth during the growing season (Oren et al. 1988a) to the measured CO_2 uptake, models of the carbon uptake by Norway spruce in the Fichtelgebirge were constructed through multiple regression. These were for each needle age-class and month based on the measurements of diurnal courses from May to October, 1986. In this stepwise regression, driving variables were light, temperature, the water vapor pressure difference, and external CO_2 concentration in different combinations. Sums of CO_2 uptake were calculated for 5-min intervals from a data set for the declining stand for the 1986 growing season. Data on climatic conditions within the canopy were not available for 1985 and a complete data set of daily assimilation was not available for the healthy stand.

Figure 1 shows that there were some climatic differences between the 2 years which, based on the models, would result in somewhat lower annual CO_2 uptake in 1985 than in 1986. Because of similar climate (Fig. 1) and a similar leaf-area index (Table 1) and because there were no appreciable differences in gas exchange by green needles at the two sites, annual CO_2 uptake was calculated for both sites using the models derived for green foliage in the declining stand. As a result, estimated monthly (Fig. 8) and annual (Table 2) carbon uptake was identical, at 4100 g m^{-2} CH_2O for the two sites (Table 2). The carbohydrates produced during the growing season were supplemented by using some of the stored carbohydrates (Fig. 3 and Table 2).

Carbohydrates are used during the growing season in maintenance respiration, tissue production, and production respiration (Landsberg 1986). Maintenance respiration is a function of temperature and the living biomass (Amthor 1984). The temperature for calculating aboveground respiration was the same as that used in the CO_2 uptake models. Root respiration was calculated from soil temperature measured during the 1985 growing season in the organic- and mineral-soil layers at the two sites (Fig. 6).

The average living biomass in a given component was used when there was no significant difference between the sites (Table 1). Otherwise, the component biomass of each site was used. The equations for the relationships between respiration rate and temperature were recalculated from Kinerson's (1975) equations for pine. Branch surface area was estimated by multiplying branch weight per ground area by 9.7 cm^2 g_{dw}^{-1}. Stem surface area was estimated from the dimensions of the average tree, assuming that it had a cylindrical shape to the

Table 2. The carbon balance of healthy and declining Norway spruce stands located in the Fichtelgebirge, northeast Bavaria, FRG[a]

		Healthy		Declining
Photosynthetic uptake:			4100	
Change in carbohydrates:		+ 163		+ 81
Total carbohydrate input:		4263		4181
Maintenance Respiration:	Branches		160	
	Trunk	140		108
	Roots[b]	95		95
	Roots[c]	65		60
Production:	Stemwood	1073		693
	Bark (Stem)		21	
	Branches		73	
	Twigs		99	
	Needles		199	
	Roots[d]		15	
Growth Respiration:	Stemwood	720		340
	Bark (Stem)		26	
	Branches		42	
	Twigs		57	
	Needles		121	
	Roots[e]		10	
Total of respiration and growth:		2915		2118
Input − (respiration + growth) :		1348		2063
Estimated belowground production[e] :		917		
belowground growth respiration :		431		

[a] All values are in g CH_2O m^{-2} ground. See text for explanations and assumptions.
[b] Roots larger than 10 mm in diameter were assumed to be the equivalent of 6-m-long roots with a diameter of 20 mm m^{-2}.
[c] Roots less than 10 mm in diameter from Schneider et al. (Chap. 4-E, this Vol).
[d] Estimated production of roots smaller than 10 mm in diameter based on insufficient number of samples (Table 1).
[e] Belowground production and respiration (assuming 0.47 g CH_2O g^{-1}_{dw}) estimated from the difference between input and use of carbohydrates at the healthy site (see text for explanation).

base of the crown and a cone shape within the crown. Needle and twig respiration was accounted for by measuring net CO_2 uptake of intact twigs. Respiration of main roots was estimated by assuming an equivalent 6-m-long root with an average mean diameter of 20 mm per m^2 ground using the equation for branches and stem. Fine-root respiration was estimated from root weight per ground area in each soil layer and site, using an equation relating respiration rate to root weight (Kinerson 1975).

Production of the different tree components where no significant site difference was found was averaged for the two sites (Table 1). Only stemwood

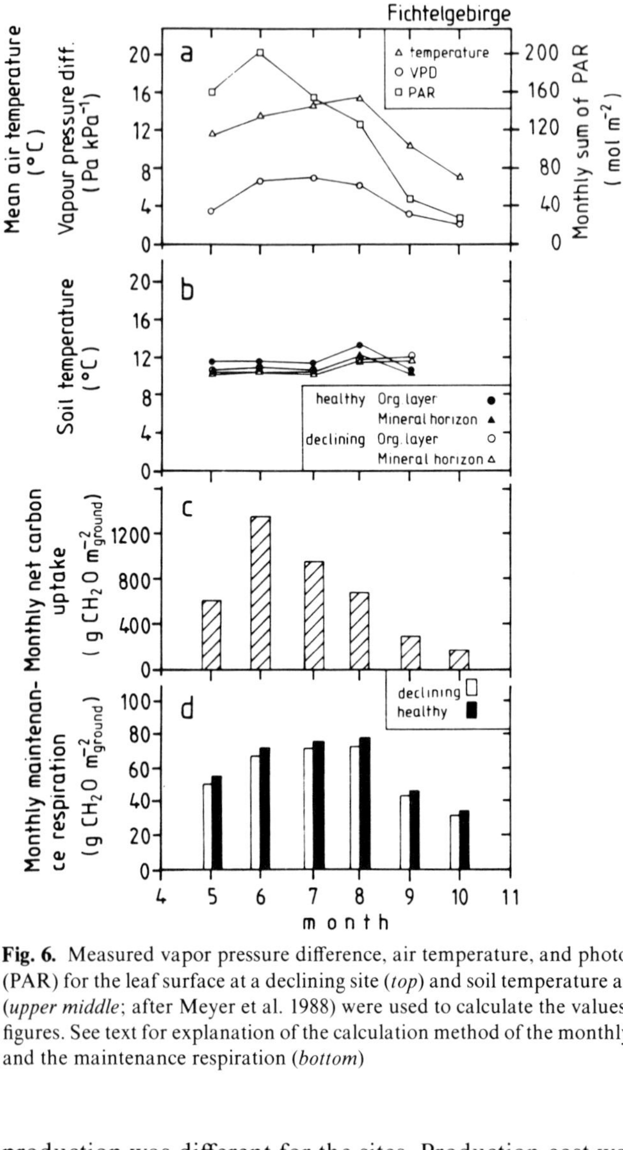

Fig. 6. Measured vapor pressure difference, air temperature, and photosynthetic active irradiation (PAR) for the leaf surface at a declining site (*top*) and soil temperature at declining and healthy sites (*upper middle*; after Meyer et al. 1988) were used to calculate the values presented in the lower two figures. See text for explanation of the calculation method of the monthly CO_2 uptake (*lower middle*) and the maintenance respiration (*bottom*)

production was different for the sites. Production cost was calculated based on values given for the different organs of pine (Chung and Barnes 1977).

The net carbon gain combined with carbohydrates used from storage, between the start and end of the growing season, was 1348 g CH_2O m^{-2}_{ground} more than the estimated use of carbohydrates at the healthy site. This difference may originate from three sources:

1. The first source is related to weather conditions. Weather conditions were better in 1986 than in 1985, resulting in a high estimate of carbon uptake which could not be balanced with the production which occurred in 1985.

2. The second source is related to root production estimation. The study was not designed so that root production could be estimated reliably. Thus, it resulted in a very low fine-root production and no estimate of production of roots larger than 10 mm in diameter (Table 1). The maximum root production reported in the literature for a spruce stand with similar characteristics to the stands in this study has been 700 g m^{-2} (Van Praag et al. 1988). This value is lower than our estimated root production (917 g m^{-2}) calculated from the unexplained difference between carbohydrate input and use at the healthy site and a production cost of 0.47 g CH_2O g_{dw}^{-1} (Chung and Barnes 1977). It should be noted, however, that not all the carbohydrates allocated to belowground are used for root production; up to 30% of the carbohydrates may be used by mycorrhizal fungi and microorganisms (Vogt et al. 1982).
3. The third source relates to inherently large errors associated with the calculations of the carbon budget. Based on maximum reported root production, total carbon uptake appears to be overestimated by about 8% at the healthy site. Ågren et al. (1980) also used climatic data from one year to estimate CO_2 uptake and related it to a growth estimate for another year. They found that despite some differences in the seasonal climatic pattern, the estimated annual CO_2 uptake differed by less than 10%. One should not, however, lose sight of the fact that large errors are incorporated in the model, and that by changing an assumption, e.g., root production, the proportion of unexplained difference may change appreciably (see discussion on carbon budgets and errors in them by Ågren et al. 1980). Nevertheless, assuming that CO_2 uptake at the declining site was overestimated by the same amount as that at the healthy site for similar reasons, and by subtracting the 8% from the total uptake at the declining site, the unexplained difference between the carbon input and use is 11%.

This 11% difference at the declining site may be due to two factors which could reduce actual CO_2 uptake below the predicted value: (1) the assimilation capacity might be reduced by an unknown proportion of yellow needles in this site not represented in the model; (2) the leaf area at the declining site may have been overestimated, because only apparently healthy trees were used to develop the relationship between leaf area and sapwood area used to estimate the leaf-area index (Oren et al. 1986). Clearly, both factors could be involved in creating the discrepancy, assuming that no particular errors occurred at the declining site.

4 Assimilation Rates in Relation to Needle Color

The maximum photosynthetic capacity of Norway spruce needles (0- to 4-year-old), measured at CO_2 and light saturation, decreased with the needle Mg-concentration, but only at concentration below 12 mol g_{dw}^{-1} (Fig. 7, *top*; Lange et al. 1986). The relationships seemed better when only one needle age class was depicted (e.g., 1-year-old in Fig. 7, *bottom*). Below 12 mol g_{dw}^{-1}, Mg concentration is associated with the degree of needle chlorosis (Fig. 8; Zimmermann et al. 1988).

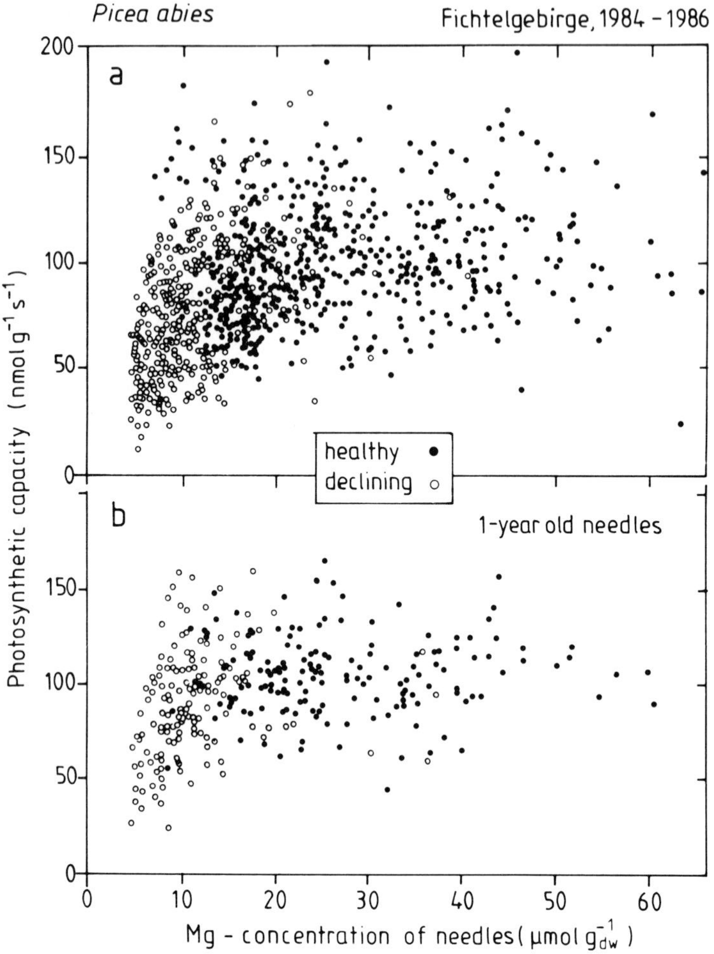

Fig. 7. Photosynthetic capacity in relation to Mg concentration in data collected from 1984 to 1986 from healthy (*closed circle*) and damaged (*open circle*) trees. Data collected at healthy and declining sites on 0-year- to 4-year-old needles are depicted together (**a**; n = 1350), and data collected at the declining site on 1-year-old foliage are depicted separately (**b**) (Lange et al., unpublished)

Under normal environmental conditions, the supply of CO_2 to the mesophyll is limited and light saturation occurs only occasionally in a relatively small proportion of the foliage. Thus, it is necessary to evaluate the effect of needle chlorosis, or Mg deficiency, on CO_2 assimilation at naturally occurring conditions. For this purpose, assimilation rates and leaf conductance were estimated for an intercellular CO_2 (c_i) of 240 ppm from CO_2-response curves (Fig. 8; Zimmermann et al. 1988). While no trend was found in leaf conductance in relation to Mg concentration, assimilation declined with Mg concentrations

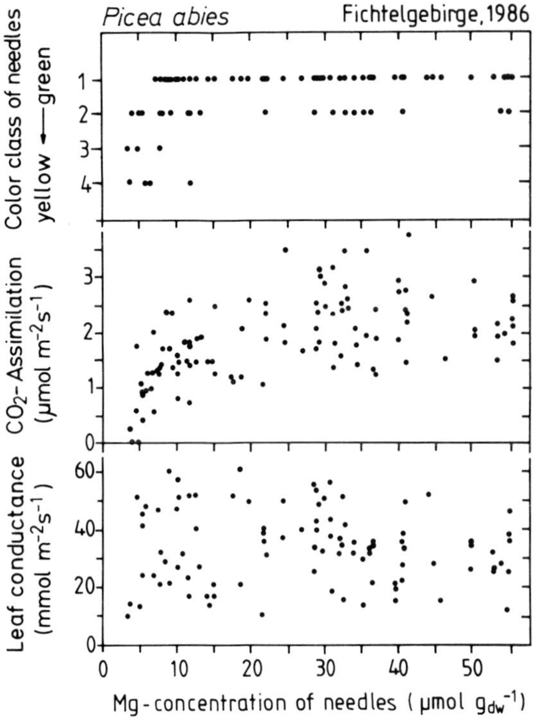

Fig. 8. The relationships between Mg concentration in needles of various ages from healthy and declining stands and (*top*) needle color, (*middle*) net CO_2-assimilation rate, and (*bottom*) leaf conductance. Values are on an all-sided needle-area basis. Measurements were made at light saturation with leaf temperature of 19 °C and various levels of intercellular CO_2 concentrations (c_i). From these CO_2 response curves, CO_2 assimilation rate and leaf conductance at c_i of 240 ppm were calculated for each twig (n = 109) (Zimmermann et al. 1988)

beginning at a concentration similar to that where the photosynthetic capacity seems affected (cf. Figs. 9 and 10).

Because some Mg is retranslocated from old needles to support the new growth of needles (Oren et al. 1988b; Chap. 3-H), the differences in Mg concentration and photosynthetic capacity between healthy and declining trees may be found in the older needles but not in those produced last (Chap. 3-D). Indeed, the only reduction in the CO_2-assimilation rate found in any of the investigations were associated with advanced stages of chlorosis of nutritionally deficient needles. Thus, stomatal function and the photosynthetic apparatus at the declining site have not been affected directly by atmospheric pollution and assimilation has probably been limited by nutrient deficiency.

That the assimilation rate at the declining site was reduced from nutrient deficiency was also shown experimentally. Severing the buds of branches with yellow or green needles before bud-break at the declining site resulted in an

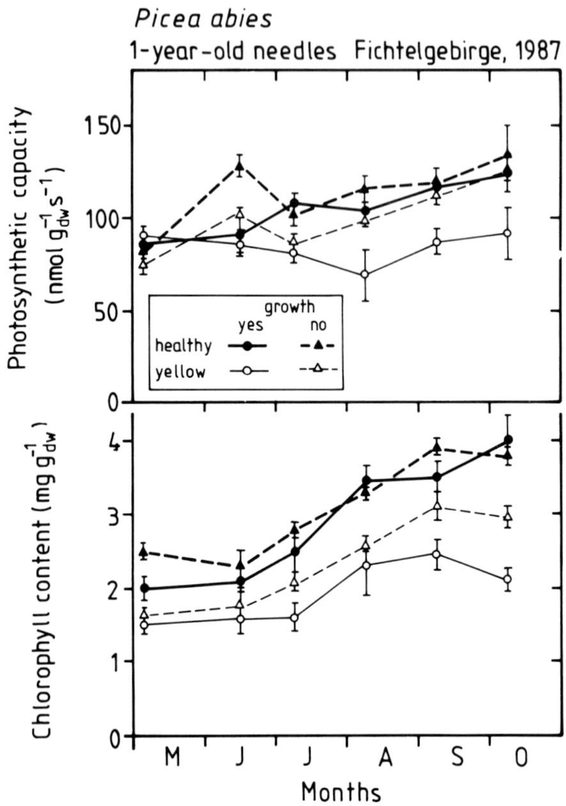

Fig. 9. Seasonal course of photosynthetic capacity (*top*) and chlorophyll content (*bottom*) of 1-year-old needles on branches with (*filled symbols*) and without (*open symbols*) the 0-year flush (Lange et al. 1987)

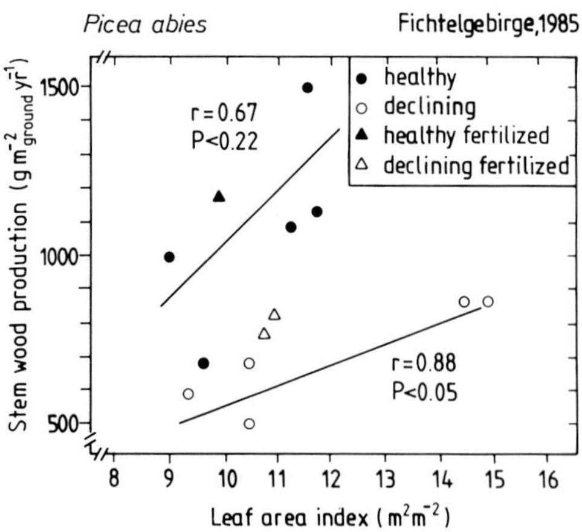

Fig. 10. Relationships between annual stemwood production and leaf-area index between plots in healthy and declining stands. *Triangles* represent three additional plots which were measured in a Mg^{2+} fertilizer experiment (see Fig. 1 in Chap. 3) but are not included in the analyses (Oren et al. 1988a)

increased concentration of several elements, including Mg, and of pigments, leading apparently to an increased photosynthetic capacity by the needles (Fig. 9; Lange et al. 1987). The increase in the photosynthetic capacity seems greater for the damaged branches, probably because of a greater relative reduction in nutrient stress.

Nonetheless, a reduction of carbon uptake due to a lower Mg concentration in a certain proportion of the needles at the declining site could not alone account for all the differences in growth at a given level of leaf-area (Fig. 10; Oren et al. 1988a) because the proportion of yellow needles at that site was small (< 10%).

5 Annual CO_2 Uptake in Relation to Canopy Leaf-Area

More than 95% of the light absorbed in spruce forests occurs when the leaf area index reaches 6 m^2 m^{-2} (Jarvis and Leverenz 1983; Waring 1983). Light absorption relates linearly to production (Linder 1987) so that the leaf-area index may be reduced by 40–50% at both sites with only a small effect on the annual carbon budget and growth. If, however, the leaf-area index was greatly overestimated at the declining site, the lower growth at that site may be explained simply as a result of lower canopy leaf-area, compounded perhaps by some needle-yellowing. This is a reasonable scenario because only about 10% overestimation of leaf area at two of the plots with low leaf-area index at the declining site could explain the lower growth at that site. It is likely, the canopy leaf area at that site was overestimated to that extent, eventhough only two of five plots at that site contained trees which had lost some foliage.

6 Conclusions

A reduction in CO_2 uptake must have accompanied the lower stemwood production at the declining site, because carbohydrates did not accumulate there. Sweet and Wareing (1966) have shown that a nutrient deficiency may reduce both photosynthesis and the sink for carbohydrates. A reduction in the sink for carbohydrates, in turn, may cause a reduction in the photosynthetic rate from, for example, starch accumulation in the needles which impairs photosynthesis. However, neither an increase in starch nor a reduction in photosynthetic rate were detected. Because of the large variation in photosynthetic rates even within one stand (Fig. 8), the sample size required to detect the small difference in photosynthetic rates (ca. 10%) necessary to account for the growth differences between the sites must be much larger than was used or was practical. Such a difference could result from a small proportion of yellow needles at the declining site in combination with an overestimation of leaf-area index in two, relatively open plots.

A nutritional effect on sink strength was found at both sites (Oren et al. 1988b; Chap. 4-H). It appears that the production of stemwood at both sites is related to the availability of Mg from the soil and to the ability of roots to develop into the mineral soil (Meyer et al. 1988).

References

Ågren GI, Axelsson B, Flower-Ellis JGK, Linder S, Persson H, Staaf H, Troeng E (1980) Annual carbon budget for a young scots pine. In: Persson T (ed) 1980 Structure and function of northern coniferous forests – an ecosystem study. Ecol Bull (Stockholm) 32:307–313

Amhtor JS (1984) The role of maintenance respiration in plant growth. Plant Cell Environ 7:561–569

Chung HH, Barnes RL (1977) Photosynthate allocation in Pinus taeda. I. Substrate requirements for synthesis of shoot biomass. Can J For Res 7:106–111

Deutscher Wetterdienst (1987) Klimatabellen für den Raum Fichtelgebirge. Wetteramt Nürnberg, FRG

Fuchs M, Schulze ED, Fuchs MI (1977) Spatial distribution of photosynthetic capacity and performance in a mountain spruce forest of Northern Germany. II. Climatic control of carbon dioxide uptake. Oecologia (Berl) 29:329–340

Guderian R (1985) Air pollution by photochemical oxidants. Formation, transport, control and effects on plants. Ecological studies 52. Springer, Berlin Heidelberg New York Tokyo

Guehl JM, Aussenac G (1987) Photosynthesis decrease and stomatal control of gas exchange in *Abies alba* Mill. in response to vapor pressure difference. Plant Physiol 83:316–322

Gross K (1987) Gaswechselmessungen an jungen Fichten und Tannen während Begasung mit Ozon und Schwefeldioxid (allein und in Kombination) im Kleinphytotron. Allg Forst Jagdztg 158:31–36

Hantschel R (1987) Water and element balance of damaged, fertilized Norway spruce ecosystems in the Fichtelgebirge with respect to physical and chemical soil heterogeneity. Bayreuther Bodenkd Ber Bd 3:1–216

Heber U, Laisk H, Pfanz H, Lange OL (1987) Wann ist SO_2 Nährstoff und wann Schadstoff? Allg Forstz 27/28/29:700–705

Höll W (1985) Seasonal fluctuations of reserve materials in the trunkwood of spruce [*Picea abies* (L.) Karst.]. J Plant Physiol 117:355–362

Jarvis PG, Leverenz JW (1983) Productivity of temperate deciduous and evergreen forests. In: Lange OL, Nobel PS, Osmond CB, Ziegler H (eds) Encycl Plant Physiology. Vol 12 D Springer, Berlin Heidelberg New York, pp 233–280

Keller T, Häsler R (1984) The influence of a fall fumigation with ozone on the stomatal behaviour of spruce and fir. Oecologia (Berl) 64:284–286

Kinerson RS (1975) Relationships between plant surface area and respiration in loblolly pine. J Appl Ecol 12:965–971

Lange OL, Gebel J, Schulze ED, Walz H (1985) Eine Methode zur raschen Charakterisierung der photosynthetischen Leistungsfähigkeit von Bäumen unter Freilandbedingungen – Anwendung zur Analyse "neuartiger Waldschäden" bei der Fichte. Forstwiss Centralbl 104:186–198

Lange OL, Führer G, Gebel J (1986) Rapid field determination of photosynthetic capacity of cut spruce twigs (*Picea abies*) at saturating ambient CO_2. Trees 1:70–77

Lange OL, Zellner H, Gebel J, Schraml P, Köstner B, Czygan FC (1987) Photosynthetic capacity, chloroplast pigments, and mineral content of the previous year's spruce needles with and without the new flush: analysis of the forest-decline phenomenon of needle bleaching. Oecologia (Berl) 73:351–357

Landsberg JJ (1986) Physiological ecology of forest production. Academic Press, Lond NY

Linder S (1987) Responses to water and nutrients in coniferous ecosystems. In: Schulze ED, Zwölfer H (eds) Potentials and limitations of ecosystem analysis. Ecol stud 61, Springer, Berlin Heidelberg New York Tokyo

MacLean DC, Schneider RE, Weinstein LH (1986) Responses of five species of conifers to sulfur dioxide and hydrogen fluoride during winter dormancy. Environ Exp Bot 26:291–296

Matyssek R (1985) Der Kohlenstoff-, Wasser-, und Nährstoffhaushalt der wechselgrünen und immergrünen Koniferen Lärche, Fichte und Kiefer. PhD thesis Univ Bayreuth FRG

Meyer J, Schneider BU, Werk KS, Oren R, Schulze ED (1988) Performance of two *Picea abies* (L.) Karst. stands at different stages of decline V. Root tip and ectomycorrhiza development and their relation to above-ground and soil nutrients. Oecologia (Berl) 77:7–13

Olszyk DM, Tibbitts W (1981a) Stomatal response and leaf injury of *Pisum sativum* L. with SO_2 and O_3 exposures. I. Influence of pollutant level and leaf maturity. Plant Physiol 67:539–544

Olszyk DM, Tibbitts W (1981b) Stomatal response and leaf injury of *Pisum sativum* L. with SO_2 and O_3 exposures. II. Influence of moisture stress and time of exposure. Plant Physiol 67:545–549

Oren R, Schulze ED, Matyssek R, Zimmermann R (1986) Estimating photosynthetic rate and annual carbon gain in conifers from specific leaf weight and leaf biomass. Oecologia (Berl) 70:187–193

Oren R, Schulze ED, Werk KS, Meyer J, Schneider BU, Heilmeier H (1988a) Performance of two *Picea abies* (L.) Karst. stands at different stages of decline. I. Carbon relations and stand growth. Oecologia (Berl) 75:25–37

Oren R, Schulze ED, Werk KS, Meyer J (1988b) Performance of two *Picea abies* (L.) Karst. stands at different stages of decline. VII. Nutrient relations and growth. Oecologia (Berl) 77:163–173

Pelkonen P, Luukkanen O (1974) Gas exchange in three populations of norway spruce. Silvae Genet 23:160–164

Rehfuess KE (1987) Perception on forest diseases on central Europe. Forestry 60:1–11

Schulze ED, Zwölfer H (1987) Potentials and limitations of ecosystem analysis. Ecol stud 61, Springer, Berlin Heidelberg New York Tokyo

Schulze ED, Lange OL, Lembke G (1972) A digital registration system for net photosynthesis and transpiration measurements in the field and an associated analysis of errors. Oecologia (Berl) 10:151–166

Schulze ED, Fuchs M, Fuchs MI (1977) Spatial distribution of photosynthetic capacity and performance in a mountain spruce forest of Northern Germany. III. The significance of the evergreen habit. Oecologia (Berl) 30:239–248

Schulze ED, Hall AE, Lange OL, Walz H (1982) A portable steady-state porometer for measuring the carbon dioxide and water vapor exchanges of leaves under natural conditions. Oecologia (Berl) 53:141–145

Sweet GB, Wareing PF (1966) The role of plant growth in regulating photosynthesis. Nature (Lond) 210:77–79

Van Praag HJ, Sougnez-Remy S, Weissen F, Carletti G (1988) Root burnover in a beech and spruce stand of the Belgian Ardennes. Plant Soil 105:87–103

Vogt KA, Grier CC, Meier CE, Edmonds RL (1982) Mycorrhizal role in net primary production and nutrient cycling in *Abies amabilis* ecosystems in western Washington. Ecology 63:370–380

Vogt KA, Grier CC, Gower ST, Sprugel TG, Vogt DJ (1986) Overestimation of net root production: a real or imaginary problem? Ecology 67:577–579

Waring RH (1983) Estimating forest growth and efficiency in relation to canopy leaf area. Adv Ecol Res 13:327–354

Waring RH, Schlesinger WH (1985) Forest ecosystems. Academic Press Lond NY

Werk KS, Oren R, Schulze ED, Zimmermann R, Meyer J (1988) Performance of two *Picea abies* (L.) Karst. stands at different stages of decline III. Canopy transpiration of green trees. Oecologia (Berl) 76:519–524

Zimmermann R (1985) Untersuchungen zur Photosyntheseleistung, Transpiration und Nadelbiomasseverteilung von zwei geschädigten Fichtenforsten in den höheren Lagen des Fichtelgebirges. Thesis, Univ Bayreuth, FRG

Zimmermann R, Oren R, Schulze ED, Werk KS (1988) Peformance of two *Picea abies* (L.) Karst. stands at different stages of decline II. Photosynthesis and leaf conductance. Oecologia (Berl) 76:513–518

4-E Root and Mycorrhizal Development in Healthy and Declining Norway Spruce Stands

B.U. SCHNEIDER, J. MEYER, E.-D. SCHULZE, and W. ZECH

1 Introduction

Roots link the soil-plant-continuum. Their health depends on both, soil environment and tree canopy function. Changes in any one of these compartments lead to a root response (Persson 1980). The health status of the root in return determines multiple aboveground plant functions, such as water and nutrient exchange (Chapin 1980), growth (Ingestad 1982), and hormonal root/shoot interactions (Schulze 1986). Figure 1 illustrates the interrelationships between the tree canopy and the root system. The canopy supplies carbohydrates for root growth (Marshall and Waring 1985) while roots supply water and nutrients to the canopy. However, root growth, mineral concentration, and the formation of root tips are determined by such soil chemical and physical properties as the nitrogen supply (Meyer 1985) and soil acidification. Soil acidification processes change molar ratios of Ca:Al and Ca:protons in the root environment (Ulrich 1981; Rost-Siebert 1985).

Because the root tips form ectomycorrhizas, feedback between ectomycorrhiza and new root tip formation may occur (Kottke and Oberwinkler 1986). Long roots generally do not form mycorrhizas, exploit new soil volume, and may develop into large roots. Such long roots differ from short mycorrhizal fine roots, which have a coralloid appearance and mainly function for water and nutrient uptake.

Changes in root growth, mineral concentration, and formation of mycorrhizas can be sensitive indicators of forest ecosystems (Meyer et al. 1985). How-

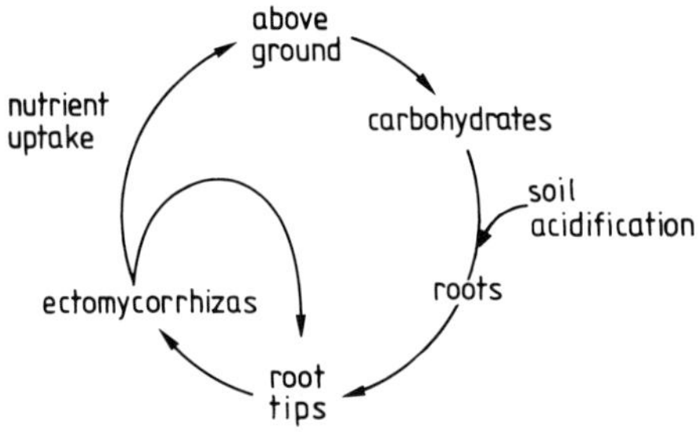

Fig. 1. Schematic diagram of the interrelationships of the carbohydrate supply from the tree canopy to the root, the soil environment and root tips and the root system and canopy via the nutrient supply (After Meyer et al. 1985)

E.-D. Schulze, O.L. Lange, R. Oren (Eds.)
Ecological Studies Vol. 77

ever, it must be shown to what extent such changes result from effects of air pollutants on the tree canopy and to what extent such changes are induced by the soil environment. Direct damage by air pollutants may affect carbon metabolism so that roots are deprived of carbohydrates, a process which secondarily affects plant nutrition.

Alternatively, increased acidification of soils (Wittmann and Fetzer 1982) may change soil chemical properties (Ulrich 1981) which affect root function, and thus, canopy nutrition. Variations in pH along the root surface in a uniform bulk soil depend on different zones of physiological activity and characterize normal root function (Marschner and Röhmheld 1985). Acidity at the soil-root interface, however, may increase by leaching of cations from damaged needles, because enhanced cation uptake leads to a nearly equivalent exchange of protons from the roots (see Chap. 3-D). Lange (see Chap. 3-E) has assumed that this mechanism is important in detoxifying the sulfuric acid produced from sulfur dioxide uptake by needles. Low No_3:NH_4 molar ratios in the soil solution, and the preferential uptake of NH_4 by roots, also lead to a release of protons from the roots and cause an acidification of the soil-root interface (Marschner et al. 1983). Ammonium uptake may also induce an efflux of cations, especially Mg (Boxman 1988), from roots.

This chapter examines root systems in healthy and damaged Norway spruce (*Picea abies* L. Karst.) stands in northeast Bavaria (see Oren et al. 1988a and Schneider and Zech 1989, for methods). In particular, it examines whether the carbohydrate supply of the canopy limits root growth, and which soil chemical factors influence root growth, the formation of mycorrhizas and mineral concentrations of roots. Finally, it describes how acidification processes influence the interrelationships between shoots, roots, and soil.

2 Distribution and Seasonal Development of Fine-Root Biomass, Number of Root Tips, and Ectomycorrhizas

2.1 Comparison Between Young Healthy and Declining Spruce Stands

Schneider (1989) has investigated the development of roots in healthy and declining Norway spruce stands and found no significant difference in total fine root ($\emptyset < 2$ mm) or large root biomass. Despite this root-biomass similarity, important differences have been found in root density and biomass in different soil horizons (Fig. 2). The density of living fine roots in the O_f horizon was significantly higher at the declining site than at the healthy site, but significantly lower in the O_h horizon and in the upper 5 cm of the mineral soil.

The seasonal development of fine-root density is illustrated in Fig. 3. At the declining site, a higher density of fine roots occurred in the O_f horizon throughout the growing season. In the healthy stand, the density of living fine roots was significantly higher in the upper mineral soil (0–5 cm depth) in April and July. Fine root density appeared to increase in the organic layer (O_f) from April

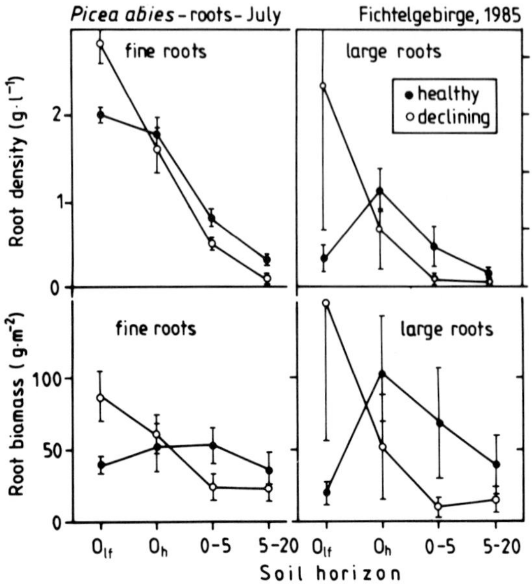

Fig. 2. Vertical distribution of fine root ($\phi < 2$ mm) and large root (2–10 mm) density and biomass at the healthy and declining sites. The data represent living fine root biomass in July. *Vertical bars* indicate standard errors (n = 5)

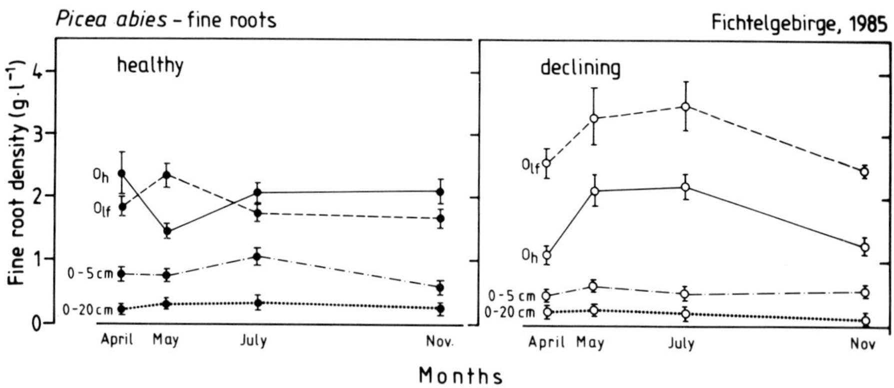

Fig. 3. Seasonal development of fine roots in different soil horizons at the healthy and declining site. *Vertical bars* indicate standard error (n = 5). The seasonal trend was not significant in any horizon ($p > 0.05$)

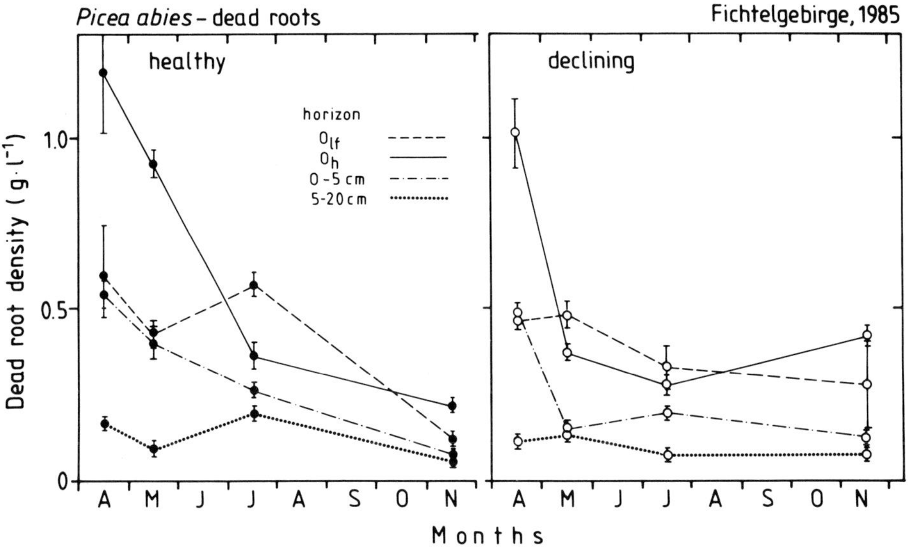

Fig. 4. Seasonal change of dead root density in different soil horizons at the healthy and declining site. *Vertical bars* indicate standard error (n = 5). Differences between sites were significant in the Oh horizon ($p < 0.05$) and in the 0–5 cm mineral soil ($p < 0.12$) in May and for the 0–20 cm mineral soil ($p < 0.05$) in July. The seasonal trend was not significant ($p > 0.1$) in the organic layer but was significant in the mineral soil ($p < 0.01$)

to July and decrease in November, but this seasonal trend was not significant ($p > 0.05$).

Differences in root biomass and density between sites should lead to differences in dead-root biomass. However, it is difficult to interpret differences between sites and horizons because dead roots represent a changing pool, depending on root growth and mortality, as well as on decomposition. Therefore, the interpretation of fine-root necromass is ambiguous, because a low density of dead roots may be expected with either low rates of mortality or high rates of decomposition. A high density of dead roots may occur with high mortality or slow decomposition. Because dead root density was similar in young healthy and declining spruce stands (Fig. 4), there was no indication of differences in mortality or in turnover between sites.

2.2 Comparison of Young and Old Trees at the Declining Site

At the declining site, the density of living fine roots in the mineral soil horizons was significantly higher in a 120-year-old stand than in a 40-year-old stand (Fig. 5). In the organic layer, the results were the opposite. The dead-root density in the mineral soil of old spruce was also significantly higher than in the young stand. The density of dead roots and of living roots in the mineral soil of the old stand

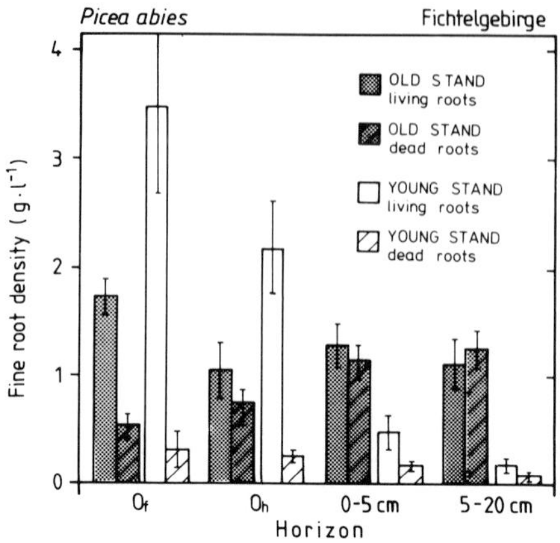

Fig. 5. Vertical distribution on living and dead fine roots in declining young (n = 5) and old stands (n = 20). *Vertical bars* indicate standard errors

was the same. The ratio of dead to living fine roots was significantly lower in the organic surface layer of the young stand. In the mineral soil of the old stand, the ratio was much higher; however, due to high variability, these differences were not significant (Table 1).

Despite the difficulties in interpreting dead-root concentrations, it may be hypothesized that the high density of dead roots in the mineral soil of the old stand was not a function of the fine-root biomass, because the concentration of dead roots did not change correspondingly. Rather, it was probably a result of accumulation from higher rates of mortality or lower rates of decomposition.

2.3 Root Tips and Ectomycorrhizas

Quantification of the functioning root system should include measurements of the formation of root tips and ectomycorrhizas. The symbiosis between root tip and fungus is essential if water and nutrients are to be made available to trees. Therefore, it was examined whether healthy and declining Norway spruce stands have different numbers of root tips, and whether declining trees would be able to compensate for a decrease in fine-root density in the mineral soil by growing more

Table 1. Ratios (mean ± standard error) of dead to living fine roots of young (n = 5) and old (n = 20) declining *Picea abies* [(L.), Karst.] spruce stands

	Young (n = 5)		Old (n = 20)	
Horizon	$\bar{x}$	SE	$\bar{x}$	SE
O_f	8	(8)	36	(24)
O_h	14	(8)	123	(135)
0–5 cm	62	(59)	250	(592)
5–20 cm	54	(29)	145	(232)

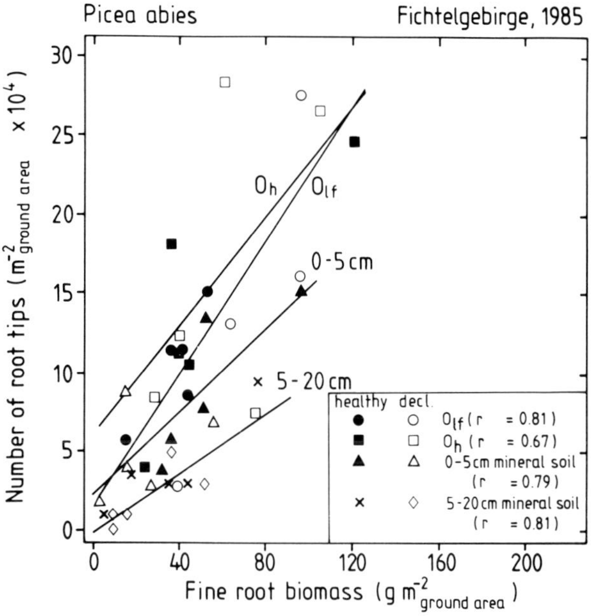

Fig. 6. Relationships between root tips and living fine root biomass during July in 4 horizons at the healthy and declining sites (n = 10; $p < 0.05$) (Meyer et al. 1988)

root tips. The number of root tips was significantly correlated with the biomass of living fine roots in each soil horizon (Fig. 6). Root density and root tips in the O_f horizon and at 0–5 cm soil depth were also positively correlated in all four samplings (Fig. 7).

The positive correlation between the number of visible ectomycorrhizas and root tips (Fig. 8) was significant in all four horizons at the healthy and the declining sites. The level of mycorrhizal infection was determined by dissecting microscopy and may be underestimated. Kottke and Oberwinkler (1986) have shown that mycorrhizas identified by electron microscopy may be present even if a full hyphal mantle has not yet developed. The availability of root tips may determine the number of mycorrhizas (Fig. 8; Chilvers and Gust 1981). Different mycorrhiza-forming fungal species may inhabit the root at different stages of decline (Haug and Oberwinkler 1987).

The distribution of root tips (Fig. 9, left) corresponds to the density of living fine roots (Fig. 2). Trees at the declining site did not compensate for lower fine-root concentrations in the mineral soil by developing more root tips. This finding has implications for plant water uptake. Periods of drought may affect young trees at the declining site much earlier, since the fine roots are concentrated in the upper organic layer rather than distributed in the mineral soil, as under healthy trees (Fig. 9, left.)

Water-uptake capacity also depends on the physiological functioning of a root (Bauch and Schröder 1982). In addition, high concentrations of dead roots

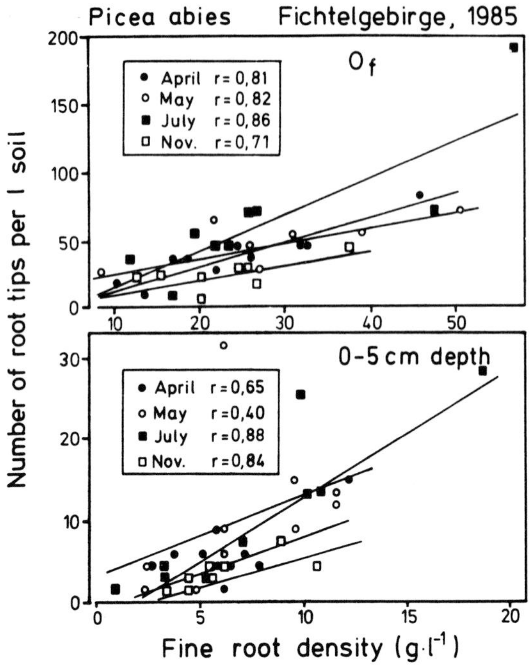

Fig. 7. Relationships between the number of root tips (l^{-1} soil) and fine root density (g l^{-1} soil) in April, May, July and November 1985 in the O_f horizon and at 0–5 cm soil depth. Data from both the healthy and declining young spruce stands are combined ($p < 0.05$ for all correlations)

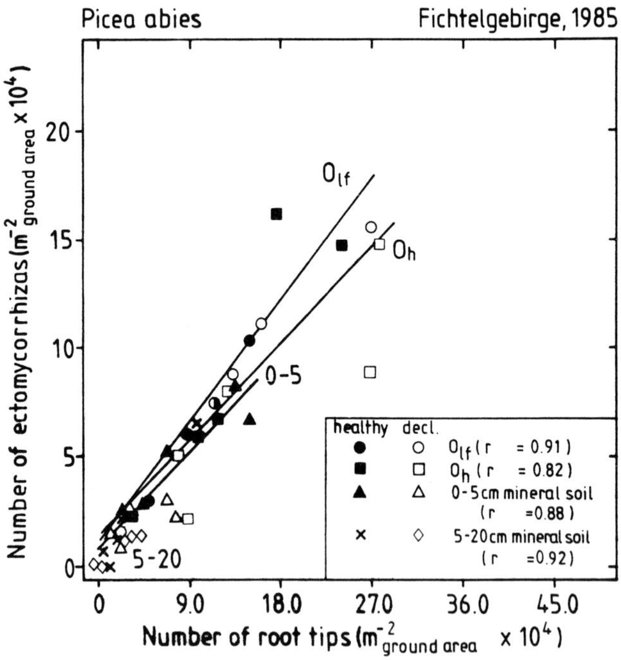

Fig. 8. Relationship between ectomycorrhizal root tips as identified by dissecting microscope and the total number of root tips in different soil horizons at the healthy and declining sites in July ($p > 0.05$) (Meyer et al. 1988)

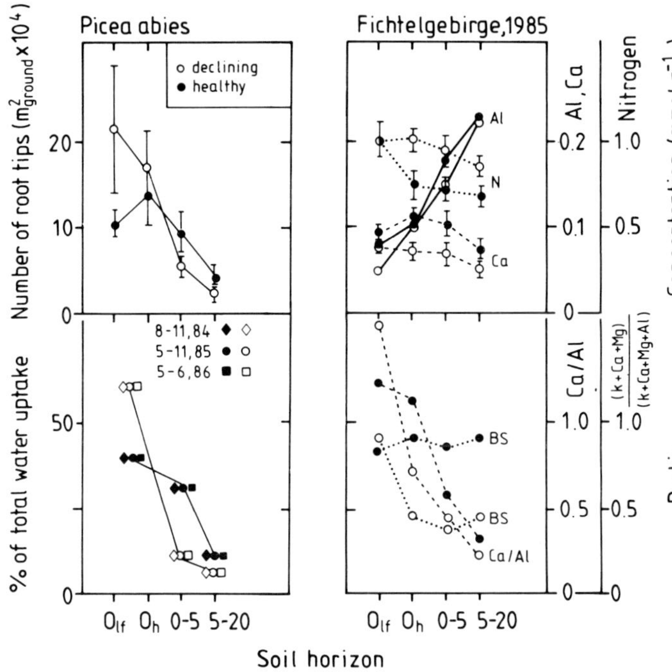

Fig. 9. Number of root tips during July (*top left*; from Meyer et al. 1988) and seasonal averages of water uptake expressed as percent of total annual uptake (*bottom left*; recalculated from Hantschel 1987) in different soil horizons at the healthy and declining site. Seasonal average Al and Ca concentration in fine roots (*top right*; from Oren et al. 1988b) and element ratios of Ca:Al and total base cations (*bottom right*) at different soil depths at healthy and declining sites

in the mineral soil of the old stand may reflect detrimental influences of the soil environment on the functioning of living roots. The higher density of living fine roots, therefore, would not necessarily mean greater availability of water to old trees.

3 Carbohydrate Concentration of Roots

Root development may be affected by a limitation in the carbohydrate supply from the canopy, e.g., as a consequence of canopy damage at the declining site. The level of nonstructural carbohydrates in roots was quite similar at both the declining and healthy sites, and the seasonal course was opposite to what would be expected if carbohydrate starvation had occurred (Fig. 10). During the period of main growth, a reduction of the sugar concentration in fine roots in the first 20 cm of mineral soil was observed at both sites. This was followed by an increase until November. The decrease in sugar concentration in roots in the organic surface layer (O_f, O_h horizons) at the healthy site stopped at bud-break and was

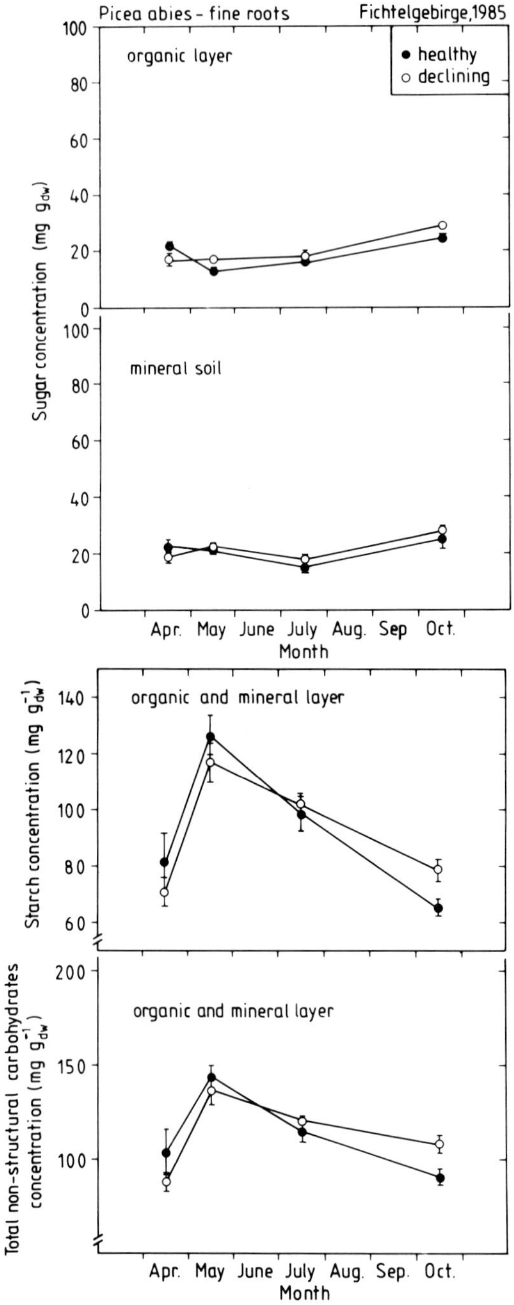

Fig. 10. Seasonal course of sugar concentration in living fine roots of the organic surface layer (*upper left*) and the mineral horizon (*lower left*), and starch (*upper right*) and nonstructural carbohydrate (*lower right*) concentrations of fine roots in the declining and healthy stands (Oren et al. 1988a)

followed by a steady increase until November. At the declining site, sugar concentrations in roots in the O-layer were slightly higher and increased throughout the growing season. Marshall and Waring have shown (1985) that the use of carbohydrates by fine roots is correlated to root growth. Figure 10 therefore suggests that the carbohydrate supply does not limit root growth. The concentration of roots in the organic surface layer and the limitation of growth into the mineral soil at the declining site must have been induced by other factors.

The carbohydrate concentrations in dead roots suggest a similar conclusion. Marshall and Waring (1985) have demonstrated that fine roots of seedlings die when all their stored carbohydrates are respired. Dead roots at both field sites, however, still contained nonstructural carbohydrates with concentrations that were significantly higher ($p = 0.027$) at the declining site (Fig. 11). This further indicates that fine roots in both stands died from causes other than a shortage of carbohydrates. In the year of observation, dry weather conditions in late summer may have contributed to the reduction of fine roots in the organic layer at the declining site.

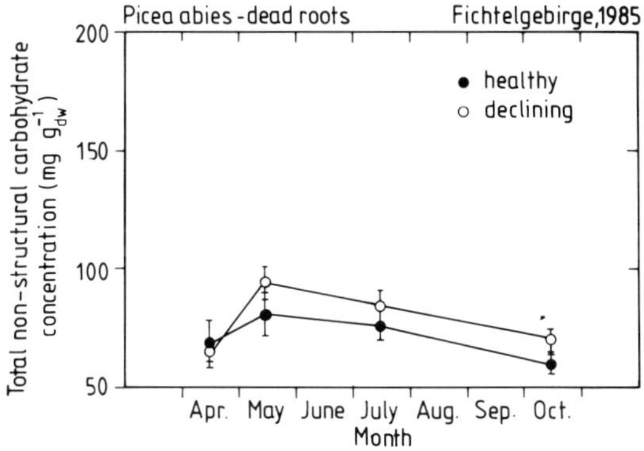

Fig. 11. Seasonal course of nonstructural carbohydrate concentration in dead fine roots in young declining and healthy stands (Oren et al. 1988a)

4 Root Growth in Relation to Soil Chemical Properties

Because carbohydrate limitation could not explain the differences in root-biomass production and distribution between healthy and declining Norway spruce stands, evidence for the influence of chemical soil properties on growth, distribution, and/or mineral concentrations of living fine roots was examined.

4.1 Comparison Between Young Healthy and Declining Spruce Stands

The Ca concentration in living fine roots from the mineral soil of young healthy spruce was significantly higher ($p = 0.05$) than in roots from declining spruce (Fig. 12). This finding corresponds to a higher Ca supply in the soil solution throughout the growing season (Fig. 13). Mg and K concentrations of living fine roots did not differ between sites (Fig. 12), and declined to a minimum in July, when reduced Mg and K concentrations in the soil solution were also detected (Fig. 13). At the same time, soil water potentials (Fig. 14) increased, probably increasing the Mg and K sorption at the exchange complex. Close correlations between Ca and Mg concentrations of both living fine roots and needles, and the fine root density in the O_f horizon of young healthy and declining stands (Fig. 15), suggest that the Ca and Mg supplies of spruce in this study depended highly on Ca and Mg availability in the organic surface layer. Studies by Süsser (see Chap. 4-B) on buffering processes in the humified material at the declining site have shown that high proton concentrations cause a very rapid displacement of mainly Mg and Ca from binding sites of the organic matter. This finding may explain differences in tissue nutrient concentrations between the sites, because due to an unfavorable fine root distribution drought may affect nutrient supply at the declining site much earlier.

In the course of a season, Al concentrations as well as molar ratios of Ca or Mg:Al did not differ between young healthy and declining spruce (Fig. 12). The increase of the Al concentration in living fine roots with soil depth (Fig. 9, right) was the same at both sites. In contrast, the Ca concentration of roots decreased with increasing soil depth only at the declining site (Fig. 9, right). At the healthy

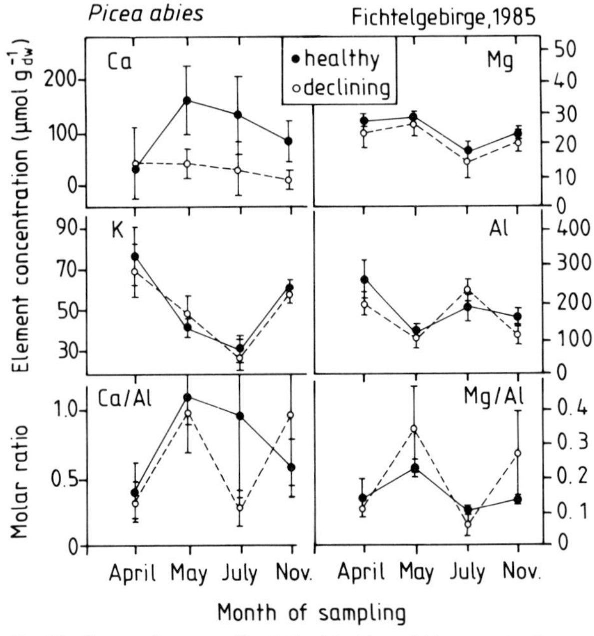

Fig. 12. Seasonal course of both Ca, Mg, K, and Al concentrations and Ca:Al and Mg:Al molar ratios in living fine roots in 0–5 cm soil depth in young healthy and declining *Picea abies* stands. *Vertical bars* indicate standard errors (n = 5). Differences in Mg and K concentrations between May and July were significant ($p < 0.05$). The seasonal trend was not significant ($p > 0.05$)

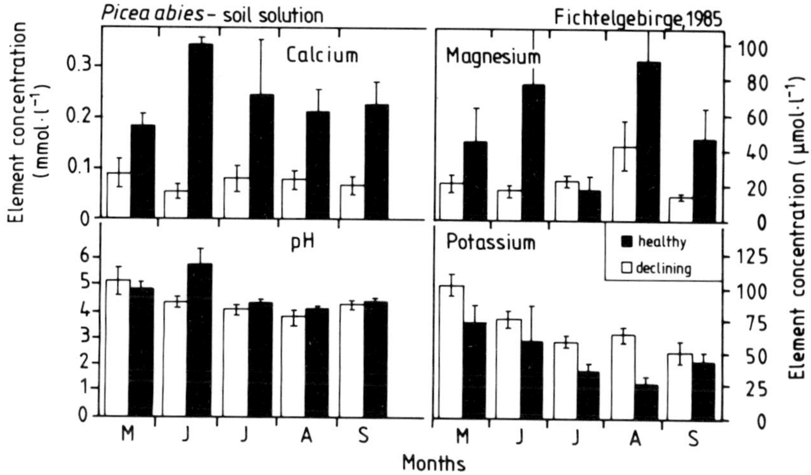

Fig. 13. Seasonal course of Ca, Mg, H^+ and K concentrations in soil solution from suction cups at 20 cm soil depth of heatlhy (n = 4) and declining (n = 5) young *Picea abies* stands. *Vertical bars* indicate standard errors. Differences of element concentrations between months were significant for Mg and K from May to July and from July to November ($p < 0.05$)

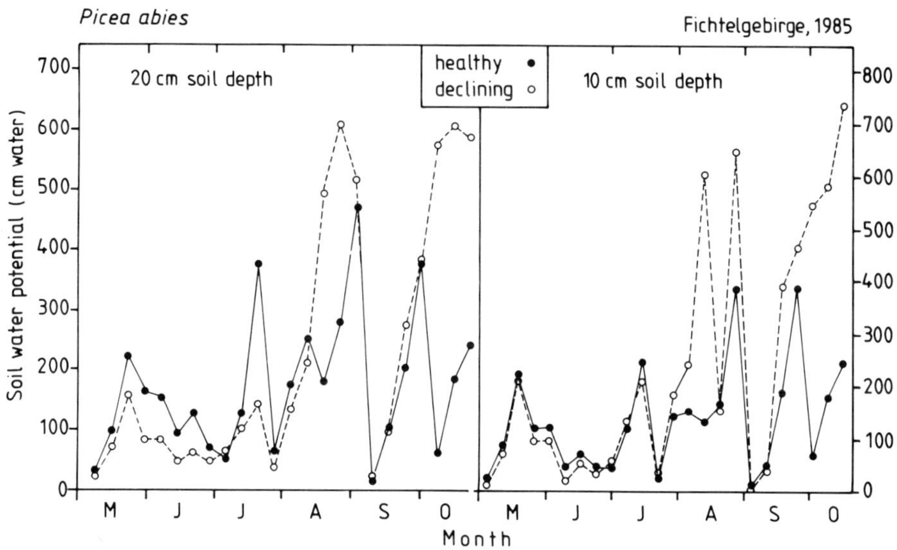

Fig. 14. Seasonal course of the soil water potential at young healthy (n = 5) and declining (n = 5) *Picea abies* stands. The soil water potential was measured once a week with tensiometers (Hantschel 1987)

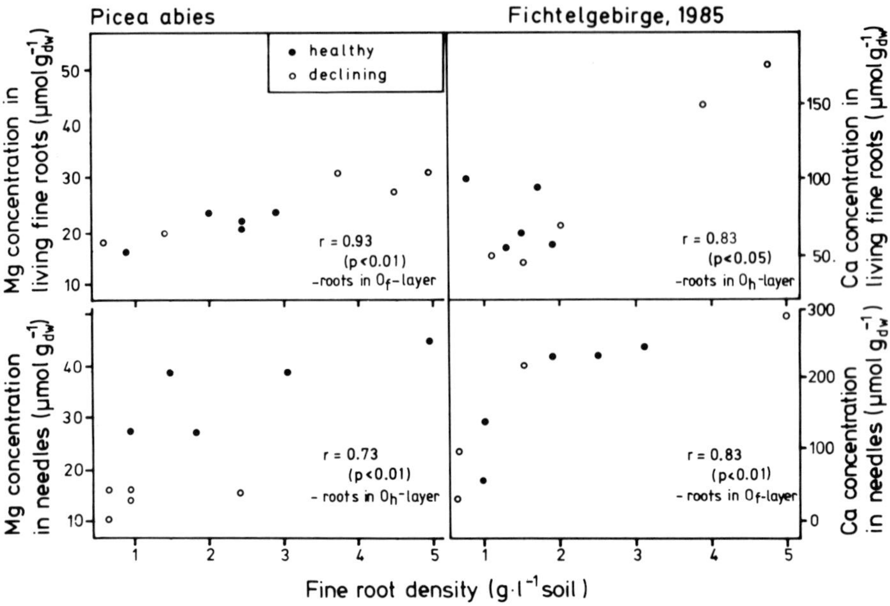

Fig. 15. Correlations between Mg and Ca needle and root concentrations and the fine root density of the O_f horizon in young healthy and declining *Picea abies* stands in July

site, root Ca-concentration in the O_f horizon was the same as at the declining site. It decreased only slightly in the lower mineral soil at a concentration which was significantly higher ($p < 0.05$) than at the damaged site. It is interesting to note that the N concentration of fine roots was higher and decreased less with soil depth at the declining site.

The Ca:Al ratio (Fig. 9, right) in fine roots decreased with soil depth at both sites. It was higher in the O_f horizon but lower in the mineral soil at the declining site. Differences within the horizons, however, were not significant and arose mainly from the better Ca supply to roots at the healthy site. The sum of K, Ca, and Mg as a proportion of total base cations plus Al (Fig. 9, right) at the declining site decreased to less than 50% of the total sum base cations beneath the O_f horizon at the healthy site.

Extracts from undisturbed soil samples of the young stand at the declining site were treated with sulfuric acid of pH 3.5 (for an estimate of the average proton input into young forest stands of the Fichtelgebirge, see Chap. 1-B). These extracts should have reflected the actual mineral composition of the soil solution in undisturbed soils as found in the root environment (Fig. 16). At the healthy site, the exchange of Ca and Mg in the course of the buffering process was higher (85 and 11 μmol l^{-1} for Ca and Mg respectively) than at the declining site (4 and 2 μmol l^{-1}). Aluminum concentrations in soil extracts from the healthy site were also significantly higher than at the declining site (12 and 5 μmol l^{-1} for the two sites,

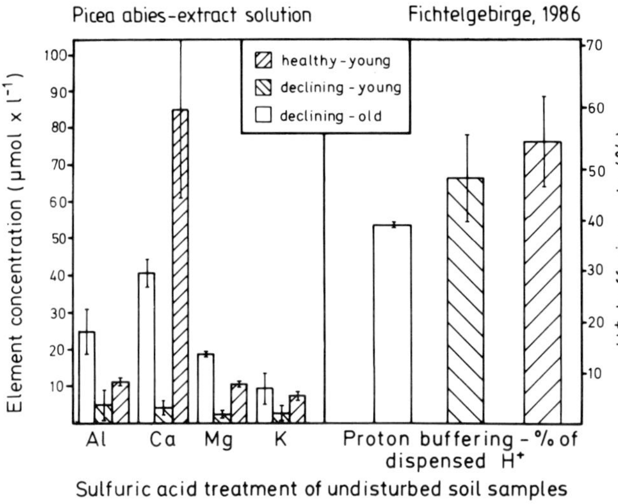

Fig. 16. Element concentrations in extracts of undisturbed soil samples from young healthy and both young and old declining Norway spruce stands (technique described by Kaupenjohann and Hantschel 1987). Soil samples were treated with sulfuric acid of pH 3.0 (old stand) and pH 3.5 (young stands) which was similar to the average proton inputs calculated from field data. The proton buffering is expressed as the % of dispensed protons

respectively). This finding may explain why molar ratios of Ca to Al in living fine roots at both sites were nearly the same, and why trees at the declining site suffer mainly from a Mg deficiency.

In addition to site comparisons, in which mean population values were calculated from five plots at each site (Oren et al. 1988a), each plot was also used as a single data point in correlation analyses ($n = 10$). The number of root tips at different soil depths as well as the total number of root tips per meter ground area was correlated with nutrient concentrations in roots and with soil chemical variables. In large roots, the correlation between root-tip number and Ca concentration of large roots was not significant ($r = 0.35$, $p = 0.08$), but the correlation between root-tip numbers and Al concentrations ($r = 0.40$; $p = 0.039$) was significant. A somewhat stronger relationship ($p < 0.001$) was found, however, between the numbers of root tips and the molar ratio of Ca:Al in large roots (Fig. 17, left). The nutrient concentration in large roots appeared to reflect the soil chemical conditions at different soil depths of the healthy and declining sites and the capacity for root-tip formation. The functional basis for this relationship is an effect of the soil solution on root-tip formation. The number of root tips in the mineral soil was positively correlated with the molar ratio of Ca:Al in water extracts of the mineral soil below the organic horizon (Fig. 17, right). It is important to note that in a comparison of different horizons (Fig. 9, right), Al mainly changed with depth. However, in a comparison of plots (Fig. 17, right), Ca changed more than Al. Evidence from laboratory studies indicates that

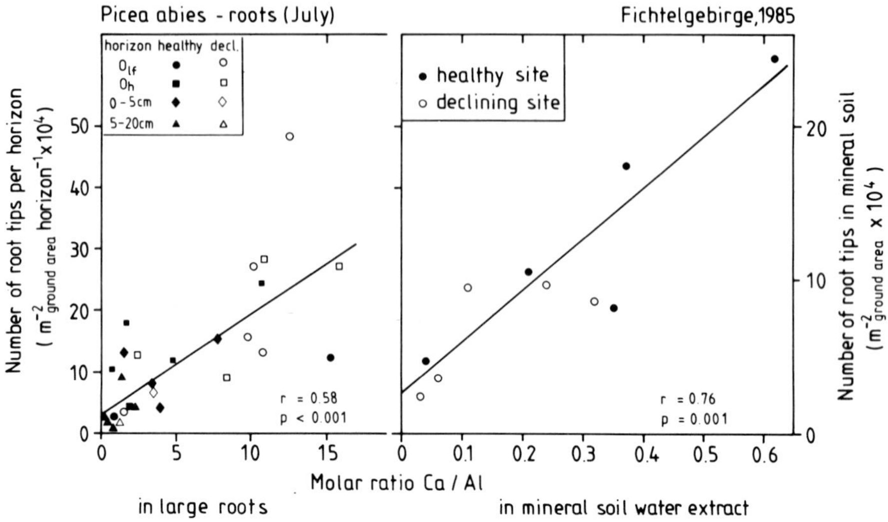

Fig. 17. *Left*: Relationship between number of root tips in July and molar ratio of Ca:Al in the large root biomass at different soil depths. The correlation between Ca/Al molar ratio and number of root tips was stronger ($p < 0.001$, r = 0.76) than the relationship between root tips and Ca concentration in large roots ($p < 0.08$, r = 0.33), or for Al concentration alone ($p < 0.04$, r = 0.40). *Right*: Relation between the total number of root tips per square meter ground area in the first 20 cm mineral soil and the molar ratio of Ca:Al in mineral soil water extracts ($p < 0.001$, r = 0.89; from Meyer et al. 1988). The correlation appears stronger than Al alone (not shown, $p < 0.04$, r = –0.66)

root growth is sensitive to the Ca or Mg:Al ratios in the nutrient solution (Rost-Siebert 1983). This finding on root tips confirms this observation for field conditions (Meyer et al. 1988).

However, the negative influence of Al or the Ca:Al ratio on fine root distribution is not the only explanation for differences in the health status of the sites. Because Ca is an essential nutrient for root growth and function (Rademacher 1986; Bauch and Schröder 1982), significantly lower Ca concentrations in living fine roots at the declining site may have reduced the vitality of both roots and shoots. In a laboratory experiment, Rost-Siebert (1985) has observed reduced root growth of spruce seedlings when the molar Ca:proton ratio in a nutrient solution is decreased. Increasing H^+ concentrations may also decrease the Ca concentration in fine roots (Stienen 1985). The density of living fine roots (Fig. 18) and the level of mycorrhizal infection (Fig. 19; Meyer et al. 1988) were positively correlated with both the pH and Ca concentration of the soil solution and the solution in humus lysimeters, respectively, in the Norway-spruce field study. The number of ectomycorrhizas was also correlated to NH_4 and Mg concentrations in the soil solution (Meyer et al. 1988). Soil pH, Mg, Ca and NH_4 are interdependent parameters. Because there were no differences in the root Mg-concentration in the mineral soil at the two sites and because NH_4 concentrations in the mineral soil were generally very low (see Chap. 4-B), these ions probably had little influence on root growth in the mineral soil.

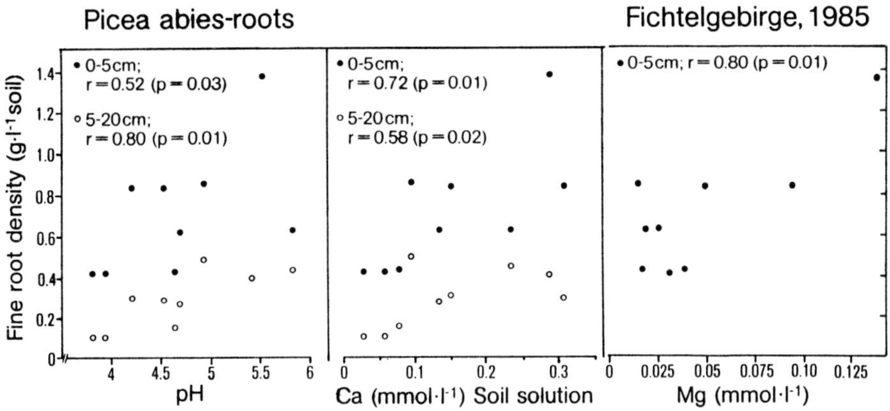

Fig. 18. Correlation between fine-root biomass in the mineral horizons and pH, Ca, and Mg concentrations of soil solution from suction cups (n = 9) in both young healthy and declining spruce stands in July 1985 (root data from the two sites is presented by horizon). Solution from the suction cups was collected at 20 cm soil depth by means of a negative pressure, which slightly exceeds the natural water tension in the soil

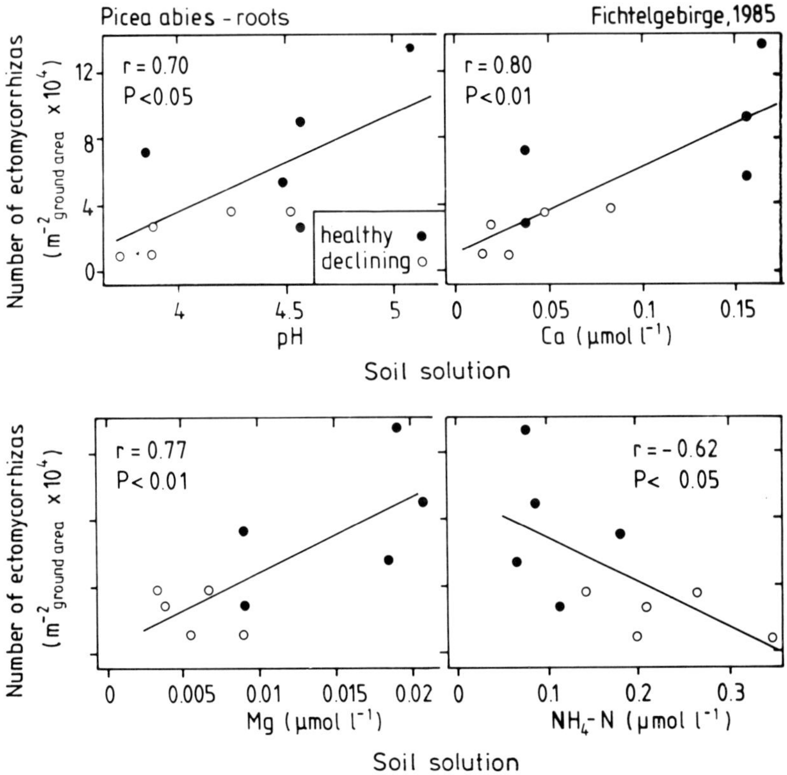

Fig. 19. Correlation between number of ectomycorrhizas in the two mineral horizons (combined) and pH, Ca, Mg and NH_4 concentrations of free-drainage soil solution (n = 10) in young healthy and declining spruce stands in July 1985 (Meyer et al. 1988). The free-drainage soil solution was collected just below the organic surface layer

4.2 Comparison Between Old and Young Spruce Stands at the Declining Site

Living fine roots of old, declining spruce stands contained less Ca (57 μmol gdw^{-1}) than those of young, declining trees (68 μmol gdw^{-1}) (Table 2). However, the Al concentrations and the Ca:Al molar ratio were not significantly different. Differences in the root biomass and necromass of old and young declining trees did not appear to have resulted from Al toxicity or an unfavorable Ca:Al molar ratio. Because H^+-input in the old spruce stands (5–6 kmol H^+) was much higher than in the young stands (2–3 kmol H^+), the comparatively low Ca and Mg concentrations in living fine roots and the high dead-root density of old declining trees may have resulted from the increased proton load of the old stand (Chap. 4-B).

The role of an increased proton load was tested by percolating sulfuric acid of pH 3.0 (corresponding to the average H^+ input in the old stands in the Fichtelgebirge; Kaupenjohann et al. 1987) through undisturbed soil samples of the old and young spruce stands. This allowed a comparison of the buffering rates of soils from both young, healthy and declining Norway spruce stands with those from old, declining stands. Soil from the old stand had a significant lower buffering rate than that from the young healthy and declining stands (Fig. 16). This indicated higher concentrations of free protons in the actual soil solution. The buffering process released more Ca and Mg from the soil of the old stand (41 and 19 μmol l^{-1} for the two elements, respectively) than from the soil of the young stands (4 and 2 μmol l^{-1}). However, while Mg concentrations in living fine roots increased, root Ca actually decreased. This suggested that free protons affect Ca but not Mg uptake (Table 2). As concentrations of foliar Mg were higher in old than in young, declining spruce growing on same soil (Kaupenjohann et al. 1987), these results indicate that old trees can incorporate more Mg than young trees.

Table 2. Ca, Mg, Al concentrations and Ca or Mg:Al molar ratios in fine roots of old (n = 20) and young (n = 5) declining *Picea abies* stands (at 5 cm soil depth, μmol/g dry weight) in July. Differences between stands were tested with a student t-test

Element	Young	SE (%)	Old	SE (%)	Significance[a]
Calcium	67.9	57	57.2	8	**
Magnesium	14.8	34	20.0	17	**
Aluminum	199.2	14	222.4	9	N.S.
			Molar Ratio		
Ca/Al	0.26	44	0.27	6	N.S.
Mg/Al	0.05	17	0.88	9	**

[a] **Significant differences (p = 0.05); N.S. no significant differences.

5 Influence of Acidifying Processes on the Soil-Root Interface

The data presented indicate that fine-root biomass, the number of root tips, and the elemental concentrations of fine roots are related to the soil chemical environment, especially to such pH-related variables as H^+, Ca, and Al. Because roots may actively change their own environment, interactions may occur between pH processes and the effects of acid deposition on the soil.

There are numerous ways by which roots change their microenvironment. Water extraction leads to soil drying, and this in turn promotes soil aggregation, which reduces the ion flux to the root (see Chap. 4-B). Cation uptake generally takes place with an exchange of protons from the root (Marschner et al. 1985). Selective ion uptake may cause pH changes near roots (Marschner et al. 1985). The demand for cations is increased under conditions of acid rain because of leaching (Hantschel 1987, see Chap. 3-D, 4-I) and acid formation in aboveground plant parts (see Chap. 3-C, D). Cations leached from the canopy will return to the humus layer, but in a different spatial distribution than from uptake. All these processes are difficult to demonstrate in the field, and only a few examples are presented below.

Norway spruce seedlings take up NH_4 over NO_3 when both are present (Ingestad 1979). The pH in the solution of a sand culture has been shown to decrease with an increasing supply of NH_4:NO_3-N (Schulze and Gebauer 1989) because of the exchange of protons from the root which occurs when the NH_4 cation is taken up by the seedlings. This acidifying process may occur when rain containing NH_4 and NO_3 (Chap. 4-B) reaches the roots, and may be of particular importance to rhizosphere processes.

Boxman (1988) has shown for *Pinus sylvestris* that cation uptake is highest with mycorrhizal infection and NO_3 nutrition (Table 3). As soon as the NH_4:NO_3 was > 1, a net efflux of base cations occurred until the NH_4 was depleted in the medium. Only then did net cation uptake occur. Conditions similar to those described by Boxman (1988) occurred in the organic surface layer at both Norway-spruce sites under investigation. The NH_4-N to NO_3-N ratio was > 1 in

Table 3. Net uptake and net efflux of Ca, Mg, K, P and protons from seedlings (*Pinus sylvestris*) grown in solutions containing only NO_3 or various molar ratios of NO_3:NH_4 with and without the mycorrhizal fungus *Rhizopogon luteolus* at pH 4 and after an incubation of 4 h (Boxman 1988)

Element:	Ca		Mg		K		H		P	
Treatment[a]	NM[b]	MYC	NM	MYC	NM	MYC	NM	MYC	NM	MYC
I	-0.5	+0.7	+0.1	+0.5	+3.2	+2.9	+1.8	-6.4	+0.6	+0.5
II	-0.6	-0.3	-1.2	-0.6	-1.5	-1.4	-3.3	-0.7	+0.9	+0.9
III	-1.6	-1.3	-6.3	-4.9	-6.6	-3.9	-11.5	-8.4	+1.0	+0.3

[a] I 90 μM nitrate; II 90 μM nitrate + 100 μM ammonium; III 90 μM nitrate + 1000 μM ammonium.
[b] NM: Nonmycorrhizal seedlings; MYC: Mycorrhizal seedlings. Fluxes in μmol g^{-1} dry roots: (+) net uptake; (-) net efflux.

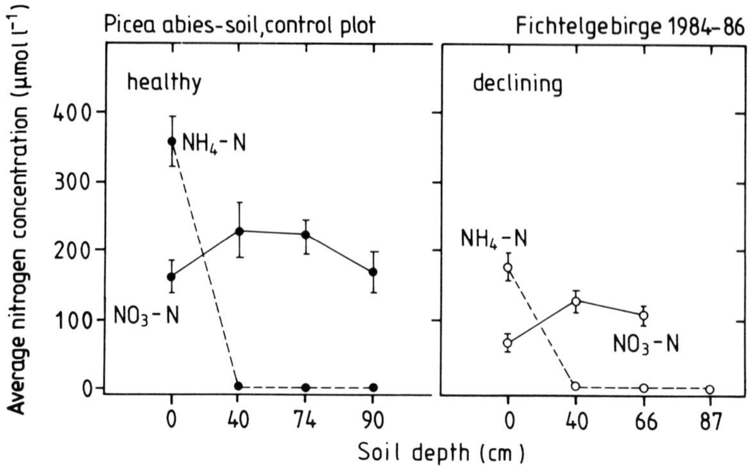

Fig. 20. Average NH_4-N and NO_3-N concentrations in the soil solution collected with lysimeters below the organic layer and with suction cups in the soil horizons at different depths (After Hantschel 1987)

the water percolating through the organic layer (Fig. 20) and < 1 at 40 cm soil depth. The conditions at 20 cm soil depth were similar to those at 40 cm soil depth. Assuming that NH_4 was assimilated in the organic layer, cations could have been released into the soil solution. This process could have contributed to the leaching of cations out of the organic layer.

Leonardi and Flückinger (1988) have demonstrated an efflux of protons at the root surface during acid treatment of the shoot of beech seedlings. If we assume that spruce stands react in a similar way, acid deposition on the canopy increases acidification of the spruce rhizosphere because of cation leaching out of damaged needles, which causes an accelerated proton release into the soil.

Schneider et al. (1989) treated 3-year-old spruce seedlings with sulfuric acid for 10 days (60 ml/tree H_2SO_4). In the course of this experiment Mg and K were leached out of the needles, and the nutrient solution became acidified to about 55 IE/tree (Chap. 3-D). This was paralleled by a significant decrease of the Ca concentrations in fine roots of stressed trees. K concentrations also declined, however, differences between treatments were less pronounced. No such changes were observed for Mg. The decline of Ca concentration in fine roots due to the acid rain treatment suggested that Ca uptake is negatively related to the proton extrusion from fine roots. This finding may well explain, why Ca concentrations of living fine roots were significantly lower for old than for young trees (s. above), because more protons are buffered in the old than in the young stand due to a higher canopy filtering capacity. These results support Ulrich's (1981) canopy-buffering hypothesis and provides evidence that acid deposition can directly influence the rhizosphere by circumventing the soil path.

Table 4. Average Ca, Mg, and K concentrations ($\mu mol\ g_{dw}^{-1}$) in fine roots of 3 year old spruce [*Picea abies* (L.) Karst.] treated for ten days with deionized water (treatment A) or sulfuric acid (treatment B) (each seedling received 60 ml water or H_2SO_4, pH 2.4, n = 16)

Element	Treatment A		Treatment B		
	x	SE	x	SE	
Calcium	129.0	(7.1)	94.4	(10.4)	***
Magnesium	93.4	(5.9)	87.9	(4.4)	N.S.
Potassium	163.6	(5.3)	144.0	(11.9)	*

*** significant differences ($p < 0.05$);
* significant differences ($p < 0.10$);
N.S. no significant differences.

6 Conclusions

Root growth of Norway spruce at the healthy and declining sites did not appear to be restricted by either carbohydrate supply or by mycorrhizal infection.

Ca and Mg concentrations in both fine roots of the O_f horizon and needles were closely related to fine-root density in the upper organic surface horizon (O_f), which suggests that roots in the organic layer supplied most of these nutrients. The Ca concentration in fine roots was not only a function of Ca availability and soil acidity but may also have increased with increasing proton loads in the tree canopy.

The distribution of fine roots, number of root tips, and degree of mycorrhizal infection within the soil profile and between the experimental plots were all related to the soil chemical environment. The Ca:Al ratio in the mineral soil and soil Ca concentration were the strongest indicators of unfavorable soil conditions. The response in root growth and distribution may have been the combined result of these and several other pH-related soil parameters, including soil NH_4, Mg, and H^+ concentrations.

The soil chemical conditions at the soil-root interface were a combination of the base saturation of the bulk soil and root activity. These conditions were strongly altered by soil aggregation (Chap. 4-B), cation leaching (see Chap. 3-D), and by the supply of NH_4 and NO_3 in rain water, and the subsequent selective uptake of NH_4 and cation exchange at a NH_4:NO_3 ratio > 1 (Boxman 1988). All these processes occur under natural conditions, but they may have been accelerated in their magnitude by acid rain and air pollution. In the Chapter on aggregation, it has been shown that the soil surface, which can be reached by roots and mycorrhizas, has been to a large extent depleted of base cations at the declining site. As shown in this Chapter and Chap. 4-F, this has implications for root and canopy nutrition.

The results suggest that the health status of roots under conditions of acid and ammonium input depends on the soil chemical conditions in the rhizosphere.

References

Bauch J, Schröder W (1982) Zellulärer Nachweis einiger Elemente in den Feinwurzeln gesunder und erkrankter Tannen (*Abies alba Mill.*) und Fichten (*Picea abies* (L.) Karst.). Forstwiss Centralbl 101:285–294

Boxman D (1988) Effects of excess nitrogen on the nutritional state of trees. In: Nilsson J (ed) Critical loads for sulphur and nitrogen. UN-ECE Workshop, Sweden

Chapin FS III (1980) The mineral nutrition of wild plants. Annu Rev Ecol Syst 11:233–260

Chilvers GA, Gust LW (1981) The development of mycorrhizal populations on pot-grown seedlings of *Eucalyptus St. Johnii* RT Bak. New Phytol 90:677–699

Hantschel R (1987) Water and element balance of damaged, fertilized Norway spruce ecosystems in the Fichtelgebirge with respect to physical and chemical soil heterogeneity (in German). Bayreuther Bodenkd Ber 3:219 p

Haug I, Oberwinkler F (1987) Some distinctive types of spruce mycorrhizae. Trees 1:172–188

Ingestad T (1979) Mineral nutrient requirements of *Pinus silvestris* and *Picea abies* seedlings. Physiol Plant 45:373–380

Ingestad T (1982) Relative addition rate and external concentration: driving variables used in plant nutrition research. Plant Cell Environ 5:443–453

Kaupenjohann M, Hantschel R (1987) Die kurzfristige pH-Pufferung gestörter und ungestörter Waldbodenproben. Z Pflanzenernähr Bodenkd 150:156–160

Kaupenjohann M, Hantschel R, Zech W, Horn R (1987) Mögliche Auswirkungen "Sauren Regens" auf die Nährstoffversorgung von Wäldern. Kali briefe (Büntehof) 18:631–638

Kottke I, Oberwinkler F (1986) Mycorrhiza of forest trees – structure and function. Trees 1:1–25

Leonardi S, Flückinger W (1988) Indirect effects of acid mist upon the rhizosphere and the leaves' buffering capacity of beach seedlings. In: Mathy P (ed) Air pollution and ecosystems. Proc Comm Eur Communities Int Symp, Grenoble, May 1987, pp 697–700

Marschner H, Römheld V (1983) In vivo measurement of root-induced pH changes at the soil-root interface: Effect of plant species and nitrogen source. Z Pflanzenphysiol 111:241–251

Marschner H, Häusslinger M, Leisen E (1985) Rhizosphere pH of Norway spruce grown under both controlled and field conditions. In: Indirect effects of air pollution on forest trees: root-rhizosphere interactions. Comm Eur Communities, COST workshop Jülich 1985:113–118

Marshall JO, Waring RH (1985) Predicting fine root production and turnover by monitoring root starch and soil temperature. Plant Soil 91:51–60

Meyer FH (1985) Einfluß des Stickstoff-Faktors auf den Mykorrhizabesatz von Fichtensämlingen im Humus einer Waldschadensfläche. Allg Forstz 40:208–219

Meyer J, Oren R, Werk KS, Schulze ED (1985) The effect of acid rain on forest roots: a review. In: Indirect effects of air pollution on forest trees: root-rhizosphere interactions. Comm Eur Communities, COST workshop Jülich 1985:16–30

Meyer J, Schneider BU, Werk KS, Oren R, Schulze ED (1988) Performance of two *Picea abies* (L.) Karst. stands of different stages of decline. V. Root tip and ectomycorrhiza development and their relation to aboveground and soil nutrients. Oecologia 77:7–13

Oren R, Schulze ED, Werk S, Meyer J, Schneider BU, Heilmeier H (1988a) Performance of two *Picea abies* (L.) Karst. stands at different stages of decline. I. Carbon relations and stand growth. Oecologia 75:25–37

Oren R, Schulze ED, Werk KS, Meyer J (1988b) Performance of two *Picea abies* (L.) Karst. stands at different stages of decline. VI. Nutrient concentrations. Oecologia 77:151–162

Persson H (1980) Fine root dynamics in a Scots pine stand with and without near optimum nutrient and water regimes. Acta Phytogeogr Suec 68:101–110

Rademacher P (1986) Morphologische und physiologische Eigenschaften von Fichten (*Picea abies* (L.) Karst.), Tannen (*Abies alba* Mill.), Kiefern (*Pinus sylvestris* L.) und Buchen (*Fagus sylvatica* L.) gesunder und erkrankter Waldstandorte. Dissertation, Fachbereich Biologie, Univ Hamburg

Rost-Siebert K (1983) Aluminum-Toxizität und -Toleranz an Keimpflanzen von Fichte (*Picea abies* (L.) Karst.) und Buche (*Fagus sylvatica* L.). AFZ 38:686–689

Rost-Siebert K (1985) Untersuchungen zur H- und Al-Ionentoxizität an Keimpflanzen von Fichte (*Picea abies* Karst.) und Buche (*Fagus sylvatica* L.) in Lösungskultur. Ber Forschungszentrums Waldökosysteme/Waldsterben, Bd 12, Univ Göttingen, 219 p

Schneider BU (1989) Wachstum und Ernährung von Feinwurzeln geschädigter und ungeschädigter Waldbestände im Fichtelgebirge. Bayreuther Bodenkd Ber 13 (in press)

Schneider BU, Kaupenjohann M, Zech W (1989) Sulfuric acid rain treatment influences mineral contents of fine roots. Water Soil Air Pollution (in press)

Schulze ED (1986) Carbon dioxide and water vapor exchange in response to drought in the atmosphere and in the soil. Annu Rev Plant Physiol 37:247–274

Schulze ED, Gebauer G (1989) Aufnalme, Abgabe und Umsatz von Stick oxiden, NH_4^+ und Nitrat bei Waldbäumen, insbesonder der Fichte. 1. Status seminar Ergebnisse PBWU-GSF, München (in press)

Stienen H (1985) Struktur und Funktion von Feinwurzeln gesunder und erkrankter Fichten (*Picea abies* (L.) Karst.) unter Wald- und Kulturbedingungen. Dissertation, Fachbereich Biologie, Univ Hamburg, 165 p

Ulrich B (1981) Theoretische Betrachtung des Ionenkreislaufes in Waldökosystemen. Z Pflanzenernähr Bodenkd 144:289–305

Wittmann O, Fetzer KD (1982) Aktuelle Bodenversauerung in Bayern. – Bayrisches Staatsministerium für Landesentwicklung und Umweltfragen. Stockert, München, 70 p

4-F Nutrient Relations of Trees in Healthy and Declining Norway Spruce Stands

E.-D. SCHULZE, R. OREN, and O.L. LANGE

1 Introduction

The effects of nutrient relations on tree vigor become complex under conditions of acid rain. The visible symptoms of needle-yellowing has been associated with magnesium deficiencies on silicate soils, potassium deficiencies on limestone, and manganese deficiencies on dolomite (Abrahamson 1980; Hauhs and Wright 1986; *Forschungsbeirat Waldschäden* 1986). In order to balance nutrition, plants should respond to a low supply of a single element by reducing growth (Ingestad 1982; Schulze and Chapin 1987). It is not yet understood which processes lead a single element deficiency to being expressed in acid-rain affected forests, so that these forests reduce their wood production while maintaining their production of new canopy components (Schulze et al. 1987; Oren et al. 1988a).

Because acid rain accelerates cation weathering and soil leaching, the reservoir of available cations is reduced on acid soils (Ulrich 1987; see Chap. 4-B). In addition to soil leaching, the chemical composition of the soil solution may change and influence root vigor (Rost-Siebert 1983; see Chap. 4-E) and nutrient uptake. Nutrients can also be lost from the foliage by canopy leaching (see Chap. 3-C). In these ways nutrient stress may develop and manifest deficiency symptoms. However, acid rain does not only adversely affect plant nutrition. Because it contains nitrate and ammonium, acid rain acts as a fertilizer and may stimulate plant growth. Thus visible deficiency symptoms could result from a nutritional disharmony (sensu Oren et al. 1988c; Chap. 4-H) rather than a simple nutrient deficiency (Schulze et al. 1987).

In this chapter, the nutrient relations of healthy and declining trees and the effect of nutrition on tree growth are examined for two stands at very different stages of forest decline. The healthy site supported trees without any visual signs of decline, while the declining site showed a high degree of variation in tree health, with patches of trees with full crowns and green needles next to patches of trees with thin crowns and yellow needles. Within the declining stand, the variation among trees was also great, with heavily damaged trees next to green ones.

In order to investigate the spatial and temporal variation in tree damage, five 80 m^2 plots containing about 20 trees were established at each site. At the declining site, the damaged and the apparently healthy trees were represented by two plots each, and another plot represented intermediately damaged trees (Oren et al. 1988a). The plots at the damaged site were selected so that major disruptions in the canopy by dead trees were avoided. In each plot, a codominant tree with green

E.-D. Schulze, O.L. Lange, R. Oren (Eds.)
Ecological Studies Vol. 77

foliage was selected for detailed above-ground observations of tree nutrition. Once a tree begins to deteriorate, many secondary functional and structural changes may occur to obscure the primary cause of decline. If decline is a general stand phenomenon, then even a codominant green tree at the declining site should be affected by the parameters which cause the decline (Oren et al. 1988a). All below-ground observations and above-ground growth and biomass estimations were made at a plot level to represent the average conditions of all the trees in the plot. For further details of sampling, and analytical and statistical methods see Oren et al. (1988a,b), Osonubi et al. (1988), Meyer et al. (1988), and Lange et al. (1987).

Since above-ground nutrient samples were taken from the healthy, green trees in all the plots, these represent the best nutritional conditions in the plots and may overestimate nutrient pools and turnover in the plot. If decline symptoms are related to a nutritional limitation, the above-ground nutrient status of yellow trees should be different from that of green trees, at least with respect to one nutrient element. Therefore, the mineral nutrition of damaged (yellow) and healthy (green) trees was also compared. This does not represent a stand-level comparison but a comparison among individual trees at various degrees of decline.

It was shown in Chapter 4-B that soils at both sites had different solution chemistry, as the healthy site had higher calcium and magnesium but lower aluminium concentrations than the declining site. The stand density in relation to mean tree biomass at the declining site was shown to be below the level of self-thinning. In addition, growth per leaf-area was lower at the declining site (Oren et al. 1988a). In this chapter, the nutrient status of trees in relation to forest decline is described, and in the following chapter (Chap. 4-H), nutrient interactions and nutritional disharmonies are analyzed in greater detail.

2 Nutrient Status of Healthy and Declining Norway Spruce Trees and Stands

In the needles of the green trees at the healthy site (Fig. 1), the magnesium content was higher than at the declining site. The changes with needle age between the sites were different on a dry wt. and on a leaf-area basis because needle dry wt. increased with age differently at each site. For instance, magnesium concentration per dry wt. decreased with needle age at the declining site, but less apparently so when expressed as content on a leaf-area basis. Similarly the magnesium content of the needles of the trees at the healthy site increased more with needle age on a leaf-area basis than as a concentration on a dry wt. basis. When single trees with green needles were compared with trees with yellow needles at the declining site (Fig. 2), the age dependent change of needle magnesium-concentration was very similar to that established for green trees on a plot basis (Fig. 1).

There was a significant seasonal trend in needle magnesium-concentration at both sites, with higher concentrations in April than in other months. This seasonal trend can be seen in greater detail in a single declining tree with yellow needles (Fig. 3). In May, the magnesium concentration of old needles decreased when the

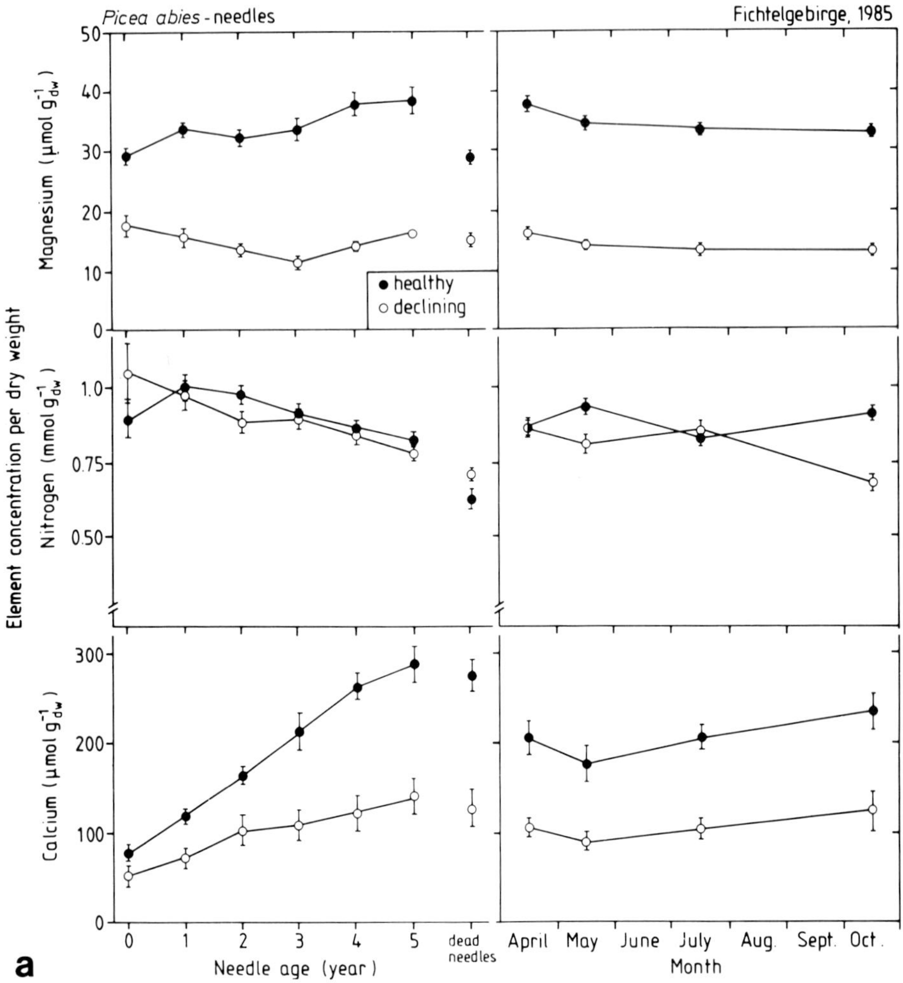

Fig. 1a,b. Average (± SE) element concentrations on a dry weight (**a**) and on a leaf area (**b**) basis in needles from 6 age classes, and dead needles from the healthy and declining sites ((*left*, n = 20) and the annual course of the average concentrations for needle ages 1-to 5-years (*right*, n = 25)

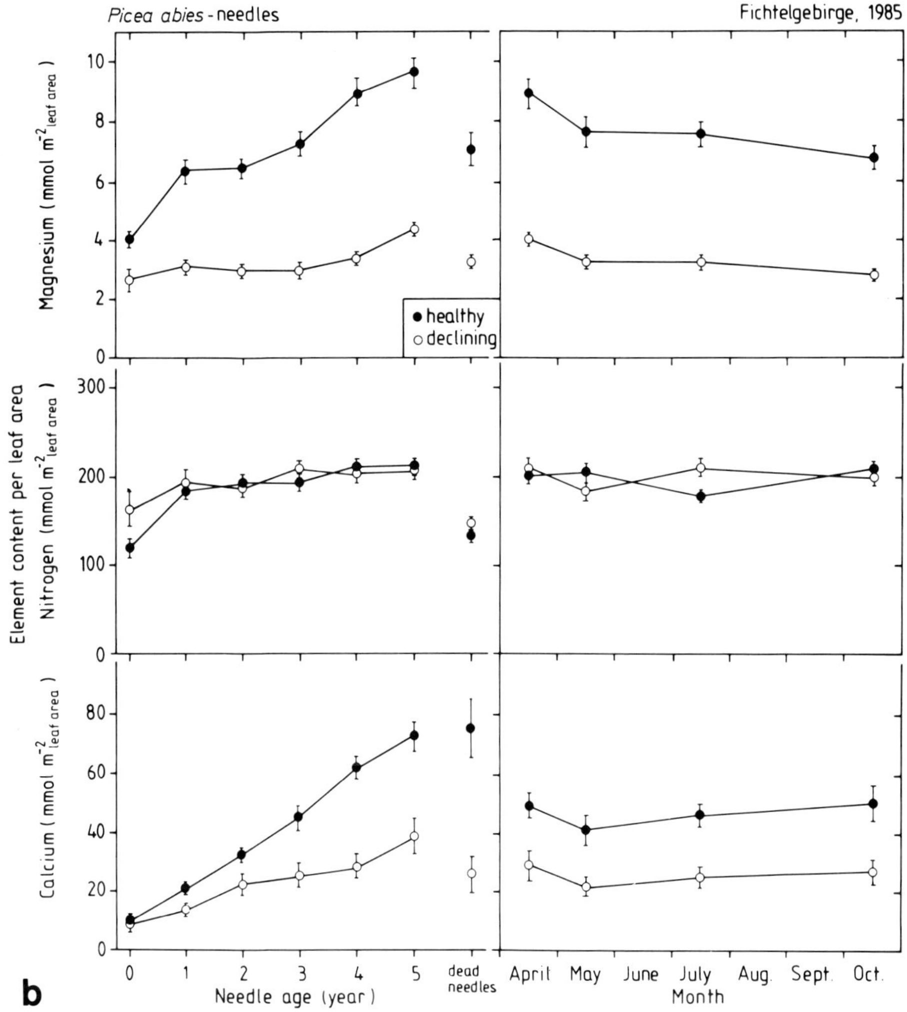

Picea abies - needles
Fichtelgebirge, 1985
Magnesium (mmol m$^{-2}_{leaf\ area}$)
Element content per leaf area
Nitrogen (mmol m$^{-2}_{leaf\ area}$)
Calcium (mmol m$^{-2}_{leaf\ area}$)
● healthy
○ declining
Needle age (year)
dead needles
April May June July Aug. Sept. Oct.
Month

b

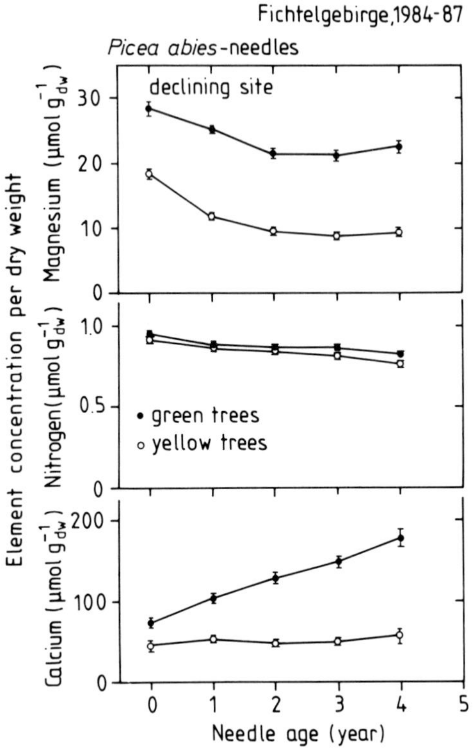

Fig. 2. Average (± SE) element concentrations on a dry wt. basis in needles from 4 age classes of green and yellow trees from the declining site

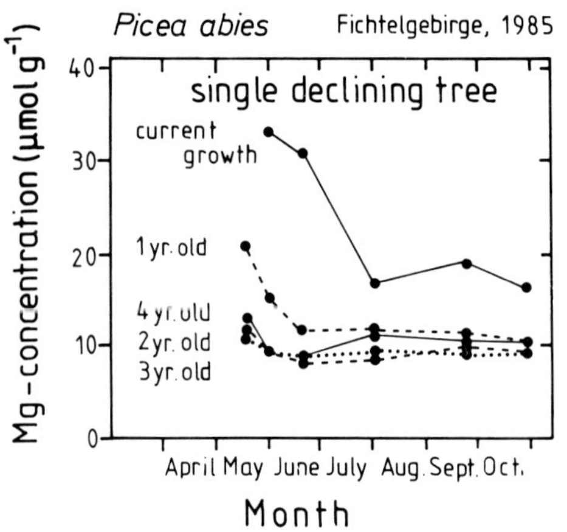

Fig. 3. Seasonal change in magnesium concentrations in needles of different ages of a yellow tree at the declining site

young needles developed. During about 4 weeks in spring the tree shown in Fig. 3 lost about 40% of the magnesium in the 1-year-old needles and became visibly yellow. The magnesium concentration of these 1-year-old needles continued to decrease until October, when the concentration reached was similar in all needle age-classes except for the current-year needles. Two- to four-year-old needles lost about 25% of their magnesium concentration in the spring, while the current year growth initially had a concentration similar to that in needles at the healthy site (Figs. 1, 3). As dry wt. increased over the season, the magnesium concentration in current-year needles had decreased by October to a lower level than that of 1-year-old needles in the spring and to a lower level than current-year needles at the healthy site (Figs. 1, 3).

Calcium concentration in the needles of green trees in the healthy stand was higher and increased with age about twice as fast as those in the declining stand (Figs. 1, 3). There was only a weak seasonal trend of needle calcium concentrations on a leaf-area basis.

Calcium concentration did not increase with needle age in yellow trees (Fig. 2). Even in trees with yellow needles, calcium should be transported with the transpiration stream into the yellow needles. The constant calcium concentration in needles of yellow trees may, therefore, indicate that there is a transport of calcium out of the needles when the calcium supply is low.

Green trees at the healthy site had a potassium concentration similar to that in green trees at the declining site. However, green trees at the healthy site had lower phosphorus, manganese, zinc and aluminum concentrations. In needles of green trees at the declining site, aluminum was significantly lower, and manganese, potassium and zinc were significantly higher than in needles of yellow trees at the same site (Fig. 4).

Except for current-year needles, all age classes had similar nitrogen concentrations per leaf area irrespective of site and season. When single green trees are compared with single yellow trees, the nitrogen concentration was the same for current and 1-year-old needles. At both sites, 25–30% of the nitrogen was retrieved from needles before shedding (Fig. 1); this proportion is smaller than that previously estimated by Schulze et al. (1987).

The element concentration in twigs (Fig. 5) is determined by the balance between the uptake from the stem and uptake from old needles, and the loss to growing tissues. A decreasing concentration indicates that demand exceeds supply as was the case for magnesium and nitrogen in the 1-year-old twigs during the main period of needle and twig growth. An increase in soluble carbohydrates during this period could only partially explain the decrease in concentration. The seasonal change in carbohydrates was even greater in 3-year-old needles (Oren et al. 1988a; Chap. 4-D), but the magnesium content per dry wt. did not decrease. This indicates that there was a net uptake of magnesium, nitrogen and calcium into the twigs during early spring and subsequent retranslocation in early summer.

Generally, site differences in ion concentrations of twigs were similar to those found for needles. The nitrogen concentrations were the same at both sites, but the magnesium and calcium concentrations at the healthy site were higher than

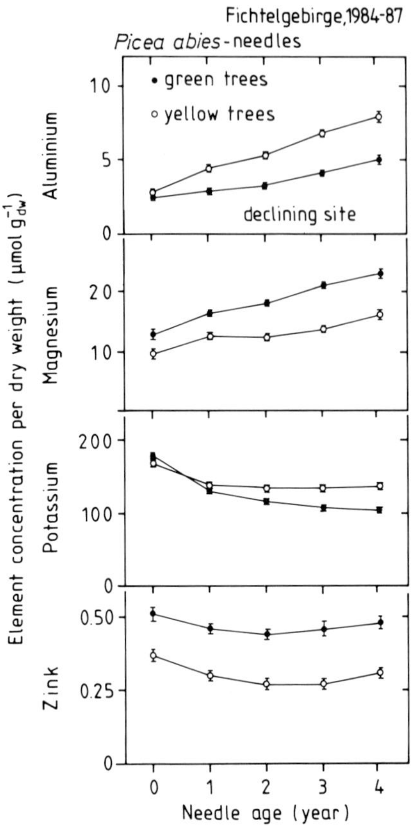

Fig. 4. Average (± SE) element concentrations per dry wt. of green and yellow trees of the declining site as dependent on needle age

those at the damaged site. The differences in magnesium concentration in twigs between the sites was probably related more to differences in concentrations in bark than in wood. Stem bark at the healthy site had higher magnesium concentrations than stem bark at the declining site. The element concentrations in wood were generally lower than in bark, and no clear differences were found between sites.

In large roots (Fig. 6; see also Chap. 4-E), magnesium concentrations in the O_{1f}-horizon were the same for both sites. This may indicate that the organic layer at both sites supplied magnesium to the roots at a similar rate. However, the magnesium concentration of roots in the lower horizons at the damaged site was lower than that at the healthy site, which reflects the differences in the supply between the sites (see Chap. 4-E). At the healthy site calcium concentrations in roots, particularly in the lower organic layer, were generally higher than in those at the damaged site. There was no difference in nitrogen concentration in large roots between the sites, but the declining site had higher concentrations in fine roots than did the healthy site at all depths, except for the lowest (Oren et al.

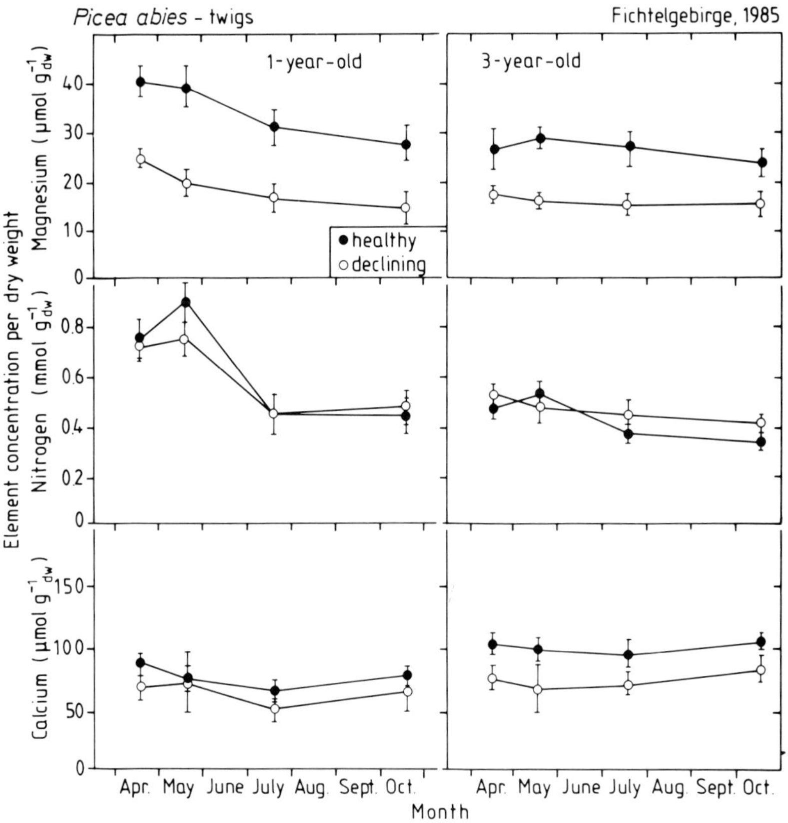

Fig. 5. Seasonal average (± SE) element concentrations in twigs from 2 age classes from the healthy and the declining sites (n = 5)

1988b). Magnesium and nitrogen concentrations in fine roots decreased during the period of growth, but the calcium concentration remained relatively constant, or even increased in large roots during summer.

3 Interactions Between Root and Shoot

In Chapter 4-E, the development of roots in the mineral horizon was shown to be strongly influenced by the soil chemical environment, which in turn is affected by the chemical conditions in the bulk soil, the degree of soil aggregation, and the proton exchange of roots.

Comparing the nutrient levels of fine roots with those of needles may provide some insight into nutritional status and root-shoot interactions, because both types of tissue contain a high proportion of living cells. The ratio between the

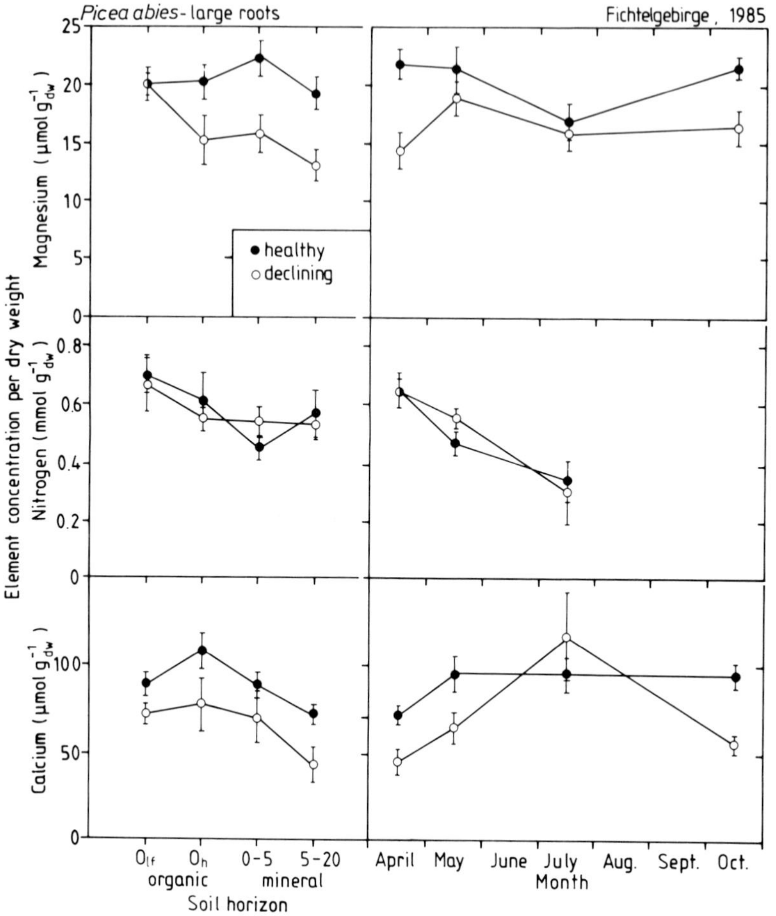

Fig. 6. Average (± SE) element concentration in 2–10 mm large roots from 4 soil horizons and for different seasons at both sites (n = 20)

magnesium concentration in 0-year needles and that in fine roots in the O_{lf} horizon was 1.32 at the healthy site and 0.74 at the declining site. The same ratio for calcium was also higher at the healthy site (0.83) than at the declining site (0.69). The ratio was only higher for nitrogen at the declining site (1.02) than at the healthy site (0.86). This may indicate that the base cation supply of trees at the declining site was low with respect to the nitrogen supply.

Because several factors determine the nutrient status of needles, it would be surprising to find a strong correlation between the element concentration in the soil solution and that in the needles. This was investigated by using the element concentrations in 3-year-old needles from the lower sun crown, which reach a relatively stable nutrient concentration. The element concentrations in October were compared with the average concentration of each element in soil water,

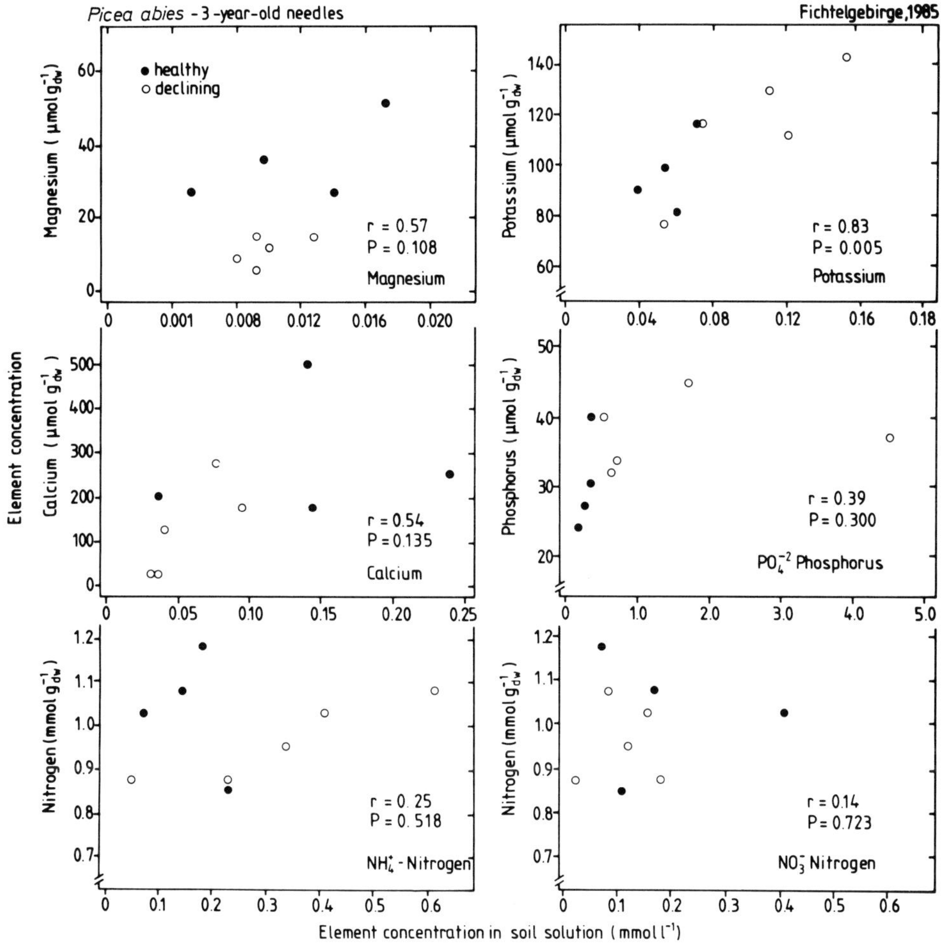

Fig. 7. Relations between the concentration of elements in needles and their concentration in the soil solution in 4 plots from the healthy site and 5 plots from the declining site, measured during October. The correlations were not very strong, probably because of the large temporal variations in element concentrations in the soil solution

which had been collected just below the organic layer during the 3 weeks prior to needle sampling in October. The correlations, with the exception of that for potassium (Fig. 7), were not very strong. In contrast to the weak correlations between soil solution and needle element content, a significant correlation between the average magnesium, calcium, and aluminum concentrations in needles from the lower sun-crown and the number of ectomycorrhizal root tips per leaf-area was found for the plots at both sites (Meyer et al. 1988; Fig. 8).

The relations between needle nutrition and root development suggest that a correlation should exist between needle nutrition and element concentration in

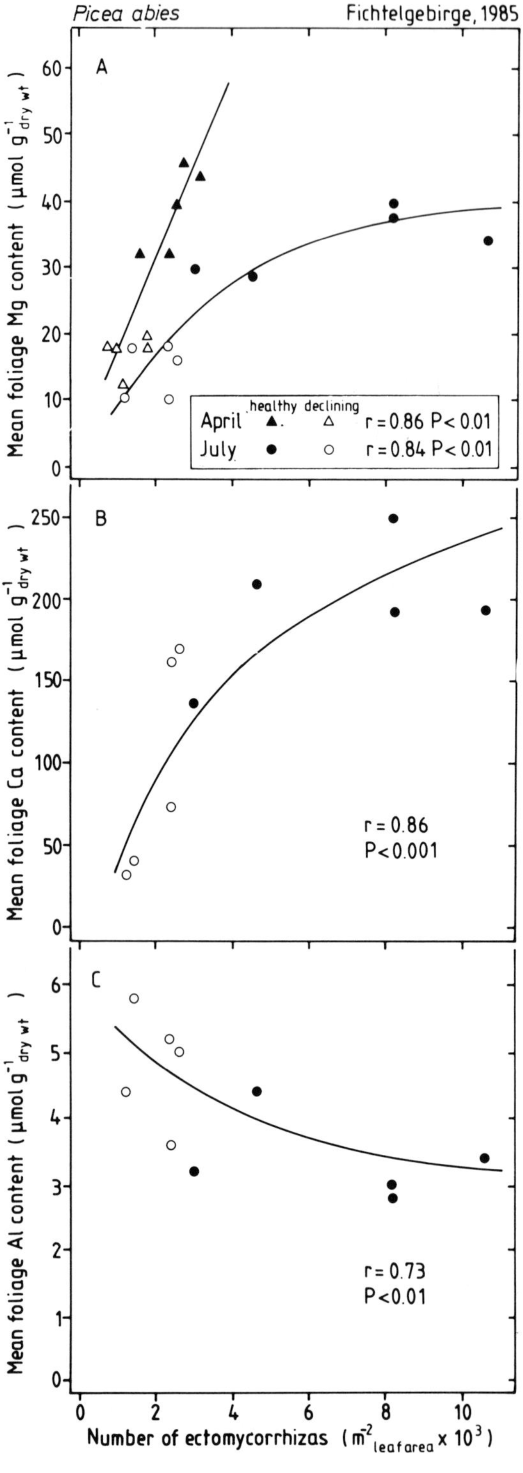

Fig. 8. Correlation between the amount of foliage magnesium in April (*top*), calcium (*middle*), and aluminium (*bottom*) in July and the number of ectomycorrhizal root tips per leaf area. The data refer to ectomycorrhizas in the first 20 cm of mineral soil (n = 10) of 2 sites

the xylem sap. Osonubi et al. (1988) measured the element concentration in the xylem during midday (10 am – 14 pm) and showed that the calcium and magnesium concentrations in the xylem of green trees were significantly higher at the healthy site, but nitrogen and potassium concentrations were similar at both sites.

The difference in xylem sap concentration between sites can be related to differences in plant water status rather than to differences in soil chemistry. When ion concentrations in the xylem were related to the plant water potential (Fig. 9), both sites had similar xylem ion-concentrations at high water potential. As the water potential decreased in trees at both sites, the xylem concentration of magnesium and potassium decreased similarly. However, Fig. 10 shows that the xylem water potentials of trees at the declining site were slightly but consistently lower than those of trees at the healthy site. This is probably because roots at the declining site were concentrated mainly in the organic layer (see Chap. 4-B) which, at both sites, was mostly dry for many summer days, whereas roots at the healthy site penetrated to a greater soil depth, which remained moist most of the season (Meyer et al. 1988; Werk et al. 1988).

Nitrogen, magnesium and potassium concentrations in the xylem sap responded differently to soil drought, as indicated from xylem water potential. While the magnesium concentration reached a low and constant level at –0.6 MPa, the potassium concentration continued to decrease with decreasing water potential. While nitrogen did not respond initially, it began to decrease when magnesium concentrations had already reached their lowest level. There is no simple interpretation for changes in xylem sap element concentrations with decreasing water potential. Due to the circulation between xylem and phloem within the trees, the xylem sap concentration may not be as closely related to the element concentration in the soil solution. It is known from experiments using ^{11}C-labeling that the unloading process of the phloem is very sensitive to water stress in roots (Goeschel et al. 1988, and Goeschel, pers. comm.). Thus, a change in xylem sap concentration at soil drought may reflect both a decrease in plant internal circulation as well as a change in uptake or nutrient availability.

Potassium is known to circulate readily within the plant between phloem and xylem, but the availability of potassium in the soil solution was higher than that of any other cation (see Chap. 4-B). The decrease in xylem potassium concentration may reflect reductions in both supply and internal circulation with increasing drought severity. The supply of magnesium from the soil solution was much lower than that of potassium (see Chap. 4-B). The initial decrease in concentration in the xylem may reflect both decrease in supply as well as in circulation. The constant concentration at low water potentials may represent that proportion which is circulated via roots in those lower soil horizons less affected by drought.

Nitrogen is also circulated in plants. The high concentration of nitrogen in the xylem sap as drought developed may indicate that the supply to the plant remained initially high, and decreased only when the mineralization rate was reduced, as the organic matter became drier. Figure 9 shows two pairs of encircled data points for nitrogen: the lower and upper pairs represent measurements

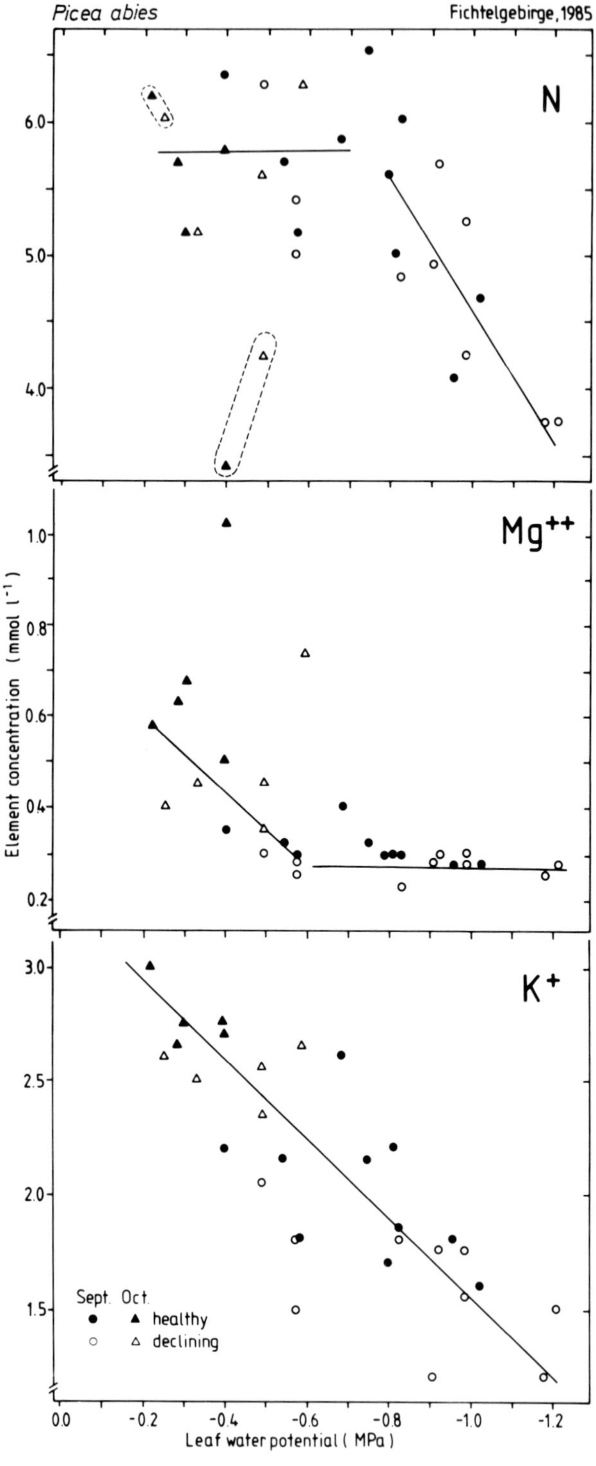

Fig. 9. Nitrogen, magnesium, and potassium concentrations in the xylem sap of trees on the healthy and damaged sites as related to the leaf water potential. Measurements and sap collection were made between 10 and 12 a.m.

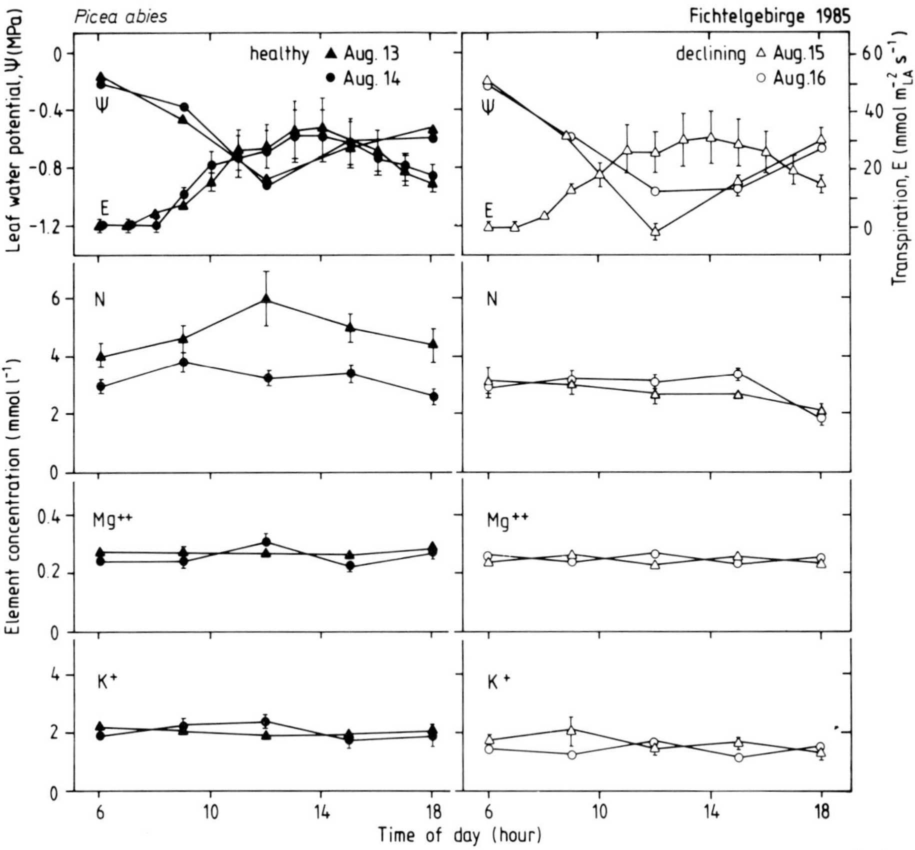

Fig. 10. Diurnal changes of nitrogen, magnesium and potassium concentrations in the xylem sap, leaf water potential and transpiration as measured by mass flow through the stem at the healthy and declining sites. Xylem sap was collected at a pressure of 1.0 MPa on August 13 and 15 and of 1.2 MPa on August 14 and 16 over the balancing water potential

before and after the end of a period of dry weather. Following soil rewetting, water potential recovered rapidly, but the nitrogen concentration remained low for an additional day and recovered by the second day after the rain event. If the nitrogen concentration is an indication of the supply from the soil, then the decrease in magnesium and potassium resulted from reduced circulation rather than a change in the supply from the soil. Indeed, the correlation between element concentrations in xylem sap and those in the soil solution was not significant.

As soil drought develops, the ratio of nitrogen to magnesium supplies will increase and may promote a nutritional imbalance in the trees. Such conditions are more likely to occur at the declining site, where most of the roots are in the organic layer, and numerous, moderate droughts could occur between rain events.

Water stress from soil drying had a different effect than that produced by low water potential in needles when the soil was wet and transpiration was high (Schulze and Hall 1982). During a daily course (Fig. 10), transpiration increased in the morning and water potential decreased accordingly until noon when both processes begin to reverse. The concentrations of magnesium and potassium in the xylem sap of green trees were not affected by this change of water potential in response to transpiration. At the healthy site there was only some variation in the diurnal course of xylem nitrogen-concentration which paralleled that of transpiration. In contrast to the absence of a diurnal effect of leaf water potential on xylem sap-concentration, the average concentrations of elements in the xylem sap as well as average xylem potential were lower in trees at the declining site.

The amount of ions transported in the xylem of twigs to the needles can be estimated from the total transpiration (see Chap. 4-C) and the xylem sap-concentration in the twigs. Table 1 shows that based on this estimation, the total amount of elements which was transported was similar at both sites. Small differences in concentrations between the sites are compensated by small, opposite differences in total transpiration. In all cases, much larger quantities of magnesium, potassium and nitrogen were transported through twigs than were demanded for growth and replacement of leached cations. At both sites only ca. 16–19% of the magnesium, 15–25% of the nitrogen, and 18–26% of the potassium transported through twigs towards the apex were actually used for growth or canopy leaching compensation. A large proportion of the quantities transported with the transpiration stream in the apical direction had to be recirculated in the phloem, since no increase in needle concentrations was measured after bud break.

Table 1. Estimate of the total mass of ions transported through the xylem during the growing season, the requirement for these ions for growth and canopy leaching during the course of the growing season, and the total nonstructural xylem water pool of the stem[a]

	Magnesium		Nitrogen		Potassium	
	Healthy	Declining	Healthy	Declining	Healthy	Declining
Xylem sap concentration mmol l^{-1}	0.25–0.44	0.17–0.35	5.4	5.1	1.6–2.3	1.5–1.9
Total element transport mmol a^{-1}	62–108	50–102	1328	1489	394–566	438–555
Element requirement in twigs for growth and leaching, mmol a^{-1}	16	12	332	218	87	129
Total nutrient pool in stem and branch water, mmol	4–7	2–4	91	51	27–39	15–19

[a] The data are calculated from the concentrations measured in the xylem sap and the total mass flow of water in the xylem.

[b] The seasonal total transpiration was 246 l m^{-2} at the healthy site and 292 l m^{-2} at the declining site (see Chap. 4-C). The total water content of the stem and branches was 17 l m^{-2} at the healthy site and 10 l m^{-2} at the declining site.

Based on the calculations shown in Table 1, the plant internal transport systems of green trees seem to have been functioning properly, unimpaired by microorganisms or viruses. This is also supported by experiments on yellow trees with labeled carbon (Kuhn and Beck 1987). It appears (not shown in Table 1) that even those trees in decline because of nutritional limitations were maintaining a large plant-internal circulation of magnesium, potassium and nitrogen. In contrast to the large amount of cations circulated through the twigs during the growing season, the total pool of ions in the water contained in stem-wood was small (3–7% of the amount transported). The results indicate that magnesium ions may be transported several times into the needles but not absorbed by the needles, even though a needle may be deficient in magnesium. This can only be interpreted to mean that the phloem cells have a higher affinity for magnesium and other elements, or that these cells are closer to the site of unloading from the xylem than the mesophyll cells of needles. Indeed, electron microscope pictures show (Parameswaran et al. 1985) that the initial damage from magnesium deficiency occurs in the Strassburger cells of the parenchyma, immediately surrounding the phloem within the needle endodermis. The data basis for calcium concentrations in the xylem sap is too limited to be conclusive, but the measurements of Osonubi et al. (1988) combined with calculations similar to those given here indicate that calcium may circulate in the plant to a larger extent than previously thought (Kirkley and Pilbeam 1984).

4 Nutrient Relations and Growth

4.1 Nutrient Pools and Nutrient Use for Growth

Assessing nutrient pools and nutrient use for growth was based on measured biomass components and element concentrations (Table 2). When both sites were compared, only those biomass components or concentrations considered to be statistically significant between the sites were used. Otherwise, the calculations were based on average numbers of both sites (Oren et al. 1988a). With respect to biomass and growth, the only differences were between the sites, stem biomass, and stem growth. All other biomass components were not significantly different.

The main difference in nutrient-elements between sites was the pool of magnesium. The healthy site contained about twice as much magnesium in the standing biomass than did the declining site. At the healthy site, the pool of calcium was 65% higher, the pool of nitrogen and potassium 20–30% higher, and the pool of phosphorus only 5% higher than at the declining site.

These differences in nutrient pools can be interpreted in conjunction with nutrient use for growth and turnover.

Since a large proportion of element uptake is returned to the soil through litter fall, the net accumulation in the biomass is almost totally determined by wood growth. At the healthy site, the nitrogen immobilization in wood was 20% higher, the calcium immobilization 40% higher, and the magnesium immobilization 75% higher than at the declining site. Thus, the amount of nitrogen available per mol of calcium or magnesium was 16 or 45% higher at the declining site, respectively.

Table 2. The biomass and growth (gm^{-2}_{ground} except for needles $m^{-2}_{project\ needle\ area}\ m^{-2}_{ground}$) and nutrient pools and nutrient use (mmol m^{-2}_{ground}) in a healthy and declining forest

a) Healthy stand-nutrient pool

Component	Biomass	Calcium	Magnesium	Nitrogen	Potassium	Phosphorus
Needle	11.25	559.0	100.4	2.33	330.1	114.5
Twig	683	67.4	21.7	0.37	83.9	28.9
Branch	772	81.7	21.1	0.39	71.3	25.6
Stem	13251	415.3	89.2	1.99	413.1	154.1
Bark	1135	221.7	33.4	0.49	87.7	23.4
Large root	180	10.7	3.9	0.12	8.3	4.8
Fine root	180	13.5	4.8	0.17	12.0	5.4
Total		1369.3	274.5	5.86	1006.4	356.7

Healthy stand-growth and nutrient use

Component	Growth	Calcium	Magnesium	Nitrogen	Potassium	Phosphorus
O-yr needle	1.79	39.4	11.31	371.1	52.5	18.26
Element accumulation		95.4	1.14	0	0	0
O-yr twig	99	10.5	2.69	50.5	9.1	3.28
O-yr branch	73	7.7	1.98	37.2	6.7	6.74
Wood stem	1073	33.8	7.22	161.0	33.5	12.55
Wood bark	43	8.4	1.27	18.5	3.3	0.89
Root	700	52.5	16.22	630.0	31.1	21.07
Total		247.7	41.83	1268.3	136.2	62.79
Retranslocation						
Needle fall	1.79	0	0	109.7	31.9	5.01
Uptake		247.7	41.83	1158.6	104.3	57.78
Litter						
Needle	1.79	134.8	12.45	249.9	20.6	13.25
Twig	20	2.0	0.54	10.2	1.8	0.66
Root	700	52.5	15.54	609.0	31.1	21.07
Total		189.3	28.53	869.1	53.5	34.98

b) Declining stand-nutrient pool

Component	Biomass	Calcium	Magnesium	Nitrogen	Potassium	Phosphorus
Needle	11.25	329.1	44.7	2.33	330.1	61.8
Twig	683	51.1	13.4	0.37	83.9	30.4
Branch	772	60.1	12.8	0.39	71.3	30.0
Stem	7288	228.4	49.1	1.09	227.2	84.8
Bark	693	135.3	20.4	0.30	53.6	21.9
Large root	180	10.7	2.6	0.12	5.9	3.1
Fine root	180	12.6	3.0	0.11	7.8	5.3
Total		827.3	146.0	4.71	779.8	337.3

Table 2. *Continued*

Declining stand-growth and nutrient use

Component	Growth	Calcium	Magnesium	Nitrogen	Potassium	Phosphorus
O-yr needle growth	1.79	25.1	5.55	371.1	52.5	25.78
Element accumulation		22.2	2.15	0	0	0
O-yr twig	99	7.3	1.64	50.5	9.1	3.84
O-yr branch	73	5.7	1.21	37.2	6.7	2.83
Wood stem	693	21.8	4.66	104.0	21.6	8.04
Wood bark	43	8.4	1.27	18.5	3.3	1.36
Root	700	46.8	14.74	630.0	31.1	31.71
Total		137.3	31.22	1211.3		
Retranslocation						
Needle fall	1.79	0	1.97	109.7	31.9	8.06
Uptake		137.3	29.25	1106.6	92.4	65.56
Litter						
Needle	1.79	47.3	5.73	249.9	20.6	17.72
Twig	20	1.6	0.33	10.2	1.8	0.78
Root	700	46.8	15.54	609.0	31.1	31.71
Total		95.7	21.60	869.1	53.5	50.21

4.2 Relations Between Element Concentrations and Growth

In order to understand the relationships between nutrient status and growth, it is essential to know which plant components grew more slowly at the declining site than at the healthy site (see Chap. 3; Oren et al. 1988a). The declining site had a leaf-area index similar to that at the healthy site ($P < 0.05$; 11.9 ± 1.1 and 10.6 ± 0.6 $m^{-2}_{projected\ LA}$ m^{-2}_{ground}, respectively). Both stands produced similar amounts of needles, twigs, branches, and roots. Indeed, differences in growth between the sites were only due to differences in stemwood production, which was 35% lower at the declining site.

For the individual plots, growth efficiency (Fig. 11), and annual above-ground production (Fig. 12), were higher at the healthy site (Oren et al. 1988a). The relationship between growth efficiency and magnesium concentration of 3-year-old needles appeared stronger than that for calcium, while above-ground production was more highly correlated with needle calcium than with needle magnesium. No relationships for either growth variable were found for nitrogen. Therefore, it appears that wood growth is mainly related to the supply of magnesium and calcium.

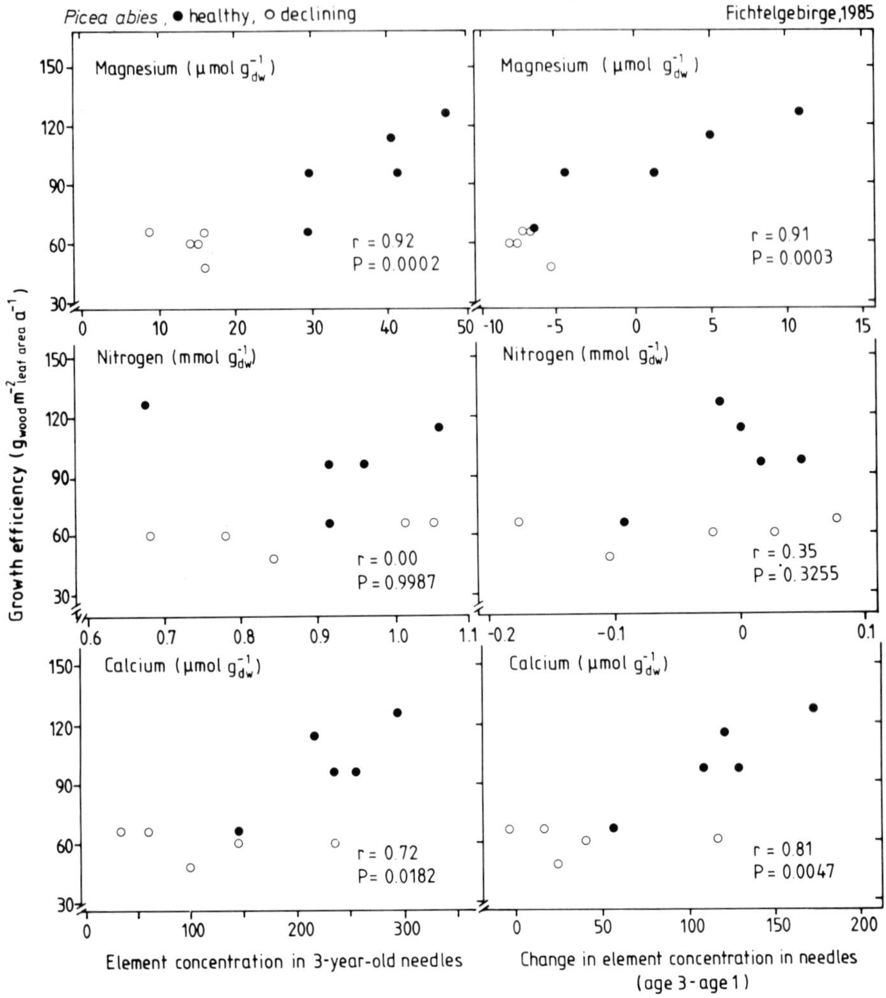

Fig. 11. Growth efficiency (annual biomass production per leaf area) in 10 plots from the declining and healthy sites as related to the element concentration in 3-year-old needles in April (*left*) and to the change in element concentration between 1- and 3-year-old needles (*right*)

4.3 Growth as Related to an Imbalance of Magnesium and Nitrogen

If stem-growth is indeed limited by the supply of calcium and magnesium, one may wonder why canopy growth is not reduced to a level where the magnesium deficiency symptoms in needles are avoided (Schulze et al. 1987). Reduced growth generally occurs when the nitrogen supply is so limiting that deficiency symptoms do not develop (Hällgren and Näsholm 1988). An indication that canopy growth and nutrient deficiencies in needles are the result of an imbalanced nutrient supply was shown by Ingestad (1959, 1982). Maximum growth of spruce

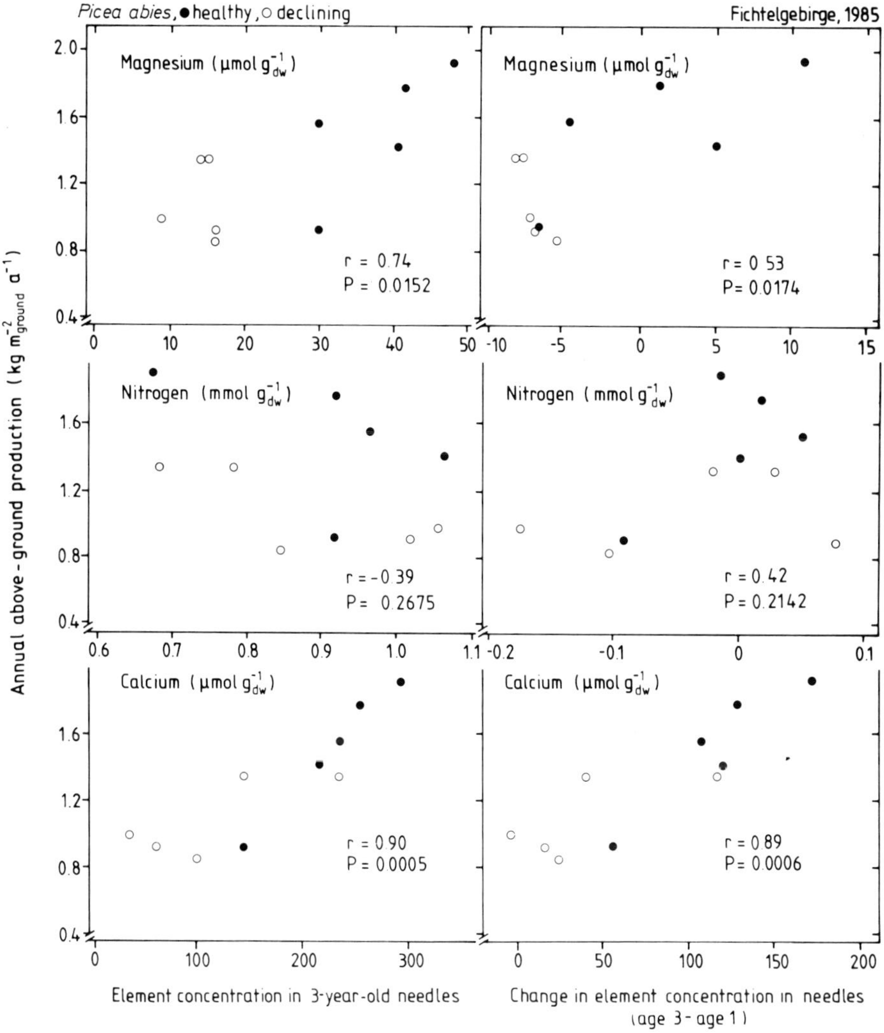

Fig. 12. Annual above-ground production in 10 plots from the declining and healthy sites as related to the element concentration in 3-year-old needles in April (*left*) and to the change in element concentration between 1- and 3-year-old needles (*right*)

seedlings at any level of nitrogen supply can be expected at or above a molar ratio of nitrogen to magnesium of 28 (mmol mol^{-1}). All plots from the declining site were below this magnesium to nitrogen level (Fig. 13), even though only green trees on the study plots were used. This may indicate that an imbalance of nitrogen to magnesium is involved in the development of deficiency symptoms.

Despite large differences between the sites in magnesium concentration in needles of all ages, the seasonal retranslocation pattern of magnesium from needles is similarly related at both sites to the nitrogen content in needles (Oren

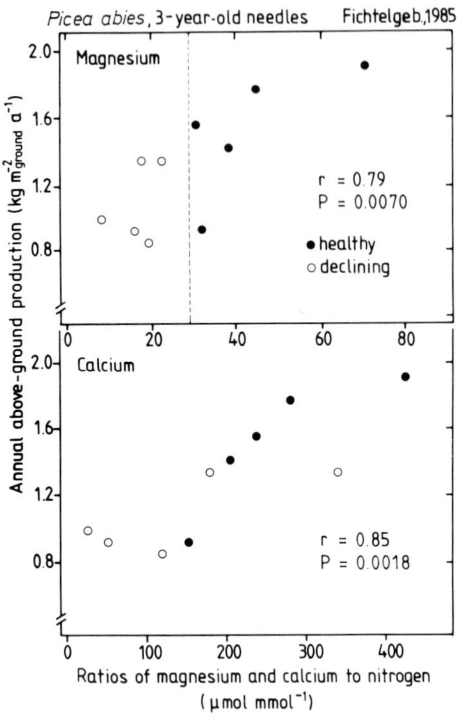

Fig. 13. The relationship between annual above-ground production and the ratio of magnesium to nitrogen in 3-year-old needles from 10 plots from the healthy and declining sites in April. The *dashed line* indicates the ratio for maximum growth according to Ingestad (1959)

et al. 1988c; see Chap. 4-H). Regardless of the absolute level of magnesium and the health status, needles with more nitrogen exported more magnesium, even if their own magnesium content dropped below the level where yellowing begins (Ingestadt 1959).

Following this retranslocation pattern, one would expect that there would be a tendency towards supplying current-year needles with magnesium or calcium as needed for uninhibited growth. The strongest relationships between nitrogen supply and growth exist when all available nitrogen is used for growth and the tissue concentration remains constant. This condition is expected in a range where nitrogen limits growth. Only at a higher supply will some of the nitrogen not be used for new tissue growth (as other growth parameters will be limiting) but rather used to increase the tissue concentration, probably for storage. At the healthy site, all of the nitrogen was used for growth of new needles so that the nitrogen content remained constant (Fig. 14). In this case, the stand could be regarded, with respect to canopy growth, as nitrogen deficient. The retranslocation of magnesium from old needles met the demand for growth based on the supply of nitrogen (see Chap. 4-H). Therefore, the current-year needles did not show symptoms of cation deficiency. This was true for all but one plot at the damaged site, where the canopy

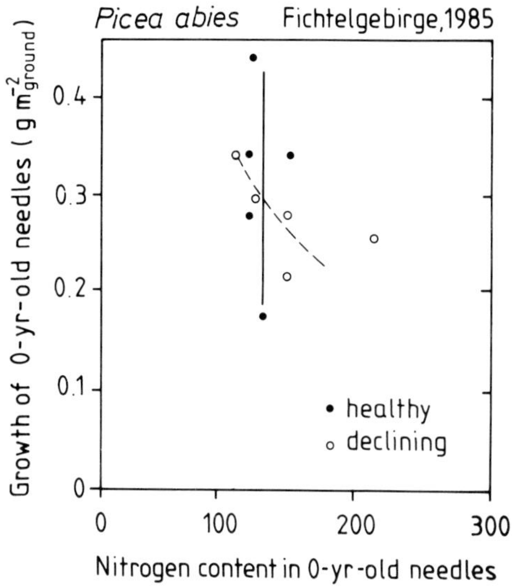

Fig. 14. Growth of needles as related to the nitrogen concentration in new needles

no longer grew but accumulated nitrogen instead. This plot exhibited strong decline symptoms (i.e. needle chlorosis and loss) and had the most severe magnesium deficiency.

The process of magnesium retranslocation from mature needles and the stimulation of new needle growth may not operate in concert. Growth stimulated by nitrogen may, in fact, lead to a dilution of the elements even in the current-year needles (Oren et al. 1988c; Chap. 4-H). The current-year needles of the green trees in the study plots showed decreasing magnesium and calcium concentrations, with an increasing nitrogen concentration as a result of dilution. This indicates that here, nitrogen promoted a greater proportion of the growth of needles and twigs than did the supplies of calcium and magnesium from uptake by the roots and from retranslocation from mature needles. In new needles, the concentration was generally maintained above the level at which yellowing occurs, even at a very high nitrogen concentration.

The effect of growth on the magnesium retranslocation from mature needles has been experimentally demonstrated by pruning vegetative buds (Lippert 1988) before new growth begins (Fig. 15). In green trees, the magnesium content of 1-year-old needles was even lower than in yellow trees. As soon as new needles began to expand, the magnesium content in older needles of intact twigs of the declining trees decreased, reaching a minimum in autumn. With removal of the bud from twigs on the same tree, the concentration in the needles initially declined, but increased during summer and reached the level of healthy twigs with intact buds by autumn. The same trend emerged for calcium and manganese. Photosynthesis of needles on pruned 1-year-old twigs recovered on the damaged tree parallel to the increase in cations (see Chap. 3-D). Pruned, healthy twigs also showed similar trends for calcium and manganese, but not for magnesium. It is

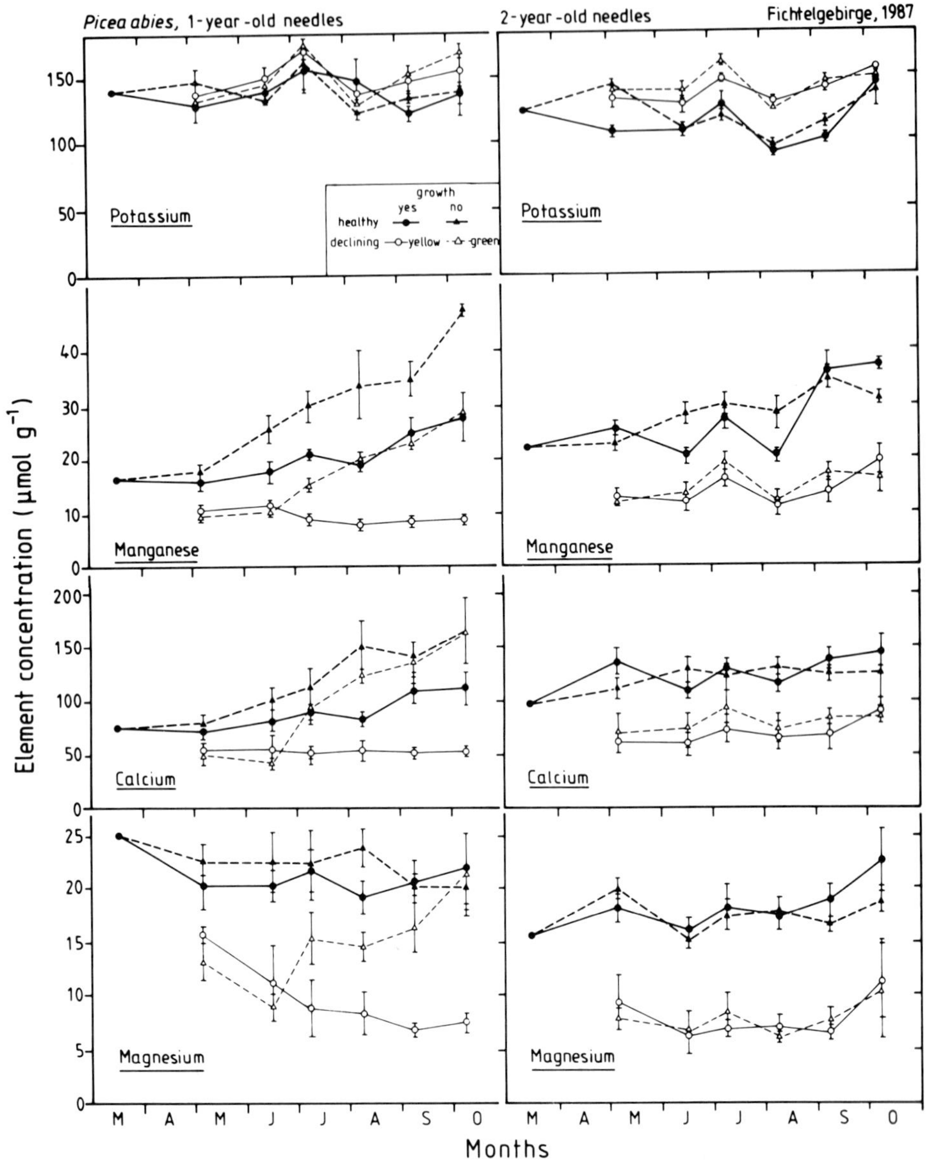

Fig. 15. Change in needle element concentrations with growth of a new twig and without growth due to removal of the bud (Lippert 1988)

interesting to note that no such trend occurred in 2-year-old and older (not shown) needles. These results indicate that the growth of new needles is the prime factor which causes retranslocation of elements from older tissues. All cations, even calcium, showed a similar response. The difference in the calcium content of pruned and nonpruned twigs may indicate that in fact calcium is phloem mobile in *Picea abies,* which has previously been disputed (e.g., Kirkby and Pilbeam 1984). In addition, the lack of response in 2-year-old and older needles is an indication of a strong plant-internal circulation and retranslocation, since calcium is transported into these needles via transpiration and at the concentration of xylem sap. Removal of the plant sink for cations causes the cation contents and the physiological rates to recover to a level of undamaged plants. As indicated by chlorosis, the location of the nutrient deficiency is not where growth occurs, but is rather in the old needles, which are the main source of elements for new needle growth. Stemwood growth, on the other hand, is limited by the low availability of magnesium at a time during the growing season following needle growth.

5 Conclusions

The present study confirms earlier observations that the visible symptom of needle – yellowing is associated with magnesium deficiency. In addition to this decline symptom, it is shown in Chap 4-D that both stands were quite similar in their canopy production, but significantly different in their wood production. The declining stand produced less wood than the healthy stand at a similar leaf area index.

In seeking those mechanisms which lead to the decline symptom of yellowing in old needles, it became clear that the plant-internal nutrient circulation rotates a small pool of available nutrients between xylem and phloem. This pool leads to retranslocation of elements between tissues. The resulting tissue concentrations of growing tissues, e.g., of 0-yr-old needles and fine roots, indicate that the declining stand is equally supplied with nitrogen, but less supplied with magnesium and calcium, than is the healthy stand. It follows that decline symptoms result from a complicated interaction of acid rain acting as nitrogen fertilizer, on the one hand, and as anion for soil cation-leaching, on the other. Needle production is related to nitrogen availability and the required magnesium for new foliage production is supplied from old foliage in proportion to the nitrogen supply. Both stands do not differ in this response, but retranslocation of magnesium from mature needles reaches the level of magnesium deficiency only at the declining site. Thus, needle-yellowing is a consequence of nitrogen-stimulated canopy growth, when magnesium supply is limited. In contrast to the healthy site, stem growth was lower at the declining site than would be expected from the nitrogen supply. In this case, it appears that stem growth is related to the supplies of calcium and magnesium, which are lower at the declining site.

References

Abrahamson G (1980) Acid precipitation, plant nutrition and forest growth. In: Drablos D, Tollan A (eds) Ecological impact of acid precipitation. Norw Interdisciplinary Res Program, SNSF Project As Norway, pp 58–63

Forschungsbeirat Waldschäden/Luftverunreinigungen der Bundesregierung und der Länder (1986) 2. Ber, Elser Druck, Mühlacker, 229 pp

Goeschl JD, Fares Y, Magnuson CE, Scheld HW, Strain BR, Jaeger CH, Nelson CE (1988) Short-lived isotope kinetics: a window to the inside. In: Beecher GR (ed) Research Instrumentation for the 21st Century. Beltsville Symp Agric Res, pp 21–52

Hällgren JE, Näsholm T (1988) Critical loads for nitrogen, Effects on forest canopies. In: Nilsson J, Grennfelt P (eds) Critical loads for sulfur and nitrogen. Nordic council of Ministers and UN-ECE. ISBN 91-7996-096-0, pp 323–342

Hauhs M, Wright RF (1986) Regional pattern of acid deposition and forest decline along a cross section through Europe. Water Air Soil Pollut 31:463–474

Ingestad I (1959) Studies on the nutrition of forest tree seedlings. II. Mineral nutrition of spruce. Physiol Plant 12:568–593

Ingestad T (1979) Mineral nutrient requirements of *Pinus sylvestris* and *Picea abies* seedlings. Physiol Plant 45:373–380

Ingestad T (1982) Relative addition rates and external concentrations: driving variables used in plant nutrition research. Plant Cell Environ 5:443–453

Kirkby EA, Pilbeam DJ (1984) Calcium as a plant nutrient. Plant Cell Environ 7:397–405

Kuhn U, Beck E (1987) Conductance of needle and twig axis phloem of damaged and intact Norway spruce [*Picea abies* (L.) Karst.] as investigated by application of ^{14}C in situ. Trees 1:207–214

Lange OL, Gebel J, Schulze ED, Walz H (1985) Eine Methode zur raschen Charakterisierung der photosynthetischen Leistungsfähigkeit von Bäumen unter Freilandbedingungen. Forstwiss Centralbl 104:186–198

Lange OL, Zellner H, Gebel J, Schramel P, Köstner B, Czygan F-C (1987) Photosynthetic capacity, chloroplast pigments and mineral content of previous year's spruce needles with and without the new flush: analysis of the forest-decline phenomenon of needle bleaching. Oecologia (Berl.) 73:351–357

Lippert M (1988) Photosynthesekapazität, Chlorophyllgehalt und Nährstoffgehalte von Nadeln geschädigter und ungeschädigter Fichten mit und ohne Konkurrenz des diesjährigen Triebs: Experimenteller Ansatz zur Analyse der Waldschäden im Fichtelgebirge. Diplomarbeit Univ Würzburg

Meyer J, Schneider BU, Werk KS, Oren R, Schulze ED (1988) Performance of two *Picea abies* (L.) Karst. stands at different stages of decline. V. Root tips and ectomycorrhiza development and their relation to above-ground and soil nutrients. Oecologia 77:7–13

Oren R, Schulze ED, Werk KS, Meyer J, Schneider BU, Heilmeier H (1988a) Performance of two *Picea abies* (L.) Karst. stands at different stages of decline. I. Carbon relations and stand growth Oecologia 75:25–37

Oren R, Werk KS, Schulze ED, Meyer J, Schneider BU, Schramel P (1988b) Performance of two *Picea abies* (L.) Karst stands at different stages of decline. VI. Nutrient concentrations. Oecologia 77:151–162

Oren R, Schulze ED, Werk KS, Meyer J (1988c) Performance of two *Picea abies* (L.) Karst. stands at different stages of decline. VII. Nutrient relations and growth. Oecologia 77:163–173

Osonubi O, Oren R, Werk KS, Schulze ED, Meyer J, Heilmeier H (1988) Performance of two *Picea abies* (L.) Karst. stands at different stages of decline. IV. Xylem sap concentrations of magnesium, calcium, potassium and nitrogen. Oecologia 77:1–6

Parameswaran N, Fink S, Liese W (1985) Feinstrukturelle Untersuchungen an Nadeln geschädigter Tannen und Fichten aus Waldschadensgebieten im Schwarzwald. Eur J For Pathol 15:168–182

Rehfuess KE, Bosch C, Pfannkuch E (1983) Nutrient imbalances in coniferous stands in southern Germany. Commun Inst For Fenn 116:122–130

Rost-Siebert K (1983) Aluminum toxidity and tolerance in spruce (*Picea abies* Karst) and beech (*Fagus silvatica* L.) seedlings. Allg Forstz 26/27:686–689

Schulze ED, Chapin FS III (1987) Plant specialization to environments of different resource availability. Ecol Stud 61:120–148

Schulze ED, Hall AE (1982) Stomatal responses, water loss, and CO_2 assimilation rates of plants in contrasting environments. Encycl Plant Physiol NS vol 12B:181–230

Schulze ED, Oren R, Zimmermann R (1987) Die Wirkung von Immissionen auf 30-jährige Fichten in mittleren Höhenlagen des Fichtelgebirges auf Phyllit. Allg Forstz 27/28/29:725–730

Ulrich B (1987) Stability, elasticity, and resiliance of terrestrial ecosystems with respect to matter balance. Ecol Stud 61:11–49

Werk KS, Oren R, Schulze ED, Zimmermann R, Meyer J (1988) Performance of two *Picea abies* (L.) Karst. stands at different stages of decline. III. Canopy transpiration of green trees. Oecologia (Berl.) 76:519–524

Zech W, Popp EL (1983) Magnesiummangel, einer der Gründe für das Fichten- und Tannensterben in NO-Bayern. Forstwiss Centralbl 102:50–55

Zöttl HW, Mies E (1983) Nährelementversorgung und Schadstoffbelastung von Fichtenökosystemen im Südschwarzwald unter Immissionseinfluß. Mitt Dtsch Bodenkd Ges 38:429–434

4-G Effects of Fertilization

M. KAUPENJOHANN, W. ZECH, R. HANTSCHEL, R. HORN, and B.U. SCHNEIDER

1 Introduction

Following the general observation that cation deficiencies are strongly correlated with forest decline, fertilizations were carried out in order to experimentally determine the involvement of nutrition in the process of forest decline. Three types of fertilization experiments were conducted.

1. The experiments attempted to diagnose whether decline phenomena were specific symptoms of nutrient deficiency. $MgSO_4$ and $CuSO_4$ were directly applied to the foliage by sprinkling salt solutions. The effects of these nutrients on Norway spruce have been studied at the Schneeberg site in the Fichtelgebirge, where trees showed severe needle-yellowing and crinkling of the top branches.
2. The experiments tried to determine whether nutritional deficiencies could be improved via soil melioration, and how quickly and effectively soil fertilization with various fertilizers of different solubility would work. Different fertilizers were applied to the soils under Norway spruce, with different degrees of needle-yellowing. In addition, the effects of the fertilizers on soil and soil solution chemistry were studied (see Kaupenjohann 1989).
3. The experiments examined the possibility of buffering atmospheric acids already in the canopy by applying lime and studying the effects of this fertilization approach on tree-nutrient status (see Kaupenjohann 1989). Lime was applied to those canopies in older spruce stands which showed needle loss. As noted in Chapter 3-D, acidic inputs may be buffered in tree canopies by increased "cation leaching." Increased cation leaching from the canopy forces the tree to accelerate cation uptake from the soil. This results in increasing acidification of the rhizosphere.

2 Sprinkling of Norway Spruce Trees with Nutrient Solutions

Table 1 shows needle elemental-analysis 6 months after sprinkling salt solution on 10-year-old Norway spruce with severe needle-yellowing. Each tree was sprayed with 200 ml of a solution containing 100 g $Mg(NO_3)_2$ l^{-1} at the beginning of May 1984. In control trees, N, K, and Cu concentrations were low, Ca and Mg contents were deficient, while P and S levels were sufficient to high. Treatment

E.-D. Schulze, O.L. Lange, R. Oren (Eds.)
Ecological Studies Vol. 77

Table 1. Mineral contents of current-year needles of Norway spruce, 6 months after sprinkling with Mg at the Schneeberg site[a]

Treatment	N	P	S	K	Ca	Mg	Cu
	mg g_{dw}^{-1}						mg kg_{dw}^{-1}
–	11.7	1.9	1.6	5.5	1.8	0.22	4.3
$Mg(NO_3)_2$ sprinkling	12.6	1.9	1.6	6.3	2.2	0.32[b]	3.8

[a] 200 ml of a solution containing 100 g $Mg(KO_3)_2$ l^{-1} per tree.
[b] $p = 0.05$.

Table 2. Mineral contents of current-year needles of Norway spruce after repeated sprinkling with Mg (see Table 1) and after applying $CuSO_4$ (100 ml of a solution containing 39 g $CuSO_4 \cdot 5\,H_2O$ l^{-1} per tree)

Treatment	N	P	S	K	Ca	Mg	Cu
	mg g_{dw}^{-1}						mg kg_{dw}^{-1}
–	12.5	1.6	1.5	6.2	1.3	0.24	3.7
$Mg(NO_3)_2$ + $CuSO_4$ sprinkling	13.4	1.7	1.7	6.8	1.9[a]	0.44[b]	7.1[b]

[a] $p = 0.05$.
[b] $p = 0.01$.

with $Mg(NO_3)_2$ did not significantly affect the contents of N, P, S, K, Ca or Cu. On the other hand, Mg concentrations increased by about 50% (0,22 mg Mg g_{dw}^{-1} in control trees and 0,32 mg Mg g_{dw}^{-1} in treated ones). However, the increase in the Mg levels was not so high that deficiency symptoms disappeared (deficiency symptom levels at about 0,35 mg Mg g_{dw}^{-1}). Therefore, the treatment was repeated in the following year. Cu was also applied to test whether the observed crinkling of the top whorl of the trees could be attributed to a Cu deficiency. Analysis indicated that there was an increase in the Mg contents of the needles by about 100% due to fertilization (Table 2). Similarly, the deficiency symptoms disappeared to a great extent in four of five treated spruce trees. Cu treatment also increased the Cu concentrations of the needles from 3.7 to 7.1 mg kg^{-1}. However, this did not affect the symptom of crinkled branches.

3 Soil Fertilization Experiments

An overview on fertilization experiments in the Fichtelgebirge is given by Kaupenjohann (1989). As one representative example, results of the experiments on spruce with needle-yellowing at the declining site near Oberwarmensteinach are shown in Table 3. Six months after fertilization, the control plot showed

Table 3. Influence of different fertilizers on the nutrient contents of current-year spruce needles in the declining stand near Oberwarmensteinach 6 months after fertilization (fertilizer was applied in May 1983)

Treatment[a]	Weight of 100 needles	N	P	S	K	Ca	Mg
	(g)	mg g_{dw}^{-1}					
I	0.45	14.0	1.9	1.8	6.1	1.7	0.6
II	0.47	14.4	1.8	2.0	6.6	2.8	1.0
III	0.49	17.2	1.6	2.0	8.0	2.1	0.7
IV	0.39	16.9	1.7	1.8	7.6	2.7	0.9
V	0.46	16.2	1.7	1.7	7.1	1.7	0.9

[a]I = Control
II = $MgSO_4 \cdot H_2O$, 1000 kg ha^{-1}
III = $MgSo_4 \cdot H_2O$, 1000 kg ha^{-1} + KNO_3, 500 kg ha^{-1}
IV = $MgSO_4 \cdot H_2O$, 1000 kg ha^{-1} + KNO_3, 500 kg ha^{-1} + P_2O_5, 90 kg ha^{-1}
V = $MgO \cdot CaCO_3$, 10000 kg ha^{-1}.

adequate N, P, and K levels and high S, low Ca, and deficient Mg contents in the needles.

$MgSO_4$ application increased the Mg concentrations of the needles from 0.6 to 1.0 mg g_{dw}^{-1}. Treatments containing KNO_3 or half-calcined dolomite (Treatments III, IV, and V; see Table 3) did not work as quickly because of either a K/Mg-antagonism or relatively slow solubility.

Two years after fertilization untreated spruce contained only 0.46 mg Mg g_{dw}^{-1} in current-year needles, and the values even decreased with needle age (Fig. 1). $MgSO_4$ strongly increased the Mg levels of all needles. The effect of the dolomite fertilizer was retarded.

Even 4 years after fertilization, $MgSO_4$ had induced at minimum a duplication of the Mg levels of all needle age-classes compared with the control plot. The high Mg supply in the dolomite treatment induced a threefold increase in the

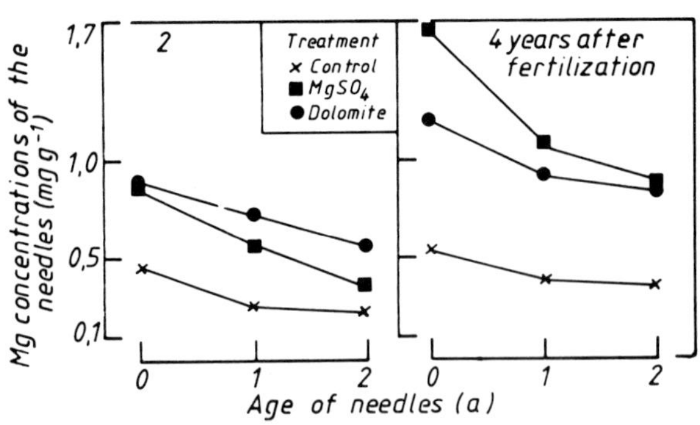

Fig. 1. Influence of different fertilizers on the Mg contents in spruce needles of different age classes at the declining site 2 and 4 years after fertilization

Table 4. Influence of different fertilizers on the nutrient contents of current-year spruce needles in the declining stand near Oberwarmensteinach 5 years after fertilization (fertilizer was applied in May 1983)

Treatment[a]	Weight of 100 needles (g)	N	P	S	K	Ca	Mg
				mg g_{dw}^{-1}			
I	0.67	14.1	2.8	1.6	10.2	2.1	0.7
II	0.64	13.8	3.0	1.8	9.8	1.7	1.1
III	0.62	14.0	2.1	1.6	9.9	1.3	1.2
IV	0.68	14.2	2.2	1.6	9.2	1.2	1.1
V	0.77	15.3	1.77	2.3	8.2	2.7	1.4

[a] See Table 3.

Mg content in current-year needles in comparison to control trees. In addition, 5 years after fertilization, the effects on Mg levels of the needles continued (Table 4).

$MgSO_4$ and dolomite increased the Mg concentrations in soil solutions to at least 75 cm of soil depth, as well as the NH_4Cl exchangeable amounts of Mg in the 0–30 cm mineral soil layer (Tables 5 and 6). In contrast, K and NO_3 concentrations in the soil solutions strongly decreased in the $MgSO_4$ treatment, while the NO_3 release from the soil was increased by up to one order of magnitude with dolomite (Table 5). The latter result can be attributed to the reinforced mineralization of the organic soil material as a result of liming.

Besides the improvement in the nutrient status of the fertilized trees, a visible recovery also occurred 1–2 years after fertilization in plots II-IV, and 3 years after fertilization in plot V.

An increase in needle weight was not induced by the fertilizers, not even by the N, P, K, and Mg treatment (Tables 3 and 4), nor was the number of needles within the fourth whorl increased as previously reported by Kaupenjohann (1989).

Table 5. Fertilizer effects on ion concentrations of soil solutions sampled using P80-suction cups at the declining site near Oberwarmensteinach[a]

Fertilizer[b]	I	II	V	I	II	V	I	II	V	I	II	V
		Mg			Ca			K			NO	
Soil depth (cm)												
25	0.3	2.6	3.3	1.6	0.9	1.7	2.3	0.3	3.2	1.8	0.3	6.0
50	0.2	6.4	5.3	1.0	1.4	1.8	1.5	0.6	2.0	1.2	0.4	10.0
75	0.2	4.5	2.6	1.1	1.4	1.3	2.3	0.5	3.0	2.1	0.2	8.7

[a] Mean values from Jan. 7, 1984 to June 30, 1985, mg l^{-1} (Kaupenjohann et al. 1987).
[b] See Table 3.

Table 6. NH_4Cl soluble Mg (kg ha^{-1}) in various soil depths at the declining site near Oberwarmensteinach 2 years after fertilization (Kaupenjohann and Zech 1987)

Soil depth (cm)	Fertilizer[a] I	II	V
0– 5	13	31	240
–10	5	20	135
–20	13	26	126
–30	11	18	79
Sum (0–30)	42	95 (33)[b]	580 (29)[b]

[a] See Table 3.
[b] Percent of the applied Mg.

4 Fertilization of Canopies

At the declining site of Oberwarmensteinach, 100-year-old spruce stands with severe needle loss were fertilized by helicopter with a lime suspension containing $CaCO_3$, KSO_4, $MgCO_3$, and $ZnSO_4$ (Table 7). In order to improve fertilizer absorption by the canopy of the trees, molasses was added to the suspension (Kaupenjohann 1989).

Six months later control trees contained rather low K levels in current needles, while the supply of other nutrient elements (P, Ca, Mg, Zn, Cu) was sufficient to high (Table 7). The fertilizer tended to increase K, Ca, and Zn contents. For example, K concentrations of current-year needles were increased by ca. 23%. Two years after fertilization, the effects on nutrient status followed the same tendency for K and Zn even more distinctly (Table 8).

Any revitalization effects of fertilization should result in decreasing needle loss. To study possible effects of recovery of trees, litterfall was collected during the period from 01.08.87 to 31.12.87. The results indicate clear reduction of litterfall from 31.9 g m^{-2} under control spruce to 19.3 g m^{-2} in fertilized stands which may be attributed to fertilization.

Table 7. Mineral element contents of spruce needles of different age classes, 6 months after fertilization (Fert.)

Needle age	Treatment	P	K	Cu	Mg	Zn	Mn
		mg g_{dw}^{-1}					
Recent	Control	2.03	5.07	3.30	1.22	0.030	0.45
	Fert.	1.95	6.23	9.26	1.39	0.038	0.74
2 years old	Control	1.65	4.97	3.80	0.97	0.025	0.42
	Fert.	1.60	5.95	4.07	1.11	0.031	0.65
3 years old	Control	1.68	5.31	3.70	0.73	0.021	0.39
	Fert.	1.56	5.60	3.98	0.91	0.028	0.52

Table 8. Mineral element concentrations of current-year needles of spruce 2 years after fertilization (Fert.)

Treatment	K	Ca	Mg	Zn
			mg g_{dw}^{-1}	
Control	4.49	3.39	1.05	0.039
Fert.	5.73	3.26	1.04	0.045

5 Conclusions

In sprinkling experiments with Mg solutions, Mg concentrations in spruce needles were increased and needle-tip yellowing disappeared from the treated trees. This experimentally indicates that the yellowing of older needle tips in declining spruce stands is due to a Mg deficiency (Zech and Popp 1983). The same results have been achieved from soil fertilization as also has been shown by Hüttl (1985), Isermann (1985), and Kaupenjohann et al. (1987). The most rapid increase in foliar Mg resulted from the $MgSO_4$ application, while the dolomite effects were retarded as a result of slower solubility. The effects of both treatments continued even 5 years after fertilization. This may be attributed to both the high amounts of applied fertilizer, as well as to the storage of Mg fertilizer in the plant available fractions in the soil. These results show that at the experimental site under current environmental conditions, Mg concentrations in the soil are too low to satisfy tree demand without fertilization. However, these results do not indicate whether the shortage of plant available Mg in the soil is (1) a result of Mg leaching from the soil due to soil acidification or (2) reflects a relative deficiency of Mg as a result of a N-induced demand increase for growth. Because decreasing NO_3 concentrations induced by $MgSO_4$ in the soil solution might be explained by increasing N uptake of Mg fertilized trees, declining stands might be saturated with N.

On the other hand, decreasing NO_3 concentrations may simply be the result of leaching after SO_4 application, which is supported by the results of several authors. Lee (1985) and Mochoge and Beese (1986) found that NO_3 was released from undisturbed soil columns after an SO_4 application. The concept of increasing leaching of ions after fertilization could also explain the decreasing K concentrations within our $MgSO_4$ treatment.

Liming increased the NO_3 release into the soil solution up to 10 mg NO_3-N l^{-1}. This relatively high concentration may be attributed to the high dose of half-calcined dolomite (10 t ha^{-1}) which had been applied. Indeed, the results of several other authors who had applied lower quantities of lime have shown that NO_3 mineralization is not that high (Matzner 1985; Beese 1985; Evers 1986).

Surprisingly, needle weight did not increase with fertilization, not even from applying nearly the total set of macronutrients (N, P, S, K, and Mg). From this result, one might conclude that another factor is growth-limiting.

Because Mg application induced clear revitalization of formerly yellow trees, some conclusions for practical forest fertilization management may be offered to revitalize and stabilize affected forest ecosystems as much as possible until reduction of air pollution [which is supposed to be the main cause of forest decline (FBW 1986)] takes effect (Ulrich 1986). Because of possible ecological effects of fertilization on a large scale, the dose of fertilizer should be kept to a minimum as much as possible. In an attempt to derive a proposal for practical forest fertilization Kaupenjohann and Zech (1987) have reviewed several fertilization trials. Where nutrient deficiency was severe, for example, these authors suggest the application of 400–600 kg ha^{-1} of a soluble fertilizer like $MgSO_4$ which may be combined with dolomite (20–30 t ha^{-1}).

The results of lime application to the canopy of trees seems quite promising. Theoretically, by this method part of the acid input into ecosystems would be neutralized before it enters the soil. However, as the results are still preliminary, more experience with this type of fertilization is necessary.

References

Beese F (1985) Wirkungen von Meliorationskalkungen auf podsoliger Braunerde in einem Buchenwaldökosystem. Allg Forstz 43:1161–1162

Evers FH (1986) Forstdüngung und Nitratbelastung von Sicker-, Oberflächen- und Quellwasser. KFK-PEF Bericht 5:57–63 (ISSN 0931-2749)

FBW (Forschungsbeirat Waldschäden/Luftverunreinigung) (1986) 2. Ber Kernforschungszentrum Karlsruhe 1–229

Hüttl RF (1985) "Neuartige" Waldschäden und Nährelementversorgung von Fichtenbeständen (*Picea abies* Karst.) in Südwestdeutschland. Freiburger Bodenkd Abh 16:1–195

Isermann K (1985) Diagnose und Therapie der "neuartigen Waldschäden" aus der Sicht der Waldernährung. VDI-Bericht 560:897–920

Kaupenjohann M (1989) Chemischer Bodenzustand und Nährelementversorgung immissionsbelasteter Fichtenbestände in NO-Bayern. Bayreuther Bodenkd Ber 11:1–202

Kaupenjohann M, Zech W (1987) Walddüngung und neuartige Waldschäden – Ergebnisse aus Düngungs – und Kalkungsversuchen. In: Glatzel G (ed) Möglichkeiten und Grenzen der Sanierung immissionsgeschädigter Waldökosysteme. Selbstverlag der Österreichischen Gesellschaft für Waldökosystemforschung und Experimentelle Baumforschung Wien, 82–98 (ISBN 3-900-491-02-3)

Kaupenjohann M, Hantschel R, Zech W, Horn R (1987) Ergebnisse von Düngungsversuchen mit Magnesium an vermutlich immissionsgeschädigten Fichten [*Picea abies* (L.) Karst.] im Fichtelgebirge. Forstwiss Centralbl 106:78–84

Lee JJ (1985) Effect of simulated sulfuric acid rain on the chemistry of a sulfate-adsorbing forest soil. Water Air Soil Pollut 25:185–193

Matzner E (1985) Auswirkungen von Düngung und Kalkung auf den Elementumsatz und die Elementverteilung in zwei Waldökosystemen im Solling. Allg Forstz 43:1143–1147

Mochoge BO, Beese F (1986) Leaching of plant nutrients from an acid forest soil after nitrogen fertilizer application. Plant Soil 91:17–29

Ulrich B (1986) Die Rolle der Bodenversauerung beim Waldsterben: Langfristige Konsequenzen und forstliche Möglichkeiten. Forstwiss Centralbl 105:421–435

Zech W, Popp E (1983) Magnesiummangel, einer der Gründe für das Fichten- und Tannensterben in Nordostbayern. Forstwiss Centralbl 102:50–55

4-H Nutritional Disharmony and Forest Decline: A Conceptual Model

R. OREN and E.-D. SCHULZE

Heavenly tune which none can hear
Of human mould with gross unpurged ear
Milton (Arcades)

1 Introduction

A base cation deficiency has been implicated in the decline of European forests (Abrahamsen 1980; Ulrich 1983; Franz 1983; Zech and Popp 1983; Tomlinson 1985; Hauhs and Wright 1986; Rehfuess 1987; Cape et al. 1988). However, it is puzzling to find that a tree, dying from a lack of a cation such as Mg at one site, may have more of that cation than a healthy tree at another site (see Bosch et al. 1983). Plant nutrition studies have revealed that for maximum growth to occur, requisite elements must be supplied to trees in a certain ratio which differs between species (Mulder 1956; Ingestad 1959, 1979a,b; see review by Powers 1984). The harmony in the supply of elements to trees, or in the uptake by trees, is the core of the following conceptual model of forest decline.

The term nutritional disharmony was chosen with care. To use a metaphor, when one instrument in a large orchestra is played too early or late or too hard or soft, the instrument is not in full harmony with the rest of the orchestra and the music is spoiled. To detect the disharmoniously played instrument may be difficult, particularly when the orchestra is large, the music complex and the listener untrained. After listening to only a small fraction of an unfamiliar musical piece, isolating such an instrument may not be possible. An element which sometimes appears in excess may be equated with that instrument played precisely to such a degree that, without its timely contribution, full harmony cannot be achieved.

Like the degree to which an instrument is required by an orchestra, the same nutrient level may not be optimal for all growth processes throughout the growing season (Linder and Rook 1984). An assessment of nutritional status at a given point in time cannot account for the dynamic nature of nutrient relations and may only result in a qualitative, perhaps erroneous, explanation of tree performance.

The mechanisms of nutrient stress leading to forest decline can be described by dividing the year into three periods: (1) a needle and twig growth period (*crown growth phase*) when retranslocation occurs from the old tissues; (2) a stemwood production period (*stem growth phase*) when uptake from the soil is required to support growth; (3) a recharge period (*recharge phase*) during the nongrowing season when nutrients are accumulated in mature tissues. An equivalent division has been proposed by Nambiar and Fife (1987; see also Powers 1984).

Our conceptual model of forest decline from nutritional disharmony is based on several concepts of plant nutrition which may explain certain, seemingly

E.-D. Schulze, O.L. Lange, R. Oren (Eds.)
Ecological Studies Vol. 77

contradictory, observations of healthy and declining Norway spruce (*Picea abies* L. Karst.) stands (see Chaps. 2, 4-D). In particular, we seek to explain why two stands produced similar amounts of crown components at a similar leaf-area index, even though the declining stand displayed yellow crowns and produced 35% less stemwood than the healthy stand (Oren et al. 1988a,b,c; see Chap. 4-D).

2 Canopy Growth Phase

In growing tissues such as young needles or new stemwood, nutrient stress is caused by one or more elements being supplied at a rate lower than the potential relative growth rate under a given set of conditions (Ingestad 1982). In the suboptimal range of N concentration, growth rate at steady state is linearly related to the uptake rate and the internal concentration of N (Ingestad 1979b, 1982; see Chap. 4-F). This has been shown regardless of irradiation, species, or the supply of other elements (Ingestad 1982). [Recently, Ingestad and Kähr (1985) have reported that conifers take up elements to levels above those needed to support their growth potential, including N if it is present in high concentrations.] In general, it appears that when growth rate becomes either limited or enhanced by any resource variable, a transition period follows in which the uptake rate of N does not match the growth rate. At this point the concentration of N may progressively increase or decrease before a new steady-state of N concentration in relation to growth is reached. During this transition period, a temporarily higher photosynthetic rate under standardized conditions (Linder and Rook 1984), or chlorosis (Ingestad and Kähr 1985), depending on foliar N concentration, may occur. When plants are chlorotic, they experience a deficiency which causes an impaired functioning of existing tissues. Once plants reduce the growth rate to match the uptake rate of the limiting element, the chlorosis is alleviated, but the growth-reducing stress remains (Linder and Ingestad 1977).

If N is nonlimiting to growth, a function-impairing deficiency of other nutrients can be caused by a retranslocation of elements in low supply from mature tissues which have already completed their growth. Just as the soil supplies nutrients to annual plants, such mature tissues as the needles and the twigs in a spruce crown can be considered the nutrient source for expanding buds. Therefore, a tree crown with limited stores of elements cannot supply the expanding twigs. This results in either reduced growth or resource exhaustion, so that the trees functions may be impaired (Glatzel 1981).

Because the growth of single young twigs after bud break is exponential, the supply rates of elements from the soil may be too low to meet growth-rate requirements. This lack of elements can result in a decrease in both the canopy growth rate and the period in which the tree benefits from its new foliage (Linder and Rook 1984). Therefore, it is not surprising that retranslocation of certain nutrients from mature tissues appears to be normal behavior for trees, regardless of the soil nutrient supply (Chapin and Kedrowski 1983). Retranslocation of elements, including N, can thus increase with increased N input to the soil

(Nambiar and Fife 1987). In fact, a high growth rate in conifers has been found to stimulate retranslocation and resorption before needles are shed (Birk and Vitousek 1986). Both processes seem to be responses to the seasonal pattern of growth rate, rather than to the availability of elements in the soil solution.

In many cases, needle growth appears to be related to N availability, but maximum growth rate, and thus annual production, can be determined by other factors, such as the supply of Mg or other elements. Because many temperate coniferous forests respond to N fertilization, particularly when N concentrations in the foliage are low (Taylor 1987), trees are most likely suffering N stress (Mahendrappa et al. 1986). Many times an early response to fertilization is an increase in total leaf biomass and the size of individual needles produced after the supply has increased (Brix 1981; Linder and Rook 1984; Binkley and Read 1985; Nambiar and Bowen 1986; Nambiar and Fife 1987; Madgwick and Tamm 1987; Hüttl and Wisniewski 1987). In terms of change in their weight and element concentrations, the responses of newly produced individual needles may be used to evaluate the status of forest stands with respect to any element (Fig. 1; Timmer and Morrow 1984; Binkley 1986). Therefore, if both needle weight and concentrations of elements have increased after fertilization, a previous deficiency can be recognized. If only concentration or only weight has increased, luxury consumption or sufficient supply are implied, respectively, because the function of the needles has not been impaired. However, if needle weight has increased and concentration has decreased, a dilution process is indicated. A deficiency has occurred, and may still be present, only when both needle weight and element concentration have increased in response to a greater element supply.

In Fig. 2 an interpretation of these relationships in ten plots representing the two sites is presented. No clear relationship between Mg concentration and needle growth existed. Based on site average, the relationships indicated that an increasing Mg supply simply resulted in greater accumulation in the needles, with no change in needle size. In contrast, an increased N supply resulted in increased needle weight until, in a single plot, the limited supply of Mg created growth-reducing stress so that instead of further increasing needle weight, N became

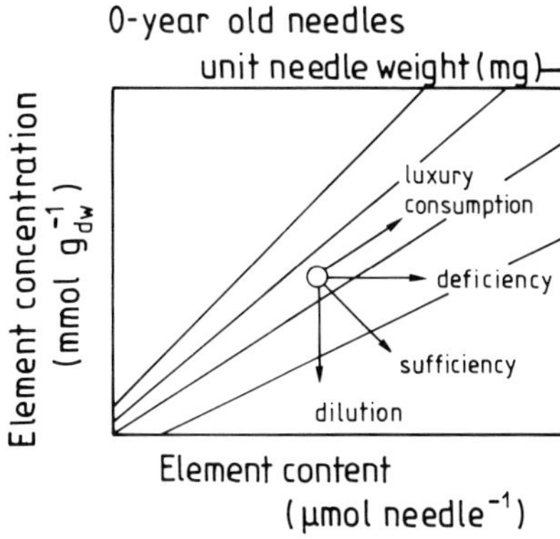

Fig. 1. Diagnosis of nutrient limitation based on the responses of single needles produced after the nutrient supply was altered (see text for explanation; after Timmer and Morrow 1984)

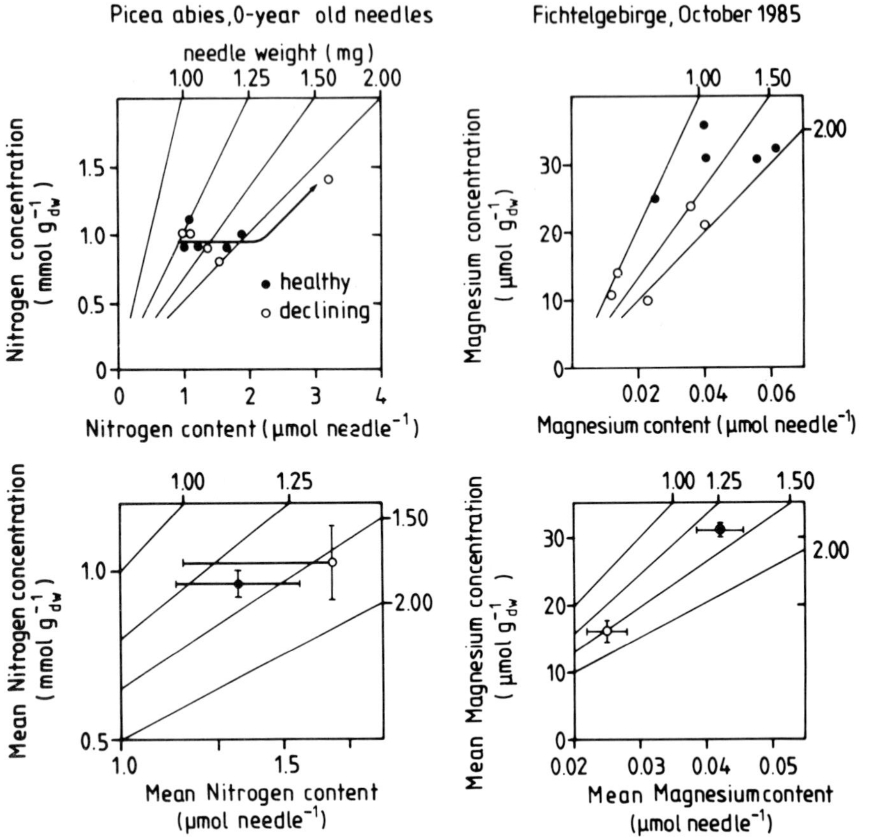

Fig. 2. Relationships between needle N and Mg concentrations and contents and needle weight in young needles from ten plots at declining and healthy *P. abies* sites (*top*) and the site average of the same data (*bottom*). Based on Fig. 1, N seemed to be sufficient until, in one plot, Mg stress prevented further needle growth and N accumulated luxuriously (the plot with high needle weight and low Mg concentration, *top left*). Magnesium seemed to be in excess (Oren et al. 1988c)

more concentrated. Thus, the N supply was first sufficient and then in excess (cf. Figs. 1 and 2). The relationship of weight changes with respect to N concentration changes should not be confused with the relationships with growth rate discussed above. Foliar diagnosis permits an evaluation of stand nutrition when conditions become increasingly limiting (i.e., not at steady state).

Foliar diagnosis may also be used to examine the effect of changes in the supply of one element, particularly N, on the concentration of other elements (Timmer and Morrow 1984; Binkley 1986; Timmer and Armstrong 1987; Timmer and Ray 1988). In many fertilization trials with fertilizers containing either only N, or N in a relatively high concentration compared to the other elements, a dilution of the other elements in growing needles often occurs (Gill and Lavender 1983; Timmer and Morrow 1984; Linder and Rook 1984; Rehfuess et al. 1983; Nambiar and Fife 1987; Weissen et al. 1988). This dilution process may also occur when other elements are supplied in appropriate proportions to N (Aronsson and Elowson 1980). Indeed, even atmospheric N input has been implicated in causing a similar effect in highly polluted areas (Nhilgård 1985; Mohren et al. 1986; Schulze et al. 1987; Glatzel et al. 1987; see review by Johnson et al. 1982). This claim has been supported by the results of long-term N treatments at rates similar to those of the input from pollution (Tamm 1968; Tamm and Aronsson 1988). Such effects are particularly noticeable during times of peak demand (e.g., bud break) in trees growing on poor soil (Zöttl and Hüttl 1986). It is apparent from Fig. 3 that the Mg concentration in young needles declined with increasing N concentration similarly at both sites, but did not fall below the level at which needle chlorosis begins (Ingestad 1959). This was not only accomplished by increasing the supply of Mg to match the growth promoted by the N, but as we showed earlier, also by preventing a further increase in needle weight once the Mg concentration became so low.

During the *crown growth phase* in the spring, most of the Mg required for growth was retranslocated from older canopy components (Table 1) in direct proportion to the demand promoted by the N concentration. This may represent a general relationship in the seasonal retranslocation of a mobile element from mature needles.

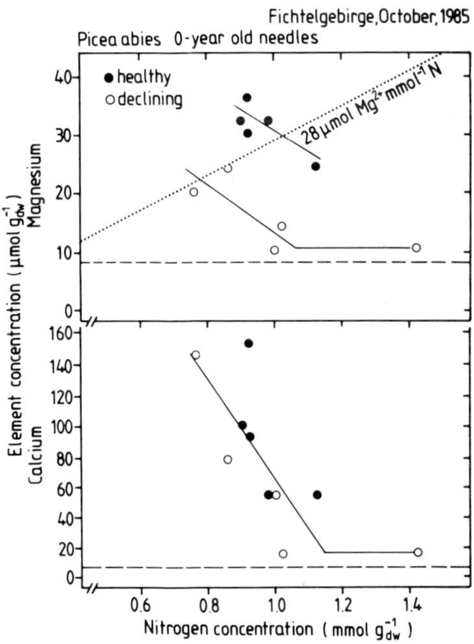

Fig. 3. Concentrations of Mg and Ca in new needles in October declined with increasing N concentration from an increase in needle weight (see Fig. 2). Levels did not reach those at which chlorosis begins (*horizontal dashed lines*, after Ingestad 1959; from Oren et al. 1988c). The *diagonal dotted line* represents the Mg to N ratio above which production is within 95% of maximum (Ingestad 1959)

Table 1. The amounts of Mg and N in newly produced tree components, and retranslocated from and recharge of mature tissues in three periods at the healthy and declining sites (mmol m^{-2} ground)[a]

	Magnesium							Nitrogen					
Phase	Canopy growth		Stem growth		Recharge			Canopy growth		Stem growth		Recharge	
stand	Healthy	Declining	Healthy	Declining	Healthy	Declining		Healthy	Declining	Healthy	Declining	Healthy	Declining
Growth							Growth						
Needle	6.89	4.78	0.64				Needle	256.7	256.7				
Twig	3.08	1.64					Twig	45.6	45.6				
Branch	2.27	1.21					Branch	33.6	33.6				
Stem							Stem						
Wood			7.22	4.66			Wood			161.0	104.0		
Bark			1.27	1.27			Bark			18.5	18.5		
Root	8.97	8.58	7.25	6.16			Root	322.0	322.0	308.0	308.0		
Total	21.21	16.21	16.38	12.09			Total	657.9	657.9	487.5	430.5		
Retranslocation/Recharge							Retranslocation/Recharge						
Needle							Needle						
Dying	−2.41	−0.99	−2.41	−0.99			Dying	−54.8	−54.8	−54.9	−54.9		
Mature	−12.01	−7.47	−8.33	−3.97	20.34	11.45	Mature						
Current					9.74	2.92	Current					114.4	114.4
Twig							Twig						
Mature							Mature	−52.6	−52.6	−23.4	23.4	76.0	76.0
1-yr	−0.90		−0.39				1-yr	−28.7	−28.7			4.9	4.9
Current			−0.39		1.29		Current					28.7	28.7
Branch							Branch						
1-yr							1-yr	−9.5	−9.5			9.5	9.5
Current			−0.29				Current					3.6	3.6
Root							Root	−54.0	−54.0			54.0	54.0
Total	−15.32	−8.46	−11.80	−4.96	31.37	14.37	Total	−199.6	−199.6	−73.3	−73.3	291.1	291.1
Net uptake	5.89	7.75	4.58	7.13	31.37	14.37	Net uptake	458.3	458.3	414.2	357.2	291.1	291.1
Annual uptake:		Healthy	41.83				Annual uptake:		Healthy	1163.6			
		Declining	29.25						Declining	1106.6			

[a] Calculated uptake rate from the soil is also presented. Detailed account of the assumptions underlying the budget is presented in Oren et al. (1988c).

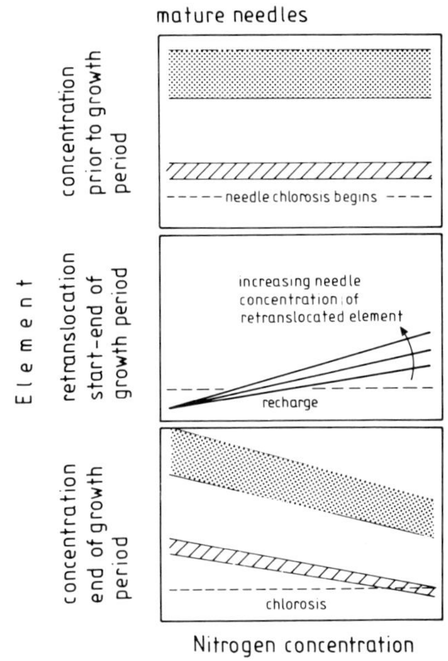

Fig. 4. A general description of seasonal retranslocation of a mobile element from mature needles in relation to the N concentration in those needles. Initial conditions at the start of the growing season are presented (*top*) for needles which were able to accumulate (recharge) large or small amounts of the element before the growth phase began (during the *recharge phase*). When the *canopy growth phase* begins, the mobile element is retranslocated out of mature needles in relation to the N concentration (*middle*). More N promotes faster growth of new tissues and greater demand for other elements. As a result, the element concentration at the end of the growth phase is inversely related to the N concentration (*bottom*). If the element is not recharged sufficiently and the needles have high N concentration, retranslocation may result in needle nutrient deficiency and chlorosis

Figure 4 (top) illustrates the conditions in trees which begin the season with high and low levels of e.g. Mg concentration (Mg represents here any mobile element) in their mature foliage. The higher variation in the Mg concentration in trees with higher Mg availability represents a real phenomenon (Oren et al. 1988b).

The mobile element moves out of these needles in proportion to the N concentration (Fig. 4, middle). New needle growth is determined by the N concentration in the mature tissues (Fagerström and Lohm 1977; Ågren 1985) as N is taken out of mature needles to support the new growth proportionally. The figure shows that at very low levels of N, uptake from the soil actually exceeds retranslocation for new needle growth, and the element concentration can increase. In addition, the variation in retranslocation increases at high N levels because other factors, such as the actual concentration of the mobile element, are also involved in affecting retranslocation (Oren et al. 1988c). Another factor which may contribute to the variation in retranslocation in relation to N is the form of N in the needles. Only part of the N in the older needles can be retranslocated and is used to promote growth of new tissues (Powers 1984). Only some of this mobile portion of the N can be reusable in various forms (i.e., storage N) while the rest can be in nonreusable forms which accumulate when growth rate is slower than the uptake rate (Millard 1988). Therefore, total N concentration may only be a crude index of how much N promotes growth and the resulting dilution of other elements.

As a result of retranslocation, the concentration of elements in needles with high N concentration decreases (Fig. 4, bottom; see also Aronsson and Elowson

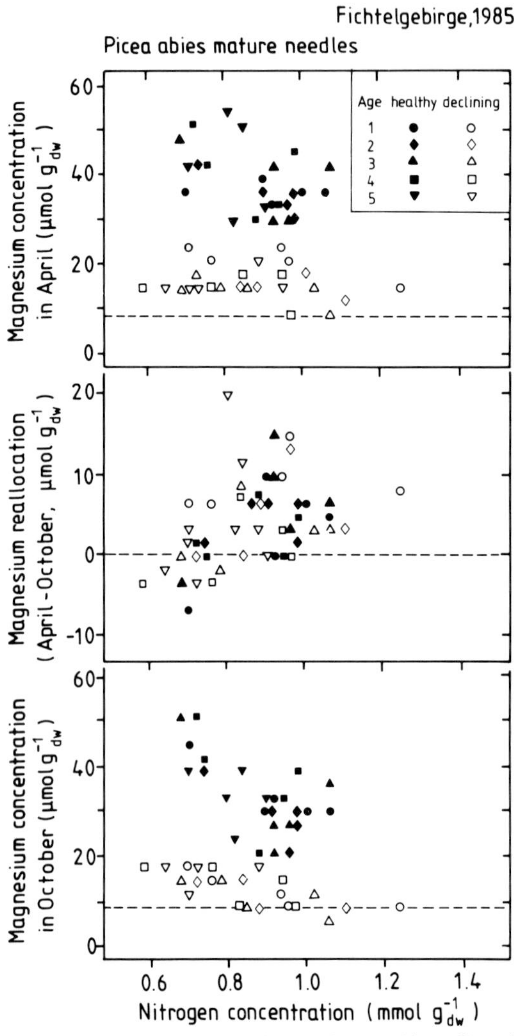

Fig. 5. The Mg concentration in 1- to 5-year-old needles prior to the growth phase in April was higher at the healthy site (*top*; from Oren et al. 1988c). The retranslocation of Mg out of the mature needles from April to October increased with the April N concentration in the needles (*middle*) and caused Mg concentrations in some needles at the declining site to decrease below the level at which chlorosis occurs (*dashed horizontal line*; after Ingestad 1959, *bottom*)

1980; Bender et al. 1986). If the concentration is low at the start of the season, it may fall to the level at which chlorosis occurs.

Figure 5 presents the data on which this model is based. All age classes of needles at the declining site began the season with low Mg and high N concentrations. They then lost much of their Mg and became chlorotic. Although their function was probably impaired, the growth of new needles was not limited, with the exception of a single plot. While needle chlorosis is a nonspecific indicator of nutrient deficiency (Linder and Rook 1984; Hüttl and Wisniewski 1987), its correlation with only one element in a given stand often indicates that this element is in short supply (Forschner and Wild 1988; Rehfuess 1987).

The growth of needles and twigs is not curtailed as long as enough Mg can be retranslocated (see also Chap. 4-D). To evaluate the contribution of retranslocation to the production of new needles and twigs, we calculated a budget for both Mg and N for each phase during the year (Oren et al. 1988c). Trees in the healthy and the declining stands used about 21 and 16 mmol Mg m^{-2} to produce their new canopy components and roots, of which retranslocation accounted for 75 and 50% in the two stands, respectively (Table 1). If we assume that retranslocation of elements in the canopy occurs mostly within the canopy (Nambiar and Bowen 1986), Mg retranslocation could completely satisfy the demand to produce canopy components at both sites. In both stands, the N required for growth was approximately 660 mmol m^{-2}, of which 30% was satisfied by retranslocation (60% if only the canopy is considered). In comparison, Nambiar and Fife (1987) estimated that N retranslocation for new foliage production in *Pinus radiata* D. Don. contributed 57 and 32% in fertilized and unfertilized trees, respectively, and P retranslocation contributed 70% in both treatments.

Because twigs grow at a higher rate and reach larger final size when they are more exposed to light, the demand for retranslocation should be greater at the top of the crown, particularly in dominant trees. However, the ratio of twigs produced to mature twig biomass is higher at the top of the crown (Schmidt-Vogt and Makkonen-Speicker 1986) and therefore a greater relative demand for Mg may be exerted on the mature needles at the top of the tree than at the lower crown. In addition, needles of all age classes may require higher Mg concentration to remain green and function properly at high light environment. Together, these processes may explain the repeated observation that mature needles exposed to higher radiation are the first to become chlorotic (Zöttl and Hüttl 1986; Hauhs and Wright 1986). Retranslocation in Norway spruce growing in regions of cold winters may be in part a passive process which reflects a change in membrane properties and an increase in membrane leakiness after the winter frost (Feiler 1985) which is more severe at the top. Alternatively, when the demand by expanding buds exceeds the supply from root uptake, a depletion of nutrient ions near the outer surface of cell membranes in mature needles may occur, stimulating movement of ion out of cells. Regardless of the mechanism of retranslocation, chlorosis can be caused by growth. When growth in the canopy is restricted by pruning (Nambiar and Fife 1987) or by severing the buds (Lange et al. 1987), the concentration of elements in mature foliage increases and chlorotic needles become green (see also Glatzel 1981).

3 Stem Growth Phase

This phase is characterized by increased competition among tissues for available nutrients for retranslocation (Fife and Nambiar 1982, 1984) and by factors which affect the nutrient supply from the soil.

Some processes and tissues in a plant are preferentially supplied by an element when its uptake rate is less than the potential growth rate. For example, when N is in short supply relative to the growth rate, leaf photosynthesis is given a higher priority for usage of N, and possibly of other nutrients, compared to growth processes (Ågren 1985; Waring et al. 1985; McDonald et al. 1986a,b). This results in a higher starch concentration in plants grown at low nutrient availability, although the starch accumulates at a lower rate (McDonald 1986b). The seasonal status and changes of carbohydrates in our stands fit this pattern. Although the instantaneous photosynthetic rates of green foliage were similar in both stands, the declining stand had a lower starch accumulation rate before bud break (Zimmermann et al. 1988; Oren et al. 1988a). At the declining site, however, growth was reduced relative to photosynthesis, as indicated by the higher starch concentrations in most tissues of the declining stand at the end of the growing season. Sweet and Wareing (1966) speculated that a change in nutrient availability from fertilization affects both photosynthesis and sink strength for carbohydrates. We conclude (Oren et al. 1988a) that the production of carbohydrates is probably less affected by the mechanism causing decline than is the sink for carbohydrates.

Under limited nutrient supply, supply to the roots is preferred over that to the leaves (Ingestad 1959; McDonald et al. 1986a). This preference is also associated with higher carbon allocation to the roots, perhaps in an attempt to increase nutrient uptake (Ågren and Ingestad 1987). At the declining site, Mg concentrations in the roots were higher than those in the needles. The opposite was true at the healthy site (Oren et al. 1988b; see Chap. 4-F). No such trend was found for N or Ca. It is conceivable that under conditions of such limitation, the allocation of nutrients for stemwood growth would also be curtailed. In this case, one would expect a strong relationship between needle Mg concentration and growth. Because both needle and stemwood growth are also highly related to the N supply, they are also reflected in the needle concentration of N. Thus, the combined effect of both the N and Mg supplies on stemwood growth can be described as a family of curves (two of which are depicted in Fig. 6). A reduction in growth would be reflected in a lower than optimal concentration of an element, such as Mg, relative to N. We assume, for lack of other information, that a limited supply of a nutrient would result in the same proportion of growth reduction at any level of N and, therefore, the reduction in growth can be interpolated from the reduction of growth at optimal N concentration.

Data from Ingestad (1959) on *P. abies* seedlings were recalculated and plotted as stem growth in relation to the N and Mg concentrations in the needles (Fig. 7; see Chap. 4-F). We incorporated the data from our ten plots in relation to the plotted curve by positioning on the curve the plot with the highest growth above the concentrations found in the young foliage in that plot (Fig. 7, left). The growth

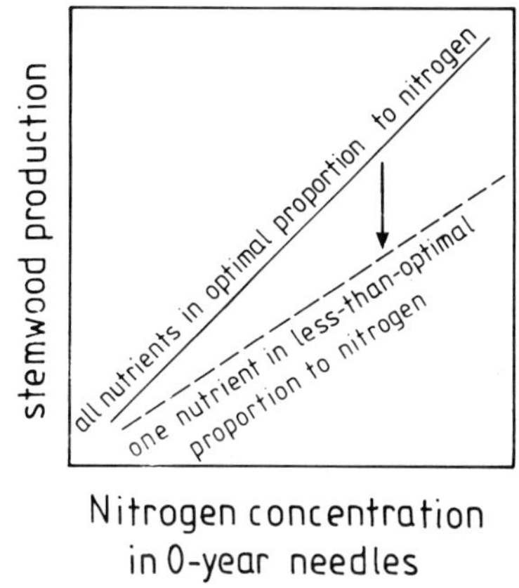

Fig. 6. Production of stemwood by trees of similar size and leaf area and subjected to similar competition should increase in relation to the N concentration in their young foliage. However, if the supply of another element were in less-than-optimum proportion to that of N, growth would be proportionally reduced

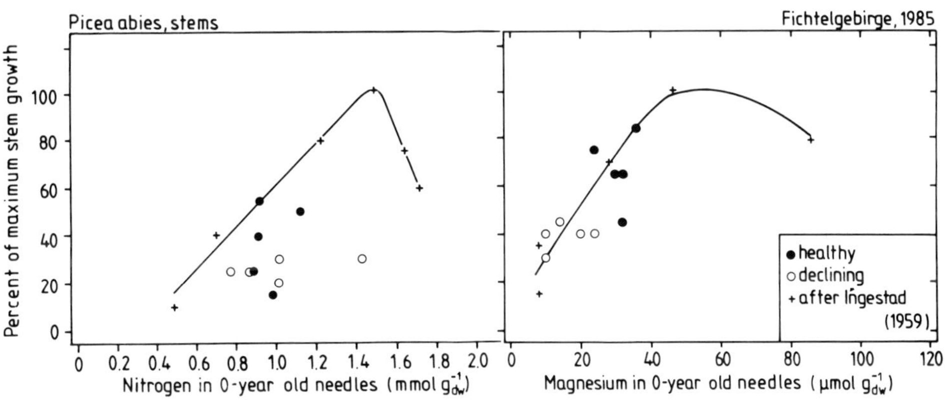

Fig. 7. The relationships between the Mg and N concentrations in new needles and the percent of maximum growth of stem recalculated from a *P. abies* seedling experiment (+ ; Ingestad 1959). A line was fitted by eye to the seedling data. Data from the plot with the highest growth of stemwood per unit of leaf area were drawn on the two lines according to the October concentrations of N and Mg in new foliage at the plot. Growth in all other plots was expressed as a percent of the growth in that plot and drawn according to the foliar concentrations (Oren et al. 1988c)

in all other plots was expressed as a proportion of the growth in the first plot and positioned according to foliar concentrations. The growth per unit of leaf area was used to account for the large differences in standing biomass between the plots which would affect growth directly (Waring et al. 1981; Oren et al. 1988a). The relationships between growth and N is very weak; the data points may be viewed as if they are positioned on many lines and represent various concentrations of Mg (Fig. 6). Indeed, the plots of both sites nearly converged on the line, indicating the direct relationship between growth and Mg concentration (Fig. 7, right).

During tne *stem growth phase*, 65% of the 16 mmol Mg m^{-2} needed for stem and root growth in the healthy stand was supplied by retranslocation from mature tissues. In comparison, only 40% was supplied from the same source in the declining stand to satisfy its lower (12 mmol m^{-2}) Mg requirement (Table 1). Therefore, in both phases the declining stand had to rely more heavily than the healthy stand on uptake of Mg from the soil, while the healthy stand relied more on retranslocation. Retranslocation of various nutrients throughout the growing season has been reported (Mahendrappa et al. 1986). Although conifer seedlings are able to take up sufficient amounts of nutrients even if the concentrations in the soil solution are very low (Ingestad and Kähr 1985), such uptake may be too low to meet the demands of large trees during periods of potentially rapid growth. Therefore, the growth in the declining stand during the *stem growth phase* was probably limited because of less ability to retranslocate. This limitation was exacerbated by the greater difficulties of the declining stand in obtaining Mg from the soil.

Element uptake from the soil depends on the concentration available from the soil solution and the characteristics of the root system (see Chap. 4-E). The soil solution at the declining site had lower concentrations of Ca and Mg than that at the healthy site. The differences in cations between the two soil solutions were probably due to (1) less available Ca and Mg from the weathering of the parent material at the declining site and (2) greater leaching, due to higher precipitation and proton input (Hantschel 1987). Under such conditions, the cation leaching rate from the soil generally exceeds the weathering rate (Johnson et al. 1982).

Although the concentration of N in the soil solution was similar at both sites, at the declining site a greater proportion was in the NH_4^+ form, while at the healthy site a greater proportion was in the NO_3^- form (Meyer et al. 1988; Oren et al. 1988b). The difference in the main form of N between the sites is probably due to a lower base saturation and pH (Meyer et al. 1988; Chap. 4-B) at the declining site. *P. abies* seedlings prefer NH_4^+ over NO_3^-, although they can grow almost as well when only NO_3^- is available (Ingestad 1979a). However, a balance of NH_4^+ and NO_3^- is more nutritionally desirable because it ensures the uptake of the accompanying ions, e.g., Ca and Mg with NO_3^- and P with NH_4^+ (Mulder 1956). The higher P concentrations, found in all tissues of the plants at the declining site, indicates that more NH_4^+ was probably being taken up than NO_3^- (Oren et al. 1988b). The lower Ca and Mg uptake at the declining site may relate not only to the low concentration of these elements in the soil solution, but also to the high proportion of N as NH_4^+ (Roelofs et al. 1985), and perhaps to a higher proton concentration in the soil (Mulder 1956).

Given favorable physical and chemical soil conditions (Sutton 1969), the expanding root system exponentially explores a larger volume of soil, while the climate promotes an exponential growth phase above-ground (Ingestad 1982). However, the soil at the declining site was likely to have a higher solution concentration of Al because of its lower base saturation, soil pH, and acid neutralizing capacity (Chap. 4-B), combined with the high ionic strength of the soil solution from the sulfate input of precipitation. This resulted in a Ca:Al molar ratio within the range of impairing root development (Rost-Siebert 1983; Tomlinson 1985; Entry et al. 1987; Matsumoto and Yamaya 1986). The number of root tips in the mineral soil of plots at both sites was related to this ratio (Meyer et al. 1988). Generally, the more favorable ratio at the healthy site may explain the more even root distribution with depth when the drier period began in July and perhaps the higher proportion of mycorrhizal root tips (Meyer et al. 1988). Mycorrhizal development (Entry et al. 1987) and ability to absorb or retain Ca and Mg (Boxman et al. 1985) is reduced at low ratios of Ca:Al. Because Ca and Mg leached by acidic deposition from the surface soil may accumulate in deeper horizons (Kelly and Strickland 1987), it is conceivable that the restriction of root growth to the upper, organic horizons reduces root access to soil Mg.

The combination of drying soil and low pH has a very negative effect on the rate of *P. abies* root growth (Bartsch 1987), a rate which is normally low in this species (Ingestad and Kähr 1985; see Chap. 4-E). Upon soil rewetting, root growth recovers much more slowly at low pH, a factor which may exacerbate the effect of soil drying on element uptake (Ingestad 1987). This phenomenon most likely affects a large proportion of the root system at the declining site which is concentrated in the more drought-prone organic layer, and probably reduces even further the capacity of the trees to take up nutrients (see Chap. 4-F).

Transpiration was similar at both sites (Werk et al. 1988; see Chap. 4-C). Because the concentration of elements in the sap of trees at both sites decreased with decreasing water potential (Osonubi et al. 1988), water was probably taken up from the volume of soil already depleted of nutrients (Clark 1984). As Mg and Ca are taken up passively (Clark 1984; Kirkby and Pilbeam 1984), their uptake is more affected at higher water potentials than is that of N. Particularly in its NO_3^- form, N should be even less affected by soil drying than NH_4^+. An uneven distribution of roots at the declining site probably caused its lower water potential and worsened its imbalanced nutrient supply even more (Osonubi et al. 1988).

A limited rate of Mg uptake will reduce the growth rate below the growth potential, as indicated by the N content in the stem bark. Even a small difference in the growth rate causes a large difference in the biomass accumulated during a growing season (Ågren 1985), as has been found for healthy and declining *P. abies* stands (Oren et al. 1988a). In addition to the seasonal discrepancy between the potential wood growth rate and the Mg uptake rate, which limits wood production, an annual discrepancy may also be present. The annual growth rate of stems in forests follows an S-shaped curve (Ingestad 1982). If the supply of Mg in one stand is too low to support an unimpaired production of wood, a large difference in the standing biomass relative to a stand with no such limitation results, even before deficiency symptoms become apparent. Differences found between the 30-year-old healthy and declining spruce stands (Oren et al. 1988a) indicate that Mg stress had occurred at the declining site many years before needle chlorosis had begun. A reduced uptake of Ca associated with a low Ca:Al ratio in the soil and a subsequent limitation in cambial activity has been implicated elsewhere in the reduction of wood production, and it has been hypothesized that foliage loss will follow (Shortle and Smith 1988).

4 Recharge Phase

From the end of one growing period to the beginning of the next, tree tissues recharge with nutrients to levels which are largely determined by the concentrations in the soil solution. The recharge is limited when the soil base-cation content is low and becomes increasingly more limited as the soil base saturation decreases from leaching.

If an unusually warm late season, promoting wood production, is followed by an early frost which limits transpiration, or if a very long period of soil frost is followed by conditions promoting early bud break, the recharge before the growing season is inadequate. With inadequate recharge, the withdrawal of Mg may even cause needles of green trees to become yellow during the period of a high canopy production rate. These symptoms, also found in healthy stands (Nambiar and Fife 1987), are reversible and may disappear during the next *recharge phase*, if Mg is not so depleted that the recharge rate is inadequate for regreening the needles. This may explain those various observations of apparent recovery as well as apparent damage of forests occasionally reported. Should the quantities of nitrogenous compounds deposited in forest ecosystems increase, adequate recharge may become increasingly difficult as canopy growth is enhanced and cations are leached from the soil.

We calculated (Oren et al. 1988c) that the Mg uptake during the *recharge phase* at the healthy site was about 31 mmol m^{-2}, compared to 14 mmol m^{-2} at the declining site (Table 1). However, because conifers with adequate nutrient cation supply absorb elements beyond their needs (Ingestad and Kähr 1985), it is reasonable to assume that the trees at the healthy site with a better supply of Mg took up at least as much Mg as those at the declining site. We can assume that the ratio of uptake rates between the stands at any phase during the year is the same as the ratio of the annual uptake rates. Based on this assumption, the trees in the healthy stand must have absorbed Mg from the soil at a rate 30% higher than the trees in the declining stand throughout the year. A higher Mg absorption rate would be likely, because the Mg concentration in the soil solution at the healthy site was more than twice that at the declining site, and the root system was more evenly distributed through the soil profile (Meyer et al. 1988). This uptake is not accounted for in the periodic balance (Oren et al. 1988c) but may have been added to storage in the stem or in a xylem-phloem cycling pool. The large amount of recharge of various elements that must be made after the *stem growth phase* is completed (Table 1; Powers 1984) may be facilitated mainly by the lack of growth during that phase (Nambiar and Fife 1987) and not by increased uptake rate.

5 Conclusion: Nutritional Disharmony and Decline

To understand how nutritional disharmony is related to the decline in forest health and productivity, the appropriate temporal resolution for sampling should coincide with distinct phenological events during the growing season, and the study should last 2 or more consecutive years. Several processes seem to operate in concert to cause nutrient stress.

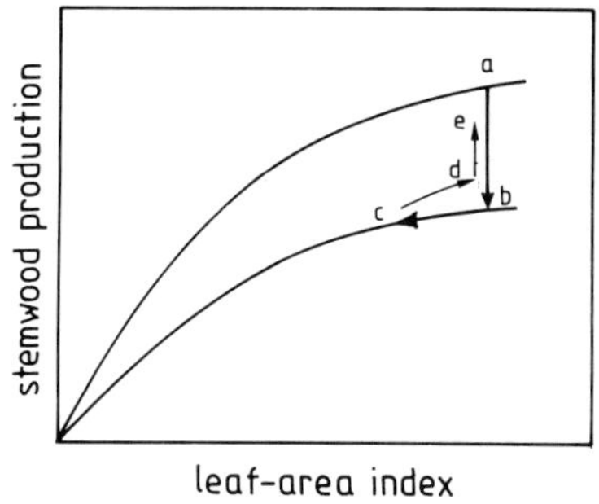

Fig. 8. Stemwood production increases with increasing leaf-area index, but at a decreasing rate due to the mutual shading within canopies. At any level of leaf-area index, limited supply of any element will proportionally reduce stemwood production (*a* to *b*). At a high leaf-area index, a reduction in leaf-area caused by nutrient limitation (*b* to *c*) is accompanied by a small reduction in stemwood production. The situation may reverse at a low leaf-area index. Increased supply of the limiting element would probably increase production by increasing both the leaf-area index (*c* to *d*) and wood production at any level of leaf-area (*d* to *e*; supporting evidence is presented in Oren et al. 1988a)

The conceptual model of forest decline from nutritional disharmony (Fig. 8) proposes that decline can be quantified by the reduction in stemwood production at a given leaf-area index (i.e., a growth-limiting nutrient stress) and by the reduction in the leaf-area index (i.e., a function-impairing nutrient stress of mature needles resulting in needle loss). A reduction in leaf-area index presumably from air pollution has been reported (Perkins et al. 1987). Needle-yellowing from a disharmony in the supply of N and cations (see Johnson et al. 1982 results in a cation deficiency and is an a posteriori indicator of decline. The effect of Mg stress on stemwood production should be more pronounced than the effect of needle chlorosis or the reduction in leaf area. This is because only very chlorotic needles show a reduction in photosynthetic rate, and at such a high leaf-area index, the productivity of older, shaded foliage is very low (Fig. 8; Jarvis and Leverenz 1983; Waring 1983). It is possible, however, that the dense canopy of spruce is not only the result of evolutionary development aimed at sequestering more essential resources in order to suppress competing vegetation (Waring 1988), but that the large crown is designed as a storehouse for elements. If so, a reduction in leaf area and accompanying storage may start a catastrophic spiraling decline (Gatto and Rinaldi 1987).

Increased N availability may promote larger canopies. However, any increase in leaf area will be short-lived. Without a comparable increase or perhaps even a decrease in the Mg supply, needles will suffer a severe Mg deficiency for a longer part of the growing season to the point that their annual CO_2 uptake may become negative (Zimmermann et al. 1988) and they will be prematurely shed. This may increase the turnover rate of needles and may adversely affect the carbon balance of the tree (Schulze et al. 1977).

Moreover, if the Mg retranslocation rate cannot meet the canopy production rate, a reduction in the production rate will result. As the needle production rate gradually falls short of replacing needle fall, leaf area will decrease. A decrease in leaf area is directly related to reduced stemwood production (Oren et al. 1986, 1987, 1988a, see Waring and Schlesinger 1985).

The concepts we have presented here may be used, in combination, to explain the frequent changes in needle color observed in declining forests, the effects of weather conditions, the stresses caused by the limiting supply of various elements, and the possible responses of the forest to alleviating these stresses (Zöttl and

Hüttl 1986). These concepts could be applied, without major modifications, to elements which are mobile in plants (e.g., phosphorus), and to trees in which new twig growth is supported by a much larger biomass of mature crown components (e.g., *Abies sp.*)

Although the particular synthesis of nutritional principles in the model described here has emerged from a comparative, nonexperimental study, the principles have already been tested on different species under different conditions. Nevertheless, before the model or nutritional disharmony can be applied to a range of soils and pollution levels, it must be verified in areas where low base-cation availability occurs across a range of N availability and proton input. The authors of this Chapter propose nutritional disharmony as a composition of hypotheses, some of which are currently being tested.

References

Abrahamsen G (1980) Acid precipitation, plant nutrients and forest growth. In: Drablos D, Tollan A (eds) Ecological impact of acid precipitation. Norw Interdisciplinary Res Program, SNSF Project, Ås, Norway, pp 58–63

Ågren GI (1985) Theory for growth of plants derived from the nitrogen productivity concept. Physiol Plant 64:17–28

Ågren GI, Ingestad T (1987) Root:shoot ratio as a balance between nitrogen productivity and photosynthesis. Plant Cell Environ 10:579–586

Aronsson A, Elowson S (1980) Effects of irrigation and fertilization on mineral nutrients in Scots pine needles. In: Persson T (ed) 1980 Structure and function of northern coniferous forests – an ecosystem study. Ecol Bull (Stockholm) 32:219–228

Bartsch N (1987) Responses of root systems of young *Pinus sylvestris* and *Picea abies* plants to water deficit and soil acidity. Can J For Res 17:805–812

Bender J, Jäger HJ, Seufert G, Arndt U (1986) Untersuchungen zur Einzel- und Kombinationswirkung von SO_2 Und O_3 auf den Stoffwechsel von Waldbäumen in Open-top- Kammern. Angew Bot 60:461–479

Binkley D (1986) Forest nutrition management. Wiley, NY

Binkley D, Read P (1985) Long-term increase of nitrogen availability from fertilization of Douglas fir. Can J For Res 15:723–724

Birk EM, Vitousek PM (1986) Nitrogen availability and nitrogen use efficiency in loblolly pine stands. Ecology 67:69–79

Boxman AW, Sinke RJ, Roelofs JGM (1985) The effect of airborne ammonium sulphate deposition on pine forests. Effects of NH_4^+ on the growth and potassium (86Rb) uptake of various ectomycorrhizal fungi in pure culture. Presented at the Muskoka Conference. Water Air Soil Pollut 32:517–522

Brix H (1981) Effects of nitrogen fertilizer source and application rates on foliar nitrogen concentration, photosynthesis, and growth of Douglas-fir. Can J For Res 11:775–780

Cape JN, Paterson IS, Wellburn AR, Wolfenden J, Mehlhorn H, Freer-Smith PH, Fink S (1988) Early diagnosis of forest decline. Inst Terrestrial Ecol Rep, 68 pp

Chapin FS III, Kedrowski RA (1983) Seasonal changes in nitrogen and phosphorus fractions and autumn retranslocation in evergreen and deciduous taiga trees. Ecology 64:376–391

Clark RB (1984) Physiological aspects of calcium, magnesium, and molybdenum deficiencies in plants. Agron Monograph 12:99–170

Entry JA, Cromack K JR, Stafford SG (1987) The effect of pH and aluminum concentration on ectomycorrhizal formation in *Abies balsamea*. Can J For Res 17:865–871

Fagerström T, Lohm U (1977) Growth in Scots pine (*Pinus sylvestris* L.). Mechanism of response to nitrogen. Oecologia (Berl) 26:305–315

Feiler S (1985) Einflüsse von Schwefeldioxid auf die Membranpermeabilität und Folgen für die Frostempfindlichkeit der Fichte [*Picea abies* (L.) Karst.]. Flora 177:217–226

Fife DN, Nambiar EKS (1982) Accumulation and retranslocation of nutrients in developing needles in relation to seasonal growth of young radiata pine trees. Ann Bot 50:817–829

Fife DN, Nambiar EKS (1984) Movement of nutrients in radiata pine needles in relation to the growth of shoots. Ann Bot 54:303–314

Forschner W, Wild A (1988) Comparative investigation on the nutrient composition of healthy and injured spruce of different locations. In: Mathy P (ed) Air pollution and ecosystems. Proc Comm Eur Communities Symp, Grenoble, France, May 1987, Reidel Publ, Dordrecht, pp 604–608

Franz F (1983) Auswirkungen der Waldkrankheiten auf Struktur und Wuchsleistung von Fichtenbeständen. Forstwiss Centralbl 102:186–200

Gatto M, Rinaldi S (1987) Some models of catastrophic behavior of exploited forests. Vegetatio 69:213–222

Gill RS, Lavender DP (1983) Urea fertilization and foliar nutrient composition of western hemlock [*Isuga heterophylla* (Raf.) Sarc.]. For Ecol Manage 6:333–341

Glatzel G (1981) Mineral nutrition and tree growth. Mitt Forstl Bundesversuchsanstalt Wien 142:245–257

Glatzel G, Kazda M, Grill D, Halbwachs G, Katzensteiner K (1987) Ernährungsstörungen bei Fichten als Komplexwirkung von Nadelschaden und erhöhter Stickstoffdeposition – ein Wirkungsmechanismus des Waldsterbens? Allg Forst Jagol Ztg 158:91–97

Hantschel R (1987) Water and element balance of damaged, fertilized Norway spruce ecosystems in the Fichtelgebirge with respect to physical and chemical soil heterogenity (in German). Bayreuther Bodenkd Ber 3:1–219

Hauhs M, Wright RF (1986) Regional pattern of acid deposition and forest decline along a cross section through Europe. Water Air Soil Pollut 31:463–474

Hüttl RF, Wisniewski J (1987) Fertilization as a tool to mitigate forest decline associated with nutrient deficiencies. Water Air Soil Pollut 33:265–276

Ingestad I (1959) Studies on the nutrition of forest tree seedlings. II. Mineral nutrition of spruce. Physiol Plant 12:568–593

Ingestad T (1979a) Mineral nutrient requirements of *Pinus sylvestris* and *Picea abies* seedlings. Physiol Plant 45:373–380

Ingestad T (1979b) Nitrogen stress in birch seedlings. II. N, K, P, Ca, and Mg nutrition. Physiol Plant 45:149–157

Ingestad T (1982) Relative addition rate and external concentration: Driving variables used in plant nutrition research. Plant Cell Environ 5:443–453

Ingestad T (1987) New concepts on soil fertility and plant nutrition as illustrated by research on forest trees and stands. Geoderma 40:237–252

Ingestad T, Kähr M (1985) Nutrition and growth of coniferous seedlings at varied relative nitrogen addition rate. Physiol Plant 65:109–116

Jarvis PG, Leverenz JW (1983) Productivity of temperate deciduous and evergreen forests. In: Lange OL, Nobel PS, Osmond CB, Ziegler H (eds) Encycl Plant Physiol. 12D:233–280, Springer, Berlin Heidelberg New York

Johnson DW, Turner J, Kelly JM (1982) The effects of acid rain on forest nutrient status. Water Resour Res 18:449–461

Kelly JM, Strickland RC (1987) Soil nutrient leaching in response to simulated acid rain treatment. Water Air Soil Pollut 34:167–181

Kirkby EA, Pilbeam DJ (1984) Calcium as a plant nutrient. Plant Cell Environ 7:397–405

Lange OL, Zellner H, Gebel J, Schramel P, Köstner B, Czygan FC (1987) Photosynthetic capacity, chloroplast pigments, and mineral content of previous year's spruce needles with and without the new flush: analysis of the forest-decline phenomenon of needle bleaching. Oecologia (Berl) 73:351–357

Linder S, Ingestad T (1977) Ecophysiological experiments under limiting and non-limiting conditions of mineral nutrition in field and laboratory. In: Bicentenary Celebration of C.P. Thunberg visit to Japan. R Swed Embassy Bot Soc Jpn, pp 69–76

Linder S, Rook DA (1984) Effects of mineral nutrition on carbon dioxide exchange and partitioning of carbon in trees. In: Bowen GD, Nambiar EKS (eds) Nutrition of plantation forests. Academic Press, Lond NY, pp 211–236

Madgwick HAI, Tamm CO (1987) Allocation of dry weight increment in crowns of *Picea abies* as affected by stand nutrition. Oikos 48:99–105

Mahendrappa MK, Foster NW, Weetman GF, Krause HH (1986) Nutrient cycling and availability in forest soils. Can J Soil Sci 66:547–572

Matsumoto H, Yamaya T (1986) Inhibition of potassium uptake and regulation of membrane-associated Mg^{2+}-ATPase activity of pea roots by aluminium. Soil Sci Plant Nut 32:179–188

McDonald AJS, Ericsson A, Lohammar T (1986a) Dependence of starch storage on nutrient availability and photon flux density in small birch (*Betula pendula* Roth). Plant Cell Environ 9:433–438

McDonald AJS, Lohammar T, Ericsson A (1986b) Growth response to step- decrease in nutrient availability in small birch (*Betula pendula* Roth). Plant Cell Environ 9:427–432

Meyer J, Schneider BU, Werk KS, Oren R, Schulze ED (1988) Performance of two *Picea abies* (L.) Karst. stands at different stages of decline. V. Root tip and ectomycorrhiza development and their relation to above-ground and soil nutrients. Oecologia (Berl) 77:7–13

Millard P (1988) The accumulation and storage of nitrogen by herbaceous plants. Plant Cell Environ 11:1–8

Mohren GMJ, Van Den Burg J, Burger FW (1986) Phosphorus deficiency induced by nitrogen input in Douglas fir in the Netherlands. Plants Soil 95:191–200

Mulder EG (1956) Nitrogen-magnesium relationships in crop plants. Plant Soil 7:341–376

Nambiar EKS, Bowen GD (1986) Uptake, distribution and retranslocation of nitrogen by *Pinus radiata* from 14N-labelled fertilizer applied to podzolized sandy soil. For Ecol Manage 15:269–284

Nambiar EKS, Fife DN (1987) Growth and nutrient retranslocation in needles of radiata pine in relation to nitrogen supply. Ann Bot 60:147–156

Nihlgård B (1985) The ammonium hypothesis – an additional explanation to the forest dieback in Europe. Ambio 14:2–8

Oren R, Schulze ED, Matyssek R, Zimmermann R (1986) Estimating photosynthetic rate and annual carbon gain in conifers from specific leaf weight and leaf biomass. Oecologia (Berl) 70:187–193

Oren R, Waring RH, Stafford S, Barrett JW (1987) Twenty-four years of Ponderosa pine growth in relation to canopy leaf area and understory competition. For Sci 33:538–547

Oren R, Sculze ED, Werk KS, Meyer J, Schneider BU, Heilmeier H (1989a) Performance of two *Picea abies* (L.) Karst. stands at different stages of decline. I. Carbon relations and stand growth. Oecologia (Berl) 75:25–37

Oren R, Werk KS, Schulze ED, Meyer J, Schneider BU, Schramel P (1988b) Performance of two *Picea abies* (L.) Karst. stands at different stages of decline. VI. Nutrient concentration. Oecologia (Berl) 77:151–162

Oren R, Schulze ED, Werk KS, Meyer J (1988c) Performance of two *Picea abies* (L.) Karst. stands at different stages of decline. VII. Nutrient relations and growth. Oecologia (Berl) 77:163–173

Osonubi O, Oren R, Werk KS, Schulze ED, Heilmeier H (1988) Performance of two *Picea abies* (L.) Karst. stands at different stages of decline. IV. Xylem sap concentrations of magnesium, calcium, potassium and nitrogen. Oecologia (Berl) 77:1–6

Perkins TD, Vogelmann HW, Klein RM (1987) Changes in light intensity and temperature as a result of forest decline on Camels Hump, Vermont. Can J For Res 17:565–568

Powers RF (1984) Estimating soil nitrogen availability through soil and foliar analysis. In: Stone EL (ed) Forest soils and treatment impacts. Univ Tennessee, Knoxville, pp 353–379

Rehfuess KE (1987) Perception on forest diseases on central Europe. Forestry 60:1–11

Rehfuess KE, Bosch C, Pfannkuch E (1983) Nutrient imbalancies in coniferous stands in southern Germany. Commun Inst For Fenn 116:122–130

Roelofs JGM, Kempers AJ, Houdijk ALFM, Jansen J (1985) The effect of air-borne ammonium sulphate on *Pinus nigra* var. *maritima* in the Netherlands. Plant Soil 84:45–56

Rost-Siebert K (1983) Aluminum toxicity and tolerance in spruce (*Picea abies* Karst.) and beech (*Fagus silvatica* L.) seedlings. AFZ 26/27:686–689

Schulze ED, Fuchs MI, Fuchs M (1977) Spacial distribution of photosynthetic capacity and performance in a mountain spruce forest in northern Germany. I. Biomass distribution and daily CO_2 uptake in different crown layers. Oecologia (Berl) 29:43–61

Schulze ED, Oren R, Zimmermann R (1987) Wirkungen von Immissionen auf 30-jährige Fichten in mittleren Höhenlagen des Fichtelgebirges auf Phyllit. AFZ 27/2:725–730

Shortle WC, Smith KT (1988) Aluminum-induced calcium deficiency syndrome in declining red spruce. Science 240:1017–1018

Sutton RF (1969) Form and development of earlier root systems. Commonw For Bur Tech Commun 7:131 pp

Sweet GB, Wareing PF (1966) The role of plant growth in regulating Photosynthesis. Nature (Lond) 210:77–79

Tamm CO (1968) An attempt to assess the optimum nitrogen level in Norway spruce under field conditions. Stud For Suec 61:1–67

Tamm CO, Aronsson A (1988) Experiments simulating increased nitrogen deposition to forest ecosystems. A poster presented in COST workshop on Ecosystem Experiments, June 1988, Mitwitz, FRG

Taylor CMA (1987) The effects of nitrogen fertilizer at different rates and times of application on the growth of Sitka spruce in upland Britain. Forestry 60:87–99

Timmer VR, Morrow LD (1984) Predicting fertilizer growth response and nutrient status of jack pine by foliar diagnosis. In: Stone EL (ed) Forest soils and treatment impacts. Univ Tennessee, Knoxville, pp 335–351

Timmer VR, Armstrong G (1987) Diagnosing nutritional status of containerized tree seedlings: comparative plant analyses. Soil Sci Soc Am J 51:1082–1086

Timmer VR, Ray PN (1988) Evaluating soil nutrient regime for black spruce in the Ontario Claybelt by fertilization. For Chron 64:40–46

Tomlinson GH (1985) Acid deposition and the loss of nutrients from forest soils. Tappi J 68:54–58

Ulrich B (1983) A concept of forest ecosystem stability and of acid deposition as driving force for destabilization. In: Ulrich B, Pankrath J (eds) Effects of accumulation of air pollutants in forest ecosystems. Reidel Publ, Dordrecht, Boston, Lancaster, Tokyo, pp 1–29

Von Bosch C, Pfannkuch E, Baum U, Rehfuess KE (1983) Über die Erkrankung der Fichte (*Picea Abies* Karst.) in den Hochlagen des Bayerischen Waldes. Forstwiss Centralbl 102:167–181

Von Schmidt-Vogt H, Makkonen-Speicker K (1986) Unterschiede im Nahrelementspiegel von Nadeln verschiedener Fichtenherkünfte (*Picea abies* (L.) Karst.). Allg Forst Jagdztg 157:145–152

Waring RH (1983) Estimating forest growth and efficiency in relation to canopy leaf area. Adv Ecol Res 13:327–354

Waring RH (1988) Responses of evergreen trees to multiple stresses. Paper presented at a Symposium on Multiple Stress Effects on Vegetation. 10–12 Feb. 1988, Asilomar, California

Waring RH, Schlesinger WH (1985) Forest Ecosystems. Academic Press, Lond NY

Waring RH, Newman K, Bell J (1981) Efficiency of tree crowns and stemwood production at different canopy leaf densities. Forestry 54:15–23

Waring RH, McDonald AJS, Larsson S, Ericsson T, Wiren A, Arwidsson E, Ericsson A, Lohammar T (1985) Differences in chemical composition of plants grown at constant relative growth rates with stable mineral nutrition. Oecologia (Berl) 66:157–160

Weissen F, Van Praag HJ, Marechal P, Delecour F, Farcy CH (1988) Les causes de la degradation sanitaire de forets en Wallonie: Le point de la situation. Extrait Bull Soc R Foretière Belg 2:57–68

Werk KS, Oren R, Schulze ED, Zimmermann R, Meyer J (1988) Performance of two *Picea abies* (L.) Karst. stands at different stages of decline. III. Canopy transpiration of green trees. Oecologia (Berl) 76:519–524

Zech W, Popp E (1983) Magnesiummangel, einer der Gründe für das Fichten- und Tannensterben in NO-Bayern. Forstwiss centralbl 102:50–55

Zimmermann R, Oren R, Schulze ED, Werk KS (1988) Performance of *Picea abies* (L.) Karst. Stands at different stages of decline. II. Photosynthesis and leaf conductance. Oecologia (Berl) 76:513–518

Zöttl HW, Hüttl RF (1986) Nutrient supply and forest decline in southwest Germany. Water Air Soil Pollut 31:449–462

4-I Nutrient Balance and Element Cycling in Healthy and Declining Norway Spruce Stands

R. Horn, E.-D. Schulze, and R. Hantschel

1 Introduction

Forest ecosystems are open systems which gain elements from the atmosphere through wet and dry deposition, and from bedrock through weathering, and lose elements through leaching to groundwater. Element flow may be partitioned into three major paths (Fig. 1): (1) atmospheric input, which is modified by canopy processes and results in a throughfall, and by soil processes, which result in seepage out of the system. This flux interacts with the other two paths. (2) Plant uptake is balanced by growth and net canopy exchange, i.e., the result of canopy uptake and leaching. (3) Nutrient uptake for growth is the sum of nutrients required for biomass increment and litterfall if the canopy is at steady state. Elements are returned to the soil via decomposition of the organic material. This cycle loses elements to the immobilized element pool in woody biomass and receives elements through mineral weathering.

The internal fluxes of elements in the ecosystem are generally large, relative to the input and output of elements. Under such conditions the ecosystem may be regarded as a "closed system with a coupled element cycle" in which photosynthesis and ion uptake equals respiration and mineralization (Ulrich 1987). The balance may change if the internal nutrient cycles are perturbed by relatively large additions of nutrients or acids from the atmosphere. These external nutrient inputs may change the nutrient status of the plants and the chemical composition of the soils, as well as increase the leaching of elements from the soil. When acids are added to ecosystems, soil leaching may accelerate soil and groundwater acidification and may result in a reduction of the total pool of base cations in the soil (see Chap. 4-B; Hauhs 1985).

The nutrient fluxes in an ecosystem provide a framework for assessing the probable consequences from perturbations to the ecosystem in terms of chemical changes (De Vries and Breeuwsma 1987). In the following chapter, nutrient balances and fluxes of healthy and declining *Picea abies* stands are compared in order to link the health status of the stands to the ecosystem nutrient cycles. The ecosystem element fluxes are based on data discussed in detail in previous chapters.

E.-D. Schulze, O.L. Lange, R. Oren (Eds.)
Ecological Studies Vol. 77

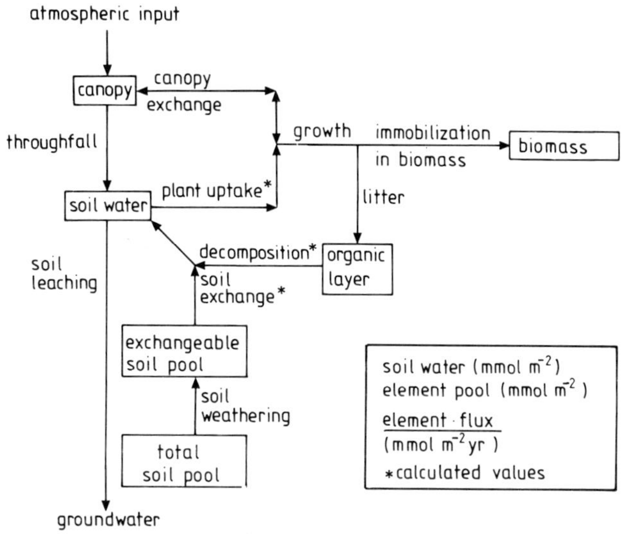

Fig. 1. Element pools and fluxes in forest ecosystems. The same diagram used in Fig. 3 for various elements. The dimensions are explained in the *lower corner* of Fig. 1

2 Assessment of the Ecosystem Balance

Two *Picea abies* stands were studied in 1985; the results are presented as an example of element cycling in a forest ecosystem with both healthy and declining stands.

2.1 Quasi-Stationary Pools

The total nutrient pool in the organic and mineral soil was based on a soil chemical analysis (Chap. 4-B). The element content of the gravel was calculated based on a gravel density of 2.65 g cm^{-3} and a measured stone content in each horizon, which varied from 20–65% (Chap. 4-B). The exchangeable soil pool was separated from the total pool because of its importance relative to actual element fluxes. The soil pools were based on a soil depth of 40 cm; most of the roots in these forests are found at a depth of 25–30 cm (see Chap. 4-E; Meyer et al. 1988). There was no nutrient recharge from deeper soil horizons by capillary rise, since the water balance was always positive during the period of observation, regardless of season (see Chap. 4-C). The nutrient pool in the biomass was divided into two parts: (1) the standing tree biomass, which was produced prior to the 1985 growing season (see Chap. 4-F); (2) the biomass produced during that season in terms of needles, twigs, branches, stem, and roots (Chap. 4-F; Oren et al. 1988a,b).

To calculate the amount of nutrients available for uptake by plants from the soil solution, each soil layer was assumed to have the same concentrations of nonpercolating and percolating water. The mean available nutrient pool in the soil was calculated by multiplying the volumetric water content in the root zone by the ion concentration of soil solution at that zone (collected once a week at different depths.)

2.2 *Element Input to the Soil*

2.2.1 *Precipitation and Canopy Processes*

Soil input (canopy throughfall, D_t) was determined by precipitation input (D_p), interception deposition (D_i), and canopy exchange processes (C_e):

$$D_t = D_p + D_i + C_e. \tag{1}$$

Precipitation input (D_p) consisted of rain, snow water, and fog water. Precipitation was collected weekly by rain and snow gauges, positioned at one location per site outside the canopy (Chap. 4-C). Due to chemical and physical processes in the canopy (see Chap. 4-B), the chemical composition of water in the canopy throughfall (D_t) differed from that of precipitation. The ionic composition of the throughfall was analyzed from water collected every week below the trees on a single control plot with six sampling units. The amount of elements in canopy throughfall was determined by precipitation input, canopy interception, and canopy ion uptake or release.

Interception deposition (D_i) cannot be measured directly. It includes very different types of particulate and gaseous deposition, such as the deposition of aerosols and dew and gas adsorption on wet surfaces. Thus, interception deposition can only be estimated by difference:

$$D_i = D_t - D_p - C_e. \tag{2}$$

In this equation, the term for canopy exchange of ions, C_e, by canopy leaching or by direct uptake remains unknown. Canopy leaching of chloride, Cl, is very small and can be neglected. Therefore, the deposition of chloride in the throughfall is assumed to equal the sum of chloride deposited by precipitation and interception (Ulrich 1983):

$$D_t^{Cl} = D_p^{Cl} + D_i^{Cl}. \tag{3}$$

Based on $f^{Cl} = D_i^{Cl}/D_p^{Cl}$, the amount of any ion, ii, which is intercepted by the canopy surfaces and washed off with rain water without being absorbed into the canopy may be approximated as:

$$D_i^{ii} = f^{Cl} \cdot D_p^{ii}. \tag{4}$$

Sulfur is deposited on the canopy as aerosol (i.e. particulate deposition) and gaseous adsorption and is taken up by the canopy as gases. The particulate deposition may be calculated using Eq. (4), assuming that the ratio f^{Cl} (or using Na instead of Cl) is the same as that of sulfur and that no aerial uptake of sulfate takes place (Chap. 3-C; Ulrich 1983). Thus:

$$D_i^{S\ part} = f^{Cl} \cdot D_p^S. \quad (5)$$

The gaseous adsorption of sulfur, $D_i^{S\ ads}$, was calculated as the remaining difference in sulfur deposition between throughfall and precipitation:

$$D_i^{S\ ads} = D_t^S - D_p^S - D_i^{S\ part}. \quad (6)$$

Sulfur dioxide assimilation was calculated from the concentration of SO_2 in the air (S_{conc}) and canopy conductance, g_{canopy}^S (see Chap. 4-C).

$$A^S = g_{canopy}^S \cdot S_{conc}. \quad (7)$$

The acidity of deposited sulfur compounds is buffered in part by cation exchange through plant surfaces. The exchange of cations canopy, C_e, should be washed from the canopy by rain and can be measured as part of the throughfall:

$$C_e^{ii} = D_t^{ii} - (D_p^{ii} + D_i^{ii\ part}). \quad (8)$$

where, in this case, ii represents base cations, e.g., potassium, magnesium, and calcium.

The interception deposition of nitrogen compounds is derived from sulfur deposition and the concentration of nitrogen in the gas phase and in aerosols (Schulze 1989). Particulate nitrogen deposition (D_i^{part}) and adsorption of nitrogen gases ($D_i^{N\ ads}$) were calculated on the basis of sulfur interception and the ratio of S/N in aerosols or gases in the atmosphere (see Chap. 1-B).

$$D_i^{N\ part} = D_i^{S\ part} \cdot (N/S)^{aerosol}; \quad (9)$$
$$D_i^{N\ ads} = D_i^{S\ ads} \cdot (N/S)^{gas}. \quad (10)$$

Nitrogen assimilation from the gas phase was calculated from the atmospheric concentration of NO_x and canopy conductance (see Chap. 1-B).

$$A^N = g_{canopy}^N \cdot N_{conc}. \quad (11)$$

Table 1. Dry deposition (mmol m^{-2} a^{-1}) of sulfur and nitrate and ammonium-nitrogen[a]

	Sulfur		NO_x nitrogen		NH_x nitrogen	
	Healthy	Declining	Healthy	Declining	Healthy	Declining
Aerosol deposition	11.6	20.7	6.5	11.6	19.1	34.1
Gas adsorption	19.7	51.4	54.1	112.6	–	–
Gas uptake	7.4	7.6	18.6	18.6	–	–

[a] Calculated from the equations in the text, measured values of NaCl in precipitation and throughfall, canopy transpiration (Chap. 4-C), and aerosol and gas concentrations (Chap. 1-B).

The calculation of the dry deposition of sulfur and nitrogen compounds was higher for the declining site (Table 1). Gas adsorption on wet surface and gas uptake of sulfur and nitrogen exceeded aerosol deposition.

2.2.2 Soil Processes

Percolating soil solution in three sampling units at a single control plot at each site was collected weekly with ceramic cups at three depths (Hantschel 1987). The solution at each depth was continuously taken up with a suction that was 50–150 hPa more negative than the soil matrix suction at that depth. This was done to minimize alteration of ionic concentration of soil water (mobile fraction). The ion concentration at each depth was determined from bulk samples from the three sampling units. The concentrations of Ca, K, Mg, Na, Al, H, NH_4-N, NO_3-N, SO_4-S, and Cl were determined in all the collected soil water samples.

Leaching of soil nutrients was calculated by multiplying the amount of seepage water with the ion concentration at a depth of 85 cm.

Plant nutrient uptake was calculated from the growth measurements of different tissues and the nutrient concentrations in the tissues from five control plots per site (see Chap. 4-F). Elements are transferred from standing biomass to the soil not only from canopy leaching, but also in the form of litter fall and root mortality. Litter fall was measured in the same five control plots per site with two 1 m^2 litter traps per plot; the element concentration in the litter was determined four times during the growing season. It was assumed that annual nutrient input by litter fall would be similar to nutrient release by decomposition (Binkley 1986). Weathering was calculated from decomposition, soil leaching, plant uptake, and throughfall based on the assumption that the soil exchangeable pool would not change much annually.

2.3 Output of Elements

Element output to the groundwater was assumed to equal the leaching of soil nutrients at 85 cm soil depth as calculated above.

The total proton surplus in the soil due to anion and cation input, plant uptake, and groundwater recharge was calculated according to Bredemeyer (1987). Several processes affect the ecosystem proton budget, including: (1) accumulation of a cation surplus in standing biomass and litter during stand development; (2) the turnover of organic acids; (3) the proton release during ammonium absorption in the soil; and (4) proton buffering in the canopy. Additionally, protons are released during gaseous and particulate sulfur and nitrogen absorption (see Chap. 3-E).

3 Seasonal Variations of Precipitation, Throughfall, and Seepage of Elements

The input of elements in precipitation appeared similar at the healthy and declining sites, except for sulfate which was higher at the declining site (Fig. 2). The seasonal inputs of most elements at both sites appeared greater during the growing season than during the rest of the year. The input of nitrogen, both as ammonium and nitrate, during the growing season was about twice that during the rest of the year. The total wet deposition of nitrate and sulfate was similar to

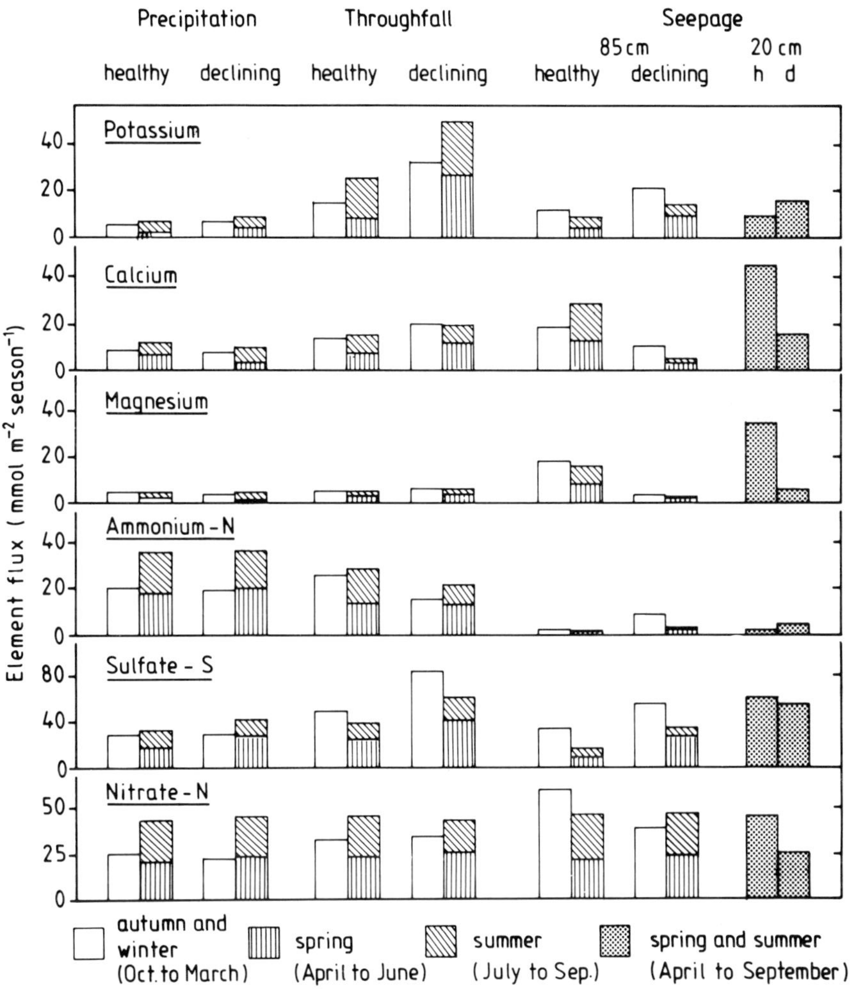

Fig. 2. Seasonal fluxes of elements (mmol m^{-2} yr^{-1}) into and out of healthy and declining stands via precipitation, throughfall, and seepage (at 20 cm and 85 cm soil depth)

values reported in other regions of southern Germany (see Chap. 4-A). The total proton input (Table 2) was 50 mmol m^{-2} yr^{-1}, which was lower than that in northern Germany (e.g., Solling: 79 mmol m^{-2} yr^{-1}; Ulrich 1983).

Throughfall element input is different from precipitation input because of the effects of dry deposition and cation exchange processes in the canopy. The potassium, calcium and sulfate fluxes in canopy throughfall seemed much higher than those in precipitation. These increases appeared greater (particularly for potassium and sulfate) at the declining site. In comparison to fluxes in precipitation, the ammonium flux in the throughfall decreased over all seasons, while the nitrate flux did not change during the growing season, and increased only in

Table 2. Annual fluxes of ions (mmol m^{-2} yr^{-1}) into and out of healthy and declining *P. abies* stands in the Fichtelgebirge, 1985 (*b.d.*: below detection limits)

	H	K	Na	Ca	Mg	NH_4-N	Zn	Al	NO_3-N	PO_4-P	SO_4-S	Cl
Healthy												
Precipitation	46	19	20	25	10	54	2	4	64	2	71	32
Throughfall	84	44	18	35	11	53	2	7	74	2	109	30
Seepage	11	21	43	45	33	1	2	11	95	b.d.	39	30
Declining												
Precipitation	57	25	31	26	10	52	4	4	63	2	76	23
Throughfall	104	92	31	43	13	34	4	12	73	6	170	37
Seepage	42	34	45	15	5	2	1	77	76	b.d.	96	19

winter. This indicated canopy uptake of nitrogen. An increase in throughfall water pH as a result of canopy buffering (Cronan and Reiners 1983) was not observed in this study (cf. proton input and throughfall in Table 2).

The element flux in seepage water was different between sites and at different soil horizons. Calculated according to Ulrich (1987), soil leaching at the declining site of M_a-cations (Al and H: 119 mmol m^{-2} yr^{-1}) was twice that of M_b-cations (Ca, K, Mg: 54 mmol m^{-2} yr^{-1}) or about two-thirds the total M_a plus M_b-cations output (173 mmol m^{-2} yr^{-1}). In contrast, soil leaching at the healthy site of the M_a-fraction (22 mmol m^{-2} yr^{-1}) was only about 20% of the leaching of the M_b-fraction (99 mmol m^{-2} yr^{-1}) or less than one-fifth of the total output (121 mmol m^{-2} yr^{-1}). Potassium losses seemed higher at the declining site. In contrast, calcium and magnesium losses appeared much higher at the healthy site.

The loss of calcium and magnesium was already apparent at the depth above which most of the roots in both stands were found (20 cm; Meyer et al. 1988) and was higher than the loss from 80 cm soil depth. The differences in fluxes between the two depths indicated that these cations moved through the upper mineral soil horizon at both sites, but mostly at the healthy site, and accumulated to some extent in the lower horizons (Kelly and Strickland 1987). At the declining site, the total loss of calcium and magnesium at 80 cm depth was similar to the precipitation input. This indicated that, at the declining site, all the magnesium and calcium released by decomposition and weathering was used for growth. The data from the healthy site indicated that internally released magnesium and calcium exceeded the amount required for growth, despite the greater growth at that site compared to the declining site (Oren et al. 1988a; Chap. 4-D).

Ammonium losses were very small at all soil depths and lower in throughfall. These indicated efficient uptake by tree roots. In contrast, nitrate and sulfate losses from the soil were high at both sites. Losses of nitrate, particularly at the healthy site and losses of sulfate at the declining site seemed greater than their inputs in precipitation. Sulfate losses seemed smaller than the input by throughfall. More sulfate was leached out of the declining site since sulfate is precipitated as $Al_n(OH)_{(3n-x)}$ + $(SO_4)_{(3n-x)/2}$, which acts as a strong sulfate sink at the healthy site (see Chap. 4-B). Sulfate losses appeared higher in winter than in the growing season.

The annual sum of ion input to the sites in precipitation and throughfall and ion loss to seepage was greater for sulfate, potassium, and hydrogen at the declining site (Table 2). However, only the losses of these ions as well as of aluminum to seepage were much greater at the declining site. The losses of potassium, sulfate, hydrogen, and aluminum during the growing season at the declining site were 1.5, 2.5, 4, and 7 times those at the healthy site, respectively. Conversely, the healthy site lost nitrate, calcium, and magnesium at a rate of 1.3, 3, and 7 times that at the declining site, respectively. Thus, the movement of larger amounts of anions through the soil profile at the declining site was accompanied by a large amount of aluminum, hydrogen, and potassium, in comparison to the healthy site, where the smaller amounts of anions were accompanied by more calcium and magnesium.

4 The Ecosystem Balance

The pools of magnesium, calcium, potassium, and nitrogen (Chapter 4-F, Fig. 3) were smaller at the declining site, except for the potassium pool in the mineral soil, which was similar at both sites. The total nutrient pool of all elements in the rooting zone was one (calcium) to three (potassium) orders of magnitude larger than the corresponding exchangeable fraction.

Soil processes and litter fall, but not leaching, dominated the transfer of all elements inside the ecosystem. Magnesium input from the atmosphere was great

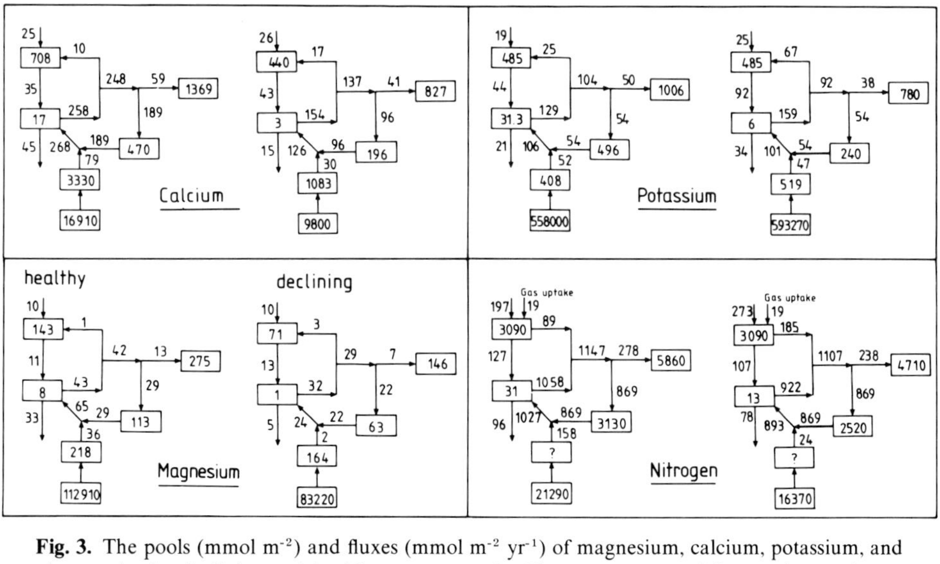

Fig. 3. The pools (mmol m^{-2}) and fluxes (mmol m^{-2} yr^{-1}) of magnesium, calcium, potassium, and nitrogen in the declining and healthy spruce stands. The components of the nutrient cycle are presented in Fig. 1

enough to account for the immobilization of growth at both sites (Fig. 3). Canopy losses from leaching accounted for up to 10% of the magnesium taken up by roots and had a small effect on the overall flux. Thus, even if there had been no canopy leaching, the magnesium taken up by the roots would still have been higher than the magnesium release through decomposition and weathering at the declining site. It was lower than that magnesium release at the healthy site. This resulted in less soil leaching at the declining site. Stem growth at the declining site immobilized as much magnesium taken up from the exchangeable pool as was lost in seepage.

The healthy site could draw magnesium from the total soil pool for another 825 years before it would be depleted to the same level as the declining site. The exchangeable magnesium pool at both sites, however, was small, relative to the flux from exchange sites into the soil water. If one assumed that the declining stand, in which growth is limited by the magnesium supply (Oren et al. 1988c; Chap. 4-F), takes all the magnesium it can from the exchangeable pool, it is clear that only a small proportion of that pool is available for uptake. Therefore, it is probable that the amount of magnesium in the exchangeable pool at the healthy site is also not completely available for uptake. At the present rate of magnesium flux from the exchangeable pool at the healthy site, the pool will be reduced to the size of the pool at the declining site in 1.5 years. However, the exchangeable pool should be replenished by weathering. It is also possible that as the pool is drawn lower, the flux out of the pool will be reduced. In the event that the release of magnesium by weathering is less than the flux out of the exchangeable pool over a series of years, the healthy stand may also experience magnesium deficiency.

The total soil calcium pool at both sites was smaller than the magnesium pool. Nevertheless, the exchangeable soil calcium pool was larger than the magnesium pool. In contrast to magnesium, immobilization of calcium in stemwood was higher than the atmospheric input. Stem growth accounted for a higher amount of calcium taken up from the exchangeable pool than did soil leaching, particularly at the declining site. Canopy leaching was again a small component of the overall flux and should have only had a small effect on the calcium cycle. In both sites, almost all calcium, which became available from weathering, decomposition, and atmospheric input was taken up by the roots. Only about 15% of the net input into the soil exchangeable pool leached from the soil into the groundwater at both sites. If the current trend in ecosystem fluxes continues, the healthy site would reach the calcium pool size at the declining site in 160 years. Due to the high stone content of the soil, weathering may be lower than estimated (cf. Fig. 3 and Chap. 4-B) and soil leaching may cause a depletion of the exchangeable pools of both calcium and magnesium. As a result, both sites will not be able to maintain these cycles at their current rate for very long.

The total soil potassium pools were similar at both sites and were much larger than the calcium, magnesium and nitrogen pools. Canopy leaching of potassium was 42 and 20% of tree nutrient uptake at the declining and healthy sites, respectively. Potassium loss by canopy leaching was larger than by immobilization in annual production of woody biomass at the declining site. Soil leaching used about 72 and 40% of the potassium flux from the exchangeable soil pool at the declining and healthy sites, respectively.

In contrast to the other elements, the outcome of the nitrogen exchange processes in the canopy was a net uptake of nitrogen from precipitation. Direct uptake from the atmosphere may have supplied 30% and 80% of the nitrogen immobilized in wood, and 6% and 18% of the nitrogen required for all growth at the healthy and the declining sites, respectively. Thus, nitrogen used in growth exceeded the amount taken up from the soil. Nitrate leaching from the soil showed seasonal variations which were unrelated to growth.

The weathering rate appeared to be a very important factor in balancing the cation fluxes of these ecosystems. Based on experiments in the laboratory (Chap. 4-B), magnesium released to a depth of 40 cm due to soil weathering was 7 and 24 mmol m^{-2} yr^{-1} in the declining and healthy sites, respectively, as compared to 2–36 mmol m^{-2} yr^{-1} estimated from the budget (Fig. 3). The estimated annual magnesium weathering rates of 0.002 and 0.03% for the declining and healthy sites, respectively, agreed with rates determined for the Solling area by Fölster (1985). The rate of calcium weathering was estimated to be 0.31–0.47%, and that of potassium to be 0.008–0.009% at the healthy and declining sites, respectively. The total weathering rate of the main base cations was 80–170 mmol m^{-2} yr^{-1}, which was similar to values reported by Fölster (1985; 10–150 mmol m^{-2} yr^{-1}) for acidic soils.

The proton balance in the soils of both sites showed a surplus of 181 mmol m^{-2} yr^{-1} at the healthy site and 285 mmol m^{-2} yr^{-1} at the declining site. The cation surplus in the annual growth of stem wood and bark at both sites includes the measured values for potassium, calcium, magnesium, manganese, and phosphorus. For sodium, iron, aluminum, and sulfur the data were taken from Bredemeyer (1987). The healthy site had a smaller proton surplus than the Solling site in northern Germany (265 mmol m^{-2} yr^{-1}; Ulrich and Summer, in prep.). The total amount of protons in forest ecosystems includes both the protons generated in the soil and the nonbuffered protons added in the canopy throughfall (Bredemeyer 1987). The total proton input was 405 mmol m^{-2} yr^{-1} at the declining site and 268 mmol m^{-2} yr^{-1} at the healthy site, which was less than that reported for the Soling area (577 mmol m^{-2} yr^{-1}).

5 Conclusions

The leaching of cations from the soil was greater at the healthy site than at the declining site. This may indicate that the declining site had reached a low level of base saturation, where the flux of nutrients to and from the exchangeable pool was restricted. The acidification process will be accelerated as sulfate is released into the soil solution when base saturation decreases. If current rates of soil leaching rates continue, the total amount of calcium and magnesium in the soil at the healthy site will be reduced to the amount found at the declining site in approximately 350 and 1300 years, respectively, assuming that the nutrients in the stems remain at the site. However, if normal harvesting is practiced, stem removal will reduce the time it takes the soil at the healthy site to reach the calcium and magnesium levels at the declining site to approximately 100 and 800 years, respectively.

Advective nutrient transports may alter some of the estimated values and the predictions, which are made based on the calculated budgets. In addition, estimation of different components in the budget are based on very short-term observations and have an inherently large error. Thus, the calculated budgets must be interpreted with care.

Despite the uncertainties in the calculated budgets, there is little doubt that, if present soil leaching rates at the healthy site continue, the soil will become too depleted of calcium and magnesium to supply the trees for growth and proper function. Even before these pools are depleted, the healthy stand may face limitations on fluxes from the exchangeable pools similar to those threatening the declining stand now. In addition its soil chemical environment will also become inhospitable to root growth, and the decline of the forest will follow. Although forest harvesting shares the responsibility for the acidification of forest soils, this analysis shows that acid precipitation plays a major role in the acidification process.

Acknowledgement. R. Oren contributed ideas to this chapter and helped in its preparation.

References

Binkley D (1986) Forest nutrition management. Wiley, NY

Bredemeyer M (1987) Stoffbilanzen, interne Protonenproduktion und Gesamtsäurebelastung des Bodens in verschiedenen Waldökosystemen Norddeutschlands. Ber Forschungszentrum Waldökosysteme Reihe A 33:1–183

Cronan CS, Reiners A (1983) Canopy processing of acidic precipitation by coniferous and hardwood forests in New England. Oecologia 59:216–223

De Vries W, Breeuwsma A (1987) The relation between soil acidification and element cycling. Water Air Soil Pollut 35:293–310

Fölster H (1985) Proton consumption rates in Holocene and present-day weathering of acid forest soil. In: Drever JI (ed) The chemistry of weathering. Reidel, Dordrecht, 197–209

Hantschel R (1987) Wasser- und Elementbilanz von geschädigten, gedüngten Fichtenökosystemen im Fichelgebirge unter besonderer Berücksichtigung von physikalischer und chemischer Bodenheterogenität. Bayreuther Bodenkd Ber Bd 3 ISSN 0931–6442

Hauhs M (1985) Wasser- und Stoffhaushalt im Einzugsgebiet der Langen Bramke (Harz). Ber Forsch Waldökosysteme Univ Gött 17:1–206

Kelly JM, Strickland RC (1987) Soil nutrient leaching in response to simulated acid rain treatment. Water Air Soil Pollut 34:167–181

Meyer J, Schneider BU, Werk KS, Oren R, Schulze ED (1988) Performance of two *Picea abies* (L.) Karst. stands at different stages of decline. V. Root tip and ectomycorrhiza development and their relation to above-ground and soil nutrients. Oecologia (Berl) 77:7–13

Oren R, Schulze ED, Werk KS, Meyer J, Schneider BU, Heilmeier H (1988a) Performance of two *Picea abies* (L.) Karst. stands at different stages of decline. I. Carbon relations and stand growth. Oecologia (Berl) 75:25–37

Oren R, Werk KS, Schulze ED, Meyer J, Schneider BU, Schramel P (1988b) Performance of two *Picea abies* (L.) Karst. stands at different stages of decline. VI. Nutrient concentration. Oecologia (Berl) 77:151–162

Oren R, Schulze ED, Werk KS, Meyer J (1988c) Performance of two *Picea abies* (L.) Karst. stands at different stages of decline. VII. Nutrient relations and growth. Oecologia (Berl) 77:163–173

Schulze ED (1989) Estimation of dry deposition in a spruce forest. Water Soil and Air Pollution (in press)

Ulrich B (1983) Interaction of forest canopies with atmospheric constituents: SO_2, alkali, and earth alkali cations and chloride. In: Ulrich B, Pankrath J (eds) Effects of accumulation of air pollutants in forest ecosystems. Reidel Publ Co, Dordrecht, Boston, Lancaster, Tokyo, pp 33–45

Ulrich B (1987) Stability, elasticity and resilience of terrestrial ecosystems with respect to matter balance. Ecol Stud 61:11–49

Ulrich B, Summer M (1989) (eds) Soil Acidity. Springer, Berlin Heidelberg New York Tokyo (in prep)

Part 5

Processes Leading to Forest Decline: A Synthesis

E.-D. SCHULZE, R. OREN, and O.L. LANGE

1 Introduction

Recent forest decline in Central Europe has been manifested by a combination of symptoms, especially needle-yellowing and loss, and reduced stand growth. Needle damage first became apparent in the late 1970's on Norway spruce [*Picea abies* (L.) Karst.], the dominant coniferous tree species planted today. Decline was observed in the early 1980's over a large forest area on several major forest species. It was the large geographic distribution of visible change in conifers and in broadleaf species, as well as the large number of opposing hypotheses on possible causes, which led to the present case study. In the mountain range of the Fichtelgebirge, most of the forests at elevations above 750 m currently display different degrees of visual decline symptoms, although some forests at lower elevations have partially recovered. At 900–1000 m elevation, a stand which 10 years earlier first displayed visual symptoms is now totally dead. In this book we analyze possible pathways which have led to the decline of forests in this area and which may explain the observed temporal symptomatic variations.

Studying whole plant functioning and relating it to an ecosystem is complicated. To help explain the results of our study, we start by describing some of the general features of the water, nutrient and energy cycles of forest stands. Forests are systems which are connected to the atmosphere and to the mineral soil by several interconnected cycles in which organic and inorganic compounds are exchanged (Fig. 1). From the atmosphere, gases, aerosols, and rain are added to the forest once partially filtered by the canopy. In closed canopy forests, most of the rainfall enters the soil as throughfall and percolates through the soil profile into the groundwater. In addition to the ions added to the forest from the atmosphere, forest ecosystems receive ions from the weathering of primary minerals. Figure 1 illustrates the difficulties of relating the observed symptoms of forest decline to a particular cause, since very different chemical substances are exchanged; the fluxes of a substance in one part of the system may affect other fluxes in seemingly remote parts. How the system responds to a change in one factor depends on how large the change is and on the intensity of connectedness among all the factors. Identifying a particular cause of decline is also difficult, in that similar symptoms can be induced by very different combinations of stress-inducing external factors.

Figure 1 includes many plant functions, but we know neither the flux rates of the different cycles, nor the details of the process by which different chemical

E.-D. Schulze, O.L. Lange, R. Oren (Eds.)
Ecological Studies Vol. 77

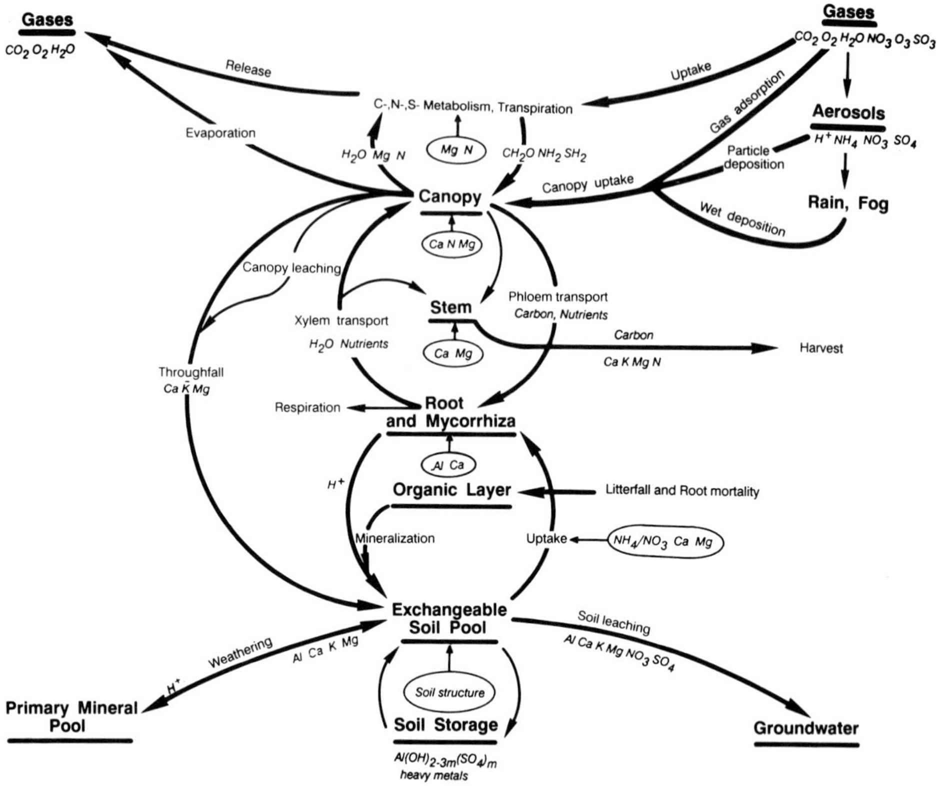

Fig. 1. A schematic model of element fluxes (*arrows*), major compartments (*underlined*), and regulating parameters (*ovals*) in a forest ecosystem under the influence of air pollution

substances are exchanged in order to maintain a balanced system at the whole plant level. This complicates the study of such systems and the interpretation of the results. The time scales of studies and their measured parameters should relate to the flux rate at each exchange point. Using the analogy of a clockwork motor, the study of a large, slow-rate cycle may overlook the fact that the system may be seriously perturbed if even a small, fast-rate cycle is not operating synchronously, and vice versa. Figure 1 represents the forest ecosystem under investigation in terms of feedback cycles which can be simplified into less realistic but easily understandable sequential effects. Figure 2 illustrates the sequence of events that appeared to lead to the apparent decline symptoms at our study sites. The reader may choose to follow Fig. 2 as we present the following integrating analysis.

2 Direct Effects of Pollutants on the Canopy

Episodic concentrations of ozone, sulfur dioxide, and nitrous oxide in the Fichtelgebirge have surpassed short-term maxima (see Chap. 1-B) known to harm conifers in controlled environment studies (Anonymous 1988).

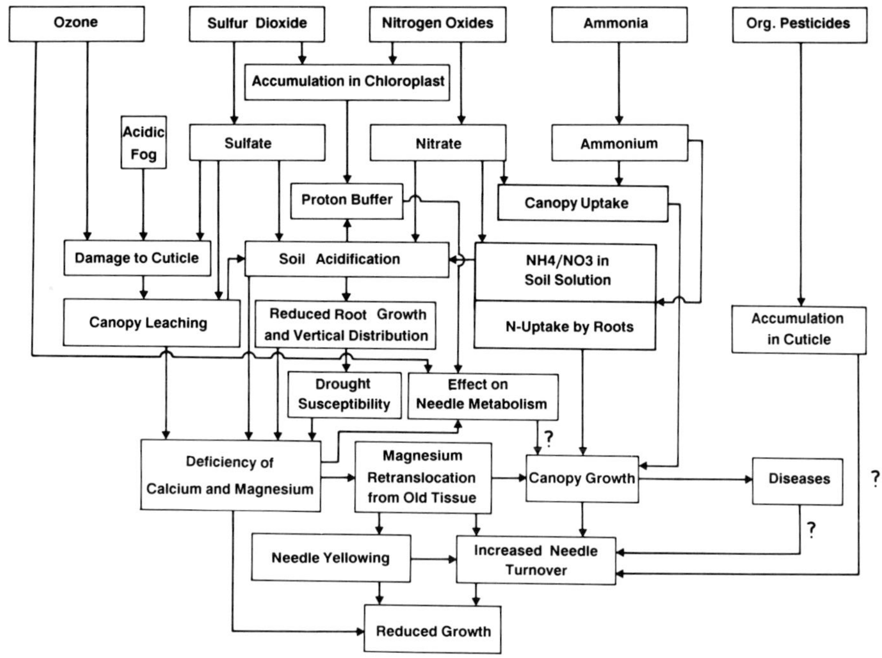

Fig.2. A schematic drawing of sequential processes which lead to the forest decline symptoms at the study site in the Fichtelgebirge

Initially, it was thought that the Fichtelgebirge was the one area in Germany where sulfur dioxide would have its greatest direct effects on trees, because of the high concentrations found in that region. Indeed, the estimated canopy uptake of sulfur dioxide was potentially sufficient to surpass the buffering capacity of the leaf cytoplasm (Chap. 3-D). However, no dose response of pollution load on the rate of photosynthesis could be observed in green or yellow needles (Lange et al. 1985; Zimmermann et al. 1988; Chaps. 3-D and 4-D). According to these studies, gaseous pollutants did not reduce photosynthesis or cause chlorosis. This was most convincingly demonstrated by an experiment in which twigs were not allowed to grow, by removing the buds (Chap. 3-D; Lange et al. 1987). One-year-old needles remained green when adjacent new growth was removed, despite exposure to the same aerial environment on the same tree in which other 1-year-old needles turned yellow when new growth was permitted to proceed as normal. This was presumably because of the absence or presence, respectively, of nutrient sink. The lack of a detrimental response to direct effects of sulfur dioxide by trees in the Fichtelgebirge might relate to the seasonal distribution of its concentrations. High levels of sulfur dioxide occur mainly in winter when stomata are closed and the canopy is protected by snow and ice. In summer, when sulfur dioxide concentrations are low, the cytoplasm is sufficiently buffered by various mechanisms, such as the exchange of protons for cations in the soil, or the consumption of protons in the reduction of nitrous oxide (Chap. 3-D).

The notion that the direct effects of sulfur dioxide on above-ground plant parts is not a primary cause of decline even in areas of very high sulfur dioxide input is not new. As early as 1912, Wieler demonstrated that local forest decline in regions of high sulfur dioxide input close to smelters was related to changes in soil chemical factors, rather than to direct atmospheric effects of sulfur dioxide. He was able to demonstrate either recovery or continuous health problems when seedlings were grown on soils exchanged from healthy and declining sites, respectively.

Initially, sulfur dioxide has been suspected of damaging Norway spruce by directly affecting such metabolic processes as photosynthesis. However, several additional ways exist, e.g., changes in cuticular waxes and stomatal behavior, by which sulfur dioxide, as well as other gaseous pollutants, might directly damage tree foliage (Fig. 2). In our study, variations in cuticular waxes (Chaps. 3-A, B) and in stomatal response to environmental factors such as light and air humidity (Chap. 3-D) were not related to decline symptoms.

Leaching of elements from the canopy has also been suggested as another mechanism leading to forest decline (Krause 1988). Chapters 3-C and 4-I show that while leaching did occur, it did not appreciably change the nutritional status of the trees. Similarly, despite accumulation of organic pesticides in the cuticle of the needles (Chap. 3-B), no apparent harm to needle metabolism was detected. It is not known, however, whether this accumulation of organic pesticides affect soil organisms and litter decomposition (see Chap. 2-C). In summary (Fig. 2), field measurements of needle physiology did not reveal any harmful effects of direct exposure to gaseous pollutants or pesticides.

3 Effects of Pollutants Via the Soil

The gaseous pollutants sulfur dioxide, nitrous oxide, and ammonia are converted into acids or ammonium in the atmosphere. This leads to acid rain or fog in which the acidity and ammonium may reach millimolar concentrations (see Chap. 1-B).

Sulfate, nitrate, and ammonium may improve plant performance because they provide macronutrients. Thus, unlike pollutants, they may have no detrimental effects. Indeed, these substances are commonly used as fertilizers to enhance plant growth. Nitrogen and sulfur compounds are continuously released from dead organic matter during the process of mineralization, even under conditions in which pollution does not occur. Thus, it seems that inputs of sulfate, nitrate, and ammonium should benefit rather than harm tree performance. However, if the rate of sulfate and nitrate deposition exceeds the rate of uptake by the vegetation, soil acidification may occur (Ulrich 1987). This is accompanied by a change in the pH buffering range of soils and thus, a reduced soil base saturation. An increased concentration of aluminum ions and the molar ratio of aluminium to base cations in the soil solution results (Ulrich 1987; Ulrich and Meyer 1987). The direct effects of gaseous pollutants on the forest canopy could be quantified in terms of dose responses. However, changes in soil chemistry are unrelated to a momentary concentration of the atmospheric input, but are related to the integrated acid load over long periods of time (Chap. 4-A) and to the mineral and organic composition of the soil since acidic deposition began.

Soil acidification affects plant performance in several ways (Fig. 2). Principally it influences root growth and reduces the supply of base cations. At the site with declining trees the supply of exchangeable base cations, such as magnesium and calcium, became very low (Chap. 4-B). The actual uptake of these ions by roots might be further reduced by cation-antagonisms in nutrient uptake (Chap. 4-E). As soil acidification progresses and the deposition of ammonium from the atmosphere continues, the concentration of ammonium may exceed that of nitrate in the soil solution. Ammonium will then be preferentially taken up by tree roots (Ingestad 1979; Boxman et al. 1988). This in turn leads to a reduced uptake of competing cations, particularly the uptake of magnesium (Jacob 1955; Mulder 1956). A suppression of magnesium uptake by this mechanism may have occurred at the study sites, as a high ammonium to nitrate ratio was found in the soil solution of the A_0 horizons. The higher concentrations of phosphorus in needles at the declining site may indicate a higher ammonium uptake at that site. Additional suppression of magnesium uptake may also relate to the high aluminum concentrations in the soil solution (see Chaps. 4-B, E). Uptake of magnesium by trees at the declining site may be low because of a low supply from the soil, an interaction with other ions in the soil solution, and an impaired root system.

In this study, changes in soil solution chemistry affected root growth and development. The number of root tips and mycorrhiza in the mineral soil correlated with the calcium to aluminium ratio in the soil (see Rost-Siebert 1985; Meyer et al. 1988; Chaps. 2-A, 4-E). Reduced in the most severely acidified upper mineral horizons at the declining site, root biomass increased in the organic layer, which probably provided a better environment for root development. The change in root distribution with soil depth affected the pattern of water uptake and increased the susceptibility of trees at the declining site to short-term drought, thereby exacerbating the difficulties in nutrient uptake (Chap. 4-C).

However, reduced cation uptake alone (Fig. 2) cannot explain the observed nutrient deficiency, which is the result of a disharmony between the uptake and demand for nutrients (Chap. 4-H) and was demonstrated by the bud-removal experiment (see Chaps. 3-D, 4-F). Laboratory studies have demonstrated that if all nutrients are similarly deficient for optimal growth, plants will adjust their growth to match the supply and show no deficiency symptoms (Ingestad 1987). Clearly, the amount of cations taken from the soil would have been sufficient to prevent deficiency symptoms if the trees had grown at a lower rate (Schulze et al. 1987).

In contrast to the reduced supply of base cations associated with soil acidification, the supply of nitrate at both study sites was higher than the actual uptake by the trees (see Chap. 4-I). Nitrate leached into the groundwater at rates equivalent to the atmospheric input. This was surprising because at both study sites, the nitrogen concentrations of needles were not in the range of optimal nitrogen nutrition (Chap. 4-H) which suggested a nitrogen deficiency. Needles did not respond to any addition of nitrogen into the tree by increasing concentrations. The apparent contradiction of nitrate leaching to the groundwater with a sub-optimal concentration of nitrogen in the tree may relate to the preferential ammonium uptake by spruce and the higher concentration of ammonium compared with nitrate in acid soils (see above). The nitrogen to cation uptake balance may be further perturbed by aerial uptake of nitrous oxide, nitrate, and ammonium from the atmosphere into the canopy (Johnson and Richter 1984; Chap. 4-I). This aerial nitrogen cannot be balanced by uptake of cations from the soil when availability is relatively low and uptake is impeded by restriction on root distribution. The imbalance in the uptake of nitrogen and magnesium appeared to cause needles to become yellow.

Yellowing of old needles was initiated by magnesium retranslocation from old into new needles. The amount of magnesium retranslocation seemed to correlate with the nitrogen concentration in mature needles (Oren et al. 1988b; Chap. 4-H) which suggests a magnesium to nitrogen interaction. A high concentration of nitrogen in mature needles appeared to stimulate the growth of new needles, and in this case, a large amount of magnesium was retranslocated to support this new needle growth. If the concentration of magnesium in mature needles was high, retranslocation did not cause old needles to turn yellow. However, at low magnesium concentrations in mature needles, retranslocation decreased the concentration of magnesium to a level at which yellowing occurs.

The interaction of nitrogen and magnesium with respect to plant performance is further complicated by the seasonal separation of needle and stem growth. Cations stored in needles are used during the period of intensive canopy growth. These cation pools seemed to be depleted at the declining site later in the season when stem growth began (Oren et al. 1988b; Chaps. 4-F, H). Thus, stem growth in summer relied mostly on the supply of magnesium from the soil. At the healthy site, stem growth was sufficiently supplied with magnesium to yield high stemwood biomass but at the declining site, the low magnesium supply from the soil resulted in reduced growth (Chap. 4-D).

In summary, nitrogen pollutants seemed to promote canopy growth and thus exhausted most of the magnesium stored in the trees before stem growth began. Stem growth was then limited by the supply of magnesium from the soil. Atmospheric nitrogen deposition seemed, therefore, to enhance the deficiency of cations which were already marginally limiting plant growth and function. It appears that this model of a nutritional disharmony is general and may explain a number of other nutritional deficiencies which are known to be associated with forest decline (Oren et al. 1988b; Chap. 4-H), such as potassium deficiencies on limestone or manganese deficiency on dolomitic soils. It could also explain the periodic occurrence and disappearance of such decline symptoms as needle-yellowing, since both growth and supply of cations from the soil solution, and thus the degree of an imbalance, may differentially change from year to year depending on climatic conditions at each site.

The ecosystem balance at the declining site (Chap. 4-I) also revealed that at the present level of very low base saturation and weathering of primary minerals, such forest practices as clear felling could remove as much base cations from the already low nutrient pool as was deposited from the atmosphere. Soil cation leaching was about the same amount as that released by mineral weathering. Cation leaching into the groundwater at the healthy site was large, compared to the size of the available nutrient pool. If leaching at the healthy site continues at the present rate, the nutrient pool there may decrease in the near future to a level similar to that at the declining site. This suggests that conditions at the healthy site are less stable than first appeared. Indeed, yellowing of needles on some trees at the healthy site, which consisted of only green trees in 1985, was detected in 1988.

4 Conclusion

A low soil calcium and magnesium concentration at the declining site in the Fichtelgebirge has probably been caused by a long period of acidic input into the soil. While stemwood growth is most closely related to the supply of magnesium from the soil, the symptom of needle-yellowing reflects a complex balance between the uptake of base cations from the soil, and nitrogen compounds from both the soil and the atmosphere. The combination of ammonium and nitrate deposition seems to have made nitrogen detrimental to forest health. Selective uptake of ammonium may cause nitrate to be leached out of the soil, accompanied with nutrient cations. The net result is a reduction in calcium and magnesium for plant uptake. At the present level of soil buffering, nitrate seems more effective than sulfate in further reducing the base saturation at both sites. Information about dry deposition and the fate of nitrogenous compounds in the humus layer is required to better evaluate these processes.

5 Historic Considerations

Historical records indicate that substantial soil acidification has occurred throughout central and northern Europe during the last 30 years (Hellbäcken and Tamm 1986; Flakengren-Grerup 1986; Ulrich and Meyer 1987). This change in soil pH is related to anthropogenic deposition of sulfur and nitrogen at higher loads than can be used by the vegetation or stored in the soil profile (Chap. 4-I). Forest use has probably also contributed to the present forest decline. The change from a mixed *Fagus-Abies-Picea* forest to a monoculture of Norway spruce may have reduced the base saturation of the soils at a time before acid rain started. In addition, litter-raking during the last century was more intensive at some sites than at others. This, however, could not have led to the large-scale forest instability which has been observed in recent history.

The historical progression may thus be summarized as follows: initially, atmospheric deposition predisposed soils to acidification, initiated soil acidification, and subsequently promoted forest decline. At first, sulfur dioxide was the major atmospheric pollutant. Sulfate inputs in northern Germany were already high in 1915 and had almost reached our present levels by 1940 (Fig. 3, after Ulrich and Meyer 1987). Information on such inputs in the Fichtelgebirge is lacking, but we can assume that the trend is fairly similar throughout Germany.

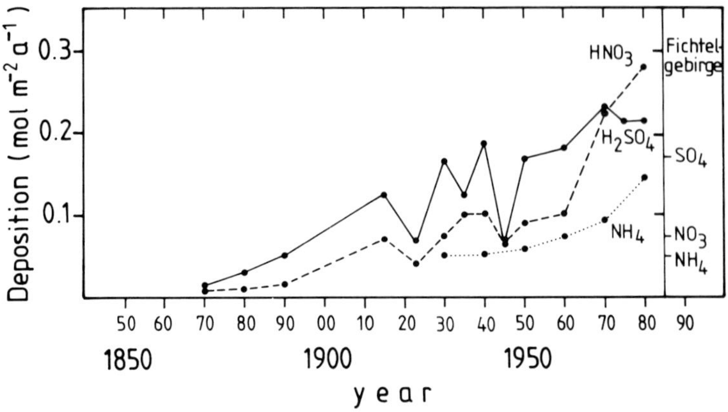

Fig. 3. The deposition of sulfate, nitrate, and ammonium since 1850. The deposition rates of sulfate and nitrate were calculated by Ulrich and Meyer (1987) as equivalents from emission data on an area basis of Germany. The ammonium deposition data are form the Netherlands (Boxman et al. 1988). Actual deposition rates of sulfur, nitrate, and ammonium in the Fichtelgebirge are lower than the calculated deposition rates (see also Epstein and Oppenheimer 1986; Derwent and Nodop 1986), or the deposition of ammonium in the Netherlands. However, for the present analysis only the time trend of deposition of the different compounds is of importance and the trends presented are probably representative also of those in the Fichtelgebirge

Sulfuric acid predisposed soil to acidification; however, sulfate was probably stored as aluminum-hydroxy-sulfate when the soils reached low base saturation (Ulrich 1987; Chap. 4-B). In the period since roughly 1915, nitrate deposition was added to the high sulfur load. The initial increase of nitrate deposition occurred when ammonium inputs were still low, and it is likely that nitrate was at that time an effective fertilizer, so that it increased forest productivity. Recently, a period has begun which is characterized by a strong increase in ammonium input. The increase in ammonium input has probably resulted in the trees taking up nitrogen mostly as ammonium (see Chap. 4-E). This may have led to excess nitrate in the soil solution, which has become a major contributor to soil acidification (see Chap. 4-B), and in turn, has also increased the rate of decline in soil pH and thus mobilized stored sulfate. Ammonium nitrate depositions, nitrate leaching from the soil, and soil acidification have probably all caused a strong nutritional imbalance in the study area since about 1980.

6 Future Perspectives

Trees that have survived the initial shift in the ecosystem balance may be presently healthy because of reduced competition between fewer trees for base cations in the soil. This is supported by the observation that decline symptoms have disappeared in recent years in some places, although climatic variations which affect the ability of needles to replenish magnesium may also cause such apparent recovery (Oren et al. 1988b; Chap. 4-H). If reduced competition among individual trees is accompanied by reduced stand growth, less nitrogen will be taken up from the soil and soil cation leaching is likely to increase. The decline symptoms may, therefore, emerge again as soon as the soil is further depleted of base cations, so that even the demand for magnesium by a reduced population of trees displaying a smaller leaf area cannot be met. The effect of a limited supply of nutrient cations may be even worse in dry years, because rooting depth has been restricted by soil acidification. The imbalance between the demand for and the supply of nutrient cations reflected in needle-yellowing will probably be even more apparent in relatively young stands as they enter the phase of maximum growth and demand for nutrients. If the inputs of nitrogen and the supply of cations were to stabilize (Linder and Ingestad 1977; Linder 1987), the forests might reach a new steady state at low growth rates without visual decline symptoms. However, as nitrogen inputs increase and continue to perturb the forest ecosystem, it is likely that stand vigor will continue to decline and tree mortality will increase in relation to variations in yearly weather.

Forests on soils which are deficient in one element will probably not die out from nutritional imbalance alone but will become unstable. Declining vigor may render the trees susceptible to a host of biological agents that could further reduce tree health and cause mortality (see Fig. 2). Following tree death and canopy opening, other factors, e.g., wind and temperature extremes, will increasingly contribute to the decline (Klein and Perkins 1988). Such a sequence of events, which can denude forests completely, temporarily at least, has been observed in a few, high elevation areas in the Fichtelgebirge.

References

Anonymous (1988) UN-ECE Workshop on critical levels of gaseous pollutants. Bad Harzburg 14–18.3.88 Final Draft Report, Umwelt Bundesamt, Berlin, 146 pp

Boxman D, Van Dijk H, Roelofs J (1988) Critical loads for nitrogen with special emphasis on ammonium. In: Nilsson I, Grenafelt P (eds) Critical loads for sulfur and nitrogen. Nordic Council of Ministers and UN-ECE. ISBN 91-7996-096-0 pp 295–322

Derwent RG, Nodop K (1986) Long-range transport and deposition of acidic nitrogen species in northwest Europe. Nature (Lond) 324:356–358

Epstein CB, Oppenheimer M (1986) Empirical relation between sulfur dioxide emissions and acid deposition derived from monthly data. Nature (Lond) 323:245–247

Falkengren-Grerup U (1986) Soil acidification and vegetation changes in deciduous forests in southwest Sweden. Oecologia (Berl) 70:339–347

Hellbäcken L, Tamm CO (1986) Changes in soil acidity from 1927 to 1982–1984 in a forest area of south-west Sweden. Scand J For Res 1:219–232

Ingestad T (1979) Mineral nutrient requirements of *Pinus sylvestris* and *Picea abies* seedlings. Physiol Plant 45:373–380

Ingestad T (1987) New concepts on soil fertility and plant nutrition as illustrated by research on forest trees and stands. Geoderma 40:237–352

Jacob A (1955) Magnesia, der fünfte Pflanzenhauptnährstoff. Enke, Stuttgart

Johnson DW, Richter DD (1984) Effects of atmospheric deposition on forest nutrition cycles. Tappi J 67:82–85

Klein RM, Perkins TD (1988) Primary and secondary causes and consequences of contemporary forest decline. Bot Rev 54:1–43

Krause GHM (1988) Impact of air pollutants on above-ground plant parts of forest trees. In: Mathy P (ed) Air pollution and ecosystems. Reidel Publ Co, Dordrecht pp 168–216

Lange OL, Gebel J, Schulze ED, Walz H (1985) Eine Methode zur raschen Charakterisierung der photosynthetischen Leistungsfähigkeit von Bäumen unter Freilandbedingungen – Anwendung zur Analyse "neuartiger Waldschäden" bei der Fichte. Forstwiss Centralbl 66:484–494

Lange OL, Zellner H, Gebel J, Schramel P, Köstner B, Czygan FC (1987) Photosynthetic capacity, chloroplast pigments, and mineral content of previous years spruce needles with and without the new flush: analysis of the forest-decline phenomenon of needle bleaching. Oecologia (Berl) 73:351–357

Linder S (1987) Responses to water and nutrients in coniferous ecosystems. In: Schulze ED, Zwölfer H (eds) Potentials and limitations of ecosystem analysis. Springer, Berlin Heidelberg New York Tokyo, Ecol Stud 61:164–179

Linder S, Ingestad T (1977) Ecophysiological experiments under limiting and non-limiting conditions of mineral nutrition in field and laboratory. In: Bicentenary Celebration of CP Thunberg visit to Japan. R Swed Embassy Bot Soc JPN, pp 69–76

Meyer J, Schneider BU, Werk KS, Oren R, Schulze ED (1988) Performance of two *Picea abies* (L.) Karst. stands at different stages of decline. V. Root tip and ectomycorrhiza development and their relation to aboveground and soil nutrients. Oecologia (Berl) 77:7–13

Mulder EG (1956) Nitrogen-magnesium relationships in crop plants. Plant Soil 7:341–376

Nihlgård B (1985) The ammonia hypothesis – an additional explanation to the forest dieback in Europe. Ambio 14:2–8

Oren R, Schulze ED, Werk KS, Meyer J, Schneider BU, Heilmeier H (1988a) Performance of two *Picea abies* (L.) Karst. stands at different stages of decline. I. Carbon relations and stand growth. Oecologia (Berl) 75:25–37

Oren R, Schulze ED, Werk KS, Meyer J (1988b) Performance of two *Picea abies* (L.) Karst. stands at different stages of decline. VII. Nutrient relations and growth. Oecologia (Berl) 77:163–173

Rost-Siebert K (1985) Untersuchungen zur H- und Al-Ionen-Toxidität an Keimpflanzen von Fichte (*Picea abies*) und Buche (*Fagus sylvatica*) in Lösungskultur. Univ Gött, Ber Forschungszentrum Waldökosysteme 12:1–29

Schulze ED, Oren R, Zimmermann R (1987) Wirkungen von Immissionen auf 30-jährige Fichten in mittleren Höhenlagen des Fichtelgebirges auf Phyllit. AFZ 27:725–730

Schulze ED, De Vries W, Hauhs M, Rosen K, Rasmussen L, Tamm CO, Nilsson J (1989) Critical loads for nitrogen deposition on forest ecosystems. Water Soil Air Pollution (in press)

Ulrich B (1987) Stability, elasticity, and resilience of terrestrial ecosystems with respect to matter balance. In: Schulze ED, Zwölfer H (eds) Potentials and limitations of ecosystem analysis. Springer Berlin Heidelberg New York Tokyo, Ecol Stud 61:11–49

Ulrich B, Meyer H (1987) Chemischer Zustand der Waldböden Deutschlands zwischen 1920 und 1960, Ursachen und Tendenzen seiner Veränderung. Univ Gött, Ber Forschungszentrum Waldökosysteme, Reihe B 6:1–133

Waring RH (1987) Nitrate pollution: A particular danger to boreal and subalpine forest. In: Fujimora T, Kimura M (eds) Human impact and management of mountain forests. Forestry For Prod Res Inst, Ibaraki, JPN, pp 93–105

Werk KS, Oren R, Schulze ED, Zimmermann R, Meyer J (1988) Performance of two *Picea abies* (L.) Karst. stands at different stages of decline. III. Canopy transpiration of green trees. Oecologia (Berl) 76:519–524

Wieler A (1912) Pflanzenwachstum und Kalkmangel im Boden – Untersuchungen über den Einfluß des Bodens durch Hüttenrauch und über die giftige Wirkung von Metallverbindungen auf das Pflanzenwachstum. Bornträger, Berl

Zimmermann R, Oren R, Schulze ED, Werk KS (1988) Performance of *Picea abies* (L.) Karst. Stands at different stages of decline. II. Photosynthesis and leaf conductance. Oecologia (Berl) 76:513–518

Subject Index